Case Study: Dodds Group

How can a company distribute its goods quickly and cost-efficiently? The Dodds Group provides the answer.

A hypothetical manufacturing company is suffering in the recession and turns its attention to its logistics function to cut costs. The company, based in the south, has a core fleet of vehicles and contracts out the balance of its tonnage. It is also in local outside warehousing, and has two-way traffic with its sister company in France. In addition it also has two-way local traffic, and a contract shunter with trailers that both loads trailers for the core fleet and moves stock to and from the outside store. The client's own staff run the core fleet on a day-to-day basis.

SPECIALIST EQUIPMENT

The customers' work bulked out before meeting maximum weights and so super cube step frames were used by the core fleet. Because of their payload these were sent occasionally on long distance journeys, only to return empty because it was not possible to find general haulage contractors running the same equipment. However, the recent legislation in vehicle lengths had not been taken into account, nor had the latest technology regarding lowering chassis and fifth wheel heights in order to increase payloads while staying within 4.2 metres for UK deliveries.

SMALL CHANGES – BIG IMPACT

After analysing the core fleet's work for three months, it was obvious that some of the work could be undertaken more cost-effectively by the contractor's trunking operation, as some of the core fleet is either under-utilised or running excess empty miles.

The solution was to reduce the core fleet, in order to cover only work within 100 miles of the factory, and also to use as many vehicles on the night shift as possible for deliveries to customers and clients able to take night deliveries.

As the vehicle costs were met by the contract, the only cost involved in night running was the additional wages and extra mileage. However, as the vehicles were being kept locally, this was offset by a reduction in the daytime mileage.

The French shipping route was changed so that the driver could take rest breaks overnight on the ferry, both to and from France. This reduced the journey time by a full day, thus increasing the working time available by the vehicle concerned by a whole shift. It now became possible for the vehicles to work locally all day, then journey to the port, travel overnight, land in France, deliver and load, and return to the French port, travelling overnight to land early enough to deliver the load and then work locally again for the whole of the following day.

It was also found that much of the work from the factory is for clients in the north of the country. By moving that work directly to the north where the contractor also had a warehouse, the company could benefit from the lower northern rents. As there was no longer a local outside store shunt operation required, and the contractor was awarded more trunking work, the shunt vehicle and trailers were taken off the contract and were replaced with an off-road shunter and sufficient trailers to cover the work by the contractor without charge.

The storage work was trunked overnight to the northern warehouse, and deliveries from the warehouse were then undertaken on local vehicles. This enabled the contractor to cost the deliveries very close to the charge for a straight delivery to the north from the factory, so the savings from the warehouse charges and the shunter were not lost. The lead time for a delivery was also reduced, thus enhancing customer service.

This customer had a first class service from its contractor, at competitive rates. However, the customer's business had changed over the years, and because the customer controlled the fleet the contractor was unaware and unconcerned with what the fleet was used for – providing of course, that it complied with legislation regarding drivers' hours, weights, etc. Substantial savings were made for the client, not by reducing the contractors costs but by increasing efficiency. Although he lost some of the core fleet, the contractor actually expanded the business as, having an increased amount of one way traffic, he was able to accept more return traffic from other clients.

THE SOLUTION? – DODD'S TRANSPORT LTD

Dodd's Transport Ltd, a family run haulage company with an 80-year history.

0181-571 5750

SEE ADVERT ON PAGE ii

The Transport Manager's and Operator's Handbook 1995

'It's become the operator/transport manager's bible . . . I can't imagine anyone carrying on his business without David Lowe's splendid guide at his side.'   **Road Law**

'What can one say of a reference book that gets better and better except that it gets better and better!'   **Roadway**

'There is none better . . . the transport man's bible'   **Headlight**

'Any transport manager would be well-advised to keep this at his elbow ready for instant use. It is to be thoroughly recommended.'   **Transport**

The 25th edition of this definitive guide provides, as ever, valuable, reliable guidance for anyone involved in the road transport industry. Clearly structured for quick and easy access, three main areas are covered:

- Transport legislation
- Technical standards
- Goods vehicle operations.

While focused on national transport operations in Britain, the *Handbook* also covers developments in the EU.

Packed with essential information, this comprehensive guide aims to build awareness of all the legal, operational and environmental factors which are of utmost importance in today's transport industry. It presents complex legal requirements in a coherent and easily digestible format, as well as offering practical operational advice and solutions to some of the environmental issues facing the industry.

Fully updated to include the most recent developments, this edition includes information on the latest UK and EU legislation, 'O' Licence and driving licence changes, new vehicle weights, UK road network developments, VED tables and VED exemptions.

As useful as when it was first published in 1970, this 25th anniversary edition of *The Transport Manager's and Operator's Handbook* is an essential reference source for any transport manager, fleet operator, owner-driver haulier, student taking transport examinations, and anyone involved in any aspect of the industry.

David Lowe is an independent transport consultant. He has spent his entire career in the transport industry and writes extensively on the subject (especially legal matters). He is the author of many popular transport books, including *A Study Manual of Professional Competence in Road Transport Management*, *The European Road Freighting Handbook* and *The European Bus and Coach Handbook*, all published by Kogan Page.

The

TRANSPORT MANAGER'S &OPERATOR'S

Handbook 1995

David Lowe

KOGAN
PAGE

Please note: From 16 April 1995 the digit '1' will be added after the initial '0' prefix to all area telephone codes. The following cities will have completely new area codes: Leeds (0532) becomes (0113 2); Sheffield (0742) becomes (0114 2); Nottingham (0602) becomes (0115 9); Leicester (0533) becomes (0116 2) and Bristol (0272) becomes (0117 9). These changes should be taken into consideration when using the telephone numbers in this book.

Formerly published as *The Transport Manager's Handbook*

This twenty-fifth edition first published in 1995

Kogan Page Limited
120 Pentonville Road
London N1 9JN

British Library Cataloguing in Publication Data

A CIP record for this book is available from the British Library.

ISBN 0 7494 1491 X
ISSN 0958 1551

Typeset from disk by Paul Stringer, Oxford
Printed and bound in Great Britain by Clays Ltd, St Ives plc

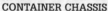  

POWER FOR A NEW GENERATION

M SERIES. Huge torque and power in a highly compact and fuel efficient package... the new M Series 10.8 litre engine from Cummins with the top rating providing a huge 1825 Nm of torque at just 1200 rpm and 380 PS at 1900 rpm.

Purpose-designed with CELECT – a highly intelligent electronic management system providing a platform for instant diagnostics, trip data and fleet management software.

The new M Series is part of a complete range of EURO 2 engines from Cummins offering both lower emissions and higher torque all the way from 3.9 to 14 litres. Including natural gas engines with exceptionally low emission levels.

All part of a new generation of power developed for the truck industry by Cummins – leading the world in diesel technology.

Cummins Engine Company Ltd.
Yarm Road, Darlington, Co. Durham DL1 4PW
Tel: 0325 460606   Fax: 0325 360196

WORLD CLASS DIESEL TECHNOLOGY

Cummins "Euro 2" line-up revealed

Cummins has announced its full Euro 2 engine line-up including the new M Series range of 11-litre engines. A Compressed Natural Gas (CNG) engine based on the popular six cylinder B series engine has also been unveiled.

100,000 CELECT

Cummins introduced Europe's first diesel engines featuring a full authority electronic engine management system in the first half of 1993. More than 100,000 Cummins CELECT engines are now in service worldwide. The pace of development, made possible by electronic systems, is dramatically illustrated by two areas where Cummins is leading the way: i) the availability of a sophisticated in-cab display (RoadRelay) to help a driver to maximise fuel consumption, and ii) the development of powerful fleet management data systems through the integration of a newly-available PC-based (Intelect) software package providing the kind of accurate engine data previously dreamed of.

No less than 24 Euro 2 ratings have been announced across four engine families, B Series 4/6-litre, C Series 8.3-litre, M Series 11-litre and N Series 14-litre. The new range, which will be progressively introduced in line with the product plans of truck/bus manufacturers using Cummins' engines, spans from 105PS to 525PS and 360Nm of torque to 1733Nm of torque.

All ratings are air-to-air aftercooled and all M & N Series ratings will be fitted with Cummins CELECT electronic engine management system as well as articulated pistons.

Tractive Effort

Peter Griffen, Manager - Market Support, spells out a warning to those who buy engines on horsepower alone, "The ability to keep increasing torque levels on engines, rather than just horsepower, can be used as a rough guide as to how technologically advanced these engines are. Some engine manufacturers barely raised torque figures for Euro 1 and many will not keep pace for Euro 2 either. This will make some quite considerable competitive torque gaps, for the same nominal horsepower. Operators who traditionally purchase on horsepower alone would do well to consider tractive effort and driveability also."

He added: "Our current L Series Euro 1 engines offer torque levels substantially in excess of many heavier and larger displacement competitors which has made them the payload conscious choice for trucks grossing up to 38/40 tonnes. The L350E CELECT rating offers a class leading 1695Nm of torgue with the L325 at 1559Nm. The new M Series builds on this strength offering from 1550Nm of torque at 305PS, to 1700Nm at 340PS and a journey time reducing 1825Nm at 380PS."

"With the M380E producing the same torque as the current N410, while weighing in at 940kg, only 12kg more than the L10 and with similar external dimensions including the same height, the M Series demonstrates what a truly advanced engine it is."

Premium over-the-road performance levels have been matched to the fuel consumption advantages of Cummins' CELECT electronically-controlled fuel management system.

1.5 Million Kilometres

For ultra long haul and extremely arduous duty cycles Cummins N Series products enjoy an unparalleled reputation. The latest N Series range offers just three high horsepower ratings (425, 475 and 525PS) with excellent torque levels to match. The N14 weighs 1256Kg, 124kgs heavier than the NTE 14-litre engine fitted to the current non-CELECT ratings, and is now

able to cover up to 1.5 million kilometres before requiring other than routine attention.

Cummins' latest compressed natural gas (CNG) engine, based on the popular six cylinder B Series, will be made available in two ratings, at 150PS and 195PS.

Model		Maximum power				Maximum torque			
		kW	(PS)	@	rpm	Nm	(lb ft)	@	rpm
B Series									
B105	(4cyl)	77	(105)	@	2500	360	(265)	@	1500
B135	(4cyl)	99	(135)	@	2500	410	(302)	@	1500
B130	(6cyl)	96	(130)	@	2500	450	(332)	@	1500
B145	(6cyl)	107	(145)	@	2500	500	(369)	@	1500
B160	(6cyl)	118	(160)	@	2500	550	(406)	@	1500
B190	(6cyl)	140	(190)	@	2500	675	(498)	@	1500
B215	(6cyl)	158	(215)	@	2500	700	(516)	@	1500
B235	(6cyl)	173	(235)	@	2500	800	(590)	@	1500
C Series									
C215*		158	(215)	@	2200	800	(590)	@	1400
C245		180	(245)	@	2200	1025	(756)	@	1400
C260*		191	(260)	@	2400	1025	(756)	@	1400
C260**		191	(260)	@	2200	950	(701)	@	1300
C280		206	(280)	@	2200	1125	(830)	@	1400
C300		221	(300)	@	2200	1125	(830)	@	1400
M Series									
M215E*		158	(215)	@	1900	950	(701)	@	1200
M250E*		184	(250)	@	1900	1025	(756)	@	1200
M280E*		206	(280)	@	1900	1166	(860)	@	1200
M305E		224	(305)	@	1900	1550	(1143)	@	1200
M305ESP		224- 261	(305- 340)	@	1900	1450- 1700	(1069- 1254)	@	1200
M340E		250	(340)	@	1900	1700	(1254)	@	1200
M380E		279	(380)	@	1900	1825	(1346)	@	1200
N Series									
N425E		313	(425)	@	1900	2100	(1549)	@	1200
N475E		349	(475)	@	1900	2250	(1660)	@	1200
N525E		386	(525)	@	1900	2350	(1733)	@	1200

*bus & specialist applications only
**high altitude rating

– 24 ratings. All air-to-air aftercooled
– Euro 2 ratings will be introduced progressively in line with product plans of truck manufacturers using Cummins engines
– All M & N Series engines feature Cummins CELECT
– M Series engines will replace current L Series ratings at "Euro 2"
– All M & N Series engines feature articulated pistons
– Factory-fitted high performance compression brake optional on all M & N Series
– B190, B215, B235 and all C Series ratings feature Holset wastegate turbochargers

New engine range announced

Cummins M Series

- four truck ratings (M305E, M340E, M380E, M305ESP)
- high torque 1550Nm, 1700Nm, 1825Nm and 1450-1700Nm respectively.
- Peak torque on M380E equal to N410 and 8% higher than L350E
- all models feature Cummins CELECT
- all models feature articulated pistons
- HX50 Holset turbocharger optimised specifically for the M Series to give greater fuel economy and excellent transient response
- oil cooler design similar to latest N Series, with no hose connections and the latest gasket technology, for a leak free design with optimum durability
- factory-fitted high performance compression brake optional on all ratings
- 2 year/unlimited mileage base engine warranty and optional 5 year/800,000km extended warranty package

The M Series is a new heavy duty engine family designed specifically for productivity-conscious operators. This ultra-modern 11-litre engine offers the performance of higher displacement engines with fuel economy that testing has demonstrated is 4% better than the fuel efficient L Series that it will replace.

Advances in fuel efficiency have been made alongside impressive torque levels which climb to 1825Nm for the top-of-the-range M380E. This torque is equal to the Euro 1 14-litre N410 and 8% higher than the class leading L350E. These levels of torque are normally associated with heavier and less fuel efficient engines in the 12-litre class, nominally rated at up to 315kW (430bhp), which do not benefit from the advanced technology evident in Cummins new M Series engines.

Fuel economy has been maximised by design-matching the Holset HX50 turbocharger to the M Series engines, the introduction of articulated pistons on all ratings and high injection pressures (15% greater than L Series) to enable fuel to be injected and burned more cleanly and efficiently.

Extensive testing including 3.2 million miles of field tests and 60,000 hours of bench testing preceded production of the M Series and over 10,000 M Series engines are now in service in world markets.

The M Series shares many of the premium features of the latest N Series within the same external dimensions as the L Series which it will replace at "Euro 2". The M Series features the most modern proven technology such as multi-valve heads with CELECT centrally-located electronically controlled unit injectors and articulated pistons. Cummins has been using multi-valve heads for 30 years on the N Series and on the L Series since it was introduced in 1982. The L350E was Cummins first engine to offer the premium combination of CELECT electronic engine management and articulated pistons which are standard on all M Series ratings.

The M Series is Cummins' first all-electronic engine family from launch. Over 100,000 engines have been manufactured using CELECT, Cummins' fully authority electronically-controlled engine management system. CELECT allows the latest engines to be optimised for fuel economy and performance, according to rating, duty cycle and application, within statutory emissions limits, in a three dimensional way not possible with less sophisticated "in-line" fuel pump systems.

Fleet Management

CELECT also offers new levels of fleet management and diagnostic information available to both the driver, in the form of the RoadRelay "in-cab" unit, and to the fleet manager. This information allows the fleet manager to maximise operating parameters and the driver to have a visual fuel consumption target to aim at.

The heavy duty pedigree of the M Series is obvious from the generous crank pin and main bearings, large camshaft and patented mid-stop liners. It is not surprising that projected average life to overhaul in long distance use exceeds current NTE 14-litre engines and is projected to be nearly 1.3 million kilometres (800,000 miles).

The M Series block while compatible with the L Series is stronger with a wide skirt. The massively strong crankshaft has partial counterweight machining for improved balance and a new crank pulley and damper design. The connecting rods are of a robust design illustrating the emphasis on reliability and durability.

The one piece cylinder head incorporates premium alloy cast iron material and advanced swirl port technology for greater durability and improved fuel economy.

The oil cooler design is very similar to the latest N Series with no hose connections and the latest gasket technology for greater durability, leak-free design and improved fuel economy. Testing has shown that oil control on the M Series is equal to the N Series.

The attention to detail on this heavy duty engine is evident throughout the specification. Microfinished camshafts increase lobe life up to a factor of three providing a substantial tolerance margin for the toughest duty cycles. While chrome valve stems reduce valve guide wear and nickel barrier rod bearing overlay increases rod bearing durability.

The productivity-conscious M Series is less than 2% (14kg) heavier than the L Series, while the projected life-to-overhaul exceeds the durable NTE 14-litre engine. Testing has also shown the M Series to have 4% superior fuel economy than the frugal L Series and with class leading torque, equal to the current N410 on the M380E, the M Series is set to add a new dimension for heavy duty truck, bus and coach operators.

ESP "Smart Power"

M305ESP

The M Series line-up includes the new dual rated "smart power" M305ESP model which represents a pioneering advance in heavy-duty engine technology.

Known as ESP, the new Cummins "smart engine" technology uses advanced and innovative computer logic to sense road and load conditions so engine computers are able to adapt output to deliver optimum horsepower and torque for climbing hills, while maintaining overall fuel efficiency on level ground.

The M305ESP engine delivers optimum fuel economy at 224kW (305bhp) when the going is easy and when running into a strong headwind or up a steep gradient the engine automatically increases torque (by 17%) from 1450Nm (1069 lb ft) up to 1700Nm (1254 lb ft) and raises horsepower by 26kW (35bhp) to improve hill climbing and reduce gear changing.

ESP engine technology was the product of a customer-led development process designed to bring together rapidly evolving electronic engine technologies in a way that provides immediate value for both operators and drivers. ESP engines use a form of artificial intelligence programmed into the CELECT electronic control module. The technology is a synthesis and extension of Cummins' long standing commitment to advanced fuel/engine management systems. Computer hardware and software for ESP, and all Cummins electronic components are developed and manufactured by Cummins with the key support of Cummins Electronics Company, Inc., (CEL), a wholly-owned subsidiary.

ESP has already provided Cummins with the annual Technical Achievement Award, presented by the Truck Writers of North America (TWNA), for introduction of a pioneering advance in heavy duty electronic engine technology.

NATIONAL BREAKDOWN
– KEEPING YOU ON THE MOVE

National Breakdown has been providing commercial vehicle operators with a nationwide recovery and breakdown service since 1977.

Since then the Company has continuously developed the services it provides, now offering its customers the best service levels ever, with over 300,000 commercial members now covered throughout the UK and Europe.

The Company's specialist Truckcall and Trucklink services provide a great deal of flexibility, members can choose either a comprehensive insured breakdown and recovery service, or a more flexible "pay-on-use" service.

The agents that are chosen by National Breakdown Commercial Recovery are all handpicked experts on every type of commercial vehicle, including tail-lifts, trailers and "reefers". Over 800 agents are on call 24 hours a day, 365 days a year. Each year the Company handles over 30,000 commercial incidents on British roadsides.

The Company maintains an emphasis on adaptability, and are capable of handling virtually any situation, quickly making arrangements for commercial vehicles carrying livestock, STGO, "just in time delivery" and vehicles carrying hazardous substances.

Commercial Fleet customers already include some of the biggest names in trucking, such as Iveco, Foden, Scania, Seddon Atkinson and ERF.

National Breakdown offer extensive commercial cover throughout the E.U., service in Europe being provided by National Breakdown's sister company Green Flag S.A.

In terms of reliability, National Breakdown was voted the best value for money recovery service by "Which" magazine and remains the fastest growing motoring organisation in the UK with over 2.5 million members – and rising.

Evidence of the organisation's commitment to quality is that it was the first motoring organisation to be awarded the British Standard Institution BS EN ISO 9002.

National Breakdown's peripheral companies including U.K. Insurance, Travellers Medical and Home Emergency Services are all success stories in their own right.

TOTAL QUALITY – TOTAL CONFIDENCE

The Eagle TX could have no better pedigree. Perkins is a world leader in the design and manufacture of diesel engines in the 5-2500bhp power range, and since the company was established in 1932 it has produced a staggering 13 million power units.

Today some 400,000 Perkins diesel engines are manufactured each year, involving three plants in the UK and 13 overseas. But impressive though these statistics might be, it is quality rather than quantity to which Perkins addresses its priorities.

Ever mindful that there is always room for improvement, Perkins has invested in the biggest training programme ever undertaken as part of its **Total Quality Programme**. This goes beyond the traditional areas of quality and involves everyone and every process from the shop floor to the Board of Directors. It is already paying dividends in advances in the quality of products and services provided, and increased customer satisfaction.

The Eagle TX is produced in Perkins Shrewsbury factory which employs a unique cell method of assembly; each engine is caringly built by a single craftsman from start to finish, before undergoing a final rigorous test cycle.

Developed from a tradition of engineering excellence ensuring the highest possible standards. It is complemented by the support of Perkins technology which is totally dedicated to developing and improving diesel engine economy and performance.

Using highly advanced equipment like the electron scanning microscope, Perkins engineers and metallurgists are constantly seeking improvements in performance and economy while guarding the environment.

With such commitment to **Total Quality**, it's hardly surprising that Perkins engines are used by over 600 producers of powered equipment in 160 countries.

Millions of Perkins engines help keep the world moving day after day.

Right across the globe, major companies trust the world's greatest power in diesel engines.

Perkins is currently the power that millions of road users rely on because of our reputation for always providing quality and reliability.

These high standards have been applied throughout our range of vehicle engines which have a reputation for outstanding power delivery and economy. They have each been specifically designed to meet the needs of market leaders and such engines as Prima, Phaser and Eagle TX cover every need from cars to 40 tonne trucks.

Furthermore, across 160 countries, we provide unparalleled customer service with Perkins Powerpart™ total after-market support.

And for the future, Perkins' commitment to develop and innovate ensures that we'll be anticipating and satisfying individual market and legislative needs the world over.

Find out more about our range of engines from 5 to 2500 bhp by phoning our worldwide headquarters on 01733 67474.

Perkins

A business of Varity Corporation. **VARITY**

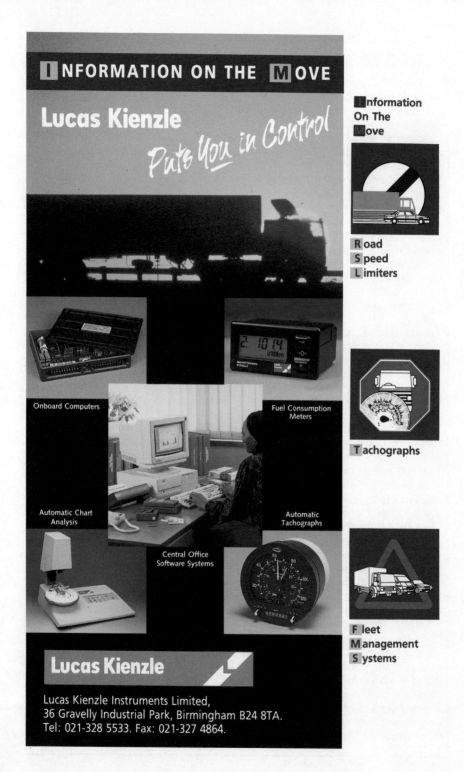

TRUCKLINE SETS THE STANDARDS ON THE WESTERN CHANNEL

Back in 1973, when Great Britain joined the Common Market, a co-operative of French farmers was deciding to start a modest ferry service to bring their produce from Roscoff in Brittany to market in Plymouth. This was the humble beginnings of what is now the Brittany Ferries Group, which quickly grew and prospered, within just a few years adding new routes of Portsmouth to St. Malo; Portsmouth to Caen; Plymouth to Santander in Northern Spain; and Cork to Roscoff.

By coincidence, also in 1973, Truckline was also offering an exclusive service for those engaged in carrying freight to the Continent using two unencumbered ports of Poole and Cherbourg. Twelve years later, in 1985, Brittany Ferries acquired Truckline and the two companies became one, with the Truckline name becoming synonymous with any freight moved on any of Brittany Ferries, seven routes or 10 ships, whether multi-purpose or dedicated freighters.

Consequently, freight carryings have grown from 20,000 movements in 1978 to over 170,000 movements crossing the Channel with Brittany Ferries/ Truckline. The reason for this expansion and continued growth in freight business is the result of offering freight customers increased frequency, good on-board services, and the most cost-effective way of crossing the channel to Central and Western France, Italy, Spain and Portugal.

Truckline's routes provide not only relaxing on-board rest periods for drivers but also saves mileage and fuel. For the long and short of it is geography with the shortest Channel crossing not necessarily the quickest. For example, the Spanish border is miles closer when choosing Truckline's Poole-Cherbourg route than Calais.

Truckline/Brittany Ferries has set the standards that others have tried to follow, building new super ferries and investing in their ports to the value of £150 million. During the last three years they have built two new ships, the 20,500 tonnes Barfleur on Poole-Cherbourg and the 27,000 tonnes Normandie for the Portsmouth-Caen route. Last year the company introduced the Val de Loire to its Plymouth to Santander Spanish route, one of the top twenty largest ferries in the world. This vessel also operates a weekly sailing from Plymouth to Roscoff and from Cork to Roscoff.

Further investment in port infrastructure in St. Malo also has been undertaken to allow Truckline/Brittany Ferries' other super ferry, Bretagne, to operate from Portsmouth to St. Malo. This is the climax of a three year plan to set the standards and make Brittany Ferries and Truckline number one on the Western Channel.

This has been recognised by the AA who have awarded 5 Star awards for the high standards of on-board catering and facilities on its super ferries. All have luxurious Drivers Clubs, with airy restaurants and lounge area offering television and video and newspapers.

On shore a central computerised freight reservations system now operates in Poole, with one call giving an instant overview of sailings and space availability. The whole operation is backed by BS5750 accreditation for its ro-ro freight, marketing, reservations and handling services at Portsmouth, Poole and Plymouth.

FOR FURTHER INFORMATION ON TRUCKLINE'S SERVICES RING FREIGHT SALES ON 01202 441133.

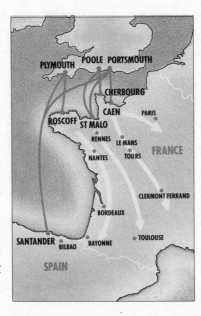

Glass's Guide to Valuations

Glass's Guide is the market leading valuation service, selling well over twice as many copies as any other available, creating a true industry standard.

Glass's Guide to Car and Commercial Vehicle Values are available in a range of formats, including books, DOS, windows and raw data to suit your business needs.

"The great thing about Glass's is that it gives a single value for a well defined, clean car. It worries me that other guides, giving three values, could encourage managers to let vehicles go for less than their true worth"

David Turner, Fleet Manager, London Borough of Lambeth

To find out more about the whole range of services available from Glass's, simply call **01932 823 823** and ask for a copy of Glass's 1995 catalogue.

Glass's Information Services Limited, St Martin's Court, 37 Queens Road, Weybridge, Surrey KT13 9TU

FODEN TRUCKS

ABERDEEN · C. Lawie Commercials · 01224 782255
ABINGDON · R.P. Cherry & Son Ltd. · 01235 531004
ALFRETON · Cotes Park Commercials Ltd. · 01773 521211
BARNETBY · Gallows Wood Trucks Ltd. · 01652 688259
BEDFORD · Banks of Sandy Ltd. · 0345 626681
BICESTER · Curtis Mechanical Services · 01869 345504
BIRMINGHAM · Millward of West Bromwich · 0121 525 5226
BOGNOR Regis · Eurotek Commercials · 01243 822014
BRISTOL · Brian Kellow Ltd. · 01275 874355
BUXTON · Barlow & Hodgkinson Ltd. · 01298 84407
CARDIFF · B.R.T. International Ltd. · 01222 598018
CARDIFF · Fairwood Diesel Engineers (Killay) Ltd. · 01222 522000
CARLISLE · Cumbria Truck Centre · 01228 36405
CARMARTHEN · O.J. Williams & Son · 01994 230355
CRAWFORD · Weston Recovery Ltd. · 0186 42 668
DARTFORD · Acorn Truck Sales Ltd. · 01322 556415
DRIFFIELD · Garton Commercials Ltd. · 01377 241120
DUBLIN · Frank Boland Ltd. · (010 3531 from UK) 4508 644
ELGIN · Baillie Brothers (Truck Services) Ltd. · 01343 87312
EXETER · Devon County Commercials Ltd. · 01392 57737
FAREHAM · Foden Fareham · 01329 827854
GATESHEAD · Albany Motors · 0191 4770501
GLASGOW · The Colin Hutton Group · 0141 336 3371
GREAT YARMOUTH · L.G. Perfect (Engineering) Ltd. · 01493 657131
HARTLEPOOL · Parsons Truck Centre · 01429 864876
HEATHROW · Foden Heathrow · 0181 893 2288
HEREFORD · Burgoyne Bros. Ltd. · 01544 327441
HODDESDON · Valley Trucks Ltd. · 01992 441551
IPSWICH · R. & G. Commercials · 01473 744800
KINGS LYNN · Lynn Commercials (1989) Ltd. · 01553 763122
LEEDS · Pelican Engineering Co. Ltd. · 0113 282 2181
LEIGHTON BUZZARD · Osborn Transport Services · 01525 383548
LIVERPOOL · Steve Gray (Motor Engineers) Ltd. · 0151 523 3393
LIVINGSTON · Caledonia Commercials Ltd. · 01506 430000
LOUGHBOROUGH · Charnwood Trucks · 01509 502121
MANCHESTER · Manchester Truck Centre · 0161 873 8048
NORTHAMPTON · Franklin Services · 01604 494208
NOTTINGHAM · K. & M. Hauliers · 01602 639630
PETERBOROUGH · Peterborough Engineering Co. Ltd · 01733 66666
POOLE · A.J.D. Commercials · 01202 736083
ROMSEY · Tom H. Andrews · 01794 512250
ROYSTON · Foulgers Garage · 01763 248332
SANDBACH · M.T.C. Sandbach · 01270 763291
SHEFFIELD · John Owen (Aggregates) Ltd. · 01909 564215
SHEFFIELD · South Yorkshire Trucks Ltd. · 01909 610055
SKIPTON · Willis Commercial · 01756 799563
SPALDING · W.A. Burdall & Company · 01775 820528
STOKE-ON-TRENT · H. & H. Commercial Truck Sevices Ltd. · 01782 575522
SWANSEA · Fairwood Diesel Engineers (Killay) Ltd. · 01792 203830
TORKSEY · R. Eastment · 01427 718638
TRABBOCH · John Maitland & Sons · 01292 591277
TUNBRIDGE WELLS · Kent & Sussex Truck Centre Ltd. · 01892 515333
WELSHPOOL · J. & D. Bowen · 01938 810726

Moss Lane . Sandbach . Cheshire CW11 9YW . Tel: 01270 763244 . Fax: 01270 762758

LESS WEIGHT . . . MORE PROFIT!

Foden tractors and rigids are renowned for their driveability, comfort and performance. But the key area in which Foden excels is saving weight. Spec. for spec. Foden trucks weigh less than the competition and that means bigger loads and bigger profits for you!

A Foden tractor unit is typically 300-500 kg lighter than its competition. That's 300-500kg more payload every trip. That's more profit per trip for you and less trips for your customer.

To demonstrate the advantage of a Foden, consider a tractor unit delivering six loads a week. For a Foden carrying an extra 380 kg at £13/tonne, you receive an extra £30 per week. 50 weeks a year for 6 years yields an **extra** £9,000 per truck for the same effort. Double shift your Foden and the benefits rise to £18,000!

Contents

LEYLAND DAF TRUCKS: BRITAIN'S FAVOURITE TRUCK COMPANY

Leyland Trucks, which has chosen to sell all its European models through Leyland DAF in the UK and DAF Trucks on the Continent, has been reborn as the largest British truck producer and one of the most efficient suppliers in the commercial vehicle business. Recently it turned in excellent trading figures, with a profit of £8.4 million in its first 11 months in business. The management-owned business employs nearly 700 people at the state-of-the-art Leyland Assembly plant and is the principal supplier of logistics vehicles to the British armed forces.

DAF Trucks is a fully independent truck builder with a pan-Continental sales and service network of more than 1,000 outlets in both west and east Europe. In the first six months of 1994 the company made a profit of £24 million and is continuing to invest heavily in research and development.

Recently Leyland DAF Trucks experts have calculated that, in real terms, the cost of operating Leyland DAF trucks has fallen by well over 50% in the last ten years a remarkable achievement and one which has consolidated the company's position in the UK market. Leyland DAF customers can expect to see further falls in operating costs in future years as more advanced technology stretches every litre of diesel still further and what it terms "the ownership experience" – the whole relationship between customer and supplier for the life of the vehicle – is further improved.

The company's excellent UK dealer network and industry-beating after sales back up service, Leyland DAFAid (which has more than 80% of broken down trucks back on the road in an average of well under two hours), reflect the dedicated approach to customer care which is the watchword at Leyland DAF Trucks. No wonder it's a favourite with British truck buyers!

Defect Repair Sheets 331   Service Records 333   Retention of
Records 333   Wall Planning Charts 334   Vehicle History Files 334

20: Safety – Vehicle, Loads and at Work   336

C&U Requirements 336   The Safety of Loads on Vehicles 336
Safety Report 338   Health and Safety at Work 340   Health and Safety at
Work etc Act 1994 343   Electric Storage Batteries 346   Control of
Substances Hazardous to Health (COSHH) 348   Notification of
Accidents 348   First Aid 351   Safety Signs 353   Vehicle Reversing 353
Safe Tipping 354   Safe Parking 354   Fork-Lift Truck Safety 355   Freight
Container Safety Regulations 355   Operation of Lorry Loaders 356   Safety
in Dock Premises 356

21: Loads – General, Animals, Food etc   357

Distribution of Loads 357   Length and Width of Loads 358   Animals 360
Food 361   Sand and Ballast Loads 363   Solid Fuel Loads 364
Container Carrying 364   Fly Tipping 364

22: Loads – Abnormal and Projecting   366

Abnormal Indivisible Loads 366   Special Types Vehicles 366   High
Loads 373   Projecting Loads 373   Lighting on Projecting and Long
Loads 375

23: Loads – Dangerous, Explosive and Waste   376

United Nations Classes and Packing Groups for Dangerous Goods 376
Packaging and Labelling 377   Dangerous Loads in Bulk Tankers/Tank
Containers 378   Conveyance of Dangerous Substances by Road in
Packages 388   Carriage of Explosives 391   Radioactive Substances 395
Petroleum 396   CIA Voluntary Code 396   Controlled and Hazardous
Waste 397

24: Fleet Car and Light Vehicle Operations   401

Excise Duty 401   Insurance 402   Company Cars and Income
Tax 404   Construction and Use Regulations 406   Drivers' Hours and
Records 407   Speed Limits 408   Seat Belts 408   Fuel Consumption
Tests 409

25: Rental, Hiring and Leasing of Vehicles   413

The Vehicle User 413   Rental 414   Hire and Contract Hire 415
Leasing 417

26: Vehicle Fuel Economy   419

Fuel and the Vehicle 419   Fuel and Tyres 421   Fuel and the Driver 421
Fuel and Fleet Management 422   Fuel Economy Checklist 425

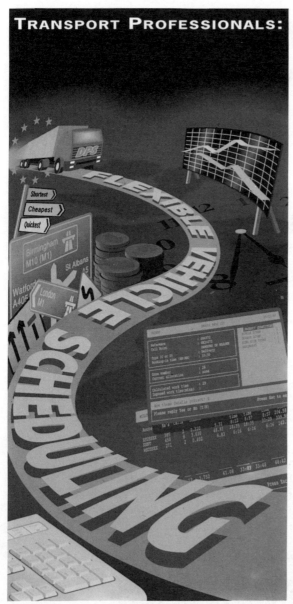

TRANSPORT PROFESSIONALS:

Are You Realising the Full Potential of Your Fleet ?

With LogiX you can.

LogiX is a complete range of powerful and versatile software products designed by logistics professionals to provide operators with both strategic and tactical information for dealing with the wide range of situations encountered within the logistics arena.

The operation of the LogiX software systems provide the less experienced user with a powerful range of features, while the advanced LogiX user can interact with the system to fine tune the decision making process.

Flexibility and ease-of-use make the system ideal for:

❑ Planning day-to-day multi-drop or trunking routes
❑ Producing detailed quotations
❑ Assessing alternative logistic strategy
❑ Fleet management analysis

LogiX is very effective with 'semi-fixed' routes, providing an optimised route schedule to reflect the reality and cost of your situation.

Logix also features:

● Detailed route itinerary based on order volume
● Vehicle Utilisation
● Driver Efficiency
● Performance Monitoring
● Journey Profitability
● Pan-European Map Databases
● File import & export with other systems in ASCII.
  Lotus* or DBase** formats

LogiX already benefits many large companies involved in grocery distribution, frozen foods, contract services, parcel freight, containers, trunking, electrical goods, and also sales representatives & service engineers.

The software uses the low cost hardware platform of the IBM PC*** or a compatible micro and comes with full documentation and training is also available.

 DPS

Distribution Planning Software Limited
Lygon Court, Hereward Rise, Halesowen, West Midlands B62 8AN
Tel: 021-585 6633 Fax: 021-585 6303

THE LOGISTICS PROFESSIONAL'S CHOICE

*Lotus is a registered trademark of Lotus Software **DBase is a registered trademark of Ashton-Tate *** IBM PC is a registered trademark of International Business Machines Corporation.

Foreword

I congratulate the *Transport Manager's and Operator's Handbook* on attaining its 'Silver Jubilee'. This is no mean feat for a specialist publication, and is, I think, an indication of the status it has achieved. I commend the contribution it has made to the development of high operating and safety standards among haulage operators.

The past 25 years have seen enormous change in the haulage industry as it has responded positively, and profitably, to the challenges and opportunities of the market both at home and abroad. An efficient, competitive haulage industry is vital to our economic prosperity.

Quality of service and operation, as perceived by the customer and by society at large, will be essential elements in the future development of the industry. Tough standards will increasingly be demanded and those engaged in it will need ready access to relevant information. I am confident, therefore, that there will be a place for this *Handbook* on the bookshelves of all transport managers and operators.

The Rt Hon Dr Brian Mawhinney MP
Secretary of State for Transport

Preface to the 25th Edition

Who would have thought in 1970, when the very first edition of this *Handbook* was published, that, 25 years later, not only would it still be in existence, having been updated and revised year by year, but that it would have been expanded over the intervening years from the original 192 pages to 538 pages (for the 1994 edition), and, more significantly, that the amount of legal gobbledegook that the transport operator has to contend with would have swollen to such proportions that it needs nearly 600 pages in which to explain it in simple terms?

For those who do not remember back that far, the *Handbook* was conceived and first produced at the time when the forerunner to the present system of 'O' licensing came into being. This was before the days of professional competence, but at a time when ideas for a 'transport manager's licence' were being propounded following the inclusion of provisions for such a licence in the 1968 Transport Act, along with the change from A, B and C carriers' licensing to 'O' licensing and, among other things, proposals for the introduction of tachographs. No wonder there was an industry-wide campaign to 'Kill the Bill' prior to that Act receiving the Royal Assent. (Some may even have wished to kill the then Minister of Transport – Mrs Barbara Castle – along with it!)

In those days the maximum weight for an artic was 32 tons (not tonnes) and with VED based on unladen weight the most an operator would have had to pay in annual duty for such a vehicle was £513. Drawbar combinations were still restricted to 30mph but artics could run at 70mph on the motorway. Now we have up to £5000 per year VED rates, 44 tonne vehicles for combined transport and a 56mph maximum speed (with speed limiters) for heavy trucks. Fines were not mentioned much in the 1970 edition but then, in those days, they did not set the offender back very much – fixed penalties were just £2. Nowadays serious road traffic offenders could find themselves paying up to £5000 for each conviction and facing imprisonment if they falsify tachograph records.

A lot has changed since 1970. The industry has become much more regulated and more strictly monitored. While the levels of professionalism have improved substantially, those who go out of their way to bend the rules (and the industry in the UK still has a lot of these) face very severe penalties if they are caught. The many changes in the law (and in the penalties for breaches of the law) and in operating practice since that time have been regularly documented in subsequent editions of this *Handbook* and, as from the very start, the object has been not merely to regurgitate the law but to provide readers with a clear explanation of what operators or commercial vehicle drivers must do, or must not do – after all, this is what they mainly want to know so they can get on, legally, with the job in hand.

The special section on the Single European Market (SEM) has been retained at the front of the text in updated form. The purpose of this is to provide readers with a general picture of the effects of the SEM on road haulage and road freight distribution. Since this remains the most important, and possibly the most momentous, event in contemporary road haulage history and for the foreseeable future, both for domestic operators as well as those operating internationally, the subject could not be ignored, even in a handbook now devoted solely to covering national transport operations within the UK. It remains a fact that many transport and distribution operators, who may have never previously contemplated doing so, are now likely to send their vehicles abroad under the pressure to expand markets across the whole Union. For this reason it is hoped that the information provided here in the usual *Handbook* format will be found to be both interesting and a useful precursor to a more detailed study of the subject to be found in the new *European Road Freighting Handbook.*

This year, as in preceding years, the text has been fully revised and updated to include new information as necessary. As always, I have tried to provide useful advice as well as chapter and verse on the legalities of operation, and to warn where possible of impending legislative changes. The intention is to provide readers with the best and simplest possible explanations of complex legal requirements to help them run their businesses or manage their transport departments with the maximum of efficiency and within the law. It is an important responsibility of goods vehicle management to comply with legal requirements. Failure to conform to the law can have very serious and very expensive consequences, as many vehicle operators and large goods vehicle drivers frequently find to their dismay. Regular scanning of the transport trade press will indicate the severe penalties imposed on lgv drivers, fleet managers and vehicle operators by the courts or the Licensing Authorities following their misdemeanours.

The purpose of this *Handbook*, as it has been from the very outset in 1970, is to help operators, managers and others responsible for road transport operations to avoid the risks of financial penalty and, worse, penalty against their Operator's licence. I sincerely hope its contents provide interesting and intelligible reading, and prove to be a useful and ready source of reference to the many and complex legal requirements relating to goods vehicle ownership and use. More importantly, I hope it helps readers to avoid problems with the enforcement authorities and the courts, to keep their Operators' (and lgv drivers') licences, and to protect their own livelihoods and the many others which may depend on these licences.

David Lowe
Tormos
August 1994

NATIONAL BREAKDOWN
– KEEPING YOU ON THE MOVE

National Breakdown has been providing commercial vehicle operators with a nationwide recovery and breakdown service since 1977.

Since then the Company has continuously developed the services it provides, now offering its customers the best service levels ever, with over 300,000 commercial members now covered throughout the UK and Europe.

The Company's specialist Truckcall and Trucklink services provide a great deal of flexibility, members can choose either a comprehensive insured breakdown and recovery service, or a more flexible "pay-on-use" service.

The agents that are chosen by National Breakdown Commercial Recovery are all handpicked experts on every type of commercial vehicle, including tail-lifts, trailers and "reefers". Over 800 agents are on call 24 hours a day, 365 days a year. Each year the Company handles over 30,000 commercial incidents on British roadsides.

The Company maintains an emphasis on adaptability, and are capable of handling virtually any situation, quickly making arrangements for commercial vehicles carrying livestock, STGO, "just in time delivery" and vehicles carrying hazardous substances.

Commercial Fleet customers already include some of the biggest names in trucking, such as Iveco, Foden, Scania, Seddon Atkinson and ERF.

National Breakdown offer extensive commercial cover throughout the E.U., service in Europe being provided by National Breakdown's sister company Green Flag S.A.

In terms of reliability, National Breakdown was voted the best value for money recovery service by "Which" magazine and remains the fastest growing motoring organisation in the UK with over 2.5 million members – and rising.

Evidence of the organisation's commitment to quality is that it was the first motoring organisation to be awarded the British Standard Institution BS EN ISO 9002.

National Breakdown's peripheral companies including U.K. Insurance, Travellers Medical and Home Emergency Services are all success stories in their own right.

Acknowledgements

I am grateful to all the many individuals and organisations who have provided helpful information for inclusion in this *Handbook*. Over the years I have been writing the *Handbook* I have approached so many people for information, guidance and assistance that it really is impossible to thank them all individually. Some people are consulted regularly year after year; some have been newly sought out for guidance. There is no doubt that without such help the *Handbook* would not have achieved its prominence as the so-called 'bible' for transport managers and others. To all these people, once again, I express my sincere thanks for their help.

Reference has been made to many sources for information – to the legislation itself, both British and EU; published guides and notes; the Department of Transport, the Licensing Authorities and their Traffic Area Office staff and the enforcement authorities. Many other relevant bodies have been consulted and both the national press and the transport trade press have provided a continuous reminder of changing legislation and developments in the industry. Similarly, a constant flow of comments and questions from clients and seminar delegates has provided the source of many leads for inclusion of information and the inspiration to continue revising the contents of the *Handbook* to include more and more legislative detail. To all these publications, organisations and nameless individuals I am greatly indebted.

Acknowledgement is made to the controller of Her Majesty's Stationery Office for permission to reproduce forms and diagrams from Statutory Instruments and from the *Highway Code*.

I would like to thank especially my wife Patricia and many friends in the industry who are a constant source of encouragement and, as always, too, the editorial staff of Kogan Page whose help ensures yet another edition of the *Handbook*.

It would not be inopportune at this time to say a special word of thanks to Philip Kogan, founder and managing director of Kogan Page Limited. Twenty-five years ago he had sufficient foresight and faith to invest in a suggestion for a book which was intended to do little more than explain quite simply the implications of the then proposed scheme of transport managers' licences. Without his early cajoling and continuous encouragement over the years I have to admit that I may never have managed to produce the very first edition, which was published in 1970, let alone the revised and updated, to say nothing of expanded, editions in each of 24 successive years.

Every effort has been made to ensure that the material contained in this book is as correct and as up to date as possible. While I accept full responsibility for any errors which may be found, I cannot stem the progress of legislation which inevitably occurs during the life of the current edition of this annual publication.

David Lowe
July 1993

Introduction

Historically, the road transport industry in Great Britain has had to contend with a mass of legislation which has imposed considerable restriction on operations, to say nothing of the burdens of high cost and considerable worry for the owners, operators and managers of goods vehicle fleets.

The past 25 or so years in particular have seen more than their fair share of such legislation. Looking back to the days when the Road Safety Act 1967 (long defunct) and the Transport Act 1968 were about to be introduced, it would have been difficult then to foresee the mass of legal requirements which would have to be faced over the years ahead. Who could have anticipated the construction of a tunnel under the channel and the subsequent development of combined road and rail transport networks all over Europe?

In 1970 we had changes in drivers' hours and records which were eventually understood and accepted; and in place of carriers' licensing, a new system of operators' licensing was also introduced in 1970 bringing own-account operators and professional hauliers under one common scheme for the control of goods vehicle operations. This, too, was accepted and largely appreciated by the industry which recognised the need for increased road safety measures.

By 1978, however, many changes in these items alone, as respective chapters in the book show, had taken place and in 1984 further significant changes were made. The hours regulations have become those of Europe; the records requirements changed to meet European law late in 1976 and subsequently have been largely replaced by tachographs on a mandatory basis from 1981; and the one-for-all system of 'O' licensing was torn apart to satisfy the demands for establishing standards of professional competence. In 1986 major changes were made to both the European and the British drivers' hours rules and the European tachograph rules, with the object of making the law simpler and more flexible for operators to apply.

Besides these particular items of legislation there have been many more changes which earlier editions of this book have charted year by year. In fact, each year has seen something new in legal responsibilities for the vehicle operator and transport manager to digest. If it has not always been significant in transport terms it has been significant in other terms, significant Acts being the ill-fated Industrial Relations Act, the Equal Pay Act, the Health and Safety at Work, etc Act, the Employment Protection Act and the Employment Protection (Consolidation) Act 1978, then the Employment Act 1980 and the Employment Act 1982, and others since then.

In previous years this introduction has referred to the great difficulty which confronts those engaged in the industry, no matter what their capacity or title, in keeping track of what must, and must not, be done (and the penalties involved for non-compliance). The pile of transport, social and environmental legislation is constantly mounting, and as it does so the problems of keeping up to date with it all – and putting it into practice – become increasingly complex and time-consuming. Now, the industry has to look towards the effects of the Single European Market with all that this entails by way of changing administrative and operational practices as well as the increased competition. From a legislative viewpoint, international transport operators may find the way eased as barriers are removed and restrictive measures, such as the requirement for road haulage permits and certain customs procedures, are eliminated. On the other hand, the domestic road haulage industry may find such steps as the legalisation of cabotage presenting them with unwelcome competition on their own doorsteps from marauding Euro-hauliers looking to pay their running costs home after delivering international loads into Britain.

So it seems that the tide of legislation cannot, and will not, be stemmed even for a brief period to allow the industry some respite from pressure and change, from new restrictions and new burdens.

The purpose of this handbook, therefore, is to gather together as much of this legislative material as can reasonably be squeezed between the covers, to explain what it is all about in lay terms which are both easy to read and to understand and apply, and thereby to provide the hard-pressed vehicle operator or transport manager with one accessible, intelligible source of information on the responsibilities laid on him by law emanating from both the British government and the EU.

The handbook is intended for the fleet operator (whatever the size of his fleet, be it large or very small), the transport manager, the owner-driver haulier or anybody else, whatever his or her title, whose responsibilities include the day-to-day control, administration or operation of goods vehicles, and the small operator who has found that there is a lot more to running a goods vehicle besides just taxing and insuring it. In addition to its function as a ready source of reference for the main legal requirements affecting goods vehicle operation and other useful information, the book may also be found, by those studying for transport examinations, to be an additional means of acquiring a detailed knowledge of the relevant legislation currently applicable to the UK road freight industry.

The major items of legislation affecting the operator and driver of goods vehicles have been covered, together with some of the other legal requirements which are not necessarily new but which it is important for him to know about. Besides transport legislation the owner or manager has been increasingly confronted with social and environmental legislation which affects him both as an employer of staff and as the occupier of premises.

Apart from this there is still a great deal more information that the operator and manager should have at his fingertips. Some of it concerns particular branches of the transport industry; the carriage of dangerous goods, abnormal loads, food or livestock, for example. The person responsible for these specialised traffics should have acquired, through experience, some knowledge of the regulations concerning his particular field. The newcomer to such operations will need to obtain the appropriate regulations direct from Her Majesty's Stationery Office (HMSO) – see addresses at end of introduction – or from a bookseller who stocks Stationery Office publications, and study them carefully.

The principal Acts of Parliament which form the basis of the legislation explained in this book are supported by, and their provisions brought into effect by, a much larger number of regulations, orders and amendments. These are the means by which the Secretary of State for Transport puts into effect the legal requirements laid down in principle in an Act of Parliament. Legislation contained in an Act is not of itself effective until brought into force by regulations or orders made by the Secretary of State, for which purpose he is given the necessary powers in the Act. Many provisions contained in such Acts may, in fact, never be brought into use but they do not automatically become extinct by lack of use. They remain dormant on the statute book unless repealed by a further Act.

European legislation arises as the result of draft proposals, similar to British government white papers, which are circulated among interested parties (the trade unions, trade associations, and so on) for comment prior to legislative action being taken, and are then enforced by means of directives and regulations of the Council of the European Union.

It is most important for transport people to be aware of the regulations affecting both their own special type of operations and transport in general. To keep up with all the individual acts, regulations, amendments and modifications which Her Majesty's Stationery Office publishes on behalf of the government in connection with transport and the statutory publications of the European Communities is no mean task (see accompanying list of HMSO bookshops where these publications can be bought – or they can be ordered through other, HMSO stockist, booksellers). A more convenient method of keeping up to date on all the new measures affecting the industry is to read the trade press regularly.

The principal journals covering the road goods transport field are *Motor Transport* and *Commercial Motor*, published weekly, *Headlight* and *Trucking International*, published monthly. Contents include the latest industry news, new regulations with simple explanations and articles on subjects of special interest such as insurance, vehicle maintenance, road tests of new vehicles, costing, education and training and management topics. *Headlight* in particular is of special interest, because of its extensive reporting of both Court and Tribunal cases, which give the reader a good indication of how the law is applied, how the enforcement agencies, and subsequently the courts, interpret legal provisions and the levels of penalties imposed on offenders and those who lose out in Tribunal cases.

It must be emphasised that the fleet operator and transport manager should keep abreast of what is happening in the industry if he wants to be efficient, progressive and stay on the right side of the law. To concentrate on the job in hand to the exclusion of all that is happening in the industry at large is an attitude adopted by many operators and many of them have already suffered the consequences of not being prepared to meet some of the drastic changes that have taken place in recent years.

A valuable aid to the operator is membership of one or other of the trade associations – the Freight Transport Association, which represents the own-account operator, and the Road Haulage Association, looking after the interests of the hire and reward professional haulier – which provide a number of services to their members as well as keeping them informed about what is happening in the industry through their respective journals, *Freight Transport* and *Roadway*. Both organisations hold open meetings and training sessions, addressed by specialists on various topics of importance, at which members can hear first-hand details of current and new legislation, ask questions and air their views.

Additionally, for the operator or manager who wants to keep abreast of current legislation and practices in the industry there are commercial seminars and training sessions ranging from one day to over a week in duration. These provide an excellent means of keeping up to date with what is going on, and of meeting and talking to other people in the industry with similar interests and problems.

The road transport industry plays an important part in the life of Great Britain. Its safe operation is essential for the well-being of the people and its efficient operation is vital to the economy. Higher standards of management and control within the industry with a far greater awareness and understanding of the legal requirements, the operational demands and the economic considerations are necessary if these essential criteria of safety and efficient operation are to be achieved and upheld.

Special Note

It should be noted that this book is intended purely as a practical interpretation of legal matters for the lay reader and is only a *guide* to matters current at the time of writing. It is *not* a definitive legal work of reference and should not be used as such. Readers are advised to check the legislation itself before committing time or expenditure to any particular course of action and any operator needing detailed legal advice is recommended to seek it through normal legal channels. The author and publishers accept no responsibility whatsoever for decisions taken or other irrevocable actions based on the contents herein.

HMSO BOOKSHOPS

London   49 High Holborn, London WC1V 6HB.
Tel 071-873 0011 (telephone orders are not accepted – see below)

Belfast   80 Chichester Street, Belfast BT1 4JY.
Tel 0232 238451

Birmingham   258 Broad Street, Birmingham B1 2HE.
Tel 021-643 3740

Bristol   Southey House, 33 Wine Street, Bristol BS1 2BQ.
Tel 0272 264306

Edinburgh   71 Lothian Road, Edinburgh EH3 9AZ.
Tel 031-228 4181

Manchester   9–21 Princess Street, Manchester M60 8AS.
Tel 061-834 7201

HMSO Publications Centre (for post and telephone orders only)
PO Box 276, London SW8 5DT.
Tel 071-873 9090 (for telephone orders)
Tel 071-873 0011 (for general enquiries)
(queuing system operating on both numbers)

An investment that pays off in many ways:

Alcoa Forged Aluminium Wheels

Alcoa has been producing, in its Cleveland (Ohio) plant, forged aluminium wheels for trucks, trailers and buses for over 45 years. In Europe, they have covered many millions of kilometres under all conditions and have passed the most stringent tests.

Alcoa wheels offer competitive advantages that no transport manager can afford to ignore:

- **Extra payload capacity**

  Forged aluminium wheels are up to 53% lighter than steel wheels and up to 16% lighter than cast aluminium wheels. This means that a substantial amount of extra payload can be carried on every trip.

- **Lower operating costs**

  Our wheels are fully machined. This gives them a perfect roundness which means truer running and reduced tyre wear. Since aluminium dissipates heat faster than steel, Alcoa wheels contribute to an extended service life of the tyres and brake linings. The lower unsprung weight and better wheel/tyre balance reduce vibrations and wear on steering and suspension systems.

- **Longer life**

  Forging and heat-treating this high-strength aluminium alloy gives the wheels great durability and resistance to high-load impact. For this reason, 300-tonne airliners are fitted with similar aluminium wheels, most of them forged by Alcoa.

- **Greater safety**

  The wheels are forged from a single piece of metal. They have no welds to fail or allow air loss from the tyres. The aluminium alloy used and the manufacturing process produce truly excellent resistance to crack propagation, and a special safety hump ensures good tyre retention even if the tyres are under-inflated.

- **Less maintenance costs**

  Our forged aluminium wheels do not need painting because they are corrosion-resistant. Their simple design has no reinforcing ribs that can trap dirt so they are simple and quick to wash. Their lower weight makes them easier to handle in the garage or at the roadside.

- **Increased comfort**

  True running and absence of vibration are invaluable where passenger comfort is important. But Alcoa forged aluminium wheels are also highly appreciated by freight haulage drivers.

- **Better appearance**

  All our wheels have the same classic design and bright appearance. They give a uniform style and a high class image to any truck, trailer or bus.

- **Five-year warranty**

  Every Alcoa wheel has a five-year warranty. This covers material or manufacturing defects, independently of the vehicle guarantee and regardless of the distance covered.

Alcoa wheels are offered as options on new vehicles by every manufacturer of trucks and trailers, and successful fleet managers insist on the quality of forged aluminium wheels which ensures them a valuable advantage for being competitive in the road transport industry.

Alcoa truck wheels are forged in the same plant and with the same expertise as wheels for 300-tonne commercial jets.

Alcoa wheels are often chosen partly for their attractive appearance, which helps any vehicle to enhance the image of its company.

Transport and the Single European Market

Following many years of publicity build-up and hype, the Single European Market (SEM) opened for business on 1 January 1993 – and following implementation of the Maastricht Treaty we should now use the more politically correct terminology by referring to the SEM as being of the European Union (EU) rather than of the European Community (EC), which we do wherever appropriate in this edition of the *Handbook*. Whether the ideals of liberalisation and harmonisation, of the free movement of people, goods, capital and services across intra-Union borders, and of the concept of 'level pitch' competition in all sectors of trade, will fully materialise, only time will tell. It will be a case of 'wait and see' before we know whether the extensive legislative planning and manoeuvring of the EU bureaucracy can be (will be) effectively implemented in all member states in a manner which permits the ambitions expressed originally in the Treaty of Rome and subsequently in the Single European Act to become fulfilled. Or will these be thwarted by changes of heart in this member state or that?

The British government has had reservations about certain aspects of monetary and political union in Europe and federalism, as have other member states, notably Denmark. However, despite these, the well-published concepts of the Single European Market – liberalisation, harmonisation and free marketeering – will be implemented. Many of them, in fact, were already with us prior to 1993.

1 January 1993 was the official commencement date for the SEM legally to take effect. However, because the progress of certain aspects of the necessary legislation has been slower than anticipated, not all the steps were completed by that date. But delays in legislative progress should not be taken to mean that the whole event has been a non-starter. The necessary constitutional moves were made in accordance with the requirements of the Treaty of Rome and the UK government ratified the Single European Act as long ago as November 1986.

The abolition of barriers to trade and to the movement of goods and people between the 12 EU member states will have a significant impact on the UK road transport industry as a whole, and on road haulage and contract distribution services in particular. For firms operating in these sectors the prospects are exciting. There are new markets to be exploited and new territories to be explored. Liberalisation of all the restrictions of the past will mean the challenge of whole new horizons.

The purpose of this special section in the *Handbook* is to direct the reader's thoughts towards the implications of the SEM, and the business opportunities which it presents for those road transport firms who set themselves up to be part of this exciting event. Much of the specific legislation which has to be observed and complied with by operators taking their vehicles into Europe – and beyond – is covered in the new companion volume to this *Handbook*, namely the *European Road Freighting Handbook*.* Here the point is to show the effects of harmonisation and liberalisation. However, this is no easy task because, like trying to describe a football match, it is a moving event. Every single move is different and is largely dependent on the one which precedes it. So, by the time this is written and then published and read, events will have moved on. Then we will be totally immersed in the whole European market as a single entity, so far as trade and commerce is concerned – and so far as those individuals who may wish to move their jobs or their home are concerned. But, for the time being, this text will provide the basis of an understanding of what the whole concept is all about.

Effect of the Single European Market

With the barriers fully removed, UK firms are free to trade, uninhibited, with the whole 350 million strong consumer market of the Union, spread over some 2,253,000 square kilometres. This means there will be no restriction on the movement of goods or people, on services or on capital. Major firms operating in the EU will come to see themselves as being Euro-based rather than nationally based – obviously many do so already – and their staffs will tend to be rather more European in spirit than of specifically British, French, German or Spanish nationality, as they move between company offices scattered across Europe. These Euro-minded firms are the ones who will snap up the opportunities presented by such a large unified market, larger in fact than each of the other major world consumer markets, the United States, Japan, even the CIS Commonwealth of Independent Soviet States, formerly the Soviet Union).

Legislation

Establishing the Single Market has involved the introduction of some 300-odd separate items of legislation, many of which have already been implemented but others, such as harmonisation of vehicle weights, will not be fully in force across the EU for some years to come – the UK has a derogation (deferment) on the weights issue until 1999 for example. In transport terms, one of the most significant steps was taken long ago with the introduction from January 1988 of the Single Administrative Document (SAD) which replaced a vast number of Customs forms used in trade between member states. Previously, each member state had its own complex and extensive requirement of national documentation for Customs clearance and entry declaration, but the SAD has simplified procedures and provides a fully recognisable document which can be readily completed and interpreted, irrespective of any language barrier within the EU, or indeed outside.

In other spheres too there has been considerable progress, for example towards the mutual recognition of professional diplomas and qualifications – the UK Certificate of Professional Competence in road transport operations is recognised throughout Europe – and towards harmonisation of national technical standards, testing and certification procedures for a wide range of products from electrical appliances to pharmaceuticals, for example. Still to come is the contentious matter of an exchange rate mechanism for Union currencies, as well as harmonisation of VAT rates between member states, and the need to establish across the EU a system of fair competition for tendering to government and public authorities without national bias.

Transport Implications

The transport implications of the SEM are very significant and whole new horizons are to be opened up. In the past, road freighting across Europe has been impeded by restrictive Customs procedures and permit requirements, resulting in excessive administrative burdens, frustrating delays and inhibiting costs. In the wider arena of distribution, past inhibition has been rather more to do with the inability to trade across frontiers. Now, with all these constraints being swept away, road freighting into Europe and the setting up of Euro-wide distribution networks will be no more complex and no more fraught with bureaucracy than operating in the domestic market-place. Our present-day concepts of what is 'local' and what is 'regional' in terms of distribution are likely to undergo considerable change. In future, local distribution for some

operators may include regular trips across the channel and regional distribution may well mean delivery networks stretching right across the 12 national states with strategically located warehouses and buffer depots in many, hitherto foreign, cities and industrial conurbations.

Road Haulage Permits

Some of the most important legislative changes directly affect road transport. For instance, the complex and restrictive scheme of road haulage bi-lateral permit quotas and allocations has been abolished and is replaced by the system of 'Community Authorisations' available automatically to all UK holders of international standard 'O' licences. The UK has a limited allocation of ECMT permits covering the EU and seven further states which are party to the ECMT agreement. These are currently issued to established international hauliers and there is no chance of new allocations unless there is any unexpected move to liberalise road freight traffic between the EU and the other signatory states (ie Austria, Switzerland, Finland, Norway, Sweden, Turkey, Yugoslavia).

Cabotage

One of the most significant current issues is the matter of cabotage (see also p 204), which is now permitted on a restricted basis to cabotage permit holders from the UK and other EU member states. Cabotage is provided for under the Treaty of Rome and is now enshrined in the Single European Act 1985. It is useful consider its meaning. The word is simply defined in the *Concise Oxford Dictionary* as: 'reservation to a country of traffic operation within its territory'.

With the liberalisation of the haulage market full cabotage is to be legalised. The worry for UK operators is that foreign hauliers will arrive in Britain from the Continent with loads to deliver and will then be on the look out, not only for a return load home, but for one or more intermediate, revenue paying, haulage trips to get them to their final pick-up point. Clearly, this means business pinched from under the noses of local hauliers, it may even mean foreign hauliers poaching their established and long-term customers on a regular basis. But this is what free and uninhibited competition is all about; and there is nothing to stop the British haulier crossing the channel and doing the same in France, or Germany, or Spain, or indeed anywhere else within the Community.

From 1 July 1990 a transitional scheme (lasting until 31 December 1992) was in operation under which 15,000 cabotage permits were originally allocated annually across the EU (with the UK having 1107) with a 10 per cent annual increment. Each permit is valid for 1 month (with 2214 one-month permits replacing the original 1107 two-month permits), during which time an unlimited number of journeys can be undertaken (ie using one permit per vehicle at a time – the permit must be carried on the vehicle). Full cabotage is now to be permitted.

Entry to the Haulage Market

Access to the European haulage market is governed solely by a system of quality licences much on the lines of the present UK scheme of operators' licensing – which was amended in 1978 to align with the original EU requirement for establishment of a professional competence, qualification for those wanting access to the road haulage industry. Entrants to the haulage business need to meet standards of good repute, financial standing and professional competence, and the granting of licences will be dependent on satisfactory proof that these standards are met. Past and more recent convictions, mainly, but not exclusively, for transport and vehicle-related offences will result in licence penalty – even revocation – and inability to sustain adequate finances will have a similar effect. There are currently proposals that licence holders should be able to prove resources or guarantees equivalent to current vehicle fleet replacement value.

Vehicle Weights and Dimensions

Membership of the EU imposes a clear mandate, as part of the Common Transport Policy, to harmonise goods vehicle weights and dimensions in order to establish a regime of fair competition among transport operators of the EU. Due to our commitment to European unity, Britain agreed to conform to EU weights and dimensional requirements, and in June 1989 acceded to 40 tonne lorries – but only from 1 January 1999. (As an interim measure, 44 tonne lorries are now allowed on UK roads but solely in connection with road-rail combined transport movements.) However, the matter is more complex and extensive than just permitting 40 tonne lorries. Harmonisation requires a number of intermediate vehicle weight increases and certain dimensional changes, most notably affecting semi-trailers, and Britain introduced 16.5 metre long articulated combinations from April 1990.

It is on this issue that most controversy currently exists. The British government (and the government of the Republic of Ireland) maintains that our roads and bridges are inadequate to cope with the stresses of the proposed Euro-vehicle and axle weights. Hence, we have seen our derogation on this matter extended further and further towards the horizon. And even then the limits may only be conceded in respect of vehicles on international journeys, leaving us with a set of double standards, namely higher weights for this operation and lower weights for domestic operation. What a pickle this will present for operators and customers alike, as they try to determine which trailers may or may not be loaded for which destinations.

Principal among the EU vehicle weights proposals, and the ones that we shall not see until 1999, are 40 tonnes on five axles for articulated vehicles with an 11.5 tonne maximum loading on drive axles. (Container carrying vehicles are now permitted to operate at up to 44 tonnes.) From 1 January 1993, 35 tonnes has been allowed on drawbar vehicles (ie road trains) and four-axle artics, and 25 tonnes on three-axle rigid vehicles.

The EU requirements for harmonised vehicle and trailer weights and dimensions are to be found in EC Directive 85/3 and further in EC Directives

89/338 and 89/461 which amend the 1985 version. Recently, an increase in length for drawbar vehicles from 18 metres to 18.35 metres was implemented, but with a restriction on load space of 15.3 metres and a minimum coupling length of 0.7 metres, leaving just 2.35 metres for cab space.

Full details of current and proposed vehicle weights and dimensions are to be found in Chapter 12.

Harmonisation of Taxes and Duties

Final legislation to effect equalisation policies on vehicle taxes and fuel duties across the EU are still some way off, but in the meantime a variety of alternative proposals keep rearing their heads in the efforts to find an equitable way of achieving parity. Much work is still to be done by the Council of Ministers of Transport for the 12 states on the problem of harmonising these. An average cost formula will need to be developed that is acceptable across the EU, but this is highly dependent on relative infrastructure costs, something which is not likely to produce ready acceptance across 12 such diverse nations.

Of particular interest to international road hauliers, for example, are the proposals that diesel fuel prices across the EU should be constrained within a price band equivalent to 195 to 205ECU per 1000 litres. A survey (carried out prior to the erratic price increases brought about by the Gulf crisis) showed that prices would have to *increase* by something in the order of 27ECU per 1000 litres in Spain (ie 4 pence per litre) to as much as 97ECU in Luxembourg (ie 13 pence per litre) and an average of 65ECU (equivalent to 8.9 pence per litre) across seven member states with diesel prices currently lower than the UK.

Little further progress has been made (or at least publicly announced) on the subject of equalising Vehicle Excise Duty across the EU. Similarly, tariffication which currently controls much of mainland Europe's road freight transport will need to be replaced by competitive market pricing. This is something which Germany is certainly not keen to see – it has threatened the imposition of road taxes on foreign (ie other EU) vehicles entering its territory, despite the clear illegality of such a move under Community legislation, for which it has been warned by the European Commission of court action.

Road Haulage Tariffs

International and domestic road haulage is to be freed from the requirements of tariff controls and it will become illegal for member states to set such tariffs.

Border Controls

A fundamental aspect of the SEM is that citizens of the European Union should be able to pass freely across intra-Union borders. A step in the direction of freeing controls on such movement was made with the Schengen Agreement between the Benelux countries, Germany and France, whereby these states would allow free passage across their borders – in effect a model for adoption by the whole Union.

Italy is currently being taken to the European Court for contravention of the Treaty of Rome (Article 30) by persisting in carrying out physical checks on intra-Union goods vehicles at its borders for customs purposes.

Negotiations are still proceeding with the so-called 'transit' countries of Austria, Switzerland and Yugoslavia to ease the various restrictions (eg Austria's night-time ban) which currently hamper goods vehicle traffic through these countries *en route* to other EU destinations.

Problems on other borders have also received wide publicity, especially those between France and Spain in which many British vehicles were caught up (with violence and vandalism of vehicles being an unfortunate result).

Hired Vehicles

The EU aims to free the hiring and use of hired vehicles (ie without drivers) across borders by 1993.

Dangerous Goods Drivers

An EU Directive requires drivers of dangerous goods vehicles to be suitably trained from July 1992 (in the case of explosives and bulk loads in tankers) and January 1995 (for all other dangerous goods loads). This requirement has been implemented in the UK under domestic legislation. Further details on this subject are to be found in Chapter 23.

Driver Licensing

Major changes to the UK driver licensing system have already been implemented (see Chapter 6 for full description of the new system) in compliance with EU Directives. Further changes are due when a second Directive comes into force. In outline, the effects of this change were the introduction from 1 April 1991 of the 'unified' licence showing all driving entitlements (ordinary and vocational) in a single licence; goods vehicles being reclassified into EU vehicle categories and the present UK requirement for vocational licences for driving *heavy* goods vehicles (hgv) over 7.5 tonnes pmw being replaced by an entitlement to drive *large* goods vehicles (lgv) over 7.5 tonnes pmw (subject to certain exemptions). When the EU *Second* Directive comes into force the vocational qualification will apply to vehicles over 3.5 tonnes pmw.

Union-Wide Transport and Infrastructure Development

Finally, if there is to be a truly united Europe with no impediments to trade, then we need to see substantial improvements in infrastructure to facilitate great movement of goods and people, and especially on this side of the channel. The Department of Transport recently created considerable publicity for its planned investment in the road network, but it needs a lot more than publicity; it needs action, nothing short of prompt action. Fears have already been expressed (by the Road Haulage Association, for example) that Britain's poor road network could result in firms in northern England losing out on trade to the south-east of England. An example quoted suggested that firms in Bruges rather than Bradford may be better placed to supply shops in the south-east. The CBI has said that road congestion costs Britain £3000 million per year. Traffic densities on our roads are twice that of Germany and three times that of France, and yet no further new motorways were planned to open. The Freight Transport Association, on the same theme, has suggested – following a survey of some of its members – that reliable congestion free roads could cut the number of lorries needed in Britain by 8000.

Department of Transport figures show that in recent years motorway traffic has increased by 7 per cent annually and on all classes of road the numbers of heavy vehicles increased by 20 per cent. Many individual road sections – parts of the M6 for example – already carry 20 per cent more traffic than they were designed for and it is projected that by the end of the century there could be 40 per cent more vehicles using Britain's existing road network.

Announcements in the early part of 1994 promise £1 billion of spending on 22 major new road projects to start in 1994–95 and a radical new prioritisation scheme to get essential schemes built faster. In June 1994 a further spending promise of £500 million on road maintenance was made but there was no promise of any significant extension of the motorway network. So, despite this extra spending Britain will still have only some 2000 miles of motorway by the end of the century, against 5000 miles in France by the mid-1990s.

Germany is planning to spend DM2000 million on road improvements in the immediate future, mainly as a result of reunification, with many of the individual projects being to update the crumbling east German network. But nevertheless the overall result will be substantial improvements across the state.

The UK road building and repair programme has gone on piecemeal over the years and will continue to do so up to the end of the century and beyond. No clear strategy has been developed to cater for the expected boom in passenger and freight traffic moving between the UK and the expanded community. With new motorways taking at least 15 years to build, Britain is clearly faced with major road infrastructure problems which will inhibit the potential for UK firms to take advantage of the new trading promises of the Single Market. The RHA is right in calling for a bi-partisan study towards a strategic roads plan and the government should be looking to spend more of the £14 billion collected in taxes from road users each year on developing an infrastructure network that puts the UK on the map as a leader in Euro-infrastructure development. We should not continue to be seen as the poor relations of the Union.

Channel Tunnel

The Channel Tunnel, which opened in May 1994, besides being a major

engineering achievement, will herald a new era in transport between the UK and mainland Europe. Its opening means that Britain is no longer an offshore island, but is umbilically joined to the whole of Europe by a fixed link rail system.

While road traffic will not be able to pass through the tunnel independently, for the road freight industry it opens up yet another possibility for overcoming the main physical obstacle to trade between the UK and mainland Europe, namely the need to cross that stretch of water. The likely 'war' between the Channel Tunnel company and the cross-channel ferry service operators can only benefit the road freight industry from both competitive pricing structures and competitive levels of service (ie speed and frequency of crossing and efficiency in handling). In response to the challenge of the tunnel, the ferry-ship operators plan to introduce larger and more luxurious vessels. The talk is of 'jumbo' ferries and of 'cruising' across the channel. These operators also recognise the need for specialisation – on either freight or passenger traffic – and for speed to compete with the through-tunnel high-speed rail services. As Peter Bennett, Railfreight Director of United Transport Logistics, says: 'This ability of the rail network to guarantee express delivery time of not just one 24-tonne consignment but 50 or so TEU's on the one train is likely to have a marked impact. Such a level of service will undoubtedly attract traffic from road and command a premium price.'

The particular advantage of the tunnel for rail freighting business lies in the increased length of through trunk haul that will be possible from the UK into central Europe. For British Rail particularly, the tunnel is a key-stone in its optimism for increased traffic volumes resulting from the SEM. It has extensive plans for high-speed rail links from mainland Britain to major cities and industrial centres in Europe, as well as providing a cross-channel service for road freight traffic. On the passenger side, British Rail and the tunnel operator (Eurotunnel) anticipate being able to syphon off some 20 per cent of short-haul air traffic from the prestige London–Paris and London–Brussels routes providing a 2-hour 30-minute service for the former – centre to centre – and 2-hour 20-minute service for the latter.

The problem with rail freighting is always the disparate cost of terminal handling and delivery at each end of the link for normal merchandise traffic (as opposed to bulk traffics). With private-siding to private-siding traffic a break-even rail trunk link of some 200 miles is necessary, but with road collection and delivery the break-even extends to 300 miles, so in theory the economics of through-tunnel business will be more advantageous to British Rail than its mainland Britain business.

The tunnel company, Eurotunnel, is predicting that it will have handled an annual freight volume in the region of 15.5 million tonnes in its opening year, increasing to 23.6 million tonnes by the year 2003 (plus some 14.5 million passengers on the shuttles and another 14 million on through trains). British Rail and SNCF have signed agreements with Eurotunnel, committing them to 50 per cent of the tunnel's capacity, even though these two organisations already operate a number of train-ferry services. Eurotunnel is talking of providing two to four 100mph freight shuttles an hour (although now lower speeds in the region of only 130kph (80mph) are being suggested), depending on demand, on a 24-hour, 7-day basis and as many as 27 freight trains each way daily. The freight shuttles will each carry up to 26 heavy vehicles of up to 44 tonnes in weight and 18.5 metres in length – anticipating changes in European truck regulations within the lifetime of the rolling stock.

Rail Network Developments

Development of the high-speed rail link between London and the Channel Tunnel is only one part of the domestic rail infrastructure needed to enable British industry to reap the rewards of increased intra-Union trading in the SEM. Large sections of our other rail network extending beyond London to the south, the west and the north will need to be upgraded to ensure full advantage of high-speed passenger and freight train travel in the future. Some sections are in place such as the east coast and west coast lines to Scotland, which will eventually allow 140mph operation with the new Electra train, although for the time being operating speeds are kept to 110–125mph and it appears they will not match the maximum speeds of the London–channel tunnel link which is destined to run at 185mph, but only in certain places, until it has crossed the channel, when it will then be able to operate at its full potential – but still not at the reputed 236mph maximum speed of the French-built TGV train which already runs between major cities in France.

Across the channel, trains will link with the European high-speed rail system that with its new lines extends as far east as Vienna (in Austria – not yet part of the EU) and south to the foot of Italy, to southern Spain and across to Lisbon in Portugal on the west. Many existing lines across Europe are to be upgraded to high-speed quality and, where appropriate, new linking lines and extensions are to be built – for example to Berlin in eastern Germany.

The problem of European rail development lies in the long-standing rivalries between national rail systems. However, the European Council of Transport Ministers is working towards agreement on a fully integrated Union-wide rail network for both high-speed passenger and combined road-rail freight services.

Road/Rail Operations

With the longer-haul potential of UK–Europe movement of goods via the Channel Tunnel the concept of road-rail (intermodal or combined) operations comes into its own. In concept, the idea is to keep the expensive part of the road operation (ie the tractive unit and driver) fully utilised on the road while the trailer is sent on its long-haul journey by rail (which may be faster and certainly relieves the pressure on road traffic). As Peter Bennett explains: 'The use of an intermodal unit allows collection and delivery of consignments by trailer. This gives greater flexibility to the producer/user thereby allowing goods to be loaded in the accustomed manner without any change in current practice. The unit can then be handled rapidly from trailer to wagon so it may enjoy the benefits of rail trunking'.

Under the proposed agreement mentioned above, road-rail operators will have access to the whole EU rail network, but on an operational basis this will only work successfully when loading infrastructures have been harmonised.

Initially, the combined freight network will be concentrated on 30 key high-volume routes both within Union territories and the territories of other members (ie shareholders) in the European rail container operation – Intercontainer. Eventually it will progress to link with the rail systems of eastern European countries and the newly independent Soviet states.

Intermodal operations have been popular in Europe for some years. For example, the German Kombiverkehr and the French Novatrans systems are totally road-rail oriented. SNCF operate intermodal services for refrigerated and for hazardous chemical loads. Intermodal operations, however, do have shortcomings. In consequence of this in April 1991 the International Road Transport Union (IRU) and the International Union of Railways (UIC) concluded an agreement between the two organisations to co-operate as partners in the development of combined road-rail transport in Europe. This agreement is particularly notable, first for the way in which it intends the two normally competing modes to respect the rules of economic competition and secondly for calling on European governments to cease regulatory discrimination between modes which hampers the efficient development of combined transport.

Without such an agreement and the co-operation of governments as requested, the true development of the potential for cross-border intermodal transport will be lost. Most especially this applies where the carriage of dangerous goods is concerned, an area where regulatory confusion is rife as government agencies pursue their own vested interests. For example, Britain's promised £16 billion, 3-year investment programme in road and rail development, announced at the end of 1990, identifies plans for both modes but makes no plan for a share of the capital to encourage (through infrastructure developments) intermodal transport – possibly the most encouraging alternative solution for both the pro-road and pro-rail lobbies.

Only a year later in August 1991 did the Secretary of State for Transport announce plans to break the British Rail monopoly on rail freight services by opening up the network for use by private freight operations – the privatisation now being well under way in mid-1994. The intention is to do away with red-tape bureaucracy and invest in a programme to encourage the development of more road-rail combination schemes – principally through hauliers establishing their own freight terminal facilities.

It has to be recognised that changing from road to road-rail operations is not a simple matter of choice; there are problems of provision of suitable infrastructures (ie transfer facilities and equipment) and vehicles (road and rail), both requiring substantial capital investment. In particular, British Rail does not have the versatility to handle the many different types of road vehicle trailer and swop-body currently in common use on the Continent, and most British hauliers do not have special trailers suitable for travel 'piggyback' or 'kangaroo' style on European rail rolling stock. However, as Peter Bennett points out, the 'duality' of intermodal transport 'not only satisfies operational and economic demands but also helps satisfy the requirements of investment criteria. Operators will have an understandable reluctance to invest in such a restrictive piece of capital equipment as a rail wagon, but will find an intermodal vehicle a far lower risk . . . the equipment defies obsolescence and ensures versatility'.

Nevertheless, with the SEM now with us, more effort will undoubtedly be put on overcoming these limitations and we should see progressive development of intermodal transport, both for its economic benefits to users and environmental benefits to the Union.

Takeovers and Mergers

Many UK firms in transport and in the wider sphere of distribution have already established their Euro-links and networks. The trade press is full of stories about takeovers, mergers and inter-trading arrangements between British and Euro firms, about new UK–Europe services and about new Euro-management appointments – bi-lingual and multi-lingual capabilities are now becoming a priority qualification for young managers. All this indicates that progressive companies did not wait for the SEM to open formally. They intended being in on the ground floor while the territory was still sufficiently clear to give them choice and flexibility in where and with whom they trade. With 1993 behind us, the whole market-place looks as though it will be well and truly sewn up. Those firms who have not seen the opportunities laid bare before them and those who still pooh-pooh the whole concept will be left way behind and may even see their own domestic trade disappear down the road in a French or Belgian lorry.

The European Road Freighting Handbook by David Lowe, also published by Kogan Page, is a companion volume to *The Transport Manager's and Operator's Handbook*. It is intended to provide both UK international road transport operators and European-based haulage and distribution firms with concise and readily understandable explanations of relevant legal requirements throughout the Community and current operational practices in international road freight transport. As such, it is expected to become as much the 'bible' in Euro-freighting as is this *Handbook* in the UK domestic transport industry.

1: Goods Vehicle Operator Licensing

Operator licensing is the regulatory control system imposed by government to ensure the safe and legal operation of most goods vehicles in Great Britain. While other individual aspects of legislation also apply to such vehicles, the 'O' licensing system provides the overriding control of road freight transport operations. Failure to observe the requirements and conditions under which 'O' licences are granted will lead to severe penalty; likewise, breach of this other legislation can result in appropriate penalties as set out in respective statutes and, subsequently, will involve penalties against the operators' licence itself. Similar licensing controls apply to goods vehicle operations in Northern Ireland under the separate Road Freight Operator's Licence scheme operated in the Province, and in other EU member states where common quality standards for entry into the haulage industry are being applied under Community harmonisation rules.

Under the current British licensing scheme, trade or business users of most goods vehicles over 3.5 tonnes maximum permissible weight must hold an 'O' licence for such vehicles, whether they are used for carrying goods in connection with the operator's main trade or business as an own-account operator (ie a trade or business other than that of carrying goods for hire or reward) or are used for hire or reward haulage operations. Certain goods vehicles, including those used exclusively for private purposes, are exempt from the licensing requirements. Details of the exempt vehicles to which 'O' licensing does not apply are given on pp 2–4.

The original system of operators' licensing was established by the Transport Act 1968. Section 60 of that Act states that no person may use a goods vehicle on a road for hire or reward or in connection with any trade or business carried on by him except under an operator's licence. The basis of the present revised regulations remains in this statute.

Regulations which took effect on 1 January 1978 made substantial changes to the 'O' licensing system which had existed from 1970 to the end of 1977. These regulations, the Goods Vehicle (Operators' Licences) Regulations 1977, introduced a three-tier system of 'O' licensing in contrast to the previous single-tier system which applied similarly to both professional hauliers and own-account operators. From 1 June 1984 further significant changes to the system were introduced by the Goods Vehicles (Operators' Licences, Qualifications and Fees) Regulations 1984 (as amended 1986, 1987, 1988, 1990 and 1993) which, principally, gave Licensing Authorities (LAs) powers to consider representations to 'O' licence applications on the grounds that environmental nuisance or discomfort would be caused to local residents by the presence of goods vehicles at or operating from particular locations.

The 'O' licensing system, which is based on the concept of ensuring legal and safe operation and thus is a system of 'quality' as opposed to 'quantity' licensing, is administered on a regional (ie Traffic Area) basis throughout Great Britain. Northern Ireland's Road Freight Operators' Licensing system is dealt with separately by the Department of the Environment in Belfast. Eight Traffic Area Offices (TAOs) form the network (see Appendix I for list) providing administrative support to the LAs, one of which currently holds the post in two areas. These LAs have the statutory power to grant or refuse operators' licences, to place environmental conditions or restrictions on such licences where necessary, and subsequently to impose penalties against licences in the event of the holder being convicted for goods vehicle related offences.

Approximately 404,000 vehicles are specified on 'O' licences in Great Britain, on about 125,000 licences of which some 60,000 are restricted licences, 50,000 are standard national licences and 15,000 are standard international licences.

Proposals for Change

At the time of preparing this 1995 (25th) edition of the *Handbook*, little progress has been made towards introducing the proposed changes to the 'O' licensing system outlined in last year's edition. These proposals were included in a Department of Transport (DoT) consultation paper on the future of 'O' licensing that was in circulation in 1993, seeking industry views on major changes to the present system which is described in detail in this chapter. Publication of this document followed a wide-ranging review of operator licensing undertaken in the latter part of 1992. Most significant among the changes envisaged at that time was the raising of the threshold at which licensing starts, from the present level of 3.5 tonnes to 6 tonnes. This, the DoT says, would bring Britain broadly into line with European requirements and would remove some 18,500 vehicles from the licensing system, thus reducing the burden of 'O' licensing on road haulage. This particular idea has now been abandoned. Among the other proposals under consideration are a change from 5-yearly renewal to continuous 'O' licensing and the issue of windscreen discs to cover the number of vehicles authorised on a licence rather than just for those actually specified. Those licensing changes still being pursued by the government (mainly the continuous licensing arrangement, which it hopes to introduce over a 5-year period from 1 July 1995) are included in the Deregulation and Contracting Out Bill currently before parliament.

Exemptions from 'O' Licensing

There are a number of categories of vehicle which are exempt from 'O' licensing requirements as described below.

Small Vehicles

The principal exemption applies to 'small' vehicles identified as follows.

Rigid vehicles are 'small' if:
(a) they are plated and the gross plated (ie maximum permissible) weight is not more than 3.5 tonnes;
(b) they are unplated and have an unladen weight of not more than 1525kg.

A combination of a rigid vehicle and a draw-bar trailer is 'small' if:
(a) both the vehicle and the trailer are plated, and the total of the gross *plated* weights is not more than 3.5 tonnes;
(b) either the vehicle or the trailer is not plated, and the total of the *unladen* weights is not more than 1525kg.

However, if the unladen weight of an unplated trailer is not more than 1020kg this does not have to be added to the unladen weight of the drawing vehicle for the purposes of deciding if the combination is 'small' for 'O' licensing purposes.

Articulated vehicles are 'small' if:
(a) the semi-trailer is plated, and the total of the *unladen* weight of the tractive unit and the plated weight of the semi-trailer is not more than 3.5 tonnes;
(b) the semi-trailer is not plated, and the total of the *unladen* weights of the tractive unit and the semi-trailer is not more than 1525kg.

Also included in the exemptions are pre-1 January 1977 vehicles which have an unladen weight not exceeding 1525kg and a gross weight greater than 3.5 tonnes but not exceeding 3.5 tons.

Other Exemptions:

Regulations list the following further specific exemptions from 'O' licensing requirements.
1. Vehicles licensed as agricultural machines used solely for handling specified goods, and any trailer drawn by them.
2. Dual-purpose vehicles and any trailer drawn by them.
3. Vehicles used on roads only for the purpose of passing between private premises in the immediate neighbourhood and belonging to the same person (except in the case of a vehicle used only in connection with excavation or demolition) provided that the distance travelled on the road in any one week does not exceed 6 miles.
4. Motor vehicles constructed or adapted primarily for the carriage of passengers and their effects and any trailer drawn by them while being so used.
5. Vehicles used for the purpose of funerals.
6. Vehicles used for police, fire brigade and ambulance service purposes.
7. Vehicles used for fire fighting or rescue work at mines.
8. Vehicles on which a permanent body has not yet been built carrying goods for trial or for use in building the body.
9. Vehicles used under a trade licence.
10. Vehicles used by or under the control of HM United Kingdom Forces.
11. Trailers not constructed for the carriage of goods but which are used incidentally for that purpose in connection with the construction, maintenance or repair of roads.
12. Road rollers or any trailer drawn by them.

13. Vehicles used by HM Coastguards or the Royal National Lifeboat Institution for the carriage of lifeboats, life saving appliances or crew.
14. Vehicles fitted with permanent equipment (ie machines or appliances) so that the only goods carried are:
    (a) for use in connection with the equipment;
    (b) for thrashing, grading, cleaning or chemically treating grain or for mixing by the equipment with other goods not carried on the vehicle to make animal fodder; or
    (c) mud or other matter swept up from the road by the equipment.
15. Vehicles used by a local authority for the purpose of enactments relating to weights and measures or the sale of food or drugs.
16. Vehicles used by a local authority under the Civil Defence Act 1948.
17. Steam-propelled vehicles.
18. Tower wagons or any trailer drawn by them, provided that any goods carried on the trailer are required for use in connection with the work on which the tower wagon is used.
19. Vehicles used on airports under the Air Authority Act 1975.
20. Electrically propelled vehicles.
21. Showmen's goods vehicles and any trailer drawn by such vehicles.
22. Vehicles plated for more than 3.5 tonnes but not more than 3.5 tons and which are not over 1525kg unladen weight and which were first registered prior to 1977.
23. Vehicles used by a highway authority in connection with weighbridges.
24. Minibuses and any trailer drawn.
25. Hackney carriages.
26. Vehicles used for emergency operations by the water, electricity, gas and telephone services.
27. Recovery vehicles.
28. Vehicles used for snow clearing or the distribution of grit, salt or other materials on frosted, ice-bound or snow covered roads and for any other purpose connected with such activities.
    (*NB: This exemption is not restricted solely to local authority-owned vehicles.*)
29. Vehicles going to or coming from a test station and carrying a load which is required for the test at the request of the Secretary of State for Transport (ie by the test station).

Exemptions also apply to vehicles used privately (ie for carrying goods for solely private purposes and not in any way connected with a business activity) and by voluntary organisations.

All other goods-carrying vehicles over 3.5 tonnes gross weight, not specifically shown as exempt in the list above must be covered by an 'O' licence. This includes such vehicles only temporarily in the operator's possession, hired or borrowed on a short-term basis, where they are used in connection with a business (even a part-time business).

The Vehicle User

An 'O' licence must be obtained by the 'user' of the vehicle for all the vehicles he operates to which the regulations apply. The 'user' may be the owner of the vehicle or he may have hired it. If the vehicle was hired without a driver, the hirer is the 'user'. There is considerable importance attached to the word 'user', and its exact meaning, both for the purposes of 'O' licensing and in other regulations. It may be explained simply as follows.

1. An owner-driver who uses his vehicle in connection with his own business is the 'user' of his own vehicle.
2. If the owner of a vehicle employs a driver to drive it for him and he pays the driver's wages then the owner is the 'user' because he is the employer of the driver.
3. If a vehicle is borrowed, leased or hired without a driver and the borrower or hirer drives it himself or pays the wages of a driver he employs to drive it then the borrower or hirer is the 'user'.

From this it can be seen that, in general, the person who pays the driver's wages is the 'user' of a vehicle, and it is this person (or company) who is responsible for holding an 'O' licence and for the safe condition of the vehicle on the road and for ensuring that it is operated in accordance with the law. However, it must be remembered that the driver himself, although an employee, is still also the user of the vehicle in the context of certain legislation (eg the Road Vehicles [Construction and Use] Regulations 1986, as amended) and he, too, is responsible for its safe condition on the road and is liable to prosecution if it is not in safe and legal condition.

A situation has arisen in recent times where owner-drivers of goods vehicles who cannot themselves obtain the professional competence qualification have had their vehicles specified on the 'O' licence of another operator but have nevertheless been paid as self-employed contractors to the other operator. This practice is illegal because if the driver owns the vehicle and uses it in connection with his business then by virtue of the regulations he is the 'user' and is therefore responsible for holding the 'O' licence for it.

Agency Drivers

Dependence on agencies for the supply of temporary drivers to provide relief manpower when regular drivers are not available has caused difficulty in interpretation of the term 'user' and in deciding who should hold the 'O' licence; the vehicle owner or the agency which employs the driver. It can be seen from item 2 above that the person who pays the driver's wages is the 'user', and is therefore the person who should hold the 'O' licence.

However, the status of the vehicle 'user' in these circumstances has been determined by the agencies getting operators to sign agreements whereby the vehicle operator technically becomes the employer of the driver rather than the agency being the employer and consequently the operator remains the legal 'user' of the vehicle. Usually the agency asks the hirer to sign an agreement whereby the agency becomes the 'agent' of the operator for these purposes in paying the driver's wages. This practice has been proved in court to be legally acceptable on the grounds that the Transport Act 1968 s92(2) states that '. . . the person whose servant or agent the driver is, shall be deemed to be the person using the vehicle'. The driver is considered to be the servant of the hirer because the hirer gives instructions and directs the activities of the driver who is temporarily in his employ.

The great danger with agency drivers is that the operator has no sound means of establishing whether the driver is legally qualified to drive or whether he has already exceeded his permitted driving times on previous days and whether he has had adequate rest periods. Reputable agencies usually go to considerable lengths to ensure that drivers provided by them for their clients are properly licensed and have complied with the driving hours rules in all respects. It is worth also remembering that the use of casually hired or temporary drivers (whose backgrounds and previous experiences may not be fully known) can result in jeopardy of the contract of insurance covering the use of vehicles and there could also be serious security risks as well as possible 'O' licence penalties for infringements of the law. For this reason the operator should confine himself to obtaining drivers from reputable agencies who are known to have vetted drivers satisfactorily.

A new Code of Practice for the employment of agency drivers has been devised by the Road Haulage Association (RHA) and the Federation of Recruitment and Employment Services (FRES). The Code sets out the respective duties and responsibilities of the haulier on the one hand and the supplying agency on the other, with a check list for each to ensure that full and correct information is exchanged as to the requirement for the driver (eg the skills and personal attributes required) and the particular job to be done. It also contains a model set of instructions and procedures which should be given to drivers.

Copies of the Code (at a cost of £1) can be obtained from local RHA offices or from the FRES at 36–38 Mortimer Street, London W1N 7RB.

Restricted and Standard 'O' Licences

There are three types of 'O' licences as follows:

1. **Restricted licences:** available only to own-account operators who carry nothing other than goods in connection with their own trade or business, which is a business other than that of carrying goods for hire or reward. These licences cover both national and international transport operations with own-account goods. Restricted 'O' licence holders must not use their vehicles to carry goods on behalf of customers' businesses – even if it is done only as a favour or is seen as being part of the service provided to a customer and even if no charges are raised (see below) – such activities are illegal and could result in penalties.

2. **Standard licences (national operations):** for hire or reward (ie professional) hauliers, or own-account operators who also engage in hire or reward operations, but restricted solely to national transport (ie operations exclusively within the UK). Own-account holders of such licences may also carry their own goods (but not goods for hire or reward) on international journeys.

3. **Standard licences (national and international operations):** for hire or reward (ie professional) hauliers, or own-account operators who also engage in hire or reward carrying, on both national and international transport operations.

National transport operations in this context include journeys to and from ports with loaded trailers which were previously considered to be international journeys.

Standard Licences for Own-account Operators

Own-account operators may voluntarily choose to hold a standard 'O' licence for national or both national and international transport operations instead of a restricted licence provided they are prepared to meet the necessary additional qualifying requirements (principally the professional competence qualification – see Chapter 2). Among the reasons which may influence them to take this step is the desire to carry goods for hire or reward to utilise spare capacity on their vehicles, especially on return trips. Such a requirement may also arise because a firm is involved in carrying goods for associate companies on a reciprocal or integrated working basis which does not come within the scope of activities which are permitted under 'O' licensing between subsidiary companies and holding companies (see p 30) or firms may find themselves in the position where they carry goods in connection with their customer's, as opposed to their own, businesses. Firms holding restricted 'O' licences may not carry on their vehicles goods on behalf of customers (ie in connection with the trade or business of the customer rather than in connection with their own business) or other firms even if such operations are described as being a 'favour' or 'part of the service' to the customer and involve no payment whatsoever. This may occur, for example, when a vehicle delivers goods to a customer and the customer then asks the driver to drop off items on his return journey because he is 'going past the door' and their own vehicle is not available. Such activities would be illegal under the terms of a restricted 'O' licence and if two convictions for such an offence are made within 5 years, the licence must be revoked by the LA.

Requirements for 'O' Licensing

In order to obtain an 'O' licence, applicants must satisfy certain conditions specified in regulations.

Restricted Licences

Applicants must be:
1. fit and proper persons;
2. of appropriate financial standing.

Standard Licences (national transport operations)

Applicants must be:
1. of good repute;
2. of appropriate financial standing;
3. professionally competent, or must employ on a full-time basis a person who is professionally competent, in national transport operations.

Standard Licences (national and international transport operations)

Applicants must be:
1. of good repute;
2. of appropriate financial standing;
3. professionally competent, or must employ on a full-time basis a person who is professionally competent, in both national and international transport operations.

Other Legal Requirements

Besides the specific requirements mentioned above, licence applicants and holders have to satisfy further legal requirements relating to the suitability and environmental acceptability of their vehicle operating centres, the suitability of their vehicle maintenance facilities or arrangements and, overall, their ability and willingness to comply with the law in regard to vehicle operating as demonstrated by signing the declaration of intent on the 'O' licence application form. These matters are dealt with in detail in this chapter.

Good Repute

For an LA to be able to grant an 'O' licence, he must determine that the applicant is of 'good repute'. Without this particular requirement being well established, the fact that the applicant may meet all other relevant criteria is of no account, no licence will be granted.

The term 'good repute' is not specifically defined in regulations but for the purposes of the standard 'O' licensing scheme it means that the applicant for a licence (ie an individual) does not have a past record of serious offence convictions during the previous 5 years (excluding those which are 'spent' – see below) or of repeated convictions relating to road transport and the roadworthiness of goods vehicles on an 'O' licence. Similarly, to be a fit and proper person in order to obtain a restricted 'O' licence means that there should not be a past record of such offences. If it is a limited liability company applying for a licence and the company has relevant convictions on its record (ie for serious or repeated offences) then the LA may use his discretion in deciding whether the firm is of good repute, a facility he does not have in the case of individuals or partnership businesses.

It should be noted that the LA will not necessarily refuse to grant a licence to an applicant who has had convictions – but he must if they are for serious or repeated offences which affect the applicant's good repute – but he will consider the number and seriousness of the convictions before making a grant. He may, for example, issue a licence for a shorter period to see if the applicant has 'mended his ways', or grant a licence for fewer vehicles than the number requested. If, during the currency of a licence, an 'O' licence holder is convicted of offences related to goods vehicle operations, the LA may call the operator to a public inquiry and determine whether he is still a fit and proper person or of good repute and whether he should be allowed to continue holding an 'O' licence (see also pp 31–2).

In the case of partnership applications for licences, if one of the partners is considered not to be of good repute, then the LA will be bound to conclude that the partnership firm is not of good repute and, on that basis, he will refuse to grant a licence.

The relevant offences for which conviction damages a person's good repute are those specified in the Transport Act 1968 s69(4) and include matters relating to:

- unlawful use of vehicles;
- unroadworthiness of vehicles, maintenance and maintenance records;
- plating and testing;
- speed limits;
- weight limits, safe loading and overloading of vehicles;
- licensing of drivers;
- drivers' hours and record keeping (including tachographs);
- certain traffic offences (eg parking);
- illegal use of rebated fuel oil;
- forgery;
- international road haulage permits.

Increasingly, convictions for vehicle excise licensing offences are featuring in consideration of 'O' licence applicants' good repute by the LAs and a number of licences have been either refused or penalised on these grounds.

Additionally, in the case of applications for standard international 'O' licences, the LA must be told of convictions for any other offence besides those concerning the matters listed above.

Serious Offences

A serious offence, by an individual or by a company or its management, as referred to above, is defined as one where, if committed in the UK, a sentence of more than 3 months' imprisonment or a Community Service Order of more than 60 hours was ordered, or a fine exceeding level four on the standard scale (currently £2500) was imposed. If committed abroad the seriousness of the offence would be determined by assessing the punishment relative to UK standards.

Spent Convictions

Spent convictions, which are referred to above in the context of good repute, are those which *do not* have to be declared on the licence application form, because they were incurred sufficiently long ago to be considered legally invalid for determining a person's past record. Under the Rehabilitation of Offenders Act 1974 a person who was convicted of an offence for which they were fined by a court need not disclose that conviction after 5 years (or 2½ years if aged under 17 at the time), or if a prison sentence of 6 months or less was imposed it need not be disclosed after 7 years (or 3½ years if aged under 17 at the time). In the case of a prison sentence of between 6 months and 2½ years the rehabilitation period (after which no disclosure is necessary) is 10 years, or 5 years if aged under 17 at the time.

A table showing the full range of rehabilitation periods is given below.

(i) Fixed rehabilitation periods:

Sentence	Rehabilitation period
• For a prison sentence between 6 months and 2½ years	10 years (5 years if aged under 17 years at the time)
• For a prison sentence of 6 months or less	7 years (3½ years if aged under 17 years at the time)
• For a fine or a community service order	5 years (2½ years if aged under 17 years at the time)
• For an absolute discharge	6 months
• For Borstal	7 years
• For a detention centre	3 years

(ii) Variable rehabilitation periods:

Sentence	Rehabilitation period
• A probation order, conditional discharge or bind over	1 year or until the order expires (whichever is longer)
• A care order or supervision order	1 year or until the order expires (whichever is longer)
• An order for custody in a remand home, an approved school or an attendance centre order	A period ending 1 year after the order expires
• A hospital order (with or without a restriction order)	5 years or a period ending 2 years after the order expires (whichever is longer)

In Scotland, supervision requirements made by children's hearings attract similar rehabilitation periods to those for care or supervision orders.

Appropriate Financial Standing

This means the applicant being able to prove to, or assure, the LA that he has sufficient funds (ie money) readily available to maintain the vehicles for which he has applied for a licence to the standards of fitness and safety required by law. (While there are no laid-down standards of monetary requirement to support a national 'O' licence application, one LA stated in 1993 that a sum of around £2000 per vehicle and trailer was needed – it is known that the DoT has produced an internal 'official' document entitled *The Guide for Assessing Financial Standing* which the Transport Tribunal has recommended should be available to licence applicants in the interests of fairness.) An additional questionnaire relating to the applicant's financial resources is sometimes required to be completed in addition to the normal 'O' licence application form (see pp 15–16) when making a new application or renewal application for a standard licence.

The LA has considerable powers to inquire into the finances of applicants, including the right to ask for the production of proof of financial standing by means of audited accounts and bank statements or bank references, savings or deposit account books, or other evidence of funds stated to be available for the maintenance of vehicles. He will particularly look at the firm's balance sheet within the audited accounts and determine its liquidity (ie its capability of paying its debts as they fall due). He will examine the relevant financial ratios such as current assets to current liabilities, which ideally should not be less than 2:1, and the so-called quick ratio of quickly realisable assets (ie items which can quickly be turned into cash) to current liabilities which, in this case, should not be less than 1:1. It is believed that where a firm's accounts show a current ratio of less than 0.5:1 the LA should consult with the financial assessors he is empowered to call on.

In complex cases, usually involving companies where, perhaps, funds are moved between one subsidiary and another, and there is cross-accounting and such like, the LAs can call on financial experts (assessors) to help determine the true position of an applicant.

Where an application is for a standard international licence, there are additional financial requirements to be met in line with minimum standards imposed by the European Union. Applicants must be able to show that they either have funds (ie available capital and reserves) or have access to funds equivalent to the lesser of the following two alternative minimum values:

1. at least 3000ECU for each vehicle authorised, or requested to be authorised, on the licence; or
2. at least 150ECU per tonne for the total of the maximum weights of all vehicles authorised or to be authorised on the licence.

This particular financial requirement applies only to those applicants for international 'O' licences since 1 October 1990. It is not applicable to operators who have continuously held such a licence since before this date.

NB: The current sterling value of an ECU (European Currency Unit) is published daily in the Financial Times. *On 23 June 1994 the ECU value of £1 was 1.2838 so 3000ECUs would equate to £2337 and 150ECUs would equate to £117.*

The EC Directive allows for these financial standards to be established by means of confirmation or assurance from a bank or from other suitably qualified sources that such funds are available, or by means of an acceptable guarantee.

Professional Competence

Details of the professional competence requirements for standard 'O' licence holders are given in Chapter 2.

Wrong Licences

It is illegal to operate on the wrong type of 'O' licence. Applicants are required to specify which type of licence they require and those who specify restricted licences will be subject to severe penalties if they subsequently carry goods for hire or reward. Operators who specify standard licences covering only national operations and who engage in international operations will be similarly penalised.

Operating Centres

Under the 1984 regulations referred to earlier, a definition is given for the vehicle operating centre. It is the place where the vehicle is 'normally kept'. This is commonly taken to mean the place where the vehicle is regularly parked when it is not in use. Thus if operators regularly permit drivers to take vehicles home with them at night and at weekends then the place where the drivers park near to their home becomes the vehicle operating centre. This place then must be declared on the 'O' licence application form.

In these circumstances an operator could have to declare a number of separate operating centres in addition to his normal depot or base and he could face environmental representation against each of these places and have restrictive environmental conditions placed on his licence in respect of their use. Alternatively, he could lose his licence if he fails to declare such places as operating centres. Failure to notify the LA of new or additional operating centres is an offence. Similarly, the practice of allowing drivers to take vehicles home regularly or, for other reasons, park away from the operating centre regularly (except when on genuine journeys away from base) puts the 'O' licence in jeopardy, as well as risking prosecution.

Licence Application

Applications for 'O' licences must be made to the LA for each Traffic Area in which the operator has vehicles based. These bases will be the operating centres (see above for definition) for the vehicles. One 'O' licence will be sufficient to cover any number of vehicles operating at one centre and any number of operating centres in any one Traffic Area. If operating centres are in different Traffic Areas then separate 'O' licences will be required for each Traffic Area. (A list of Traffic Area Office addresses is given in Appendix I.)

Form GV79

Application for a licence has to be made on the appropriate form – form GV79 obtainable from the Traffic Area Office – which is straightforward and simple to answer. This form incorporates questions relating to vehicle operating centres, the previous history of licence applicants during the past 5 years and the type of licence required.

NB: Proposals to change this form are among the 'O' licence revisions currently under consideration by the DoT.

There are questions to be answered on the form relating to the name and address of the business, its partners or directors; details are requested of the applicant's previous experience in operating goods vehicles; information regarding vehicles currently owned and those which it is planned to acquire is also required and the address of their respective operating centres. Questions ask if the applicant company or individual or any partners of the business have convictions which are not 'spent' (under the Rehabilitation of Offenders Act 1974 a person is relieved of the obligation to disclose information about a conviction which is 'spent' – see p 9). Details are required of any such convictions including the date of the conviction, the nature of the offence, the name of the Court and the penalty imposed.

Further questions require information about vehicle maintenance – who is to do it, when and where is it to be done, and what facilities there are at that place. Questions are asked about the financial status of the business proprietor, his partners or the directors of the business, in particular asking whether during the past three years any of them has been made bankrupt, been involved with a company which has gone into insolvent liquidation, or been disqualified from acting as a director or taking part in the management of a company. Details about the professionally competent person supporting the application are required, where that person lives, their actual place of work and the address of the operating centre for which they are responsible.

Declaration of Intent

When the applicant signs the form he is not only declaring that the statements of fact made on the form are true but also, in effect, he is making a legally binding promise – a declaration of intent – that statements of what he intends to do will be fulfilled. The declaration of intent relates to the observation of certain aspects of the law concerned with goods vehicle operation and the maintenance of vehicles included in the licence application. If at some time during the currency of the licence the LA finds that these intentions have not been fulfilled, as evidenced by any convictions for relevant offences, he may use his powers to revoke, suspend or curtail the licence. The basis on which the applicant makes the declaration of intent is that he promises the following:

I will make proper arrangements so that:

- the rules on drivers' hours are observed and proper records are kept;
- vehicles are not overloaded;
- vehicles are kept fit and serviceable;
- drivers will report safety faults in vehicles as soon as possible;
- records are kept (for 15 months) of all safety inspections, routine maintenance and repairs to vehicles and make these available on request;

I will:

- have adequate financial resources to maintain the vehicles covered by the licence;
- tell the Licensing Authority of any changes or convictions which affect the licence;
- maintain adequate financial resources for the administration of the business (applies to standard licence applicants only).

As already stated, the application form is straightforward and simple to answer and extensive explanatory notes are provided for guidance. The Traffic Area Office also sends applicants a free booklet, *A Guide to Goods Vehicle Operators' Licensing* (GV74), to provide further help. This does not, however, mean that the form should not be carefully studied or that any answer will do in an attempt to gain a licence. While it is obvious that to get an 'O' licence the statement of intent must be signed, it should be remembered that the consequences of not fulfilling the stated intentions can lead to penalties so severe as to put a small operator out of business and even to cause hardship to a large one. A warning about this in the following terms is included in the explanatory notes on the form so that applicants are left in no doubt as to the consequences of making false statements or not fulfilling statements:

'I declare that the statements made in this application are true. I understand that the licence may be revoked if any of the statements are false or I do not fulfil the statement of intent made . . .'

It is also worth remembering that the earlier warning in the following terms is still valid:

Under Section 69(1)(c) of the Transport Act 1968, the Licensing Authority has the power to revoke, suspend, or curtail any licence he has granted if it comes to his notice that any statement of fact made is false or that any statement of intention or expectation has not been fulfilled. He may also prohibit you from holding any operator's licence for as long as he thinks fit (S.69 [5]).

Proposals have been made to extend the declaration of intent to secure a promise from operators that they will make proper arrangements to ensure that their vehicles are driven within the speed limits. This follows both the action of certain LAs in imposing penalties against operators and their lgv drivers for regularly exceeding speed limits as evidenced by tachograph chart recordings and the purge by police on speeding offenders generally.

Form GV79A

Another form is involved in making an application for an 'O' licence. This is form GV79A which is a supplementary sheet used for supplying details of vehicles for example, registration number, maximum gross weight body type – flat or sided including skeletals, box body or van, tanker or other type such as cement mixer or livestock carrier – and whether the vehicle is articulated, a tipper or refrigerated. Certain designation letters and numbers are used to indicate body and vehicle type as follows:
1. Flat or sided including skeletals
2. Box body or van
3. Tanker
4. Other type (such as cement mixer, livestock carrier):
   - T Tipper
   - R Refrigerated
   - A Articulated.

Additional Application Forms

Two supplementary application forms are used in connection with certain licence applications. These forms are GV79E (pale green in colour), dealing with environmental information, and GV79F (beige in colour), dealing with financial information. The forms are used only when the LA requires further information following the initial application on form GV79 on either or both of the relevant matters (ie environmental issues or finance).

Environmental Information
Form GV79E is sent to licence applicants if the LA receives representations from local residents following publication of details of the applicant's proposals regarding his vehicle operating centre in the local newspaper. The form must be completed and returned to the LA who will then consider the application in the light of this further information, the information given by those making the environmental representations, and as a result of making his own enquiries. The form requires details of the applicant's name and address, and the address of his proposed operating centre (see p 12 for a definition). It then requires information about the vehicles to be normally kept at the centre and the number and types of trailer to be kept there. Information must be given about any other parking place in the vicinity of the operating centre which is to be used for parking authorised vehicles (ie those authorised on the licence). If the applicant is not the owner of the premises he must send evidence to show that he has permission or authority to use the place for parking vehicles.

A number of further questions must be answered on the form about the operating times of authorised vehicles. In particular, what time lorries will arrive at and leave the centre, whether they will use the centre on Saturdays or Sundays, what times they will arrive and leave on these days, whether maintenance work will be carried out there and between what hours, and whether any of this work will take place on Saturdays or Sundays and if so between what hours? The LA also wants to know whether there are any covered buildings at the centre in which this work is carried out.

A plan showing the parking positions for authorised vehicles must be sent when returning the completed form. This should show entry and exit points, main buildings, surrounding roads with names and the normal parking area for the vehicles. The scale of the plan must be indicated and this is suggested as being 1:500 which is 1 centimetre to 5 metres or, roughly, 1 inch to 35 feet. A larger scale of 1 inch to 100 feet can be used if this is more convenient when the operating centre is large. If the proposed operating centre has not previously been used as such the LA must be given information about any application for or planning permission granted for the proposed use of the site as a goods vehicle operating centre.

Financial Information
Form GV79F is sometimes sent to new standard licence applicants when the LA requires additional information to enable him to consider whether the applicant meets the financial requirements for this type of licence. An application will be refused unless the LA is satisfied that the applicant has sufficient financial resources to set up and run his business both legally and safely. This fact is pointed out clearly on the form. Answers have to be given to questions about the vehicles, their average annual mileage and the estimated running cost for each individual type of vehicle.

Details must be given about the funds available to start up the business and where these are held (eg in the bank, in savings or as agreed bank overdraft or loan facilities or in the form of share capital), and about the start-up costs for the business including the purchase price or amount of down payments on vehicles and on premises and the sum to be held in reserve as working capital. The applicant is required to give a forecast of the annual expenditure and income for his road haulage operations for a financial year. The LA expects this information to give a clear indication of the business finances for the year ahead. In certain cases the LA may ask for monthly information.

NB: Not all Licensing Authorities currently use this form, some prefer to rely on supporting financial information in other forms – bank references for example.

Date for Applications

Application for an 'O' licence should be made at least 9 weeks before the day on which it is desired to take effect. In some Traffic Areas the time taken to process applications is much longer than nine weeks so new operators should be aware of the fact that it is illegal to start operating vehicles until their licence has actually been granted. Where there is an urgent need to start operations before a licence is granted through the normal processes, application can be made to the LA for an interim licence (see p 18).

Where an application is made to renew an existing licence prior to its expiry date, the old licence will continue in force until the LA makes his decision on the new application, even if this is some time after the expiry date of the previous licence. Where an existing licence expires without a renewal application being made, the old licence ceases to be valid immediately following the expiry date and it becomes illegal to continue operating vehicles on that licence. In such circumstances a fresh application has to be made and vehicles cannot be operated until a new licence is granted.

Offences while Applications are Pending

Applicants for licences have a duty to advise the LA if, in the interval between the application being submitted and it being dealt with by the LA, they are convicted of a relevant offence (see p 9) which they would have had to include on the application form had the conviction been made before the application was made. Failure to notify the LA is an offence and it could jeopardise any licence subsequently granted.

Advertising of Applications

Applicants for 'O' licences who are seeking a new licence, renewal of an existing licence (technically a new licence application – licences are not 'renewable') or variation of an existing licence are required to arrange for publication of an advertisement (following a specified format to contain the necessary information for potential environmental representors – see Figure 1.1) in a local newspaper circulating in the area where the operating centre is located. If more than one operating centre is specified in the application separate advertisements must be placed for each such centre in the respective local newspapers serving those locations.

How to fill in advert

Full name(s) of Company, Partners, or Sole Trader applying for licence

Put your trading name here if different
(cross out if it does not apply)

Address where you normally receive letters including postcode

Choose the one you are applying for
(cross out the others)

Put in the address of your operating centre including postcode *(Each newspaper advert must only contain a list of the operating centres within the distribution area of that paper)*

Put the maximum number of vehicles/ trailers you want including your margin of spares *(Cross out trailers if you do not have any)*

Put the number of vehicles/ trailers you have now *(Cross out trailers if you do not have any)*

Put in the new conditions you want on the licence
(Cross out if you do not want this type of change)

Put in the old conditions
(cross out if you do not want this type of change)

Put in existing conditions
(cross out if you do not want this type of change)

Put in the address of the Traffic Area Office you are sending your application to.

Goods Vehicles Operator's Licence

Trading as

of

is applying for a licence to use
is applying to renew without change a licence to use
is applying to renew with change a licence to use
is applying to change an existing licence to use

as an operating centre for

goods vehicle(s)

and

trailer(s)

instead of

goods vehicle(s)

and

trailer(s)

with the following environmental conditions applying

instead of

by removing the following environmental conditions from the licence

Owners of occupiers of land (including buildings) in the vicinity of the operating centre who believe the use or enjoyment of the land will be prejudicially affected, may make written representation to the Licensing Authority at

within 21 days following the publication of this notice.

Representors must at the same time send a copy of their representation to the applicant at the address given in this notice.

Figure 1.1 *Format which must be used for 'O' licence newspaper advertisement*

The advertisement need appear only once but it must be published during a period extending from not more than 21 days before and not more than 21 days after the licence application is made. There is no specified minimum or maximum size requirement for the advertisement but the LAs advise that it 'should not be too small and should be easy to read'. Most adverts appear in single column format extending to a few inches of text and at an average cost estimated to be in the region of £75). Normally the advertisement will appear in the public or official notices section of the newspaper. Proof that the advertisement has appeared must be given to the LA before he considers the application and failure to produce this proof (usually achieved by sending in the appropriate page torn from the newspaper showing the advertisement itself and the name and date of the paper) will mean that the LA, by law, must refuse to consider the application. Normally this would mean making a fresh application which, of course, delays the whole matter, adds to the costs by requiring another advertisement and could mean vehicles having to stand until the licence or variation is granted.

The sole purpose of the advertisement is to give local residents an opportunity (given to them under the regulations) to make representations against the licence but only on environmental grounds (see p 22).

Interim Licences

In certain circumstances the LA may grant an interim licence pending his decision on the full licence application. The circumstances under which such a licence may be granted are not specified but they may be connected with some urgent need to move goods quickly because they are perishable or for some other urgent reason. A grant of an interim licence should not be taken as a guarantee that a full-term licence will be granted by the LA. An interim licence will not be granted in any case where the main requirements for 'O' licensing appear not to be met. For example, such a licence would not be granted to an applicant for a standard 'O' licence if he has not yet passed the CPC examination nor, indeed, while examination results are being awaited, nor on the assumption that the candidate will have passed. Neither will a grant of an interim licence be considered before the statutory 21-day waiting period for environmental representations and objections has expired. Interim licences are not granted for any fixed period. Normally they remain in force until the LA has made his decision on the grant of a full licence or, alternatively, until they are revoked.

Duration of Licences

Operators' licences will normally be granted for 5 years although in some instances, where operators have had past convictions for offences relating to the operation of goods vehicles or where the LA is not fully satisfied as to the ability or intention of the applicant to operate satisfactorily within the law, licences for a shorter period may be granted to enable the LA to consider the applicant's record over such shorter period before granting a full-term licence. Licences may continue in force beyond their expiry date pending the LA's decision on a renewal application but only where the new application was submitted prior to the expiry date of the old licence.

A proposal is under consideration by the DoT for the issue of 'lifetime' or 'continuous' 'O' licences – see p 2.

Licence Fees and Discs

The fee for a 5-year 'O' licence is £185 (increased, from 1 June 1994, from £170), plus £8 per quarter per vehicle specified on the licence (Form OL1) payable on issue. On receipt of the fee the licence itself will be issued together with windscreen discs for each of the vehicles specified on the licence. There is no fee for additional trailers. For a licence variation which requires publication in *Applications and Decisions* (see p 24) a fee of £185 is charged. The fee for an interim licence or direction is £10 per authorised vehicle.

'O' licence discs must be displayed on the vehicle (normally in the windscreen) in a clearly visible position near to the excise duty disc and in a waterproof container. Licence discs are coloured as follows to differentiate between restricted and standard licences and between standard national and standard international licences:
1. Restricted Licence – orange.
2. Standard Licence, National – blue.
3. Standard Licence, International – green.
4. Interim Licence – yellow.
5. Copy Discs – word 'COPY' in red across face of disc.

Licence discs are not interchangeable between vehicles or between operators. They are valid only when displayed on the vehicle whose registration number is shown on the disc (even if it is faded almost beyond recognition) and when that vehicle is being 'used' by the named operator to whom it was issued. Heavy fines are imposed on offenders who loan and borrow discs; their own 'O' licence may be jeopardised and the vehicle insurance could be invalidated.

Standardised 'O' licence discs, non-specific to Traffic Areas and incorporating an anti-fade agent, are now being issued (starting 1 January 1993).

Issue of Community Authorisations

Following the introduction of the new system of Community Authorisations for intra-EU movements in connection with the Single Market liberalisation of road transport operations, all UK holders of standard 'O' licences covering international operations will be issued (automatically) with a Community Authorisation document to be kept at their main place of business together with certified copies equalling the total number of vehicles authorised on their operator's licence (which must be carried on vehicles when undertaking cross-border journeys within the EU).

Licence Surrender/Termination

In the case of licence surrender, vehicle fees are proportionately refunded but only on a full year basis. The licence (or variation) fee itself is not refundable. If a licence is prematurely terminated by the LA by way of penalty for contraventions of the law, there is no refund of licence fees for the outstanding period on the licence but vehicle fees will be refunded proportionately on a full-year basis. The licence itself and all the vehicle windscreen discs must be returned to the LA on surrender or termination of an 'O' licence.

The LA's Considerations

When an application for an 'O' licence is made, the LA will have certain points to take into consideration before deciding whether or not to grant any licence. Mainly he has to ensure that the basic legal requirements as described previously have been met and particularly that those relating to vehicle operation and maintenance will be complied with. These points are dealt with here.

Fit Persons and Good Repute

The first point, and one of the fundamental requirements for 'O' licensing (as already described on p 8, but a repeat here is useful), is whether the applicant is a fit person or is of good repute and is therefore fit to hold a licence. Basically, being a fit person and being of good repute are the same but the former relates to restricted 'O' licences over which the EU has no influence while the latter is the term used by the EU in setting its requirements for the holding of a licence to carry goods for hire or reward (ie the UK system of standard 'O' licences). The LA, when deciding this, will take into account any previous record which the applicant (or the partners or directors of the applicant's business) might have had as an operator in terms of their ability or willingness to comply with the law in respect of vehicle operations and particularly maintenance, drivers' hours and records, overloading and the like, and also any previous convictions they may have for offences relating to the roadworthiness of vehicles and for other relevant offences. More recently, amendment to the regulations requires that LAs take account of 'serious offences' besides just road transport offences when determining the good repute of an individual, or of company management for a corporate 'O' licence application.

Maintenance Facilities/Arrangements
The next point that the LA will consider, and one of the most important since it is at the very foundation of the 'O' licensing system, is whether the applicant has suitable facilities or has made satisfactory arrangements for the maintenance of vehicles to be specified on the licence in a safe and legal condition and for keeping suitable maintenance records (this is dealt with in more detail in Chapter 19). In particular the LA will be concerned to know that vehicles are being subjected to safety inspections at regular intervals of time or mileage.

The LA for the West Midlands and South Wales Traffic Areas, for example, currently suggests that for operators in his area inspections should be based on a time interval only with no mileage alternative and that a period of 6 weeks between inspections is the maximum that he would find acceptable. He also seeks assurances that operators are using a *written* driver defect reporting system and wall charts for planning inspection and maintenance schedules. Other LAs seek similar assurances.

Drivers' Hours and Records

The LA will consider whether there are satisfactory arrangements for ensuring that the law relating to drivers' hours and records (including tachographs) will be complied with.

Overloading

Similarly, the LA will want to be sure that arrangements are made to prevent the overloading of vehicles and that vehicle weight limits in general will be observed.

Professional Competence Requirements

The LA will want to know details of the nominated professionally competent person, who may be the applicant himself or an employee (who, since regulation changes in November 1991, no longer need be a full-time employee – see also pp 40–1) who holds a certificate of competence covering national or both national and international transport operations (or a person who qualifies by exemption or by having passed the appropriate Royal Society of Arts examination) as appropriate to the type of standard 'O' licence applied for (see Chapter 2).

Number of Qualified Persons

There is no restriction under the regulations (see Chapter 2) on the number of people in a transport department or organisation who may be professionally competent and consequently hold certificates of competence. Further, although restricted 'O' licence holders have no need to specify the name of a professionally competent person in order to obtain a licence there is no restriction on such licence holders or their employees being professionally competent if they qualify personally.

In determining how many qualified persons must be named on a standard 'O' licence the LA will take account of the management structure of applicant firms, but generally there will need to be a minimum of one qualified person per 'O' licence. The LA may require the names of more qualified persons to be specified if he considers it appropriate in view of a division of responsibilities for the operation of vehicles under the licence or if vehicles specified on the licence are located at different operating centres within the Traffic Area. The requirement for there to be at least one holder of a certificate of professional competence (CPC) at each vehicle operating centre and the need for nominated CPC holders to be 'full-time' employees of the licence holder have been abandoned.

Financial Standing

In addition to these points the LA is required under the regulations (as already stated) to establish details of the applicant's financial standing (a bank statement or a bank manager's letter of reference or an accountant's certificate of solvency may be requested, for example, or other evidence of the availability of funds) because this has a bearing on the applicant's ability to operate and maintain vehicles in a safe condition and in compliance with the law (see also pp 15–16). While considering an applicant's financial status for this purpose the LA has authority to call for the services of an assessor from a panel of persons appointed by the Secretary of State for Transport if this is appropriate due to the complexity of the financial structure of the applicant's business or affairs. The LAs watch for 'O' licence holders who are prosecuted for vehicle excise offences and those who ask for time to pay fines or request the opportunity to make payment in instalments following conviction for offences by the courts and take this as good cause for investigating their financial position.

Representations by Local Residents

Opportunities are given to local residents individually to make representations against 'O' licence applications and variations on environmental grounds. Local residents are more carefully defined in the regulations as 'owners or occupiers of land within the vicinity' (ie of the operating centre). Those residents who wish to make representation must do so individually because group action is not permitted (although a group of individual representors may appoint a joint spokesperson to put forward their case), and nor is representation by any environmental pressure group, political party or other campaigning body. Similarly, Parish Councils, which regularly feature in such matters, have no right of objection *per se* unless they *own* land in the vicinity of an operating centre featuring in an 'O' licence application, in which case they may make representation as the owner of the land.

The grounds on which such owners or occupiers can make their representations are confined purely to environmental matters such as noise, vibration, fumes and visual intrusion but could include obstruction. They do not include road safety matters which are not an environmental issue. The grounds must be stated precisely in the written representation; to state that the representation is made for 'environmental reasons' is not sufficient. The exact wording which forms the basis of representations is specified in the legislation in the following terms, 'that place (ie the operating centre) is unsuitable on environmental grounds for . . . (such use and) . . . any adverse effects on environmental conditions arising from that use would be capable of prejudicially affecting the use or enjoyment of the land' (ie the land owned or occupied by the person making the representation).

One of the facts that has been difficult to establish in connection with this is a definition of the term 'within the vicinity'. It has been shown that residents living along an access road to a vehicle operating centre can be considered to be in the vicinity and adverse environmental effects of vehicles travelling along the road could be taken account of by the LA in his consideration of any environmental representations against a licence application. Each LA is left to make his own determination of whether a representor lives 'within the vicinity', but as a general rule if a representor can see, hear or smell a vehicle operating centre from his property then he will be considered to be 'in the vicinity' for the purposes of making an environmental representation.

There is no opportunity for people living near an operating centre to make representations on grounds other than environmental matters or to use the opportunity to vent long-standing grudges against the vehicle operator. The LAs will not consider any representation which falls outside the terms described above or which is considered to be vexatious, frivolous or irrelevant.

Local residents will normally become aware of their opportunity to make representations against the grant of a licence or licence variation through the local newspaper advertisement placed by the applicant (see pp 16–18). Those people wishing to make a representation must do so in writing (or have their solicitor do so on their behalf), within a period of 21 days from the date of publication of the advertisement, to the LA at the Traffic Area Office address given in the advertisement. They must also send an exact copy of their representation (ie letter to the LA) to the licence applicant at his address which is also given in the advertisement. Their letter must clearly state the 'particulars' of the matters forming the basis of their representation so that both the LA and the licence applicant may be fully aware of the specific grounds on which the representation is made. Failure to be specific as to the facts in this letter, or failure to submit the representation within the specified time scale, will render the representation invalid.

Representors have no right of appeal should their case against use of the operating centre fail.

These rights of representation are not to be confused with the rights of objection described later.

Suitability of Premises

Licensing Authorities must inquire into and be satisfied that the place or places to be used as vehicle operating centres are both suitable and environmentally acceptable. Local residents have rights (as described above) to make representations about the environmental consequences of the use of places for transport depots or vehicle operating centres and the LAs are bound to listen to these representations and make appropriate decisions about the application depending on the weight of the argument on either side – residents or operator (see above and p 33). In particular the LA, when considering the suitability of premises, will take account of the following:

- the nature or use of any other land in the vicinity of the operating centre and the effect which the granting of the licence would be likely to have on the environment of that land;
- how much granting a licence which is to materially change the use of an existing (or previously used) operating centre, would harm the environment of the land in the vicinity of the operating centre;
- for a new operating centre, any planning permission (or planning application) relating to the operating centre or the land in its vicinity;
- the number, type and size of the authorised vehicles (including trailers) which will use the operating centre;

- the parking arrangements for authorised vehicles within and near to the operating centre;
- nature and times of use of the operating centre;
- nature and times of use of equipment at the operating centre;
- how many vehicles would be entering or leaving the operating centre, and how often.

Applications and Decisions

When an application for a new 'O' licence, a variation of an existing licence or a renewal (ie a new licence to replace an existing licence) of an existing licence is received by the LA, details of the application (ie the name of the applicant and the number of vehicles and trailers included in the application) will be published in a Traffic Area notice called *Applications and Decisions* (As&Ds). As its name implies, this notice will also contain details of licences granted by the LA and details of public inquiries to be held. The notice is published weekly or fortnightly by all Traffic Areas and may be inspected at Traffic Area offices or purchased on an individual copy or regular basis. It is by means of this notice that statutory objectors (see below) are able to know when applications have been made against which they may wish to object. They can do this within 21 days of publication of the relevant As&Ds notice.

Objections to the Application

Applications for 'O' licences are open to statutory objection by certain bodies listed below (and only by the listed bodies – no other individual or organisation has this statutory right). Potential objectors to 'O' licences become aware of pending applications for new licences or variations to existing licences through the publication *Applications and Decisions* mentioned above.

Objections to applications can only be made by the bodies mentioned on the grounds that the applicant does not meet the essential qualifying criteria for the grant of a licence, namely that the applicant is not of good repute, is not of adequate financial standing or does not meet the professional competence requirements (where appropriate), that the law in respect of those matters which the LA will be considering when he is deciding whether or not to grant a licence is not likely to be complied with, namely that the drivers' hours and records regulations will not be observed, that vehicles will be overloaded and that there are not satisfactory arrangements or facilities for maintaining the vehicles. The bodies may also object on environmental grounds (for example that the operating centre is environmentally unsuitable).

The list of bodies who may make statutory objection to an 'O' licence application are as follows:
- A chief officer of police
- A local authority
- A planning authority
- The British Association of Removers (BAR)
- The Freight Transport Association (FTA)

- The Road Haulage Association (RHA)
- The General and Municipal Workers' Union (GMWU)
- The Rail, Maritime and Transport Union (RMTU) (formerly the National Union of Railwaymen and the National Union of Seamen)
- The Transport and General Workers' Union (TGWU)
- The Union of Shop, Distributive and Allied Workers (USDAW)
- The United Road Transport Union (URTU).

These are the only sources of objection (not to be confused with an environmental representation) to an application for an 'O' licence. If any of these bodies do make a statutory objection they are required to send a copy of their objection to the applicant at his published address at the same time as sending one to the LA and this must be within 21 days of the publication of details of the application in *Applications and Decisions*. Failure to send a copy to the applicant renders the objection invalid.

Grant or Refusal of a Licence

The LA has power to grant an 'O' licence to applicants if he considers that they meet all the necessary requirements. Alternatively, he may refuse to grant a licence or he may grant a licence for a shorter period, as previously mentioned (see p 18) or he may grant a licence for fewer vehicles than the number applied for if he doubts the ability of the applicant to be able to comply with the law with more vehicles, to be able properly to maintain more vehicles or to be able adequately to finance the operation of more vehicles. He can impose environmental conditions on any licence granted and can also refuse to accept the name put forward for the professionally competent person (in the case of standard licence applications) if he believes that the person is not of good repute. The LA may be influenced in his decision by the points made by any statutory objectors or environmental representors.

Licence Grant with Conditions

The case made by those making valid environmental representations may influence the LA to either refuse the application on the grounds that the operating centre is not environmentally suitable or alternatively to grant the licence but with environmental conditions attached. Thus to prevent or minimise any adverse effects on the environment he may place conditions or restrictions on the licence granted under the following headings:

- the number, type and size of authorised vehicles (including trailers) at the operating centre for maintenance or parking;
- parking arrangements for authorised vehicles (including trailers) at or in the vicinity of the centre;
- the times when the centre may be used for maintenance or movement of any authorised vehicle; and
- how authorised vehicles enter and leave the operating centre.

As an alternative to placing environmental conditions on the licence, the LA may seek undertakings from the operator that he will or will not follow certain practices in order to reduce environmental disturbance of local residents (eg control the number of vehicle movements into and out of the centre). The licence holder should be aware that any such undertakings he may voluntarily give to the LA become legally binding upon him and could result in penalty against his licence if he subsequently fails to observe them.

Licence holders who find they have breached environmental conditions on their licence through unforeseen circumstances must notify the LA. Failure to comply with any of these conditions during the currency of a licence may result in the LA imposing penalties on the licence such as suspension or curtailment. In serious cases the licence may be totally revoked.

Additional Vehicles

Seeking a Margin

At the time of making an application for an 'O' licence the applicant is given the opportunity to request authorisation for any additional vehicles which he may need to acquire or hire during the currency of the licence. By taking this opportunity the operator saves the problems of making a fresh application when wanting to add or hire-in vehicles on a temporary basis to meet trading peaks. It also saves facing any further environmental representations or statutory objections because once the original application is granted with additional vehicles specified, extra vehicles can be added within the number authorised by completing form GV80 when they are acquired and sending it to the LA within 1 month. There will be no need for the details to be advertised in a local newspaper or published in *Applications and Decisions.*

If additional vehicles were requested and the request was granted at the time of making the original application, the operator will have a 'margin' for extra vehicles on the licence. As described above, the LA will need to be notified within 1 month of actually *acquiring* the additional vehicles – *not* the date of putting them into service – (by submitting form GV80) so that a windscreen disc can be issued for the vehicle and the appropriate fee charged (ie £32 per vehicle per year of the remaining duration of the licence).

It is useful here to clarify the terms used in connection with the numbers of vehicles for 'O' licensing purposes:

- Authorised vehicles – the maximum number of vehicles/trailers which the licence is actually granted to cover (it is illegal to operate more than this number of vehicles at any time)

- Specified vehicles – the actual vehicles which the operator has in possession and which are specified on the licence by registration number

- Margin – the difference between the number of authorised and specified vehicles on the licence, in other words the vehicles still to be acquired by the operator whether on a permanent or a temporary basis.

NB: Operate in this context means 'use' and this applies to vehicles hired without drivers. Conversely, if they are hired with drivers, then they are operated under the hire firm's 'O' licence and not within the margin of the operator's 'O' licence (see below).

Hired Vehicles

If an operator plans to hire extra vehicles without drivers (ie where he intends to have his own employee or a hired agency driver to drive the vehicle/s) during the currency of his licence, whether for a short period (a day, a few days or even 1 or 2 weeks) or on a long-term contract, they must be covered by his 'O' licence and he will need to have applied for a sufficient margin of additional vehicles on his licence to cover these. If vehicles are hired, within the margin, for less than 1 month there will be no need to notify the LA but details of vehicles hired for more than 1 month must be sent to the LA and an 'O' licence disc obtained for display on the vehicle.

It is illegal to operate (ie to have employed drivers to drive) more vehicles (ie of over 3.5 tonnes gross weight) than are authorised on the 'O' licence even for a temporary period or reason (eg when an authorised vehicle is off the road for service or repairs or to cover additional delivery requirements).

Number of Extra Vehicles

When making the request for additional vehicles on the initial application for an 'O' licence, the number which may be requested is not limited in any way but it is recommended that it should be in reasonable proportion to the number of vehicles already operated (or initially required) and, most important, it should only be of a number which the applicant can maintain, and prove he can maintain (both physically and financially) on the same basis as the remainder of his fleet. If the request for additional vehicles relates to vehicles which are to be hired rather than owned it must be remembered that the person who hires a self-drive vehicle is fully responsible for the mechanical condition of the vehicle in so far as safety and legal requirements are concerned.

An applicant specifying additional vehicles on the original application should give some careful thought to the exact number of vehicles which may be needed and the reasons for needing them because the LA will ask questions about this if he calls the applicant to a public inquiry. Evidence in the form of business forecasts and trends in trade would be most useful as would figures to indicate past growth of the business; evidence also to show the financial prospects of the applicant during the currency of the licence period will help towards convincing the LA that he would be justified in granting a licence for the additional vehicles requested.

Replacement Vehicles

If for some reason an authorised vehicle ceases to be used the LA must be advised of the fact, but if at that time or later another vehicle is acquired to replace it the operator must advise the LA on form GV80 within 1 month of *acquiring* the replacement vehicle. Vehicles which are not removed from the licence (even when standing smashed or cannibalised in a yard or workshop) are still counted as specified vehicles and cannot be replaced by others within the authorised number on the licence until they are removed by notifying the LA and the windscreen discs are returned to the Traffic Area Office.

Licence Variation

If the holder of a restricted 'O' licence wishes to change the licence to a standard 'O' licence it is necessary to apply on form GV81 and satisfy the legal requirements for standard licences, national or international. Standard national licence holders who wish to change to a licence covering international operations must satisfy the LA that they, or an employee, are professionally competent in international transport operations (apply on form GV81).

Where the licence holder finds, during the currency of the licence, a need to add vehicles to the fleet that were not specified on the original application then an application must be made to the LA by completing form GV81 and submitting this well in advance (minimum 9 weeks). This will necessitate placing an advertisement in a local newspaper as with the original application (see pp 16–18).

Unless the variation is only of a trivial nature the LA will publish details of it in *Applications and Decisions* and it may attract objectors in the same way as a new application and the public inquiry procedure will be the same as that already described.

The licence holder must never operate more vehicles or trailers than the total number specified on his licence. When extra vehicles are required the operator must wait until the application for an increase in the licence is granted before actually putting the vehicles on the road. This can take at least nine weeks – which is the minimum application period required – and as many as 12 to 15 weeks in some Traffic Areas.

Form GV81

Form GV81 is a five-page document called *Application to Change a Goods Vehicle Operator's Licence* and must be used by applicants who wish to:
1. change the number of vehicles authorised;
2. change the operating centre, add another centre or stop using a centre;
3. change the type of licence (eg restricted to standard);
4. change or remove a condition attached to the licence (including conditions on the use of operating centres).

Most of the changes involve the need to advertise the application in local newspapers (see p 16) and where this is necessary the fact is pointed out on the form. The applicant must send a copy of the published advertisement with the form or state the expected date of publication and the name of the newspaper in which it is to be published. The usual information regarding name and address, addresses of operating centres and the number of vehicles to be based there which are in possession now or to be acquired has to be given. Also required is similar information about vehicle maintenance arrangements to that required on the original GV79 application (see pp 12–13).

Section 3 of the form deals with any addition or deletion to the specified operating centres and section 4 is completed if the applicant wishes to change the type of licence (eg from restricted to standard or from standard national to standard international licence). Details of the professionally competent person must be given where appropriate, so too must details of any convictions (other than those which are 'spent' – see p 9) of the applicant or the partners or co-directors.

Section 5 of the form has to be completed if the applicant wishes to change or remove any of the environmental conditions attached to the use of the operating centres. Reasons must be given as to why this change or removal is wanted and details given of any alternative proposals if the applicant has any.

The form has to be signed and the applicant is warned that the licence may be revoked if any of the statements given are false.

NB: Proposals to change this form are among the 'O' licence revisions currently under consideration by the DoT.

Transfer of Vehicles

If a vehicle is transferred from the Traffic Area in which it is licensed to a base in another Traffic Area for a period of more than 3 months, it must be removed from the original licence and specified on a licence in the new Traffic Area. Transfers for periods of less than 3 months are permitted with no need for notification to the LA provided an 'O' licence with a sufficient margin to cover the transferred vehicles is already held for that Traffic Area.

If the operator does not hold an 'O' licence in the other Traffic Area, or holds a licence in the area but it does not have a sufficient margin to accommodate the transferred vehicles, then an application for a new licence or a variation of the existing licence must be made to the LA for that Traffic Area. It is illegal to operate vehicles (ie over 3.5 tonnes gvw) from a base in a Traffic Area unless a licence is held in that area.

Notification of Changes

Licence holders should notify the LA in writing, within *1 month* of any changes in the legal entity of their business such as a change of name, address, ownership, if a new partnership has been formed, a limited company formed or the constitution of the partnership has been changed, as this makes a material difference to the information given in answer to questions on the original GV79 licence application. The LA must also be informed if the proprietor or persons concerned in the business die or if the business becomes bankrupt or goes into liquidation.

A change of operating address as given in the original licence application must be notified to the LA within *3 weeks.* A change of operating centre (or the use of an additional operating centre) also requires a variation application using form GV81 and the need to follow the newspaper advertisement procedure.

Other changes which must be notified in writing are those in maintenance facilities or arrangements and any breach of environmental conditions which the LA placed on the licence. Failure to notify the LA of such changes can have the same result as making false statements or failing to fulfil intentions stated in the original application, namely the risk of licence suspension, curtailment or revocation. The offender could also be prosecuted with the consequent penalties which can be imposed by the courts.

Subsidiary Companies

A holding company can include in its application for an 'O' licence vehicles belonging to any subsidiary company in which it owns more than a 50 per cent shareholding. But associate companies (ie where the shareholding arrangement is less than 50 per cent), owned by the same holding company, cannot have vehicles specified on each other's licences and separate divisions of a company are not permitted to hold separate licences unless they are separate entities in law.

The vehicles of any subsidiary company acquired during the currency of the holding company's 'O' licence can, if desired, be included in the holding company's licence either within its existing licence margin or by making application to the LA, on form GV81, to vary the licence. It is not generally likely that an application to include a subsidiary company's vehicles on the holding company's licence would be published in *Applications and Decisions* or that it would attract any objections.

Under the regulations, for the purposes of determining whether goods are carried for hire or reward in order to choose between a restricted or a standard 'O' licence, goods belonging to, or in the possession of, a subsidiary company are considered to belong to, or be in the possession of, the holding company and vice versa, so that in such cases a restricted 'O' licence would be adequate even if charges for the movement of the goods were made between the holding company and its subsidiary.

Renewals

Use of the term renewal in the context of 'O' licensing is not strictly correct: licences are not renewed, application has to be made for a new licence when an existing licence expires. This applies when all existing 'O' licences expire, whether at the end of the full 5-year period or, if the original licence was for a shorter period, at the end of that period, when a fresh application has to be made for a new licence using the same forms GV79 (the application) and GV79A. This application will be considered by the LA in the same way as a new application and it will be published and liable to attract objectors and environmental representations in the same way. The LA will, of course, by this time have information from his staff about the conduct of the applicant during the currency of the previous licence and this will have a bearing on his consideration of the renewal (ie new licence) application.

Farmers' 'O' Licences

Farmers' goods vehicles over 3.5 tonnes gross weight must be specified on an 'O' licence in the same way as already described for other goods vehicles.

Goods vehicles owned by farmers which are taxed at the full goods vehicle excise duty rate and specified on an 'O' licence may be used with complete freedom to carry any goods for any person to any destination. If, however, vehicles owned by farmers are taxed at the concessionary 'F' licence excise duty rate (see Chapter 8) they will still need an 'O' licence but may only be used for the purposes which enable them to claim the concessionary 'F' licence rate. The conditions applicable to an 'O' licence application from a farmer are the same as those applying to other haulage and own-account operators.

Penalties against 'O' Licences

The LAs, as the issuing authorities for goods vehicles licences, are also given considerable legal powers to revoke, suspend or curtail an 'O' licence for a large number of reasons, of which the following are a few of the important examples.

1. Contravention by the licence holder of the provision, in the case of standard 'O' licences, regarding professional competence requirements.
2. Failure to notify the LA of changes in the business.
3. Convictions for failure to maintain vehicles in a fit and serviceable condition.
4. Contravention of speed limits, overloading or offences in connection with loading or unloading vehicles in restricted parking or waiting areas.
5. Failure to ensure that drivers are correctly licensed.
6. Convictions relating to the use of rebated (duty free) fuel oil in vehicles (see Chapter 8).
7. Failure to keep records relating to vehicle inspections and repairs and driver defect reports.
8. For falsely stating facts on applications for 'O' licences and for not fulfilling statements of intent or environmental conditions placed on the licence.
9. If the licence holder becomes bankrupt or, in the case of a company, goes into liquidation.
10. If a place not listed on the licence is used as a vehicle operating centre.

Offences are committed, for which prosecution and a court appearance may follow, if:

1. a windscreen licence disc is not displayed;
2. a change of address is not notified;
3. a licence is not produced for examination on request;
4. a duplicate windscreen disc is not returned if the original is found;

5.   a disc is not returned when a vehicle is disposed of;

6.   a subsidiary company featured on a holding company licence is disposed of and the LA is not advised.

Furthermore, the LAs are active in preventing speeding by lgv drivers, first by imposing a penalty of suspension on the lgv driving licences of offending drivers and then by penalising the 'O' licences of firms whose drivers persistently and wilfully exceed speed limits. Evidence of such matters is mainly obtained during routine enforcement checking of tachograph charts, where recordings showing frequent instances of driving above 100kph are clear evidence of breach of the 60mph maximum speed limit for vehicles exceeding 7.5 tonnes maximum laden weight.

Usually, the offending licence holder will be called to public inquiry by the LA and be required to explain why the offences occurred and what action is being taken to put matters right or to ensure they will not happen again. Depending on his reaction to such explanations, the LA may initially give a warning about future conduct and the likely consequences if there is any repetition of the contraventions of the law or he will decide that an appropriate penalty should be imposed. This will be suspension, curtailment or premature termination of the licence or revocation. If the licence holder is found no longer to comply with the basic requirements for 'O' licensing, namely good repute, financial standing or professional competence, then the LA must revoke the licence (except in the latter case where a period of temporary derogation is permitted – see below). If a restricted 'O' licence holder is convicted twice in a period of 5 years of operating outside the terms of the licence his licence must be revoked.

Curtailment is the most commonly imposed penalty and this implies removal of one or more authorised vehicles from the licence for any period up to the expiry of the licence. Suspension involves suspension of the whole licence and this may be combined with premature termination so the LA can review the whole operation under the provisions for consideration of a new licence application. As with premature termination of an existing licence, the need to apply for a new licence places the operator at risk of objection and environmental representation. The LA can direct that a vehicle on a licence which has been suspended or limited may not be used by another operator for a maximum of 6 months during the period of suspension.

When the LA revokes an 'O' licence – which is not done lightly – he may order the holder to be disqualified, for a certain period or indefinitely, from holding or obtaining an 'O' licence and the order may be limited to one or may apply to more Traffic Areas. Following an order to revoke a licence, the LA may allow the licence holder to request a 'stay' to enable the operation to continue until an appeal to the Transport Tribunal is heard.

Temporary Derogation

There are provisions in the regulations to enable a standard 'O' licence to remain in force for up to 1 year initially and a further 6 months (maximum derogation is 18 months) if the LA feels it is appropriate if the specified professionally competent person named on the licence dies or becomes legally incapacitated (ie unable to carry out his duties due to reasons of mental disorder) in order to allow a replacement person to be found and specified.

The regulations enable the LA to defer revocation of or refusal to grant a standard 'O' licence in the event of the death or incapacity of the holder of the licence, a transport manager, or a partner whose professional competence is relied upon. Further, in the event of the death or incapacity of the licence holder the LA is empowered to authorise another person to carry on the business during the changeover period as though that person was the licence holder. Also, the LA may allow time for a transport business to be transferred to another person licensed to carry it on or for a transport manager or new partner to be appointed.

Where a person who was carrying on a business as a licence holder dies, becomes mentally incapacitated, bankrupt or goes into liquidation or where a partnership is disolved the LA must be notified within 2 months. The person carrying on the business will be considered by the LA to be the holder of the licence if a new licence application is made within 1 month in the case of restricted 'O' licences and within 4 months in the case of standard 'O' licences.

Production of 'O' Licences

Operator's licence holders must produce their 'O' licence (form OL 1 plus forms OL 1(R) or OL 1(S) where appropriate) for examination when required to do so by the police, DoT examiners (ie certifying officers) or by the LA or a person with his authority. The holder has 14 days in which to present the licence either at one of the operating centres authorised on the licence or at his principal place of business in the Traffic Area. In the case of production to the police this can be at a police station of the holder's choice also within 14 days.

Licensing Courts/Transport Tribunal

Inquiries and Appeals

Licensing Authorities regularly hold public inquiries (PIs) to which 'O' licence applicants are called to explain the basis of their operations and to enable the LA to seek more information prior to determining whether he should grant a licence or not. In the event of a representation on environmental grounds or an objection being made to a licence application the LA will hold a public inquiry at which the parties (applicant, objectors or those making representations) will have an opportunity to state their case further. If the application is refused in whole or in part or if environmental conditions are attached to a licence the applicant has rights of appeal against the LA's decision to the Transport Tribunal. Normally an existing licence will remain in force while an appeal is being heard and the LA may allow a revoked or suspended licence to continue during this time. If the LA refuses this the Tribunal can be asked to allow it to do so.

Statutory objectors also have a right of appeal to the Transport Tribunal if an application for a licence is granted and they still feel that their objection is valid. Those individuals making representations on environmental grounds have *no* similar right of appeal if their case fails.

It should be noted that the Transport Tribunal is the *only* source of appeal in regard to 'O' licensing matters.

Besides public inquiries conducted for the purposes of determining 'O' licence applications, such inquiries are also held at the LA's behest where it is necessary for him to examine the conduct of a licence holder for disciplinary purposes under the powers given him by the Transport Act 1968. Section 69 of this Act empowers him to conduct such inquiries (hence reference to 'section 69 inquiries') and impose penalties of suspension, curtailment or revocation of a licence (further details of this matter are given on p 31).

Appeals are heard only in London whereas public inquiries are usually held in the town or city in which the LA's office (ie the Traffic Area Office – see Appendix I for addresses) is situated.

Public Inquiries

It is useful here to mention the way in which a public inquiry is conducted. It is presided over by the LA or his deputy and is open to members of the general public, other operators and interested persons who may sit in and listen, and to the press, who may report all that is said.

Under rules applied since 1 November 1992, LAs may restrict general attendance at a public inquiry to protect an operator's business, particularly in regard to personal matters, commercially sensitive information and other information obtained in confidence. Hitherto, the LA could only close a PI to hear financial information. Further, LAs must disclose at a PI any information or evidence received in writing prior to the inquiry if it is intended that such information is to be taken into account in reaching a decision.

Verbal evidence is given to the LA by the applicant or by his legal representative if he has one – and this is strongly advised in most cases due to the complexities of making legal presentations and arguing points of law in a court-room situation and possibly the need to cross-examine witnesses such as a vehicle examiner or an environmental representor, even though it is an inquiry *not* a court. In fact, the 'call-up' letter to operators facing public inquiries in which the LA states his reasons for calling the PI and states his powers to curtail, suspend or revoke 'O' licences, as well as giving details of the time and location for the inquiry, advises this. Such advice, if it is required, should be sought from an experienced transport lawyer who fully understands the legal basis of the whole licensing system as well as the intricacies of the PI system (see also p 455).

Most evidence at PIs will be given in response to the LA's questions – the LA effectively playing the role of 'prosecutor' – by the licence holder/applicant, the objectors and those making representations. The evidence, unlike at criminal or civil courts, is not given under oath and statements made at a public inquiry that are defamatory or libellous of other people do not have protection by privilege. The offended person can take civil action if such statements or comments come to his notice. Similarly, if an applicant or witness lies, he will not be prosecuted for perjury, but, where this is an applicant, anything he says in support of his application may be taken by the LA to be a statement of intent to which he will be bound for the duration of any licence granted. Evidence in some instances may be provided in writing and the LA may ask for certain supporting documents, in which case the applicant should have these to hand with extra copies for the objectors to examine. When he has heard all the evidence the LA will normally make a decision without conferring with anybody else. He may announce this at the time or defer his decision to be given later in writing. The entire proceedings of the inquiry will be recorded and transcripts can be obtained by interested parties.

Licensing Authorities must now give at least 21 days' written notice of public inquiries both to operators and other parties entitled to attend, and similar notice if they intend to vary the time or place of the inquiry. However, given the consent of all parties, this requirement can be varied.

Appeals to the Transport Tribunal

The Transport Tribunal, which is now under the control of the Lord Chancellor, is a fully independent body comprising at least three sitting members, one of whom is the president, and must be a lawyer. Two or more of the other members, must also be legally qualified and have experience in the transport industry. Currently, the Tribunal comprises the President, a panel of two legal members (appointed by the Lord Chancellor) and five lay members (appointed by the Secretary of State for Transport).

An appeal may be made against an LA's decision to refuse to grant an 'O' licence, if he attaches environmental conditions to an 'O' licence, if a licence is granted authorising fewer vehicles than the number applied for, if a licence is granted for a shorter period than that applied for, or where an existing licence is withdrawn, suspended or prematurely terminated by the LA.

A time limit of 28 days is allowed in which to make an appeal to the Transport Tribunal following an LA's decision, counting from the date of publication of the issue of *Applications and Decisions* in which the decision is published. Where the LA's decision is not published within 21 days an appeal can be made within 49 days of notification of the decision by the LA.

Where an LA makes a disciplinary decision against an 'O' licence (ie suspension, curtailment or revocation) and the licence holder wishes to appeal, he can apply for a 'stay' of the decision until the appeal is heard in order to keep his vehicles operating. Otherwise he would have to observe the decision irrespective of the consequences (financial and operational) on his business. An initial request for a 'stay' of the decision is made direct to the LA, but failing this an application must be made immediately to the Tribunal giving details of the decision and the reason for requesting the 'stay'. Application for a 'stay' of the decision cannot be made if there is no intention to appeal.

Appeals to the Transport Tribunal must be in writing and six copies should be sent to the Tribunal stating the decision against which the appeal is made, the grounds for the appeal, and the names and addresses of every person to whom a copy of the appeal has been sent.

Copies of the appeal must be sent to the LA and to all objectors if the appeal is being made by a licence applicant, or to the applicant if the appeal is being made by an objector to the decision.

Although the Tribunal has the powers and status of the High Court, its proceedings are conducted informally and appellants may represent themselves or be represented by any person they choose (there are no wigs and gowns even for barristers present). However, in the best interests of the applicant he should be legally represented at an appeal by a solicitor or barrister experienced in transport law to ensure that his case is fully and correctly made.

When an appeal is heard, the Tribunal examines the transcript of the public inquiry or the LA's statement of his reasons for the decision against which the appeal is lodged and then may ask further questions of the applicant or his advocate. No oath has to be taken and there is no protection by privilege. The proceedings are open to the public and the press. Tribunal appeal decisions may be announced at the hearing or later. All parties will be sent a full statement of the decision usually within 3 weeks of the hearing.

Generally, Tribunal decisions will fall into one of three categories. Either to uphold the LA's decision, to change the decision or to refer the matter back to the LA with a direction that he should reconsider his decision but taking account of legal guidance from the Tribunal. In exceptional circumstances the Tribunal may review its decision subject to a request to do so made within 14 days of the appeal hearing. Decisions of the Tribunal are binding from the date they are given; in other words they have immediate effect.

Further appeals against decisions of the Tribunal may be made to the Court of Appeal or the Court of Sessions in Scotland but only on points of law, not on the original decision of the LA or the subsequent ruling of the Transport Tribunal.

No fees are payable in respect of appeals but costs may be awarded against frivolous, vexatious, improper or unreasonable appeals.

Further details of the appeals procedure can be found in booklet GV251A available free from Traffic Area Offices. The address of the Tribunal is: 48–49 Chancery Lane, London WC2A 1JR, Telephone 0171-936 7494, Fax 0171-404 0896.

Northern Ireland

Northern Ireland has its own scheme of Road Freight Operator's Licensing administered in the Province by the Department of the Environment. This is described fully in Chapter 11.

Goods vehicle operators in the Province do not need to obtain a short-term 'O' licence prior to entry into Great Britain. Similarly, there is no need for GB operators to obtain a short-term licence prior to entry into Northern Ireland. A goods vehicle operating on a current 'O' licence issued in Great Britain or a Road Freight 'O' licence issued in Northern Ireland is permitted to carry goods throughout the UK. However, cabotage (ie picking up and delivering a load within one country by a vehicle from another) is not permitted in either case.

Goods vehicles from Northern Ireland engaging in own-account operations for which a Road Freight Operator's licence is not required in the Province must, while operating in Great Britain, carry a document showing details of their load and route in Great Britain.

Use of Light Goods Vehicles

Many existing transport operators and new entrants to the industry have sought to avoid the problems and pitfalls of 'O' licensing by using vehicles defined as 'small' vehicles – those vehicles not exceeding 3.5 tonnes maximum permissible weight. With such vehicles there is no need to obtain an 'O' licence (the threshold for 'O' licensing may be raised to 6 tonnes in due course – see comment on p 2) and consequently no need to face the LA and satisfy all the conditions previously explained. The operator is also free from the legal requirements under other legislation for his drivers to operate tachographs or to keep other written records of their hours of work and to hold lgv driving licences.

Despite this apparent freedom the operator of such vehicles does have certain obligations and responsibilities. First, if he tows a trailer with such a vehicle the combined weight of both vehicle and trailer (if over 1020kg unladen) could exceed the 3.5 tonne weight threshold above which an 'O' licence would be needed and the provisions of the EU or British driver's hours law and the relevant record keeping or tachograph requirements may apply (see Chapters 3, 4 and 5). Second, if he also operates, or plans to operate in the future, larger vehicles which are within the scope of 'O' licensing, his conduct as an operator of small vehicles will be taken into account by the LA when deciding whether to grant or renew his 'O' licence.

The LAs have made the point that when an operator applies to renew an 'O' licence they (the LAs) would take notice of any relevant convictions in respect of smaller vehicles belonging to the operator and could call the operator to public inquiry to show cause why the 'O' licence should not be revoked or curtailed. The operator of small vehicles still has to ensure that his vehicles are not overloaded and that they are kept in a safe mechanical order under other regulations; they must be tested annually after they become 3 years old. Drivers of these vehicles are required to observe the drivers' hours regulations with certain exceptions. All these individual legal exemptions and requirements are discussed in later chapters.

Foreign Vehicles in the UK

Vehicles entering Great Britain from countries with which the UK has concluded a bilateral agreement do not need an 'O' licence; they may, however, require other documents (ie permits – or own-account documents). Vehicles from countries where no such agreement exists must be covered by an 'O' licence which will be issued for a period of not more than 90 days. Such an 'O' licence does not permit the holder to engage in internal transport in Great Britain – cabotage operations may however be permitted under appropriate permits issued to the operator in his own EU member state.

Foreign vehicles entering Great Britain under an EU or ECMT permit do not need an 'O' licence provided the permit is being carried on the vehicle.

Looking to the Single European Market

The advent of a Europe-wide single free market, with the trading barriers between EU member states demolished, will have an effect on road transport operations. Since the beginning of 1993 the liberalisation of community transport has meant the scrapping of road haulage permits to allow unrestricted movement of vehicles throughout the community and the likelihood that full cabotage operations (which are currently permitted on a very restricted basis) will be permitted. In particular, on the UK haulage scene, we can expect to see EU-registered vehicles undertaking legally permitted cabotage operations within the UK and possibly making significant in-roads into traditional domestic markets at the expense of local operators. It should be noted that such practices must be accompanied by strict observance of UK law on road traffic operation (eg speed limits, traffic rules, special loads requirements and suchlike), and on EU or international law so far as vehicle technical standards are concerned. There are also proposals to implement much stricter industry entry qualifications to add safety and environmental provisions to the current good repute, financial standing and professional competence requirements, to introduce 'Community Licences' and to harmonise road tax (ie vehicle excise duty).

NB: A more detailed examination of the effects of the SEM on the UK road transport industry is provided in the special chapter added to the front of this edition of the Handbook. *The subject of road freight transport in the SEM is covered fully in David Lowe's book* The European Road Freighting Handbook, *also published by Kogan Page.*

2: Professional Competence

The legal requirement for certain people employed in hire or reward road transport operations (ie in both road freight and road passenger sectors) to be professionally competent came into effect in Britain on 1 January 1978.

On the road goods side of the industry, this scheme was the final outcome of many years' work on the development of a plan to make individuals more responsible for the safe operation of vehicle fleets in their charge and better qualified to understand the legal, economic and operational requirements for safe and efficient goods vehicle operation. Previous proposals (in the Transport Act 1968) were based on a licensing system which would require transport managers to hold a 'transport manager's licence'. As a result of EU influence via EC Directive 74/561 'On Admission to the Occupation of Road Haulage Operator in National and International Operations', the proposed British transport manager's licensing scheme was abandoned in favour of one which provides all those people who qualify, not just those fulfilling the role of transport manager, with either a certificate of professional competence or a professional competence qualification by exemption or by examination.

Under the Goods Vehicles (Operators' Licences, Qualifications and Fees) Regulations 1984 (as amended), road haulage operators (including own-account operators who wish to carry goods for hire or to reward or in connection with a business which is not their own) are required to meet the professional competence requirement in order to obtain a standard operator's licence or renew (ie apply for a new licence to replace the expired licence) an existing standard 'O' licence.

Own-account transport operators who have no desire to or intention of carrying goods for hire or reward or for any purpose which is not in connection with their own trade or business (including not doing favours for customers by carrying their goods also) can apply for a restricted 'O' licence for which there is no need to meet the professional competence requirement.

The important point for road hauliers and others who carry goods for hire or reward is that in order to obtain a new or to renew (see note p 30) an existing operator's licence they must request a standard 'O' licence covering either national or both national and international operations and specify in their application the name of a person who is professionally competent and who is responsible for the operation of the vehicles authorised on the licence – the law requires the person to have 'continuous and effective responsibility' for the management of the transport operation. This person may be the applicant himself, if suitably qualified, or it may be a person holding the title of transport manager (or some other person – the law is not concerned as to the person's actual job-title only that they should be the person who is responsible for the vehicle operations on a day-to-day basis) who is professionally competent and is employed by the applicant.

Employment in this context no longer means full-time employment with the licence holder. Part-time or casual employment of a professionally competent person would not previously have met the requirements of the regulations but a change in the regulations from November 1991 allows a part-time qualified employee to be nominated in support of a licence application. Also, whereas a professionally competent person in one firm could not previously be specified in support of the licence held by another, non-related firm, this situation may now be viewed differently by Licensing Authorities (LAs) with this change in the rules.

There has been an LAs' decision which indicates that a self-employed (ie freelance) transport manager may be acceptable as the professionally competent person specified on an 'O' licence. The employment requirement does not necessarily mean that the person concerned must devote all of their working time to the transport management function, they may have other duties and responsibilities in the firm (eg such as administration manager, works manager, company secretary etc).

Concern among the LAs over the matter of part-time (ie so-called 'proxy') transport managers has led one of their number, Mr John Mervyn Pugh (LA for the West Midlands and South Wales Traffic Areas) to suggest that seven key questions need to be considered when determining whether the name of a part-time professionally competent person would be acceptable in support of a standard 'O' licence:

1. How many hours does he work for the licence applicant?
2. How many other employers does he have?
3. How many vehicles is he responsible for?
4. What is the distance between each of the operating centres for which he is responsible?
5. What is the nature of his duties – is he also a fitter or vehicle driver?
6. Where does he live in relation to his places of employment?
7. Is he of good repute?

Another LA has said that he would want details of exactly how a 'proxy' transport manager would apportion his time between firms employing him in this role, precisely what his duties would be for each employer and how much he would be paid.

Who may become Professionally Competent?

Professional competence is available and applicable only to individuals. Under the regulations, a firm or a corporate body cannot be classed as being professionally competent. Any individual, male or female, whether employed with the title of 'transport manager' or not, may become professionally competent if they meet the necessary qualifying conditions or otherwise pass the official examination. There is no pre-qualifying standard and no requirement that the person should have any previous experience of or is, has been or plans to actually work in the transport industry in any capacity whatsoever. It is open to absolutely anybody to become professionally competent if they so wish.

However, only those who are actually responsible for the operation of goods vehicles on a day-to-day basis will need to be 'nominated' as the professionally competent 'transport manager' in support of an application for a standard 'O' licence. Such individuals must themselves be of good repute and of appropriate financial standing (as well as the applicant – see Chapter 1) otherwise their name may not be acceptable to the LA despite the fact that they are qualified as being professionally competent (see below).

Proof of Professional Competence

Proof that a person is professionally competent and is therefore able to satisfy the requirements of the 'O' licence system is dependent on holding a certificate of professional competence (CPC) issued by an LA (form GV 203) or confirmatory evidence provided by any other body approved by the Secretary of State for Transport, namely a Royal Society of Arts examination pass certificate or a membership certificate from one of the recognised professional institutes which confers exemption (see pp 43–4). No other document provides evidence of professional competence for these purposes.

Classes of Competence

Professional competence in road freight operations falls into two classes covering national operations only or both national and international operations. A parallel scheme covers road passenger operations.

All certificates of professional competence (CPCs – form GV 203) granted under the original Grandfather Rights scheme (see below) cover both national and international operations (although it does not specifically say so on them).

In cases where people qualify for professional competence under the exemption arrangements, the level at which they qualify determines whether they are entitled to be classed as professionally competent in national transport operations only or in both national and international transport operations (see pp 43–4).

Candidates who achieve professional competence by examination will obtain appropriate documentary evidence (ie a Royal Society of Arts examination pass certificate) indicating that they have passed the modular examinations covering national operations only (modules A and B freight – or C passenger) or the additional examination covering international operations (module D freight – or E passenger).

Those qualified for professional competence in national operations only will be permitted to engage in or be responsible for operations conducted on standard 'O' licences solely within the UK (ie covering national transport operations).

Holders of certificates of professional competence (form GV 203) or those qualifying for professional competence in both national and international operations, by exemption or examination, will be permitted to engage in or be responsible for the operation of goods vehicles on standard 'O' licences covering national operations within the UK and international operations outside the UK.

National and International Transport Operations

Following legal decisions made in the EU the definition of what constitutes an international journey has been changed in Britain. So, for example, where loaded trailers are taken to a port for onward movement outside the UK, while the tractive unit and driver do not leave the UK, such journeys constitute national operations (previously, these were international operations), and a standard 'O' licence covering national transport operations only is needed for such operations (and the employed manager will need to be professionally competent only in national operations).

Qualifications for Professional Competence

The qualifications needed to obtain professional competence fall into three categories as follows:
1. By experience in the industry prior to 1 January 1975 (known as Grandfather Rights).
2. By exemption.
3. By examination.

Each of these three qualifying methods is described below.

Grandfather Rights

NB: The issue of certificates of competence (form GV 203 – known as CPCs) under this scheme ended on 31 December 1979, since when the only means of qualifying for professional competence is by exemption or examination. However, since many people continue to ask about the scheme it was felt useful to retain an explanation in the current edition of the Handbook.

Transport managers and other people employed in 'responsible road transport employment' prior to 1 January 1975 were able to obtain the grant of a so-called 'Grandfather Rights' certificate of competence, without examination, as of right. Once granted, the certificate will continue to remain valid for as long as the scheme is in force.

For the purposes of the (now extinct) Grandfather Rights scheme, responsible road transport employment was defined as employment in the service of a person or a firm carrying on a road transport undertaking and was employed in a position where the individual had responsibility for the operation of goods vehicles used under an operator's licence.

Similar conditions applied to any person who was the holder of an 'O' licence prior to 1 January 1975. In other words any owner-driver or small fleet operator who held an 'O' licence in his own name or under a business name qualified, having been a licence holder. In the case of an 'O' licence held by a partnership prior to this date, all the partners qualified under the Grandfather Rights arrangements. Where a licence was held by a limited company, the person responsible for the day-to-day operation of the vehicles under the licence (eg the transport manager) qualified for the Grandfather Rights grant of a CPC.

To obtain a certificate under this arrangement, application had to be made to the LAs by 30 November 1979. Since that date, the opportunity to obtain a CPC other than by examination or exemption has ceased.

Issue of CPC Certificates
Certificates of professional competence (form GV 203) issued under the Grandfather Rights scheme were obtainable from the Traffic Area Offices on application provided that the necessary form and certification of appropriate qualifying experience by an employer was supplied. Although certificates carry the name of the issuing Traffic Area, they are valid for operations in all Traffic Areas and individuals had no need to obtain separate certificates for each area in which they were responsible for the operation of vehicles.

There was no fee for the issue of a CPC, and certificates thus granted remain valid for the life of the scheme. There is no system for revocation or disqualification of CPC holders but where the holder is also the 'O' licence holder, or is the nominated professionally competent transport manager employed by an 'O' licence holder, then there is a requirement that he must be of good repute (see pp 8–9 and p 48) and of adequate financial standing; otherwise the 'O' licence will be subject to penalty (see pp 31–2). Grandfather Rights CPCs issued on the basis of information provided which is subsequently found to be false may be withdrawn by the LA.

Exemption

People who did not qualify for a CPC based on previous experience in road transport operations as described above and new entrants to the industry may obtain the professional competence qualification if they satisfy certain exemption criteria.

The exemption qualifications are based on holding current and valid membership of one or more of a number of professional bodies at certain levels. Additionally, a new 'Certificate in Transport' qualification offered by the Chartered Institute of Transport has been recognised by the DoT and accepted as giving exemption.

There are two levels of exemption qualification; one covering both international and national operations and the other covering national operations only. These are as follows:

For both national and international operations
1. Membership of the Chartered Institute of Transport (CIT) in the grade of Fellow or Member (engaged in the road transport sector).
2. Membership of the Institute of Transport Administration in the grade of

Fellow, Member, Associate Member or Associate (by examination) (engaged in the road transport sector).
3. Membership of the Institute of Road Transport Engineers in the grade of Member or Associate Member.
4. Membership of the Institute of the Furniture Warehousing and Removing Industry in the grade of Fellow or Associate.

See also note above about CIT certificate which is to confer exemption for national and international transport operations.

For national operations only
1. Membership of the CIT in the grade of Associate Member (engaged in the road transport sector).
2. Membership of the Institute of Road Transport Engineers in the grade of Associate (by examination).
3. Membership of the Institute of Transport Administration in the grade of Graduate or Associate (must be at least 21 years old, have 3 years' practical experience and be a holder of the NEBSS Certificate) (engaged in the road transport sector).
4. Holder of the General Certificate in Removals Management issued by the Institute of the Furniture Warehousing and Removing Industry.
5. Holder of the Royal Society of Arts Certificate in Road Goods transport gained by examination since May 1984.

There are no grounds for obtaining professional competence by exemption other than those detailed above. Valid membership of the relevant body (ie subscription paid etc) is sufficient to confirm professional competence but, if required, the institutes will issue a confirmatory certificate or statement (ie not a certificate of competence of the type issued under Grandfather Rights).

Examination

A system of examinations has been established to enable new entrants to the transport (ie road freight and road passenger) industry and those people who do not qualify under the previously mentioned Grandfather Rights or exemption arrangements to study for and obtain professional competence by examination. The examination scheme is organised and conducted on behalf of the Department of Transport by the Royal Society of Arts and examinations for both goods and passenger vehicle operations are held at main centres throughout the country four times each year. The original EC Directive of 1974, has been amended by EC 438/89 resulting in the need for an extensively revised examination syllabus which was put into effect by the Royal Society of Arts from 1 January 1991 and applied to examinations from March 1991. Of particular note was the inclusion of management, marketing techniques and environmental aspects of operating goods vehicles as additional examination subjects. Further revision of the syllabus took effect from October 1993.

Likely* dates for the 1995 examinations are as follows: Friday 10 March 1995, Friday 16 June 1995, Friday 6 October 1995 and Friday 1 December 1995.

Potential examination candidates should check these dates with the Royal Society of Arts or their local examination centre.

For the national examination a modular system is adopted with a common 'core' for both goods (ie road freight) and passenger subjects and then the individual modal (freight module B or passenger module C) sections. The 'core' (module A) comprises 20 questions to be answered in 30 minutes and the remaining freight modal section (module B) comprises 40 questions to be completed in one hour. Candidates who pass either module A alone, or the freight module B, or passenger module C sections alone, will be issued with a 'credit' pending their achievement of the full certificate, which they can gain having re-sat and passed the remainder of the examination. Thus, 'failed' examinees do not have to study for and sit the whole examination again, but only the module in which they failed.

This split modular system does not apply to the international examinations (freight module D or passenger module E) which remain unchanged with the candidate taking the one hour examination at a single sitting during which 30 questions have to be answered.

The examinations can be taken at individual sittings or all at the same examination centre on the same day (providing the local centre is offering all examinations at one sitting – this usually depends on there being a sufficient number of candidate entries). The choice is open to the candidate.

Fees for the examinations are currently as follows: module A taken alone is £7.60, module B or C taken alone is £14 and module D costs £18.60. These fees are payable to the Royal Society of Arts. Additional charges may be made by examining centres for the facilities they provide. Candidates who fail examinations may apply to the RSA for details of their performance. They should state where and when they sat the examination and should enclose a remittance of £2.50.

Note: Fees are reviewed for each examination year commencing in October, so prospective candidates should check the latest fees when making application to sit the examinations.

The address of the Royal Society of Arts is as follows: RSA Examinations Board, Westwood Way, Westwood Business Park, Coventry CV4 8HS (Tel: 0203 470033. Fax: 0203 468080).

Examination Method

Examinations conducted for the purpose of providing the qualification for professional competence are based on the 'objective testing' or multiple choice method. This means that the candidate is faced with either a number of questions, each of which is provided with a choice of possible answers (usually four), only one of which is correct, or alternatively statements to which a 'True' or 'False' answer must be indicated. The candidate must select and mark the correct answer to each question or statement.

This method of examination removes the need for high standards of literacy and ensures uniform levels of marking for all candidates. The examinations are designed to be within the grasp of candidates whose educational standard corresponds to the level normally reached at school leaving age.

Examination Syllabus

Candidates for the examination are not compelled to study beforehand but a study syllabus covering the examination subjects has been formulated and is available from the Royal Society of Arts. Study may be undertaken at courses organised by a number of bodies including the RTITB Services (now Centrex), group training associations, trade associations or commercial organisations. Some local technical colleges and other teaching establishments also offer part-time, usually evening, classes on the subject. The study may involve full-time or part-time attendance at such courses or it may be by correspondence course or by training or home learning package (ie notes and cassette tapes) or with the aid of teaching manuals. David Lowe's *Study Manual of Professional Competence in Road Transport Management* (Kogan Page Ltd, 120 Pentonville Road, London N1 9JN, Tel: 071-278 0433, covering both the freight national and international examinations, ie *not* the PSV syllabus) has been used successfully by many candidates, and in some cases without resorting to attendance at study courses or evening classes.

The RSA syllabus for road freight operations effective from the March 1991 examinations onwards is detailed here:

Syllabus for National Transport Operations

MODULE A
1. **Law**
   1.1 Elements of Law
   1.2 Business and Company Law
   1.3 Social Legislation
2. **Business and Financial Management**
   2.1 Financial Management Techniques
   2.2 Commercial Business Conduct
   2.3 General Insurance
4. **Road Safety**
   4.1 Traffic Legislation

MODULE B
1. **Law**
   1.1 Taxation
2. **Road Haulage Business and Financial Management**
   2.1 Marketing
   2.2 Commercial Conduct of the Business
   2.3 Insurance of Vehicles and Goods in Transit
   2.4 Methods of Operating
3. **Access to the Market**
   3.1 Operator Licensing
4. **Technical Standards and Aspects of Operation**
   4.1 Weights and Dimensions of Vehicles and Loads
   4.2 Vehicle Selection
   4.3 Vehicle Condition, Fitness and Maintenance
   4.4 Loading of Vehicles and Transit of Goods

5. Road Safety

  5.1 Drivers' Hours and Records

  5.2 Driving Licences

  5.3 Speed Limits

  5.4 Procedure in Case of Traffic Accidents

Syllabus for International Transport Operations

MODULE D

  1. Law

  2. Control of Road Haulage Operations

  3. Practice and Formalities Connected with International Movements

  4. Operations, Technical Standards and Road Safety

The RSA states that new legislative measures will not be included in the examination for at least three months from the date of implementation.

Those wishing to study for the examination should ensure that they have an up-to-date syllabus, which covers both goods and passenger vehicle operations, from the Royal Society of Arts (see above for address).

Note: A separate syllabus is published to cover Northern Ireland requirements.

Length of Study Course

The Royal Society of Arts recommends that the study course for the national examination should involve 72 hours of direct teaching and for the international examination a further 30 hours of direct teaching. Candidates starting from 'scratch', and particularly those without any assisted tuition, may find that they need considerably more than this amount of time which, in any case, does not include time needed for private study outside the tuition periods.

Transfer of Qualifications

Provisions are made to allow an interchange of professional competence qualifications and recognition of professional competence qualification certificates between the United Kingdom, Northern Ireland and other EU member states. Thus the UK Licensing Authorities are required to take into account any certificate of professional competence or any alternative to such a certificate showing relevant experience in the road haulage industry issued either in Northern Ireland or in other member states when considering an application for an 'O' licence.

Similarly, the Licensing Authorities in the UK will issue a Certificate of Qualification (fee £20 – payable to the Traffic Area Office) confirming the good repute, professional competence and, where relevant, the financial standing of persons or companies from the UK who wish to be admitted to the occupation of road haulage operator, or to be employed to manage the transport operations of a goods haulage undertaking in Northern Ireland or in any other EU member state.

Where a person seeking a Certificate of Qualification is not, or has not been, an 'O' licence holder in the UK, the LAs will, of course, have no knowledge of their experience, good repute or financial standing and thus will be unable to issue a certificate. In these cases application can now be made to the Secretary of State for Transport who is empowered to issue a certificate to such a person on payment of the requisite fee.

Good Repute

The subject of good repute has been dealt with in detail in Chapter 1 (see pp 8–9), where it is more properly located since it is an issue that arises only when an application for an 'O' licence is made by an individual, by a partnership business or by a company. However, in recent times, a great deal of attention has been focused on the 'good repute' aspect of professionally competent persons. For this reason, and to catch the attention of those who may read only this chapter rather than the 'O' licensing chapter (Chapter 1), some of the salient facts are repeated here.

As stated above, the good repute of such persons is only called into question when their names are put forward in support of a standard 'O' licence application. At that time the LA will want to know if the person has previous convictions for offences relating to the operation of goods vehicles (see below).

Following on from the previous explanation of the interchange of qualifications between the UK and other EU member states, the LAs, in considering an application for an 'O' licence and the good repute of the nominated professionally competent person, are required to take into account any convictions incurred by the person in Northern Ireland or in a country or territory outside the UK.

The relevant convictions are those which correspond to the relevant convictions under the British regulations, namely offences relating to vehicle roadworthiness, speeding, overloading, safe loading, drivers' hours and record keeping, drivers' licensing, maintenance, illegal use of rebated fuel, certain traffic offences and to the International Road Haulage Permits Act 1975.

Further, the Transport Tribunal has ruled that the LA can consider other convictions when considering a professionally competent person's good repute. The Tribunal ruled that the LA does not have to regard only those convictions mentioned above (as identified in the 1968 Transport Act) in deciding good repute.

In another instance an LA has said that a nominated professionally competent person must be clear of past financial difficulties. This followed a case where a nominated person was shown to be an adjudged bankrupt and the LA refused to accept the person's name on this account.

However, in new moves to protect the rights of individuals, LAs are no longer permitted to examine and rule upon a transport manager's good repute or professional competence unless the individual has been notified in advance that this is to happen and has had an opportunity to make personal representation to the LA (at a public inquiry if necessary and represented by a solicitor if he so wishes) concerning any allegations made about him.

3: Goods Vehicle Drivers' Hours

Any person who drives a goods vehicle which is being used for commercial or business purposes (ie not including those used for purely private purposes and other vehicles specifically exempted – see pp 51–3) irrespective of the vehicle weight must conform to strict rules on the amount of time they may spend driving. For most goods vehicle drivers the rules also include requirements relating to minimum breaks to be taken during the driving day and to both daily and weekly rest periods.

The driving hours rules are applied for reasons of public safety, to protect road users from the dangers of having overworked and tired drivers at the wheel of heavy vehicles. They are enforced vigorously by both the police and Vehicle Inspectorate (VI) enforcement officers and offenders are being dealt with very severely by the courts to emphasise the importance with which these road safety measures are viewed.

Drivers found to be in contravention of the rules as set out in this chapter can expect to be prosecuted and fined heavily on conviction – the most serious of such offences can result in imprisonment – and those who hold lgv driving entitlements may find they have placed these, and thus possibly their job, in jeopardy.

Employers of such drivers also risk prosecution and heavy fines on conviction for similar offences. Additionally they may have penalties imposed on their 'O' licences by the LAs since they promised in the declaration of intent at the time of their 'O' licence application that they would make arrangements to ensure the drivers' hours law would be observed. It is also a specific requirement of the EU rules that employers must make periodic checks to ensure the rules are observed and must take appropriate action if they discover breaches of the law to ensure there is no repetition of offences. On the Continent, breaches of the rules may result in drivers incurring heavy on-the-spot fines which must be paid immediately, otherwise the vehicle may be impounded and the driver held until the fine is paid.

Goods vehicle driver employers and drivers themselves need to understand the hours rules clearly and especially which particular set of rules applies to them depending on the vehicle being driven or the nature of the transport operation on which they are engaged. Three main sets of rules apply, the EU rules, the British domestic rules and what are known as the AETR rules for international journeys outside the EU (now fully aligned with the EU rules). The specific requirements under each of these sets of rules are explained in the following pages.

European Union Rules

The currently applied EU rules, contained in EC Regulation 3820/85, came into effect in the UK on 29 September 1986. British regulations as follows implementing these rules and modifying previous provisions came into effect on the same date:

1. The Community Drivers' Hours and Recording Equipment (Exemptions and Supplementary Provisions) Regulations 1986 – which implemented the EU regulations in the UK and the derogations (ie exemptions) from them.
2. The Community Drivers' Hours and Recording Equipment Regulations 1985 – which made consequential changes to the 1968 Transport Act provisions.
3. The Drivers' Hours (Goods Vehicles) (Modifications) Order 1985 – which implemented changes to national and domestic driving under the 1986 Act.
4. The Drivers' Hours (Harmonisation with Community Rules) Regulations 1986 – which harmonised the rules for those who drive under both the EU regulations and the revised 1968 Act rules.

Vehicles Covered

EU rules always take precedence over national rules. Therefore, since all goods vehicles used in connection with business activities as stated above are covered by one or other of the sets of drivers' hours rules (ie EU, British or AETR), the requirements of specific EU legislation must be considered first. Mainly, this involves examination of the list of EU rules exemptions (see below) to determine whether the vehicle or the transport operation falls within scope of the rules or outside.

Where a vehicle has a maximum permissible weight not exceeding 3.5 tonnes it is exempt from the EU rules or if it is one identified on the EU exemption list or is used for a purpose which is shown in the list as exempt as mentioned previously (eg vehicles used by the public utilities and local authorities etc) it automatically comes within the scope of the British domestic rules set out in the 1968 Transport Act as described in detail on pp 59–60.

Where vehicles are over 3.5 tonnes permissible maximum weight, including the weight of any trailer drawn, and are not exempt as shown by the list or are not used for an exempt purpose, the EU hours law applies.

Exemptions to EU Rules

The EU regulations list a number of exemptions and national governments (eg the British government) are permitted to make certain other exemptions (derogations) if they so wish, as shown below.

International Exemptions (under EU regulations)
1. Vehicles not exceeding 3.5 tonnes gross weight including the weight of any trailer drawn.
2. Passenger vehicles constructed to carry not more than nine persons including driver.

3. Vehicles on regular passenger services on routes not exceeding 50 kilometres.
4. Vehicles with legal maximum speed not exceeding 30kph (approx 18.6 mph).
5. Vehicles used by armed services, civil defence, fire services, forces responsible for maintaining public order (ie police).
6. Vehicles used in connection with sewerage; flood protection; water, gas and electricity services; highway maintenance and control; refuse collection and disposal; telephone and telegraph services; carriage of postal articles*; radio and television broadcasting; detection of radio or television transmitters or receivers.

   NB: The DoT has ruled that this exemption from the EU driving hours rules applies equally to parcels carriers operating in competition with the Royal Mail (ie the Parcelforce service) as with the Royal Mail itself. Instead, these operations fall within the scope of the British domestic hours rules as described in this chapter. However, despite this exemption, the EU tachograph rules still apply to such operations.
7. Vehicles used in emergencies or rescue operations.
8. Specialised vehicles used for medical purposes.
9. Vehicles transporting circus and funfair equipment.
10. Specialised breakdown vehicles.
11. Vehicles undergoing road tests for technical development, repair or maintenance purposes, and new or rebuilt vehicles which have not yet been put into service.
12. Vehicles used for non-commercial carriage of goods for personal use (ie private use).
13. Vehicles used for milk containers or milk products intended for animal feed.

National Exemptions (under British derogations)
1. Passenger vehicles constructed to carry not more than 17 persons including driver.
2. Vehicles used by public authorities to provide public services which are not in competition with professional road hauliers.
3. Vehicles used by agricultural, horticultural, forestry or fishery* undertakings, for carrying goods within a 50km radius of the place where the vehicle is normally based, including local administrative areas the centres of which are situated within that radius.

   * to gain this exemption the vehicle must be used to carry live fish or to carry a catch of fish which has not been subjected to any process or treatment (other than freezing) from the place of landing to a place where it is to be processed or treated.
4. Vehicles used for carrying animal waste or carcasses not intended for human consumption.
5. Vehicles used for carrying live animals from farms to local markets and vice versa, or from markets to local slaughterhouses.
6. Vehicles specially fitted for and used:
   - as shops at local markets and for door-to-door selling
   - for mobile banking, exchange or savings transactions
   - for worship

      –   for the lending of books, records or cassettes

      –   for cultural events or exhibitions.

7. Vehicles (not exceeding 7.5 tonnes gvw) carrying materials or equipment for the driver's use in the course of his work within a 50km radius of base provided the driving does not constitute the driver's main activity and does not prejudice the objectives of the regulations.

8. Vehicles operating exclusively on islands not exceeding 2300 sq km not linked to the mainland by bridge, ford or tunnel for use by motor vehicles (excludes Isle of Wight).

9. Vehicles (not exceeding 7.5 tonnes gvw) used for the carriage of goods propelled by gas produced on the vehicle or by electricity.

10. Vehicles used for driving instruction (but not if carrying goods for hire or reward).

11. Tractors used after 1 January 1990 exclusively for agricultural and forestry work.

12. Vehicles used by the RNLI for hauling lifeboats.

13. Vehicles manufactured before 1 January 1947.

14. Steam propelled vehicles.

Note: In exemption 2 above relating to vehicles used by public authorities, the exemption applies only if the vehicle is being used by:

(a) a health authority in England and Wales or a health board in Scotland
     –   to provide ambulance services in pursuance of its duty under the NHS Act 1977 or NHS (Scotland) Act 1978; or
     –   to carry staff, patients, medical supplies or equipment in pursuance of its general duties under the Act;

(b) a local authority to fulfil social services functions, such as services for old persons or for physically and mentally handicapped persons;

(c) HM Coastguard or lighthouse authorities;

(d) harbour authorities within harbour limits;

(e) airports authority within airport perimeters;

(f) British Rail, London Regional Transport, a Passenger Transport Executive or local authority for maintaining railways;

(g) British Waterways Board for maintaining navigable waterways.

EC Regulation 3820/85

To apply the rules as required and appreciate their implications it is necessary to understand the definitions of certain words and phrases used as follows:

Driver

The regulations apply specifically to the 'driver' of the vehicle. For these purposes a 'driver' is any person who drives the vehicle, even for a short period, or who is carried on the vehicle in order to be available for driving if necessary. They do not apply to persons carried as mates (ie to help load and unload only) or to statutory attendants.

Driving

Driving is time spent behind the wheel actually driving the vehicle and relates to an accumulation of periods spent driving before a break is needed or a daily rest period is commenced. The maximum limit for driving before a break is taken is 4½ hours.

Driving Time
Driving time is the accumulation of driving between two daily rest periods, or between a daily rest period and a weekly rest period, and must not exceed the limits described in this text.

Vehicle Categories
Drivers of all types and weight categories of vehicles within the scope of the regulations are treated equally in regard to driving, break and rest periods.

A Day
For the purposes of the regulations a day may be taken to mean any period of 24 hours starting from the end of a daily or weekly rest period (ie when the tachograph is set in motion) – in other words, a rolling period. If a driver drives a vehicle to which the EU regulations apply on any day (no matter how short the actual time spent driving or how short the journey on the road – for example, a 10-minute drive down the road would still bring the driver within scope of the rules) then the legal requirements apply to him for the whole of that day (ie 24-hour period) and the week in which that day falls.

Fixed Week
The definition of a 'week' for the purpose of the regulations is a fixed week from 00.00 hours Monday to 24.00 hours on the following Sunday. All references in the rules to weeks and weekly limits must be considered against this fixed week. Reference to 'fortnight' means two consecutive fixed weeks as described above.

Employers' Responsibilities

Under the EU rules employers have specifically stated responsibilities as follows:
1. They must organise driver's work in such a way that the requirements of the regulations are not broken (ie on driving times, breaks and rest periods etc).
2. They must make regular checks (ie of tachograph charts – see Chapter 5) to ensure the regulations are complied with.
3. Where they find any breaches of the law by drivers, they must take appropriate steps to prevent any repetition.

Additionally, of course, it goes without saying that employers should ensure their drivers do understand how the law in this respect applies to them and how to comply with its detailed provisions. Although it is no excuse in court for a driver to say he did not know the law, the court would expect the employer to have instructed the driver in its requirements and may well convict the employer for 'failing to cause' the driver to conform to the law (or for permitting offences) if it felt that insufficient attention had been given to this matter.

Driving Limits

Goods vehicle drivers are restricted in the amount of time they can spend driving before taking a break and the amount of driving they can do between any two daily rest periods (or a daily and a weekly rest period), in a week and in a fortnight. The maximum limits are as follows:

Maximum aggregated driving before a break:		4½ hours
Maximum daily driving:	normally:	9 hours
	extension:	10 hours on 2 days in a week
Maximum weekly driving:		6 daily driving shifts*
Maximum fortnightly driving:		90 hours

NB: The High Court ruled in 1988 that drivers can exceed the maximum of six daily driving shifts within 6 days as specified in the EU rules provided they do not exceed the maximum number of hours permitted in six consecutive driving periods. It should be noted that where a driver spends the maximum amount of driving time behind the wheel in 1 week (ie 4 x 9 hours plus 2 x 10 hours = 56 hours), during the following fixed week he may drive for a maximum of only 34 hours.

Break Periods

Drivers are required by law to take a break or breaks if in a day the aggregate of their driving time amounts to 4½ hours or more. If the driver does not drive for periods amounting in aggregate to 4½ hours in the day there is no legal requirement for him to take a break during that day.

Break periods must not be regarded as parts of a daily rest period and during breaks the driver must not carry out any 'other work'. However, waiting time, time spent riding as passenger in a vehicle or time spent on a ferry or train are not counted as 'other work' for these purposes.

The requirement for taking a break is that immediately the 4½-hour driving limit is reached a break of 45 minutes must be taken. This break may be replaced by a number of other breaks of *at least* 15 minutes each distributed over the driving period or taken during and immediately after this period, so as to equal at least 45 minutes and taken in such a way that the 4½-hour limit is not exceeded. A break period which was otherwise due in accordance with this requirement does not have to be taken if immediately following the driving period the driver commences a daily or weekly rest period, so long as the 4½ hours' aggregated driving is not exceeded.

According to DoT examples, the driver could legally operate the following procedures:

1. Drive 1 hour, 15 minutes' break, drive 3½ hours, 30 minutes' break, drive 1 hour, 15 minutes' break, drive 3½ hours, commence daily rest period.

2. Drive 1 hour, 15 minutes' break, drive 1 hour, 15 minutes' break, drive 2½ hours, 15 minutes' break, drive 2 hours, 30 minutes' break, drive 2½ hours, commence daily rest period.
3. Drive 3 hours, 15 minutes' break, drive 1½ hours, 30 minutes' break, drive 3 hours, 15 minutes' break, drive 1½ hours, commence daily rest period.

Following a highly publicised court case brought by the Lancashire police (*DPP* v *Mayfield Chicks Ltd*), it was established, after appeal against the original conviction, that the 4½-hour period should be looked upon as a rolling period so that at no time during the day must the driver have exceeded an aggregate of 4½ hours' driving without having had 45 minutes' break in respect of that driving. The previous 'wipe the slate clean' interpretation of the break period requirement has now been shown to be incorrect and must not be followed.

Further legal action has been instituted as a result of continuing confusion over varying interpretations of the four and a half-hour rule described above. A Manchester Crown Court judge (considering a number of appeals before him on the issue) asked the European Court of Justice to give a ruling that will provide a single, clear interpretation of the rule. The result of this ruling (given in December 1993) is that both the 'wipe the slate clean' and the 'rolling 4½ hours' interpretations of the driving hours rules are wrong. The Court ruled that where a driver has taken 45 minutes' break, either as a single break or as several breaks of at least 15 minutes each during or at the end of the 4½-hour period, the calculation should begin afresh, without taking account of the driving time and breaks previously completed by the driver. It also ruled that the calculation of the driving period begins at the moment when the driver sets his tachograph in motion and begins driving.

In hearing the long delayed appeals (in June 1994) the Manchester Crown Court held that the interpretation of the European Court's decision on the 4½ hours driving rule means that a driver cannot count any break taken in excess of the required 45 minutes after driving for 4½ hours as part of the break period legally required in respect of the next 4½-hour driving period.

NB: It is important to note that break periods (ie especially the 45-minute period as well as the alternative minimum 15-minute periods) should not be curtailed even by a minute or two. Prosecutions have been brought for break periods which are alleged not to conform to the law even though they have been only a matter of minutes below the minimum specified in the regulations. Such matters are, of course, shown clearly on tachograph recordings which provide ample evidence for the prosecution case.

Rest periods

Rest periods are defined as uninterrupted periods of at least 1 hour during which the driver 'may freely dispose of his time'. Daily rest periods, and particularly rest periods which are compensating for previously reduced rest periods, should not be confused with, or combined with, the statutory break periods that are required to be taken during the driving day as described above.

Daily Rest Periods

Once each day drivers are required to observe either a normal, a reduced or a split daily rest period during which time they must be free to dispose of their time as they wish. Thus in each 24-hour period one or other of the following daily rest periods must be taken:

Normal daily rest: 11 hours

or alternatively,

Reduced rest: 9 hours – may be taken three times in a week but the reduced time must be compensated by an equal amount of additional rest taken with other rest periods before the end of the next following fixed week

Split rest: Where the daily rest period is not reduced (as above) the rest may be split and taken in two or three separate periods during the 24 hours, provided:
(a) one continuous period is of at least 8 hours' duration;
(b) other periods are of at least 1 hour's duration;
(c) the total daily rest period is increased to 12 hours.

Double-manned Vehicles

Where a vehicle is operated by a two-man crew, the daily rest period requirement is that each man must have had a minimum of 8 hours' rest in each period of 30 hours.

Daily Rest on Vehicles

Daily rest periods may be taken on a vehicle provided:
(a) the vehicle has a bunk so the driver (but not necessarily a mate or attendant) can lie down
(b) the vehicle is stationary for the whole of the rest period.

It follows from this that a driver on a double-manned vehicle cannot be taking part of his *daily rest period* on the bunk while his co-driver continues to drive the vehicle. He could, however, be taking a *break* at this time while the vehicle is moving or he could merely spend his time lying on the bunk with his tachograph chart recording other work.

Daily Rest on Ferries/Trains

Daily rest periods which are taken when a vehicle is carried for part of its journey on a ferry crossing or by rail may be interrupted, but *once* only, provided:
(a) part of the rest is taken on land before or after the ferry crossing/rail journey;

(b) the interruption must be 'as short as possible' and in any event *must not* be more than 1 hour before embarkation or after disembarkation and this time *must* include dealing with customs formalities;

(c) during both parts of the rest (ie in the terminal and on board the ferry/train) the driver must have access to a bunk or couchette;

(d) when such interruptions to daily rest occur, the total daily rest period must be extended by 2 hours.

Weekly Rest Period

Once each fixed week (and after six driving shifts – see note on p 54) a daily rest period must be combined with a weekly rest period to provide a weekly rest period totalling 45 hours. A weekly rest period which begins in one fixed week and continues into the following week may be attached to either of these weeks.

While the normal weekly rest period is 45 hours as described above, this may be reduced to:

(a) 36 hours when the rest is taken at the place where the vehicle or the driver is based; or

(b) 24 hours when taken the rest period is taken elsewhere.

Reduced weekly rest periods must be compensated (ie made up) by an equivalent amount of rest period time taken *en bloc* and added to another rest period of at least 8 hours' duration before the end of the third week following the week in which the reduced weekly rest period is taken.

Compensated Rest Periods

When reduced daily and/or weekly rest periods are taken, the compensated time must be attached to another rest period of at least eight hours' duration and must be granted, at the request of the driver, at the vehicle parking place or at the driver's base. Compensation in this respect *does not* mean compensation by means of payment; it means the provision of an equivalent amount of rest time taken on a later occasion but within the specified limits (ie by the end of the next week for compensated daily rest and by the end of the third following week in the case of compensated weekly rest periods and in each case added to other rest periods).

Summary of EU Rules

The following table summarises the EU rules applicable to both national and international goods vehicle operations:

Maximum daily driving:	9 hours 10 hours on 2 days in week
Maximum weekly driving:	6 daily driving periods (see p 55)
Maximum fortnightly driving:	90 hours
Maximum driving before a break:	4½ hours

Minimum breaks after driving:	45 minutes or other breaks of at least 15 minutes each to equal 45 minutes
Minimum daily rest (normally):	11 hours
Reduced daily rest:	9 hours on up to 3 days per week (must be made up by end of next following week)
Split daily rest:	The 11-hour daily rest period may be split into two or three periods – one at least 8 hours, the others at least 1 hour each: total rest must be increased to 12 hours
Minimum weekly rest (normally):	45 hours once each fixed week
Reduced weekly rest:	36 hours at base – 24 hours elsewhere (any reduction must be made up *en bloc* by end of the third following week)
Rest on ferries/trains:	Daily rest may be interrupted once only if: – part taken on land – no more than 1 hour between parts – drivers must have access to a bunk or couchette for both parts of rest – total rest must increase by 2 hours

Emergencies

It is permitted for the driver to depart from the EU rules as specified above to the extent necessary to enable him to reach a suitable stopping place when emergencies arise where he needs to ensure the safety of persons, the vehicle or its load, providing road safety is not jeopardised. The nature of and reasons for departing from the rules in these circumstances must be shown on the tachograph chart.

Prohibition on Certain Payments

The EU rules prohibit any payment to wage-earning drivers in the form of bonuses or wage supplements related to distances travelled and/or the amount of goods carried unless such payments do not endanger road safety.

British Domestic Rules

The current British drivers' hours rules contained in the 1968 Transport Act (as amended) came into effect on 29 September 1986. They apply to goods vehicle drivers whose activities are outside the scope of the EU requirements as described above (ie which are specified as exempt from the EU rules) and comprise only limits on daily driving and daily duty.

NB: Detailed provisions which previously applied in regard to daily and weekly duty, daily spreadover and daily and weekly rest period limits were completely abolished when these rules changes were introduced.

It is important to emphasise that drivers who are exempt from the EU rules, either because their vehicle does not exceed 3.5 tonnes permissible maximum weight or because the activity in which they are engaged falls within the scope of the EU exemptions list (see pp 51–2), must observe the British domestic rules.

NB: It should be noted that:

1. *Although no records of driving or working times are required to be kept, drivers of light goods vehicles (ie not exceeding 3.5 tonnes permissible maximum weight) must still conform to the legal limits on maximum daily driving and maximum daily duty.*

2. *If a trailer is attached to a vehicle not exceeding 3.5 tonnes permissible maximum weight (which itself is exempt from the EU rules on account of its weight) thereby taking the combined weight to over 3.5 tonnes then the EU rules must be followed as described in the foregoing text (unless it is exempt for other reasons – namely the nature of the operations on which it is engaged) and a tachograph must be fitted to the vehicle and used by the driver for record-keeping purposes (see Chapter 5).*

Exemptions and Concessions

The British domestic rules apply to drivers of all goods vehicles which are exempt from the EU regulations as described earlier but with the following further exceptions which are totally exempt from all hours rules control:
1. Armed forces
2. Police and fire brigade services
3. Driving off the public road system
4. Driving for purely private purposes (ie not in connection with any trade or business).

The British domestic rules do not apply:
(a) to a driver who on any day does not drive a relevant vehicle; or
(b) to a driver who on each day of the week does not drive a vehicle within these rules for more than 4 hours (Note: this exemption *does not* apply to a driver whose activities fall within scope of the EU rules).

Driving and Duty Definitions

For the purposes of the British domestic rules:

(a) driving means time spent behind the wheel actually driving a goods vehicle and the specified maximum limit applies to such time spent driving on public roads. Driving on off-road sites and premises such as quarries, civil engineering and building sites and on agricultural and forestry land is counted as duty time, not driving time;

(b)   Duty time is the time a driver spends working for his employer and includes any work undertaken including the driving of private motor cars, for example and non-driving work which is not driving time for the purposes of the regulations. The daily duty limit does not apply on any day when a driver does not drive a goods vehicle.

Summary of British Domestic Rules

Maximum daily driving:	10 hours
Maximum daily duty:	11 hours
Continuous duty:	no specified limit
Daily spreadover:	no specified limit
Weekly duty:	no specified limit
Breaks during day:	no requirement specified
Daily rest:	no specified requirement (but obviously minimum of 13 hours so as not to exceed the daily 11-hour duty limit)
Weekly rest:	no specified requirement

Emergencies

The daily driving and duty limits specified above may be suspended when an emergency situation arises. This is defined as an event requiring immediate action to avoid danger to life or health of one or more individuals or animals, serious interruption in the maintenance of essential public services for the supply of gas, water, electricity, drainage, or of telecommunications and postal services, or in the use of roads, railways, ports or airports, or damage to property. Details of the emergency should be entered by the driver on his record sheet when the limits are exceeded.

Light Vehicle Driving

As explained above, drivers of light goods vehicles not exceeding 3.5 tonnes permissible maximum weight – as with drivers of other goods vehicles which fall within scope of these rules – must observe the daily limits on driving (10 hours) and duty (11 hours). However, only the 10-hour daily driving limit applies when such vehicles are used:
1.   By doctors, dentists, nurses, midwives or vets.
2.   For any service of inspection, cleaning, maintenance, repair, installation or fitting.
3.   By a commercial traveller and carrying only goods used for soliciting orders.
4.   By an employee of the AA, the RAC or the RSAC.
5.   For the business of cinematography or of radio or television broadcasting.

Mixed EU and British Driving

It is possible that a goods vehicle driver may be engaged in transport operations which come within scope of both the EU drivers' hours rules and the British domestic hours rules on the same day or within the same week.

When such a situation arises he may conform strictly to the EU rules throughout the whole of the driving/working period or he may take advantage of the more liberal British domestic rules where appropriate. If he decides on this course of action and thereby combines both British and EU rules he must beware of the following points:

1. time spent driving under the EU rules cannot count as an off-duty period for the British rules;
2. time spent driving or on duty under the British rules cannot count as a break or rest period under the EU rules;
3. driving under the EU rules counts towards the driving and duty limits for the British rules;
4. if any EU rules' driving is done in a week the driver must observe the EU daily and weekly rest period requirements for the whole of that week.

AETR Rules

Drivers on international journeys beyond the EU which take them to or through the list of countries given below are required to observe The European Agreement Concerning the Work of Crews of Vehicles Engaged in International Road Transport – 1971 (commonly known and referred to as AETR). When on such journeys the driver must observe the AETR rules for the whole of the outward and return journey including the portion travelled in Britain and through other EU member states. These rules are now fully harmonised with the EU rules contained in EC Regulation 3820/85 as described in the preceding text of this chapter.

The countries beyond the EU referred to above where the AETR rules apply (ie on journeys to or through) are as follows: Austria, the Czech Republic and Slovakia (formerly Czechoslovakia), Norway, Poland, Sweden, Commonwealth of Independent Soviet States, (CIS) and (the former) Yugoslavia.

Tax Relief on Driver Allowances

Sleeper-cab allowances

The amount paid to drivers for overnight subsistence varies considerably from area to area – the national general figure is currently £19.90 (1994 figure). Since 1991 the Inland Revenue has agreed that lgv drivers can be paid night-out allowances on a tax-free basis (eg for use of sleeper cabs) amounting to 75 per cent of the national figure (ie £19.90, which amounts to £14.92). This applies even if a locally agreed allowance is less than the national average. Where there are existing agreements with local tax offices for higher levels of subsistence payment (up to a maximum of 2.6 per cent above the standard figure) the tax-free allowance will be 75 per cent of the higher figure.

Owner-drivers are dealt with differently for tax purposes and may NOT claim such night-out allowances against their tax liability.

Meal Expenses

In regard to repayment to drivers of meal expenses incurred while away from base, the Inland Revenue has issued a Statement of Practice as follows but it should be noted that the Statement 'has no binding force and does not affect a taxpayer's rights of appeal on points concerning his liability to tax'.

1. Employees Absent from Home and Normal Place of Employment – General

An employer is allowed to make reimbursement of certain expenses payments without deduction of tax to an employee working temporarily away from home and his normal place of employment. The payments are those which are intended to cover the extra cost of travelling and subsistence which the employee incurs because he is away on duty – but not the cost of his travelling from home to his normal place of work or his usual expenses on food or meals taken when at his normal place of work. Similarly, where the employee is not reimbursed, a deduction for expenses may be allowed where the extra expenses of the absence are incurred wholly, exclusively and necessarily in the performance of the employee's duties.

2. Travelling Appointments

Where travelling itself is an essential feature of an employee's duties, with the result that he has to spend money on meals in restaurants or cafes above what he would spend if he had a fixed place or area of work or were able to get home for meals, a deduction may be allowed for the extra expenses necessarily incurred in the performance of the duties. Drivers who qualify for consideration under this heading are those who are engaged full time in travelling in the performance of their duties. By this is meant employment as a driver throughout the full normal working hours of each day, except those days when the employee is precluded from working by reason of sickness or other reasonable cause, or which are holidays or rest days. Employees whose jobs entail only incidental travelling would not be regarded as holding travelling appointments. Even full-time drivers are excluded from consideration if they travel only in a limited area. This is because they incur no additional expenses when at work from one day to the next, as they have an established pattern of expenditure on meals in the same way as any other employee who has to work at a distance from home and who cannot return home for lunch.

Inspectors of Taxes require claimants to give full details of the nature of their duties in addition to providing evidence of the expenditure actually incurred, as described in the following paragraph. In practice, relief is not restricted by reference to amounts of expenditure saved by not having meals at home or a fixed place of work.

3. Evidence of Expenditure

The amount of relief which can be allowed depends mainly on the bills and vouchers which can be supplied by a driver in support of his claims. Employees who consider that they may have a potential claim should ensure that bills, receipts, etc are available in support of any claim for relief for the current tax year and future years; but if exceptionally they are unable on any occasion to obtain bills or receipts, they must make a note at the time of the date, place and amount spent. (Where the employer makes a contribution in cash or otherwise towards the cost of meals this must be specified and the amount received set off against the expense in arriving at the net amount for which relief is claimed.) An expenses deduction cannot be given in the absence of evidence of expenditure and tax districts will not accept estimated figures of outgoings.

Copies of this statement may be obtained by calling at or writing to the Public Enquiry Room, New Wing, Somerset House, Strand, London WC2R 1LB.

There have been recent reports about tax inspectors querying payment of night-out allowances where drivers have stayed only a few miles from base. Currently the agreement with the Inland Revenue for night-out payments is as stated above and this applies for nights out without receipts, except where alternative figures have been agreed, in which case an annual increase of 9.3 per cent on the previously agreed figure is acceptable. These payments should only be made where the employee does actually spend the night away and incurs extra expense. If he uses the bunk in a sleeper cab the allowable amount is only that required to meet the expenses he incurs, not the full night-out allowance. Where expense payments to drivers or other staff for nights away exceed the amount stated above the employer should be prepared to bear the tax on the extra amount or, alternatively, he should include it as part of the wages within PAYE.

There have been many reports in the past year of the Inland Revenue refusing claims for tax-free payments of night-out allowances above the general limit. Employers who have paid in excess of this amount (or a locally agreed rate) without deduction of tax may find themselves liable to meet the tax due on the additional amounts paid except where it can be proved that the expense was genuinely incurred (by production of a valid receipt).

Revision of EU Rules

The European Council of Transport Ministers have been discussing proposals for amendment to the current EU regulation (EC 3820/85) which came into force on 29 September 1986. At the time of preparing this 1995 edition of the *Handbook*, there is little further news of the matter other than the fact that there is strong support for imposing daily duty (ie total working) limits on drivers (as well as on workers in other industries). Currently the suggestion is that this should be a maximum of 11 hours, reducing by up to 4 hours the present maximum working (as opposed to driving) limit for drivers. On another aspect, a draft directive has been published aimed at standardising enforcement of the drivers' hours law provisions throughout the member states.

Besides these steps, there are proposals under the EU Social Chapter of the Maastricht Treaty which are designed to protect the health and safety of employee workers (but not apparently of self-employed persons), to establish minimum daily, weekly and yearly rest periods for all workers in Europe, and to restrict night-working to an average of only 8 hours in every 24, with a ban on work on two successive night shifts. A number of derogations from these provisions will apply in cases of emergency, for seasonal work and by collective agreement between employers and employees – road transport will also be derogated.

4: Goods Vehicle Drivers' Records

Drivers of goods vehicles over 3.5 tonnes gross weight must keep records of the time they spend driving such vehicles and their working times. Most drivers are also required to record breaks taken during driving periods and their daily and weekly rest periods. Where vehicles fall within the scope of the EU drivers' hours law as described in Chapter 3 (ie unless they are specifically exempt as shown in the relevant exemptions list) the record-keeping requirement is based on the mandatory use of tachographs as described in Chapter 5.

Where a vehicle is outside the scope of the EU rules then the British Domestic (ie 1968 Transport Act – as amended) driving hours rules apply (see Chapter 3) and the driver of such a vehicle is required to keep written records by means of a 'log-book' system.

To summarise the main record-keeping alternatives, these are as follows:
1. Goods vehicles not exceeding 3.5 tonnes permissible maximum weight – NO RECORDS.
2. Goods vehicles over 3.5 tonnes permissible maximum weight (including the weight of any trailer drawn) operating within EU rules – TACHOGRAPH RECORDS.
3. Goods vehicles over 3.5 tonnes permissible maximum weight exempt from EU rules – WRITTEN 'LOG-BOOK' RECORDS (but subject to further exemption in certain cases).
4. Goods vehicles over 3.5 tonnes permissible maximum weight exempt from both EU and British Domestic rules (eg military vehicles) – NO RECORDS.

This chapter describes the record-keeping requirements applying to drivers falling within item 3 above, namely, those operating under the British Domestic rules who must keep written 'log-book' records.

Exemptions from Record Keeping

Written records do not have to be kept in the following cases:
1. By drivers of vehicles which are exempt from 'O' licensing except that the exemption does not apply to drivers of Crown vehicles which would have needed an 'O' licence if the vehicle had not been Crown property.
2. By drivers of goods vehicles on any day when they drive for 4 hours or less and within 50 kilometres of the vehicle's base (NB this exemption is applicable only in the case of domestic operations – it does not apply to tachograph use – see p 60).
3. By drivers using an EU tachograph for record-keeping purposes which has been calibrated and sealed at a DoT-approved tachograph centre.

Record-Keeping System

Prior to 29 September 1986 when the EU rules and British domestic hours rules changed, the written record-keeping requirement related to the use of the EC diagrammatic-type International Control Book (ICB). From this date there was a transitional arrangement for phasing out the use of this record book and the introduction of a new-style of simplified British record which relates specifically to the British domestic driving hours rules. Regulations brought the 'simplified' record book into use from 2 November 1987 with the old-type EC diagrammatic control book ceasing to have any further legal standing from that date.

Record Books

Ready-printed record books can be purchased 'off the shelf' or firms can have their own version pre-printed with their own name and logo if desired. In the latter case it is important that the specific requirements of the regulations are observed in both the format and the printing of the book.

The book must be a standard A6 format (105mm x 148mm) or it may be larger if preferred. It must comprise a front sheet on which is entered relevant information, a set of instructions for the use of the book, and a number of individual weekly record sheets with facilities for completing these in duplicate (ie with carbon paper or carbonless copy paper) and for the duplicate sheet to be detached for return to the employer when completed.

There is no legal requirement for the numbering of record books or for their issue against an entry in a register of record book issues as previously required.

Weekly record sheets in the book must follow the format set out in the regulations with appropriate spaces for entries to be made under the following headings:
1. Driver's name.
2. Period covered by sheet week commencing . . . week ending . . .
3. Registration number of vehicle(s).
4. Place where vehicle(s) based.
5. Time of going on duty.
6. Time of going off duty.
7. Time spent driving.
8. Time spent on duty.
9. Signature of driver.
10. Certification by employer (ie signature and position held).

Issue and Return of Record Books

Employers must issue their drivers with record books when they are required to drive vehicles to which the British domestic driving hours regulations apply and where records must be kept. Before issuing the book the employer must complete the front cover to show the firm's name, address and telephone number preferably with a rubber stamp if he has one.

When a record book is issued to the driver he should complete the front cover with his surname, first name(s), date of birth and home address and the date he first used the book. When the book is completed he should also enter the date of the last entry (ie date of last use). There is space to record the name and address of a second employer.

Books issued by an employer to an employee-driver must be returned to that employer when complete (subject to the requirement for the driver to retain it for 2 weeks after use) or when the employee leaves that employment. He must not take it with him to his new employer. Any unused weekly sheets and all duplicates must be included when the book is returned.

Two Employers

Where a driver has two employers who employ him to drive goods vehicles to which the British domestic hours rules apply, the first employer must issue the record book as described, and the second employer must write or stamp his firm's name and address on the front cover of the record book with a statement that the holder is also a driver in his employment. When the driver does part-time driving work for another employer he must disclose to each employer, if requested, details of his working and driving times with the other employer. Similarly, when a driver changes to a new employer the former employer must give the new employer details of the driver's previous driving and working times if requested.

Record Book Entries

The driver must make entries on the weekly sheet for each day on which a record is required (instructions on the correct use of the book are printed inside the cover). Care must be taken to ensure that an exact duplicate of the entry is made simultaneously (ie two separately written repeat entries are *not* acceptable even if no carbon paper is available). When completing a daily sheet he must enter all the required details under each of the headings. If he changes vehicles during the day he must write in the registration number for each vehicle. He must then sign the sheet before returning it to his employer.

Completion of the record is straightforward, the driver having to enter the vehicle registration number, the time of coming on duty, and at the end of the day he must enter the time at which he went off duty and he must sign the sheet. He may enter any remarks concerning his entries, or point out corrections which should be made, in the appropriate box at the foot of the record. The employer may also use this space if required for making comments regarding the record. This space may also be used for recording the name of a second driver.

Corrections

Entries in the record book must be in ink or made by a ball-point pen and there must be no erasures, corrections or additions. Mistakes may only be corrected by writing an explanation or showing the correct information in the remarks space. Sheets must not be mutilated or destroyed.

Return and Signing of Record Sheets

On completion of the weekly sheet and after it has been signed by the driver, the duplicate copy must be detached from the book and handed to the employer within 7 days of the date of the last entry on the sheet, and then within a further 7 days the employer must have examined and signed the duplicate sheet. However, if in either case it is not reasonably practicable to do so within this time, these actions must be carried out as soon as it is possible to do so.

Retention and Production of Record Books

The driver should have his record book with him at all times when working and must produce it for inspection at the request of an authorised examiner. The book should be shown to the employer at the end of every week or as soon as possible after the week so he can examine and countersign the entries. Following completion of the book the driver must continue to keep it with him for a further 2 weeks (available for inspection by the enforcement authorities) before returning it to his employer.

Completed record books must be retained by the employer, also available for inspection by the enforcement authorities, for not less than 12 months.

WEEKLY RECORD SHEETS

WEEKLY SHEETS

1. DRIVER'S NAME

2. PERIOD COVERED BY SHEET

WEEK COMMENCING (DATE) ....

TO WEEK ENDING (DATE) ....

DAY ON WHICH DUTY COMMENCED	REGISTRATION NO. OF VEHICLE(S) 3	PLACE WHERE VEHICLE(S) BASED 4	TIME OF GOING ON DUTY 5	TIME OF GOING OFF DUTY 6	TIME SPENT DRIVING 7	TIME SPENT ON DUTY 8	SIGNATURE OF DRIVER 9
MONDAY							
TUESDAY							
WEDNESDAY							
THURSDAY							
FRIDAY							
SATURDAY							
SUNDAY							

10. CERTIFICATION BY EMPLOYER

I HAVE EXAMINED THE ENTRIES IN THIS SHEET

SIGNATURE ....

POSITION HELD ....

Figure 4.1 *Simplified record sheet for British domestic transport operations*

5: Tachographs – Fitment and Use Requirements

Tachograph instruments installed in goods and passenger vehicles provide a means of recording time and the speed and distance travelled by the vehicle. This record enables drivers' working activities and driving practices to be monitored to ensure that legal requirements – especially observance of the driver's hours rules – have been met.

The fitment of tachographs and their use in relevant vehicles (ie those operating within the scope of the EU driving hours rules – see Chapter 3) for record-keeping purposes, originally became a legal requirement in the United Kingdom on 31 December 1981. EU regulations (EC 3821/85) which came into effect on 29 September 1986 amended some of the original requirements.

The legislation requires the fitment and use of tachographs in most goods vehicles over 3.5 tonnes permissible maximum weight with certain EU-approved exemptions which are listed at the end of this chapter (see pp 81–3). Since the list of exemptions is limited, many categories of goods vehicle which may be thought to be exempt are not so. It should be stressed that the law applies to *any* goods vehicle over 3.5 tonnes gvw (or a combination of a goods vehicle and goods-carrying trailer which together exceed 3.5 tonnes permissible maximum weight) when used for business purposes unless it is specifically exempted as shown by the exemption list.

In particular, it should be noted that there is no exemption for short distance operations, infrequent-use vehicles or occasional drivers – once a relevant vehicle is on the highway the law applies in full.

This means that tachograph instruments must be installed in the vehicle and that whosoever drives it must keep a tachograph record and observe the EU driver's hours law in full for both the day on which the driving takes place and the week in which that day falls. It should also be noted that vehicles which are exempt from the tachograph rules are not necessarily exempt from record-keeping requirements (for example, those operating under the British Domestic hours rules). See Chapter 4 for details of activities where written records must be kept.

Under the EU regulations a number of specific basic requirements relating to tachograph use must be met as follows:
1. The tachograph instrument must conform to the technical specification laid down in the EU regulations (EC 3821/85 Annex I – see p 83).
2. The instrument must be calibrated and officially sealed at a DoT-approved calibration centre to ensure that accurate (ie legally acceptable) records are made (see p 72).
3. The instrument must be used in accordance with the regulations, with individual responsibilities being observed by both employer and driver.

Employer's Responsibilities

The regulations place specific responsibilities on the employer of a driver who drives within the EU rules as follows:

1. The employer must organise the driver's work in such a way that he is able to comply with both the driver's hours and tachograph rules.
2. The employer must supply drivers with sufficient numbers of the correct type of tachograph charts (ie one chart for the day, one spare in case the first is impounded by an enforcement officer, plus any further spares which are necessary to account for any charts which become too dirty or damaged to use), and he must ensure that completed charts are collected from drivers no later than 21 days after use.
3. The employer must periodically check completed charts to ensure that the law has been complied with (ie that the driver has made a chart for the day, that he has completed it fully and properly and that he has observed the driving hours rules). If breaches of the law are found the employer must take appropriate steps to prevent their repetition.
4. The employer must retain completed charts for 12 months for inspection by Vehicle Inspectorate (VI) examiners, if required.
5. The employer must give copies of the record to drivers concerned who request them.

Drivers' Responsibilities

Drivers of vehicles operating within the EU rules must observe the tachograph requirements. In particular this means understanding what the law requires and how to comply with it. The specific responsibilities of the driver in regard to the law are as follows:

1. Drivers using tachograph charts must ensure that a proper record is made by the instrument:
   (a) that it is a continuous record
   (b) that it is a 'time right' record (ie recordings are in the correct 12-hour section of the chart – daytime or night-time hours).
2. In the event of instrument failure or in circumstances where no vehicle is available when the driver is working he must make manual recordings of his activities on the chart 'legibly and without dirtying' it.
3. Drivers must produce for inspection on request by an authorised inspecting officer a current chart for that day plus the charts relating to the current week and for the last day of the previous week in which he drove.
4. Drivers must return completed charts to their employer no later than 21 days after use.
5. Drivers must allow any authorised inspecting officer to inspect the charts and tachograph calibration plaque which is usually fixed inside the body of the instrument.

Two-Crew Operation

Reference above to a driver also includes any other driver who is carried on the vehicle to assist with the driving. In this case a two-man tachograph must be fitted and both drivers must use it to produce records. The person who is driving must have his chart located in the uppermost (ie number 1) position in the instrument and use the number 1 activity mode switch to enable his activities and vehicle speed and distance recordings to be made on the chart as appropriate. The person who is riding passenger must have his chart in the rearmost (ie number 2) position and must use the number 2 activity mode switch to record his other work activities, or break or rest periods. Only time group recordings are made on this chart; driving, speed and distance traces are *not* produced on the second-man chart.

Tachograph Calibration, Sealing and Inspection

To make legally acceptable records tachograph installations in vehicles must be calibrated initially at DoT-approved tachograph centres (see Appendix VI) and subsequently must be inspected every 2 years and fully re-calibrated every 6 years or after repair at an approved centre.

NB: In recent times problems have arisen over the date when 2-yearly checks on tachographs are due. Operators of new vehicles have missed these dates having assumed that the 2 years runs from when they first took delivery of the vehicle but in fact it runs from the date of original calibration which may have been some months earlier.

The DoT specifies and approves the premises (and the display of approved signs), equipment, staff (including their training) and procedures for the installation, repair, inspection, calibration and sealing of tachographs. Such centres must be approved to the BS 5750 Part 2 quality assurance standard before they can gain DoT approval. No other workshops or individuals are permitted to carry out such work and any work carried out by unauthorised agents would render the installation incapable of producing legally acceptable records.

It is an offence for any unauthorised person to carry out work on tachograph installations. It is also an offence for a vehicle operator (maximum fine £5000) to obtain and use a tachograph instrument repaired by a firm which is not BS 5750 approved.

The calibration process requires the vehicle to be presented to an approved tachograph centre in normal road-going trim, complete with body and all fixtures, unladen and with tyres complying with legal limits as to tread wear and inflated to manufacturer's recommended pressures. At the centre, the necessary work on the installation is carried out to within specified tolerances.

The regulations specify a range of tolerances within which the tachograph installation must operate and valid for temperatures between 0°C and 40°C as follows:

	On bench test	On installation	In use
Speed	± 3kph	± 4kph	± 6kph
Distance	± 1%	± 2%	± 4%
Time	in all cases, ± 2 minutes per day with a maximum of 10 minutes per 7 days		

In the case of both speed and distance figures shown above the tolerance is measured relative to the real speed and to the real distance of at least 1 kilometre.

Calibration and Periodic Inspection Fees

The DoT's Vehicle Inspectorate Executive Agency specifies the official fees to be charged in connection with tachograph calibration and periodic inspections. Calibration currently costs £33 plus VAT and the official time for the task is $1\frac{1}{2}$ hours. The 2-yearly inspection costs £21 plus VAT. These prices are exclusive of any replacement parts used.

Sealing of Tachographs

Approved centres seal tachograph installations after calibration or 2-yearly inspections with their own official seals (each of which is coded differently) and details of all seals are maintained on a register by the DoT. The seals are of the customs type whereby a piece of wire is passed through each of the connecting points between the vehicle and the tachograph itself and then a lead seal is squeezed tight on to the wire with special pliers which imprint the centre code number in the metal. Attempts to remove any of the seals or their actual removal will show and need to be accounted for.

The purpose of sealing is to ensure that there is no tampering with the equipment or any of its drive mechanism or cables which could either vary the recordings of time, speed or distance or inhibit the recording in any way. Such tampering is illegal and once seals are broken the installation no longer complies with the law and legally acceptable records cannot be made.

Besides the seals inside the body of the instrument head, the following points are sealed:
1. the installation plaque;
2. the two ends of the link between the recording equipment and the vehicle;
3. the adaptor itself and the point of its insertion into the circuit;
4. the switch mechanism for vehicles with two or more axle ratios;
5. the links joining the adaptor and the switch mechanism to the rest of the equipment;
6. the casings of the instrument.

Seal Breakage

Obviously, there are occasions when certain of the seals have to be broken of necessity to carry out repairs to the vehicle and replacement of defective parts (eg the vehicle clutch or gearbox). The only seals which may be broken in these circumstances are as follows:

1. those at the two ends of the link between the tachograph equipment and the vehicle;
2. those between the adaptor (ie the tachograph drive gearbox) and the point of its insertion into the circuit;
3. those at the links joining the adaptor and the switch mechanism (ie where the vehicle has a two-speed rear axle) to the rest of the equipment.

With the introduction of statutory speed-limiter fitment on many heavy vehicles, the EU now permits operators to break tachograph seals for the purpose of fitting such devices, but the seals must be replaced – at an approved tachograph centre – within 7 days.

While it is permitted to break the particular seals listed above for other authorised purposes (eg in connection with vehicle maintenance), a written record must be kept of the seal breakage and the reason for doing so. The installation must be inspected or re-calibrated and fully sealed following repair or seal breakage as soon as 'circumstances permit' and before the vehicle is used again. It is illegal to remove any of the other seals and tampering with seals by drivers is tantamount to committing fraud (ie for the purposes of making fraudulent records), which is an offence liable to lead to a prison sentence.

Calibration Plaques

When a tachograph has been installed in a vehicle and calibrated, the approved centre must fix, either inside the tachograph head or near to it on the vehicle dashboard in a visible position, a plaque giving details of the centre, the 'turns count', and the calibration date. The plaque is sealed and must not be tampered with or the sealing tape removed. When an instrument is subjected to a 2 year inspection or re-calibration a new plaque must be fitted. If a vehicle is found on the road with an 'out-of-date' plaque an offence will have been committed and prosecution may follow.

The normal sequence for plaques is that one will show the initial calibration date, the next (2 years later), which is fitted alongside the first plaque, will show the date of the 2 year inspection and a third plaque will show the second 2 year inspection. After a further 2 years a 6 year re-calibration of the installation will be due and at this time all the previous plaques will be removed, the new calibration plaque will be fitted and the procedure described above starts again. In between times, following certain repairs, a 'minor work' plaque may be fitted but this does not alter the sequence of dates for 2 year inspection and re-calibration plaques.

The 2 year and 6 year periods referred to above for inspections and calibrations are counted to the day/date, *not* to the end of the month in which that day/date falls.

Tachograph Breakdown

If tachograph equipment becomes defective (or the seals are broken for whatever reason, including authorised breakage, as described above, to carry out mechanical repairs to the vehicle, or unauthorised interference) it must be repaired at an approved centre as soon as 'circumstances permit', but in the meantime the driver must continue to record manually on the chart all necessary information regarding his working, driving, breaks and rest times which are no longer being recorded by the instrument. There is *no* requirement to attempt to record speed or distance.

Once a vehicle has returned to base with a defective tachograph, it should not leave again until the instrument is in working order and it has been re-calibrated (if necessary) and the seals replaced. If it cannot be repaired immediately, the vehicle can be used so long as the operator has taken positive steps (which he can satisfactorily prove later if challenged by the enforcement authorities – see paragraph below) to have the installation repaired as soon as reasonably practicable.

If a vehicle is unable to return to base within *1 week* (ie 7 days) counting from the day of the breakdown, arrangements must be made to have the defective instrument repaired and re-calibrated as necessary at an approved centre *en route* within that time.

Defence

There is a defence in the regulations against conviction (ie not against prosecution) for an offence of using a vehicle with a defective tachograph. This has the effect of allowing subsequent use of the vehicle with a defective tachograph provided steps have been taken to have the installation restored to a legal condition as soon as circumstances permit and provided the driver continues to record his driving, working and break period times manually on a tachograph chart. In such circumstances it will be necessary to satisfactorily prove to the enforcement authorities – and to the court if they proceed with prosecution – that a definite booking for the repair had already been made at the time the vehicle was apprehended and that this appointment was for the repair to be carried out at the earliest possible opportunity. It is also a defence to show that at the time it was examined by an enforcement officer the vehicle was on its way to an approved tachograph centre to have necessary repairs carried out. However, this defence will fail if the driver did not keep written records of his activities in the meantime.

Use of Tachographs

Drivers are responsible for ensuring that the tachograph instrument in their vehicle functions correctly throughout the whole of their working shift in order that a full and proper recording for a full 24 hours can be produced. They must also ensure that they have sufficient quantities of the right type of charts (see below) on which to make recordings.

Time Changes

Drivers must ensure that the time at which the instrument clock is set and consequently recordings are made on the chart agrees with the official time in the country of registration of the vehicle. This is a significant point for British drivers travelling in Europe who may be tempted to change the clock in the instrument to the correct local European time rather than, for example, having it indicate and record the time in Britain. To re-emphasise the point, this means that for British drivers in British-registered vehicles the tachograph chart recording must accord with the official time in Britain regardless of the country in which that recording was made – the tachograph clock must *not* be re-set to show local time when travelling abroad.

Dirty or Damaged Charts

If a chart becomes dirty or damaged in use, it must be replaced and the old chart should be securely attached to the new chart which is used to replace it.

Completion of Centre Field

Before starting work with a vehicle in which tachograph charts are used the driver must enter on the centre field of the chart the following details:
1. His surname and first name (not initials).
2. The date and place where use of the chart begins.
3. Vehicle registration number.
4. The distance recorder (odometer) reading at the start of the day.

At the end of a working day, the driver should then record the following information on the chart:
1. The place and date where the chart is completed.
2. The closing odometer reading.
3. By subtraction, the total distance driven – in kilometres.

Making Recordings

When the centre field has been completed the chart should be inserted in the tachograph instrument ensuring that it is the right way up (it should be impossible to fit it wrongly) and that the recording will commence on the correct part of the 24-hour chart (day or night). The instrument face should be securely closed. The activity mode switch (number 1) should be turned as necessary throughout the work period to indicate the driver's relevant activities, namely driving, other work, break or rest periods. While some drivers find difficulty in getting into the habit of turning the switch to coincide with each change of activity, nevertheless this is what the law requires and it is an offence to fail to do so (ie not keep proper records).

Overnight Recordings
At the end of his shift the driver can leave the chart in the instrument overnight to record the daily rest period or alternatively it can be removed and the rest period recorded manually on the chart. Generally, enforcement staffs prefer an automatic recording of daily rest made by the instrument but this is not always

practicable where the vehicle may be used on night-shift work, may be driven for road testing or other purposes by workshop staff or moved around the premises by others when the driver is at home having the rest period. Also, if the driver is scheduled to start work much later on the following day there will be an overlap recording on the chart which is illegal.

Vehicle Changes

If the driver changes to another vehicle during the working day he must take the existing chart with him and record details of the time of change, the registration number of the further vehicle(s) and distance recordings in the appropriate spaces on the chart. He then uses that chart in the next vehicle to record his continuing driving, working activities and break periods. This procedure is repeated no matter how many different vehicles (except those not driven on the public highway) are driven during the day so the one chart shows all of the driver's daily activity.

Mixed Tachographs

It is important to note that the various makes and models of tachograph currently available in the UK have different charts and they cannot all be interchanged. So the driver who switches from a vehicle with one make of instrument to a vehicle with a different make during the working day will have to make fresh entries on a second or even third chart. At the end of the day all the charts used should be clipped together to present a comprehensive (and legal) record for the whole day. However, there are some charts now available on the market suitable for dual use in different makes of tachograph. It is the employer's duty to issue drivers with the correct charts (ie with matching type approval numbers to those on the tachograph instrument in use) in sufficient numbers for the schedule which the driver has to operate.

Note: Where instruments are standardised in a fleet, one chart will suffice for the whole day and the driver must take this with him from one vehicle to the next, recording changes as previously described.

Manual Records

Drivers are responsible for ensuring that the instrument is kept running while they are in charge of the vehicle and should it fail or otherwise cease making proper records they should remove the chart and continue to record their activities manually on it as previously described. They must also make manual recordings on the chart of work done or time spent away from the vehicle (for example, periods during the day spent working in the yard, warehouse or workshop). Manual recordings must be made legibly and in making them the sheet must not be 'dirtied'.

Records for Part-Time Drivers

The rules on the use of tachographs described in the foregoing text apply equally to part-time or occasional drivers such as yard and warehouse staff, office people and even the transport manager. The rules also apply fully even if the driving on the road is for a very short distance or period of time – a 5-minute drive without a tachograph chart in use would be sufficient to break the law and risk prosecution. Vehicle fitters and other workshops staff who drive vehicles on the road for testing in connection with repair or maintenance are specifically exempt from the need to keep tachograph records when undertaking such activities (see exemption list on pp 81–3) but this exemption *does not* apply to them when using vehicles for other purposes (eg collecting spare parts, ferrying vehicles back and forth, taking replacement vehicles out to on-road breakdowns, taking and collecting vehicles to and from goods vehicle test stations etc).

Retention, Return and Checking of Tachograph Charts

Drivers must retain and be able to produce, on request by authorised examiners (including the police), completed tachograph charts for each driving day of the current week and for the last day of the previous week on which they drove. Remember, charts do not have to be made for non-driving days or rest days and therefore cannot be asked for by the police or others in respect of such days.

Charts must be returned by drivers to their employer no later than 21 days after use and, on receiving the charts, the employer must periodically check them to ensure that the drivers' hours and record-keeping regulations have been complied with. They must then be retained, available for inspection if required, for a period of 1 year.

Where a driver has more than one employer in a week (eg as with agency drivers), he must return the tachograph charts to the employer who first employed him that week. This provision clearly presents a problem for those firms which regularly employ agency drivers and which may find difficulty in securing the return of charts for driving work done with their vehicles. At the present time, there is no legal solution to this issue.

Official Inspection of Charts

An authorised inspecting officer may require any person to produce for inspection any tachograph chart on which recordings have been made. Further, he may enter a vehicle (see note below) to inspect a chart or a tachograph instrument (he should be able to read the recordings relating to the nine hours prior to the time of his inspection) and the calibration plaques and detain a vehicle for this purpose. At any reasonable time he may enter premises on which he believes vehicles or tachograph charts are kept and inspect the instruments in such vehicles and the completed charts. He can require (by serving a notice in writing) charts to be produced at a Traffic Area Office at any time on giving at least 10 days' notice in which to do so. Where a chart is suspected of showing a false entry he may 'seize' the chart (but not for reasons other than evidence of a false entry, or an entry intended to deceive, or an entry altered for such purposes) and retain it for a maximum period of 6 months, after which time, if no charges for offences have been made, the chart should have been returned. If not, the person from whom it was taken can apply to a magistrate's court to seek an order for its return.

NB: It has now been held that police and enforcement authorities do not have legal powers under current UK law to actually enter vehicle cabs for the purpose of examining tachograph charts and instruments (although they do have powers to examine such instruments) – due to an apparent oversight, the relevant provision in the 1968 Transport Act giving such authority was excluded from the original Commencement Order for the Act and has never been reinstated. Moves are under way to correct this anomaly.

In practice, Licensing Authorities (LAs) regularly ask operators to provide batches of tachograph charts covering one or more of their vehicles for a short period or possibly some months either on a routine basis or following investigations into, or leads about, hours' law or tachograph infringements – and where it is suspected that drivers are regularly exceeding speed limits. These charts are then analysed for contraventions of the law.

Cases have been reported where the police have demanded that operators should send in to them, by post, tachograph charts required for inspection. This has been shown to be an illegal practice: there is no provision in the law which allows random collection of charts by the police or demands for charts to be submitted by post. Only in cases where an on-the-spot examination reveals possible offences can the police then request (on production of their authority) copies of the relevant chart for that day, other charts for that week and for the last day of the previous week. Vehicle Inspectorate enforcement officers, on the other hand, may take random selections of charts away for examination.

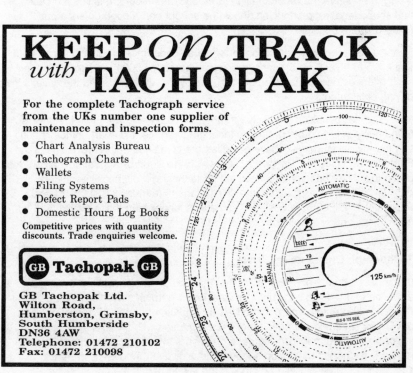

It is an offence to fail to produce records for inspection as required or to obstruct an enforcement officer in his request to inspect records or tachograph installations in vehicles.

Offences

Some tachograph related offences have already been mentioned in connection with specific requirements of the law but there are other overriding, and very serious, offences to be considered. In particular, it is an offence to use, cause or permit the use of a vehicle which does not have a fully calibrated tachograph installed; for the driver to fail to keep records by means of a tachograph (or manually if the instrument is defective); and to make false recordings. Further, it is an offence for the driver to fail to return used charts to his employer within 21 days after use or to fail to notify his first employer of any other employer for whom he drives vehicles to which the regulations apply.

In many cases offences committed by the driver result in charges against the employer for 'causing' or 'permitting' offences. For example, where tachograph charts go missing and cannot be produced for examination by the enforcement authorities, the employer may be charged with any one of three (or even all three) relevant offences; namely, failing to cause the driver to keep a record, failing to preserve the record, failing to produce the record. It is in such cases that sound legal representation should be sought and a good defence put forward where possible.

Penalties on summary conviction for offences under these regulations can be a fine of up to level 5 on the standard scale (currently £5000) and conviction for such offences can jeopardise both the employers' 'O' licence and the driver's lgv driving entitlement.

Much heavier penalties are likely for those transport operators and lgv drivers found guilty in UK courts of forging tachograph charts and chart entries relating to international journeys. These penalties are contained in the provisions of the Forgeries and Counterfeiting Act 1981 which the Court of Appeal now allows the police to use for bringing prosecutions for tachograph offences committed outside the UK. Fines may be in excess of the current level 5 standard scale maximum of £5000, and custodial sentences longer than those available to the courts under other legislation.

Tachograph Charts as Evidence

In the past it has been made clear that tachograph charts, while providing evidence of 'the facts shown' for drivers' hours purposes, would not be used by the police and enforcement authorities as evidence to bring prosecutions against drivers for speeding offences – not to be confused with the LAs' actions in imposing short-term lgv driving bans on drivers found from their charts to have regularly exceeded maximum speed limits. However, the DoT is currently considering the possibility of changing the law to permit the retrospective checking of tachograph charts for speeding and prosecution of drivers where such evidence is shown. This follows publicity surrounding a number of serious coach and lorry crashes where speeding was thought to be a contributory factor.

Exemptions

There is no requirement for the fitment and use of tachographs in the following vehicles or in vehicles used in connection with the particular transport operations specified in the exemption list.

NB: It should be noted that this is the same list of exemptions as that for the EU hours law under EC 3820/85 (see p 51–3):

National and International Exemptions
1. Vehicles not exceeding 3.5 tonnes gross weight including the weight of any trailer drawn.
2. Passenger vehicles constructed to carry not more than nine persons including driver.
3. Vehicles on regular passenger services on routes not exceeding 50 kilometres.
4. Vehicles with legal maximum speed not exceeding 30kph (approx 18.6 mph).
5. Vehicles used by armed services, civil defence, fire services, forces responsible for maintaining public order (ie police).
6. Vehicles used in connection with sewerage; flood protection; water, gas and electricity services; highway maintenance and control; refuse collection and disposal; telephone and telegraph services; carriage of postal articles; radio and television broadcasting; detection of radio or television transmitters or receivers.
   NB: Under a European Court ruling, British Gas must fit tachographs to vehicles it uses for the delivery of gas appliances, gas cylinders and meters to ensure fair competition with private transport operators. The point at issue, and contended by British Gas, was that its vehicles engaged on such deliveries were being used in connection with gas services.
7. Vehicles used in emergencies or rescue operations.
8. Specialised vehicles used for medical purposes.
9. Vehicles transporting circus and funfair equipment.
10. Specialised breakdown vehicles.
11. Vehicles undergoing road tests for technical development, repair or maintenance purposes, and new or rebuilt vehicles which have not yet been put into service.
12. Vehicles used for non-commercial carriage of goods for personal use (ie private use).
13. Vehicles used for milk containers or milk products intended for animal feed.

Further Exemptions in National Operations only
14. Passenger vehicles constructed to carry not more than 17 persons, including the driver.

15. Vehicles used by public authorities to provide public services which are not in competition with professional road hauliers.
16. Vehicles used by agricultural, horticultural, forestry or fishery* undertakings, for carrying goods within a 50km radius of the place where the vehicle is normally based, including local administrative areas the centres of which are situated within that radius.

    to gain this exemption the vehicle must be used to carry live fish or to carry a catch of fish which has not been subjected to any process or treatment (other than freezing) from the place of landing to a place where it is to be processed or treated.
17. Vehicles used for carrying animal waste or carcasses not intended for human consumption.
18. Vehicles used for carrying live animals from farms to local markets and vice versa, or from markets to local slaughterhouses.
19. Vehicles specially fitted for and used:
    - as shops at local markets and for door-to-door selling;
    - for mobile banking, exchange or savings transactions;
    - for worship;
    - for the lending of books, records or cassettes;
    - for cultural events or exhibitions.
20. Vehicles (not exceeding 7.5 tonnes gvw) carrying materials or equipment for the driver's use in the course of his work within 50km radius of base provided the driving does not constitute the driver's main activity and does not prejudice the objectives of the regulations.
21. Vehicles operating exclusively on islands not exceeding 2300 sq km not linked to the mainland by bridge, ford or tunnel for use by motor vehicles (excludes Isle of Wight).

22. Vehicles (not exceeding 7.5 tonnes gvw) used for the carriage of goods propelled by gas produced on the vehicle or by electricity.
23. Vehicles used for driving instruction (but not if carrying goods for hire or reward).
24. Tractors used after 1 January 1990 exclusively for agricultural and forestry work.
25. Vehicles used by the RNLI for hauling lifeboats.
26. Vehicles manufactured before 1 January 1947.
27. Steam propelled vehicles.

Note: In the exemption above relating to vehicles used by public authorities, the exemption applies only if the vehicle is being used by:
  (a) *a health authority in England and Wales or a health board in Scotland*
    - *to provide ambulance services in pursuance of its duty under the NHS Act 1977 or NHS (Scotland) Act 1978; or*
    - *to carry staff, patients, medical supplies or equipment in pursuance of its general duties under the Act;*
  (b) *a local authority to fulfil social services functions, such as services for old persons or for physically and mentally handicapped persons;*
  (c) *HM Coastguard or lighthouse authorities;*
  (d) *harbour authorities within harbour limits;*
  (e) *airports authority within airport perimeters;*

(f) British Rail, London Regional Transport, a Passenger Transport Executive or local authority for maintaining railways;

(g) British Waterways Board for maintaining navigable waterways.

Item 4 above includes certain works trucks and industrial tractors which have a statutory 30kph speed limit imposed upon them but this does not include fork lift trucks which may come within the scope of the rules.

Declaration of Exemption

When presenting a vehicle for the lgv annual test which the operator believes is exempt from the tachograph regulations in accordance with the list above , a 'Declaration of Exemption' form has to be completed.

The Tachograph Instrument

A tachograph is a cable or electronically driven speedometer incorporating an integral electric clock and a chart recording mechanism. It is fitted into the vehicle dashboard or in some other convenient visible position in the driving cab. The instrument indicates time, speed and distance and permanently records this information on the chart as well as the driver's working activities. Thus, the following factors can be determined from a chart:

1. The varying speeds (and the highest speed) at which the vehicle was driven.
2. The total distance travelled and distances between individual stops.
3. The times when the vehicle was being driven and the total amount of driving time.
4. The times when the vehicle was standing and whether the driver was indicating other work, break or rest period during this time.

Recordings

Recordings are made on special circular charts, each of which covers a period of 24 hours (Figure 5.1), by three styli. One stylus records distance, another records speed and the third records time-group activities as determined by the driver turning the activity mode switch on the head of the instrument (ie driving, other work, breaks and rest periods). The styli press through a wax recording layer on the chart, revealing the carbonated layer (usually black) between the top surface and the backing paper (some charts are made of Melanex and have no carbonated layer – recordings expose the translucent backing sheet). The charts are accurately pre-marked with time, distance and speed reference radials and when the styli have marked the chart with the appropriate recordings these can be easily identified and interpreted against the printed reference marks.

Movement of the vehicle creates a broad running line on the time radial, indicating when the vehicle started running and when it stopped. After the vehicle has stopped, the time-group stylus continues to mark the chart but with an easily distinguishable thin line. The speed trace gives an accurate recording of the speeds attained at all times throughout the journey, continuing to record on the speed base line when the vehicle is stationary to provide an unbroken trace except when the instrument is opened. The distance recording is made by the stylus moving up and down over a short stroke, each movement representing five kilometres travelled; thus, every 5 kilometres the stylus reverses direction, forming a 'V' for every 10 kilometres of distance travelled. To calculate the total distance covered the 'V's are counted and multiplied by ten and any 'tail ends' are added in, the total being expressed in kilometres.

When a second chart is located in the rear position of a two-man tachograph, only a time recording of the second man's activities (ie other work, break or rest) is shown. Traces showing driving, vehicle speed or distance cannot be recorded on this chart.

Precautions against interference with the readings are incorporated in the instrument. It is opened with a key and a security mark is made on the chart every time the instrument is opened.* When checking the chart it can be easily established at what time the instrument was opened and thus whether this was for an authorised reason or not. Interference with the recording mechanism to give false readings, particularly of speed, can be determined quite simply by an experienced chart analyst.

*NB: Changes to the EU tachograph specification to reduce the possibility of fraudulent recordings were introduced to apply to new instruments receiving Type Approval from 1991. These will include a provision for the chart to be marked at every interruption of the power supply (eg when a fuse is removed).

Faults

Tachographs are generally robust instruments, but listed below are some of the faults which may occur:
1. failure of the cable drive at the vehicle gearbox;
2. failure of the cable drive at the tachograph head;
3. failure of the adaptor/corrector/triplex gearbox;
4. cable braking or seizure;
5. electrical faults affecting lights in the instrument or the clock;
6. incorrect time showing on the 24-hour clock (eg day-shift work becomes shown against night hours on the charts);
7. failure of the tachograph head;
8. damage to the recording styli;
9. failure of the distance recorder;
10. damage to charts because of incorrect insertion.

Fiddles

A key feature of tachograph recordings is that careful observation will show results of the majority of faults in recordings as well as fiddles and attempts at falsification of recordings by drivers. The main faults likely to be encountered will show as follows.
1. Clock stops – recordings continue in a single vertical line until the styli penetrate the chart.
2. Styli jam/seize up – recordings continue around the chart with no vertical movement.

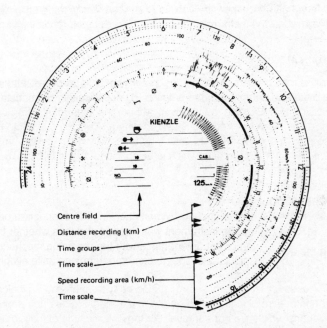

Centre field
Distance recording (km)
Time groups
Time scale
Speed recording area (km/h)
Time scale

Figure 5.1 *A typical tachograph chart showing recordings of time, distance and speed*

3. Cable or electronic drive failure – chart continues to rotate and speed and distance styli continue to record on base line and where last positioned respectively. Time-group recordings can still be made but no driving trace will appear.

Attempts at falsification of charts will appear as follows.

1. Opening the instrument face will result in a gap in recordings.
2. Winding the clock backwards or forwards will leave either a gap in the recording or an overlap. In either case the distance recording will not match up if the vehicle is moved.
3. Stopping the clock will stop the rotation of the chart so all speed and distance recordings will be on one vertical line (see item 1 above about how faults in instruments show on charts).
4. Restricting the speed stylus to give indications of lower than actual speed will result in flat-topped speed recordings while bending the stylus down to achieve the same effect will result in recordings below the speed base line when the vehicle is stationary.
5. Written or marked-in recordings with pens or sharp pointed objects are readily identifiable by even a relatively unskilled chart analyst.

NB: This is only an outline list of a large number of possible faults and attempts at falsification likely to be encountered. Some driver fiddles are one-off attempts, crudely and clumsily executed and obvious; others are much more sophisticated in their execution, often as part of an on-going violation of legal requirements. These are more difficult, but not impossible, for the transport manager or fleet operator to detect and would certainly be picked up quickly by an experienced chart analyst.

EU Instruments and Charts

Tachographs may only be used for legal record-keeping purposes if they are type-approved and comply with the detailed EU specification. Such instruments have provision for indicating to the driver, without the instrument being opened, that a chart has been inserted and that a continuous recording is being made. They also provide for the driver to select, by an activity mode switch on the instrument, the type of recording which is being made. In the UK this must be one of the following:
1. Driving time.
2. Other work time.
3. Break and rest periods.

Two-man instruments are also provided with a means of simultaneously recording the activities of a second crew member on a separate chart located in the rear position in the instrument.

The charts used for legal purposes must also be type-approved as indicated by the appropriate 'e' markings printed on them. It is illegal to use non-approved charts or charts which are not approved for the specific type of instrument being used. Care should be taken that charts used have accurate time registration – cheap and non-type approved versions, which are illegal anyway, have been found in the past to be significantly inaccurate in the way they are printed thus producing inaccurate and worthless records.

Chart Analysis

Analysis of the information recorded on tachograph charts can provide valuable data for determining whether drivers have complied with the law on driving, working, break and rest period times, and have conformed to statutory speed limits. The data can also be extremely useful as a basis for finding means of increasing the efficiency of vehicle operation and for establishing productivity monitoring and payment schemes for drivers.

Many fleet operators use and rely upon the services of tachograph analysis agencies for checking their charts for conformity with the law. However, it should be noted that should such firms fail to recognise and notify the operator of deficiencies in their records, it is the operator who is at risk and his licence, not the analysis bureau. The LAs repeatedly remind operators that responsibility for driver compliance with the hours law rests entirely with them, not outside agencies. Generally also, the checking carried out by such firms is for standard hours law infringements only, which are mainly picked up by computerised analysis and may not include identification of other irregularities or cleverly executed false entries or fraudulent recordings. Similarly, such analysis may not identify driver abuse of vehicles of frequent and excessive speeding.

Tachograph manufacturers supply accessories to enable detailed chart analysis to be carried out. A chart analyser magnifies the used chart to the extent that detailed analysis beyond the scope of a normal visual examination can be made of the vehicle's minute-by-minute and kilometre-by-kilometre progress. Journey times, average running times and speeds, delivery times, route miles, traffic delays and many other relevant factors can be readily established. With the aid of a fixed hairline cursor on the magnifier to allow precise definition of the time and speed scales and recordings on the chart, even a vehicle's rates of acceleration and deceleration can be determined.

The German company, Mannesman Kienzle GmbH (and its British counterpart Lucas Kienzle Instruments Limited of Birmingham), which is the leading tachograph manufacturer, has undertaken considerable research into chart analysis and is able to offer users the service of its analysis experts both in Germany and the UK as well as in other countries to determine the activities of vehicles and particularly the progress of a vehicle immediately prior to an accident and at the point of impact. In some instances such analysis has shown that witnesses' accounts of the speed of the vehicle and its braking force just before the accident have been far from correct.

It is also claimed that by detailed analysis of the charts and by keeping drivers aware of the information obtained, driving methods can be improved, thus saving fuel and cutting down on the wear and tear on vehicle brakes, tyres, transmission and other components.

Latest Developments

Discussions have been taking place in Brussels on the feasibility of replacing conventional tachographs with hi-tech electronic devices which are more resistant to tampering and easier for enforcement officers to read on the spot. A number of such (black-box type) devices have been designed, some of which can 'down-load' to hand-held computers. For such a change to occur, significant legislative measures would be necessary to allow legal recording of drivers' working, break and rest period times to be recorded electronically, rather than by the present stylus and chart system. Additionally, governments (and the EU) would have to be satisfied that they were not over-burdening transport operators (and indeed enforcement agencies) with unacceptably high costs in the changeover. For this reason, it is considered that the current concept of tachographs will remain the principal means of record keeping for some years to come.

In the meantime, of course, the technology – and the hardware – exists for any operator who wishes, on a voluntary basis, to equip his vehicles with on-board computers, automatic driver identification, smart-card logic and other electronic monitoring devices.

Comprehensive information on tachographs and their use, including information on carrying out detailed chart analysis is contained in The Tachograph Manual *by David Lowe and published by Kogan Page Limited.*

6: Driver Licensing and Licence Penalties

Any person wishing to drive a motor (ie mechanically propelled) vehicle on a public road in the UK or within Europe must hold a licence showing a driving entitlement (either full or provisional) for the relevant category of vehicle. Specifically in the case of goods vehicle driving, drivers must hold a current licence showing a relevant lgv vocational driving entitlement (either full or provisional). This is the legal responsibility of the individual concerned and heavy penalties are imposed on any person found to be driving without a licence, without a licence covering the correct category of vehicle or while disqualified from driving by a court. Furthermore, it is the responsibility of the employer of any person required to drive for business purposes to ensure that such employee drivers, irrespective of their function, status or seniority, are correctly licensed to drive company vehicles. The fact that a driver may be disqualified or has allowed his licence to lapse without the employer knowing is no defence for the employer against prosecution on a charge of allowing an unlicensed person to drive a vehicle.

The law states that it is an offence to drive, or to cause or permit another person to drive, a vehicle on the road without a current and valid driving licence, or to drive, or cause or permit the driving of, a large goods vehicle without a valid (ie full or provisional) category C or C+E entitlement.

Driving without a current and valid driving licence covering the category of vehicle being driven can invalidate insurance cover (which is itself an offence and is usually included among the charges for an unlicensed person driving) and could result in any accident or damage claim being refused by the insurance company under the terms of its policy contract – which is invariably conditional upon the law being complied with in full.

Recent Licensing Changes

Major changes to Britain's ordinary and hgv driving licence schemes were introduced in 1990/1 as the UK sought to harmonise with EU requirements and further changes will take place when the EU's so-called 'second' driver licensing directive (EC Directive 439/91) takes effect on 1 July 1996 – see page 117.

Since 1 April 1991 the pink European model, 'unified' driving licence (Euro-licence), which shows all of an individual's entitlements to drive (ie for motorcycle, car, light and large goods vehicles, and passenger-carrying vehicles), has been issued from the DVLA, Swansea – Licensing Authorities (LAs) no longer have responsibility for issuing heavy goods and passenger vehicle licences. The new licence carries the words 'European Communities Model' and has 'Driving Licence' printed in nine languages (besides the language of the country of issue – eg English), including Greek and Gaelic, on the front. The holder's photograph can be added to the licence, but this is not compulsory at present in Great Britain (as it is in certain EU member states and in Northern Ireland); in any case, at present the Swansea computer cannot handle photographs, but see also note on p 90.

British-issued Euro-licences also show, where appropriate, provisional driving entitlements and any endorsements of penalty points or licence disqualification made by the courts. This part of the document is called the 'counterpart' and is coloured green.

National driving licences issued in EU member states are still recognised throughout the Union and existing British 'green' ordinary driving licences continue to be valid both in Britain and in other countries which recognise British licences. Eventually, when the 'Euro-licence' is fully in use, national licences will become obsolete.

For a large proportion of British ordinary driving licence holders (ie those who have no vocational – lgv or pcv – entitlements) who currently hold 'licences-for-life', the changes mentioned above will not be noticed unless or until they apply to change the details on their licence (eg to record a new address or, in the case of women who have recently married, a new name), when they will be issued with the new style Euro-licence. Holders of vocational (ie old-type hgv/psv) licences became involved in the new scheme when they renewed their present old-type licences.

All existing entitlements to drive are maintained under the new licensing scheme so that no existing licence holder is deprived of his/her rights to drive particular vehicles, either now or in the future.

The Legislative Changes

The changes described above resulted from new and amended legislation. Principally the changes are contained in the Road Traffic Act 1988, the Road Traffic (Driver Licensing and Information Systems) Act 1989 and the following regulations; the Motor Vehicles (Driving Licences)(Amendment) Regulations 1990 (which, for example, contains the list of EU vehicle categories for licensing purposes) and The Motor Vehicles (Driving Licences)(Large Goods and Passenger Carrying Vehicles) Regulations 1990 (which deal with entitlements to drive large goods vehicles over 7.5 tonnes gross weight and passenger carrying vehicles which are used for hire and reward operations), plus a number of subsequent amendments.

The licence-scheme changes brought about by this legislation involved significant change in vocational (ie hgv/psv) licensing. Particularly, the terminology changed so that heavy goods vehicles (hgv) and public service vehicles (psv), as they were known under the old UK scheme, are now called large goods vehicles (lgv) and passenger carrying vehicles (pcv), respectively, and goods vehicles are no longer classified by the number of axles. Furthermore, where previously drivers had separate *licences* to drive such vehicles, their qualification to drive them now are referred to as *entitlements* within the unified licence scheme and are shown in the single combined (ie unified) licence document along with their ordinary (ie car, light goods vehicle and any motorcycle) driving entitlements.

For driver licensing purposes, a 'large goods vehicle' is defined as follows:
(a) an articulated vehicle (but see also exemption 17 on p 97); or
(b) a motor vehicle (other than an articulated vehicle) constructed or adapted to carry or haul goods with a permissible maximum weight (including the weight of any trailer drawn) which exceeds 7.5 tonnes.

Passenger carrying vehicles are defined as follows:
(a) large passenger carrying vehicles are vehicles constructed or adapted to carry more than 16 passengers;
(b) small passenger carrying vehicles are vehicles which carry passengers for hire or reward and are constructed or adapted to carry more than 8 but not more than 16 passengers.

For licensing purposes also, the old British system of vehicle groups (shown on ordinary licences) and classes (shown on hgv licences) has been abandoned in favour of the EU system of vehicle 'categories' (see table on p 93). Applicants for new or renewed hgv/psv licences are now issued with a 'unified' licence showing their goods and/or passenger vehicle driving entitlements under these new categories.

Photographs on Licences

The EU will eventually require all British-issued driving licences to carry a photograph of the holder as part of a campaign to eliminate misuse and fraud involving licences – currently the DoT in the UK is carrying out a feasibility study to determine how to proceed with this initiative. Northern Ireland-issued driving licences already carry a photograph of the holder. It is expected that as part of the Euro-harmonisation programme, the European Commission will insist on the UK's compliance by the year 2001.

Organ Donor Option

From 1 March 1993 UK-issued Euro-driving licences have incorporated an organ donor consent section (in the green counterpart of the licence document). Previously only provisional driving licences had this facility and separate donor cards were enclosed by the DVLC when sending out full driving licences. Completion of the section is entirely voluntary and both the DoT and the Department of Health have jointly confirmed that a consenting donor's organs would not, in any event, be removed without the prior permission of the driver's next of kin. The separate organ donor card will remain valid for existing licence holders.

The Issuing Authority

Responsibility for the *issue* of vocational driving entitlements rests with the Driver and Vehicle Licensing Agency (the Department of Transport's Executive Agency) at the Driver and Vehicle Licensing Centre (DVLC),

Swansea. However, it should be noted that the Traffic Commissioners (ie called the Licensing Authority when dealing with goods vehicle matters) still retain a disciplinary role in regard to vocational entitlements as described on page 114.

All applications in connection with driver licensing (ie for both ordinary and vocational driving entitlements) should be addressed to the Driver and Vehicle Licensing Centre, Swansea SA99 (followed by the postal code 1AD for new applications; 1AB for renewals, duplicate and exchange licences; and 1BR for all vocational entitlement applications). Further information on driver licensing can be obtained from the Driver Enquiry Unit, DVLC, Swansea SA6 7JL. Tel 01792 783838 (for general enquiries about the new licensing system) and 01792 772151 (for enquiries about specific applications).

Age of Drivers

Certain minimum ages are specified by law for drivers of various categories of motor vehicle as follows:

1.	Invalid carriage or moped*	16 years
2.	Motorcycle other than a moped* (ie over 50cc engine capacity)	17 years
3.	Small passenger vehicle or small goods vehicle (ie not exceeding 3.5 tonnes gross weight and not adapted to carry more than nine people, including the driver)	17 years
4.	Agricultural tractor	17 years
5.	Medium-size goods vehicle (ie exceeding 3.5 tonnes but not exceeding 7.5 tonnes gross weight)	18 years
6.	Other goods vehicles (ie over 7.5 tonnes gross weight) and passenger vehicles with more than nine passenger seats	21 years

*Note: The definition of moped is as follows:
1. *In the case of a vehicle first registered before 1 August 1977 a motorcycle with an engine cylinder capacity not exceeding 50cc which is equipped with pedals by means of which it can be propelled.*
2. *In the case of a vehicle first registered on or after 1 August 1977 a motorcycle which does not exceed the following limits:*
    (a) maximum design speed 30mph;
    (b) kerbside weight 250kg;
    (c) cylinder capacity (if applicable) 50cc.

If a goods vehicle and trailer combination exceeds 3.5 tonnes permissible maximum weight the driver must be at least 18 years of age; if such a combination exceeds 7.5 tonnes the driver must be at least 21 years of age (and will need to also hold an lgv driving entitlement).

Members of the armed forces are exempt from the 21 years age limit for driving heavy goods vehicles when such driving is in aid of the civil community (the limit is reduced to 17 years). Similarly, exemption from the 21 year minimum age limit applies to learner lgv drivers of vehicles over 7.5 tonnes gross weight if they are undergoing registered training by their employer or by a registered training establishment. In this case the minimum age is reduced to 18 years.

Disabled young people who receive a mobility allowance may drive cars at 16 years of age, provided they can do so safely.

Road Rollers

A person under 21 but not less than 17 years old may drive a road roller if it:
- is propelled by means other than steam;
- has an unladen weight of not more than 11,690kg*;
- is fitted with hard tyres or rollers;
- is not constructed or adapted to carry a load other than water, fuel, accumulators and other equipment used for the purpose of propulsion, loose tools, loose equipment and any object which is specially constructed for attachment to the vehicle so as to increase, temporarily, its unladen weight.

NB: If this weight is exceeded the minimum age for driving a roller is 21 years.

Agricultural Tractors

A person under 17 but over 16 may drive an agricultural tractor only if it is:
- of the wheeled type;
- not more than 2.45 metres wide including the width of any fitted implement;
- specially licensed for excise duty purposes as an agricultural machine;
- not drawing a trailer other than one of the two-wheeled or close coupled four-wheeled type which is not more than 2.45 metres wide.

A 16 year old must not drive an agricultural tractor on a road unless he or she has passed the appropriate test.

Vehicle Categories/Groups for Driver Licensing

For driver licensing purposes vehicles are defined according to specified groupings or categories which are shown on licences by means of capital letters as indicated in the following lists.

British Licence Groups
Pre-existing full, British-type ordinary (ie non-vocational) driving licences – many of which are still in existence – cover one or more of the following vehicle groups:

Group	Class of vehicle	Additional groups covered
A	A vehicle without automatic transmission, of any class not included in any other group	B, C, E, F, K and L
B	A vehicle with automatic transmission, of any class not included in any other group	E, F, K and L

C	Motor tricycle weighing not more than 425kg unladen, but excluding any vehicle included in group E, J, K or L	E, K and L
D	Motor bicycle (with or without side-car) but excluding any vehicle included in group E, K or L	C, E and motor cycles in group L
E	Moped	–
F	Agricultural tractor, but excluding any vehicle included in group H	K
G	Road roller	–
H	Track-laying vehicle steered by its tracks	–
J	Invalid carriage	–
K	Mowing machine or pedestrian controlled vehicle	–
L	Vehicle propelled by electrical power, but excluding any vehicle included in group J or K	K
M	Trolley vehicle	–
N	Vehicle exempted from duty under section 7(1) of the Vehicles (Excise) Act 1971	

EU Vehicle Categories

Under the EU licensing system, the old British system of vehicle groups as listed above has been replaced by the EU vehicle categories listed below. All future licences will specify driving entitlements against these vehicle categories.

Category	Vehicle type	Other categories covered
A	Motorcycle (with or without sidecar) – excluding vehicles in categories K, P	B1, P
B	Motor vehicle not exceeding 3.5 tonnes mass and not more than 8 seats (excl driver's), not included in any other category (incl drawing trailer not exceeding 750kg mass)	B+E, B1, C1, C1+E, D1, D1+E, F, K, L, N, P
B1	Motor tricycle not exceeding 500kg mass and max design speed exceeding 50kph – excluding vehicles in categories K, L, P	
C	Goods vehicle exceeding 3.5 tonnes mass (including drawing a trailer not exceeding 750kg mass, or a single-axle trailer not exceeding 5 tonnes pmw – *see note in italics below*)	
C1	Goods vehicles exceeding 3.5 tonnes but not exceeding 7.5 tonnes mass (incl drawing trailer not exceeding 7.5 tonnes)	B, B+E, B1, C1+E, D1+E, F, K, L, N, P
D	Passenger vehicles with more than 8 seats (excl driver's seat) (including drawing a trailer not exceeding 750kg mass or a single-axle trailer not exceeding 5 tonnes pmw – *see note in italics below*)	
D1	Passenger vehicle (not used for hire or reward) with between 8 and 16 seats (excl driver's) (incl drawing trailer not exceeding 750kg mass)	B, B+E, B1, C1, C1+E, D1+E, F, K, L, N, P

B+E	Motor vehicle in category B drawing a trailer exceeding 750kg mass	
C+E	Goods vehicle in category C drawing a trailer exceeding 750kg mass, or a single-axle trailer exceeding 5 tonnes pmw	
C1+E	Goods vehicle in category C1 drawing a trailer exceeding 750kg mass (max weight of combination not to exceed 8.25 tonnes)	
D+E	Passenger vehicle in category D drawing a trailer exceeding 750kg mass, or a single-axle trailer exceeding 5 tonnes pmw	
D1+E	Motor vehicle in category D1 drawing a trailer exceeding 750kg mass	
F	Agricultural tractor excluding any vehicle in category H	K
G	Road roller	
H	Track laying vehicle steered by its tracks	
K	Mowing machine or pedestrian-controlled vehicle	
L	Electrically-propelled vehicle	K
N	Vehicle exempted from duty under the Vehicles (Excise) Act 1977, s7	
P	Moped	

NB: In the above table the term 'mass' means the permissible maximum weight (pmw) for the vehicle/trailer – commonly referred to as the 'gross weight'. Legislation now also uses the term maximum authorised mass (abbreviated to mam) which has the same meaning as pmw mentioned above.

Examination of the table above detailing the new vehicle categories shows that for driving goods vehicles, entitlements covering vehicle categories B and C are needed. This will provide qualification to drive cars and small goods vehicles up to 3.5 tonnes (ie category B) and goods vehicles of more than 3.5 tonnes gross weight (ie category C). Sub-category C1 is designed to cover the driving of medium-size goods vehicles between 3.5 tonnes and 7.5 tonnes gross weight (in other words, vehicles below the lgv vocational driving requirement which applies above 7.5 tonnes gross weight). An entitlement to drive vehicles in category C – which cannot be obtained without first holding a category B entitlement – covers all rigid goods vehicles, irrespective of gross weight or the number of axles the vehicle has.

Where there is a requirement to drive articulated vehicles and drawbar combinations (ie rigid vehicles in combination with large trailers over 750kg maximum gross weight) an additional qualification covering category C+E is necessary – *NB: except in the case of a category C vehicle drawing a single-axle trailer not exceeding 5 tonnes pmw for which a category C entitlement alone will suffice on a temporary basis until 1 July 1996 when the facility will be withdrawn.* This C+E entitlement will indicate whether driving is 'limited to drawbar trailers'. Without this specified restriction, category C+E entitlement permits the driving of articulated goods vehicles.

NB: A provision in the regulations which permitted category C entitlement holders to drive vehicles over 3.5 tonnes gross weight drawing a single-axle trailer of any weight (in addition to the standard provision which allows the towing of trailers not exceeding 750kg gross weight) was removed from 1 April 1993. From this date a C+E entitlement is required to drive rigid vehicles drawing all trailers exceeding 750kg gross weight. A gross train weight limit of 8.25 tonnes is imposed on vehicle trailer combinations driven under a category C1+E entitlement.

Existing car drivers (ie category B entitlement holders) who wish or need to drive only light or medium-sized goods vehicles (ie up to 7.5 tonnes gross weight) may obtain an entitlement covering sub-category C1 on application since this is an automatic right for such drivers. New drivers requiring a sub-category C1 entitlement can do so by passing a test in any category B or C1 vehicle. New drivers seeking full category C or C+E entitlements must take an lgv test in a specified minimum test vehicle (see Chapter 7).

Holders of old-type Class 2 or 3 hgv licences will gain automatic entitlement to drive goods vehicles in category C and C+E ('limited to drawbar trailers') at the time of licence changeover. Class I hgv licence holders will gain automatic entitlement to drive full category C+E vehicles (ie covering articulated vehicles as well as drawbar combinations). Similar provisions will apply to those holding Class 1, 2 or 3 hgv licences with an 'A', automatic transmission, restriction, but their entitlement will specify a restriction to driving only goods vehicles with automatic transmission.

Drivers who hold existing licences (ie old-type green or new-type pink) showing a qualification to drive vehicles in Group A (or Group B for automatic transmission vehicles) may drive goods vehicles up to 7.5 tonnes permissible maximum weight if they are over 18 years of age. This right is continued under the new licensing system and will be maintained after the introduction of the further EU Directive on driver licensing from 1 July 1996. Similarly, holders of new-type licences showing entitlement to drive vehicles in categories B and C1 (with any restriction to driving automatic transmission vehicles only being shown), who are also over 18 years of age, can drive such vehicles without taking a further test and will be allowed to do so following introduction of the new Directive referred to above.

However, once this further Directive takes effect (from 1 July 1996), new drivers passing the car and light vehicle test (ie with vehicles up to 3.5 tonnes permissible maximum weight) for the first time will not be permitted to drive vehicles above this weight without securing additional driving categories on their licence. It is stressed that this restriction to 3.5 tonne driving will apply *only* to those who first pass their test after the new Directive comes into force. It will not be applied retrospectively to existing licence holders.

Towed Vehicles

It has been ruled that a person who steers a vehicle being towed (whether it has broken down or even has vital parts missing, such as the engine) is 'driving' the vehicle for licensing purposes and therefore needs to hold current and valid driving entitlement covering that category of vehicle.

Learner Drivers

Learner drivers must hold a provisional driving entitlement to cover them while driving under tuition (the new Euro-licence is issued with the provisional driving entitlement shown on the green 'counterpart').

In the case of full category C lgv entitlement holders, this can be used in place of a provisional entitlement for learning to drive vehicles in category C+E (ie drawbar combinations and articulated vehicles). But it should be noted that under the new arrangements, full entitlements in categories B and C1 *cannot* be used as a provisional entitlement for learning to drive vehicles in categories C or C+E – a proper provisional entitlement for these classes is required.

Learner drivers must be accompanied, when driving on public roads, by the holder of a full entitlement covering the category of vehicle being driven (see also below) and must not drive a vehicle drawing a trailer, except in the case of articulated vehicles or agricultural trailers. An 'L' plate of the approved dimensions must be displayed on the front and rear of a vehicle being driven by a learner driver (see Chapter 7).

Learner drivers (of category B and C1 vehicles) are not allowed to drive on motorways. However, learner lgv drivers (ie for licence categories C and C+E) who hold full ordinary entitlements (ie licence categories B and C1 – private and goods vehicles up to 7.5 tonnes gross weight) may drive on motorways while under tuition.

Supervision of 'L' Drivers
Qualified drivers who supervise learner drivers in cars and light vehicles (ie category B and C1) must be at least 21 years old and have held a full driving entitlement for at least 3 years (excluding any periods of disqualification). Contravention of these requirements could lead to prosecution of the supervising driver and, on conviction, a fine of up to £400, the imposition of 2 driving licence penalty points and possibly licence disqualification.

This provision does not apply to the supervision of learner lgv drivers (ie under provisional category C and C+E entitlements).

Exemptions from Vocational Licensing

Exemptions from the need to hold an lgv driving entitlement apply when driving certain vehicles as follows (in most cases such vehicles may be driven by the holder of a category B licence):
1. Track laying vehicles
2. Steam-propelled vehicles
3. Road rollers
4. Road construction vehicles used or kept on the road solely for the conveyance of built-in construction machinery
5. Engineering plant (which includes certain types of mobile crane)
6. Works trucks

7. Industrial tractors, that is tractors which are not land tractors, have unladen weights of not more than 3.5 tonnes, and which are designed and used primarily for work off roads, or for work on roads in connection with road construction or maintenance and which are constructed so as to be incapable of a speed of more than 20mph on the level under their own power
8. Land locomotives and land tractors (ie agricultural vehicles)
9. Digging machines
10. Vehicles used less than 6 miles per week on public roads (ie exempted from duty under the Vehicles (Excise) Act 1971, s7(1))
11. Articulated tractive units weighing not more than 3050kg unladen which have no trailer attached
12. Vehicles used for no purpose other than the haulage of lifeboats and the conveyance of the necessary gear of the lifeboats which are being hauled
13. Vehicles manufactured before 1 January 1960 used unladen and not drawing a laden trailer
14. Vehicles in the service of a visiting military force or headquarters
15. Wheeled armoured vehicles, the property of, or under the control of, the Secretary of State for Defence
16. A heavy goods vehicle when driven by a police constable for the purpose of removing it to avoid obstruction to other road users or danger to other road users or members of the public, for the purpose of safeguarding property, including the vehicle and its load, or for other similar purposes
17. Any articulated vehicle which has a permissible maximum weight of not more than 7.5 tonnes, or the tractive unit of which does not exceed 2.05 tonnes unladen weight
18. Any rigid vehicle which has a permissible maximum weight of not more than 3.5 tonnes which is towing a trailer (ie not an articulated vehicle)
19. Any vehicle other than an articulated vehicle with an unladen weight of not more than 10.02 tonnes which belongs to the holder of a psv 'O' licence when such a vehicle is going to or returning from a place where it is to give assistance to a disabled vehicle operating under a psv licence or when moving such a vehicle to prevent it causing an obstruction or to a place where it is to be repaired, stored or broken up
20. Breakdown vehicles which weigh less than 3050kg unladen, provided they are fitted with apparatus for raising a disabled vehicle partly from the ground and for drawing a vehicle when so raised, are used solely for the purpose of dealing with disabled vehicles, and carry no load other than a disabled vehicle and articles used in connection with dealing with disabled vehicles
21. Play buses
22. Fire fighting or salvage vehicles driven by a member of the armed forces
23. Vehicles driven by members of the armed forces on urgent work of national importance as ordered by the Defence Council.

Application for Ordinary and Vocational Entitlements

Applications for driving licences, showing full ordinary, any provisional entitlements required, and lgv vocational entitlements (ie for driving large goods vehicles over 7.5 tonnes gross weight) have to be made to Swansea on Form D1 (obtainable from main post offices, direct from Swansea or Local Vehicle Registration Offices).

Questions on Form D1 are concerned with details of the applicant and any current ordinary driving licence which he holds, whether he has ever been refused a driving licence or been disqualified from driving and details of any convictions against him. Also the applicant is asked to declare whether he meets the statutory eyesight requirement (see below) and give information about his health, particularly as to whether he has ever suffered any heart condition; loss of sight in either eye, cataract, double or tunnel vision; epileptic attack, stroke or loss of consciousness; drink or drug addiction; and whether he is receiving treatment for angina, diabetes, mental or nervous disorder. Any other known condition which may affect a person's driving must be declared (see also below).

The DVLA's Considerations for Vocational Entitlements

Applicants for vocational driving entitlements must meet specified conditions as follows:

- (a) they must be fit and proper persons;
- (b) they must meet laid-down eyesight requirements;
- (c) they must satisfy a medical examination and specifically must not
    - (i) have had an epileptic attack since the age of 5 years
    - (ii) suffer from insulin-dependent diabetes

The decision as to whether or not an applicant will be granted an lgv driving entitlement rests entirely with the DVLA and in making this decision it will take into account any driving convictions for motoring offences, drivers' hours and record offences, and offences relating to the roadworthiness or loading of vehicles, against the applicant in the 4 years prior to the application and any offence connected with driving under the influence of drink or drugs during the 11 years prior to the application. The applicant has to declare such convictions on the application form but the DVLA has means of checking to ensure that applicants have declared any such convictions against them.

LAs' Powers in Respect of Vocational Entitlements

Although the issue of vocational (ie lgv/pcv) entitlements is now the prerogative of the DVLA, Licensing Authorities still have powers to consider the fitness of persons applying for or holding such entitlements. This disciplinary role allows an LA to call upon applicants or entitlement holders to provide information as to their conduct (and if necessary to appear before him to answer in person), to refuse the grant of an entitlement, and to suspend or disqualify a person from holding such an entitlement. The LA's decision must be communicated to the person concerned, upon whom it is binding, and obviously to the Secretary of State for Transport (effectively the DVLA) – see also p 90.

Date for Vocational Applications

Application for an lgv driving entitlement should be made not more than 3 months before the date from which the entitlement is required to run.

Reminders will be sent out by the DVLA to existing licence/entitlement holders 2 months prior to the expiry date of their existing licence/entitlement.

Medical Requirements for Vocational Entitlements

Strict medical standards for vocational entitlement holders are legally established to determine that those wishing to drive large goods or passenger carrying vehicles are safe to do so and are not suffering from any disease or disability (especially epilepsy and heart ailments, for example) which would prevent them from driving safely.

UK applicants for lgv driving entitlements must satisfy such medical standards on first application and subsequently. To do so, they must undergo a medical examination and have their doctor complete the medical certificate portion of the application Form D4 (previously DoT 20003) not more than 4 months before the date when the entitlement is needed to commence. A further examination and completed medical certificate is required for each 5-yearly renewal of the entitlement after reaching age 45 years. After reaching the age of 65 years a medical examination is required for each annual renewal of the entitlement. Further medical examinations may be called for at any time if there is any doubt as to a driver's fitness to drive. The form DoT 20003 requests the applicant's consent to allow the DVLA's medical adviser to obtain reports from their own doctor and any specialist consulted if this helps to establish their medical condition.

Doctors charge a fee for conducting such medical examinations (approximate cost currently £35–45) which the candidate must pay himself (or his employer may pay it for him) – examinations are not available on the National Health Service in the UK. The medical fee for licence/entitlement renewal can be claimed as an allowable expense for income tax purposes.

Epilepsy

Under a change in the law, effective from 1 January 1993, the previous ban on a person who has had an epileptic attack since the age of 5 years from holding a vocational (ie lgv/pcv) entitlement has been removed. A person will now be prevented from holding an lgv/pcv entitlement *only* if they have a 'liability to epileptic seizures'. Applicants must satisfy the DVLA that:

- they have not suffered an epileptic seizure during the 10 years prior to the date when the entitlement is to take effect;
- no epilepsy treatment has been administered during the 10 years prior to the starting date for the entitlement; and
- a consultant nominated by the Secretary of State for Transport has examined their medical history and is satisfied that there is no continuing liability to seizures.

Other Medical Conditions

Other medical disabilities which may cause failure of the examination include:

(a) sudden attacks of vertigo ('dizziness');
(b) heart disease which causes disabling weakness or pain;
(c) a history of coronary thrombosis;
(d) the use of hypertensive drugs for blood pressure treatment;

(e) diabetes requiring insulin injection or oral agents (normally, insulin-dependent diabetes sufferers are barred from holding lgv entitlement but, if they held an hgv driving licence, and the LA was aware of their condition, prior to 1 January 1991, an entitlement may be granted).

A licence will be refused to a driver who is liable to sudden attacks of disabling giddiness or fainting and those who have had a cardiac pacemaker fitted are advised to discontinue lgv driving, although driving below the 7.5 tonne lgv threshold is permitted if a person who has disabling attacks which are controlled by a pacemaker has made arrangements for regular review from a cardiologist and will not be likely to endanger the public.

Suspected Coronary Health Problems

Drivers who have suspected coronary health problems are permitted (since April 1992) to retain their lgv driving entitlements while medical enquiries are made. (Previously the rule was to revoke the licence pending enquiries into the holder's health.) Such drivers no longer have to submit to coronary angiography (ie angiogram testing). The DVLA says that ECG exercise tests will be undertaken no earlier than 3 months after a coronary event and, providing the driver displays no signs of angina or other significant symptoms, he is allowed to keep his driving entitlement while investigations are made, but subject to the approval of his own doctor.

Alcohol Problems

A person with repeated convictions for drink-driving offences may be required to satisfy the DVLC (with certification from his or her own doctor) that they do not have an 'alcohol problem'.

Medical Appeals

The final decision on any medical matter concerning driving licences rests with the Medical Advisory Branch of the DVLA. However, there is the opportunity of appeal, within 6 months in England and Wales, to a magistrate's court, and within 21 days in Scotland, to a sheriff's court. In other cases the refused driver may be given the opportunity to present further medical evidence which the medical adviser will consider.

Further information on medical conditions relating to driving are to be found in a booklet *Medical Aspects of Fitness to Drive*, available from the Medical Commission on Accident Prevention, 43 Lincoln's Inn Fields, London WC2A 3PN (price £3).

Notification of New or Worsening Medical Conditions

Once a licence has been granted (ie whether ordinary or covering vocational entitlements), the holder is required to notify the Drivers Medical Branch, DVLC at Swansea SA99 1TU of the onset *or worsening* of any medical condition likely to cause them to be a danger when driving – *failure to do so is now a specific offence.* Examples of what must be reported are giddiness, fainting, blackouts, epilepsy, diabetes, strokes, multiple sclerosis, Parkinson's disease, heart disease, angina, 'coronaries' , high blood pressure, arthritis, disorders of vision, mental illness, alcoholism, drug-taking, and the loss or loss of use of any limb. In many cases the person's own doctor will either advise them to report their condition to the DVLC themselves or the doctor (or hospital) may advise the DVLC direct. In either case the driving licence will have to be surrendered until the condition clears.

There is no requirement to notify the DVLA of temporary illnesses or disabilities such as sprained or broken limbs where a full recovery is expected within 3 months.

Eyesight Requirement

The statutory eyesight requirement mentioned above for ordinary (ie car and light goods vehicle) licence holders is for the driver to be able to read, in good daylight (with glasses or contact lenses if worn), a motor vehicle number plate at 75 feet if the symbols are $3\frac{1}{2}$ inches high and from 67 feet if the symbols are $3\frac{1}{8}$ inches high. It is an offence to drive with impaired eyesight and the police can require a driver to take an eyesight test on the roadside. If glasses or contact lenses are needed to reach these vision standards they must be worn at all times while driving. It is an offence to drive with impaired eyesight. There are proposals for drivers to undergo regular eye tests.

The vision requirement for vocational (ie lgv and pcv) entitlements is that new drivers with spectacles must meet a standard of at least 6/9 in the better eye and 6/12 in the other and, without glasses, at least 3/60 in each eye separately.

Existing drivers who held an hgv driving licence on 1 January 1983 and still held it on 1 April 1991 must meet a standard of at least 6/12 in the better eye and 6/36 in the other and at least 3/60 uncorrected (ie without glasses). A person with sight in only one eye would normally be refused an lgv driving entitlement unless the following conditions apply:

(a) an hgv licence was held before 1 January 1983 and was still held on 1 April 1991 and the LA who granted the licence knew of the condition before 1 January 1991 – and the visual acuity in the eye is no worse than 6/12; or

(b) an hgv licence was not held on 1 January 1983 but was held on 1 April 1991 and the LA who granted the licence knew of the condition before 1 January 1991 – and the visual acuity in the eye is no worse than 6/9.

If they meet these requirements, new or existing drivers may drive with contact lenses or following a cataract operation. In a case where a person first held an hgv driving licence before 1 April 1991 and the LA was aware of their eye condition before 1 January 1991, an lgv entitlement will be granted if they can meet the lower standards detailed above for existing drivers. Where the wearing of contact lenses was not acceptable for hgv drivers at one time, drivers may now be permitted to drive if their vision with contact lenses meets the required standard.

Cover Against Loss of Licence/Entitlement on Medical Grounds

In view of the risk of a driver losing his lgv entitlement in later life due to the onset or worsening of a medical condition and therefore jeopardising his employment prospects, it is possible for drivers (or their employers) to insure against loss of driving licence/vocational entitlement. Special insurance schemes to cover such an eventuality are offered by, among others, the FTA, RHA and the drivers' union, the TGWU, the latter promoting a scheme provided by Unity Trust. Various other independent schemes are offered by district offices of the TGWU.

EyeCare Plan
The United Road Transport Union has launched a plan in conjunction with opticians Dolland and Aitchison (D&A) to provide drivers with an eye examination, a certificate for driver licensing purposes and a pair of spectacles at an inclusive price. The Focus Drivers Eyecare Plan is available at 500 D&A branches and at Eyeland Express branches throughout the UK.

Drug Testing of LGV Drivers

Random drug testing of lgv drivers is becoming an increasing practice in the UK, especially among tanker fleet operators, following the pattern in the USA which has had mandatory testing since 1992. There is no suggestion at this stage that the practice should become mandatory in this country. Most of Britain's major oil companies now carry out random testing for both alcohol and drug problems – Shell has produced a staff booklet identifying 11 banned substances (including amphetamines) and warning of the consequences of drink or drug abuse.

Licence Fees and Validity

The fee for a full or provisional ordinary driving entitlement (ie for vehicle categories A, B, B+E, C1, C1+E, D1 and D1+E) is currently £21. This fee also covers the conversion of the provisional entitlement to a full entitlement after passing the driving test. Full driving entitlements in these categories are valid from the date of issue until the applicant's 70th birthday (unlike vocational entitlements – categories C, C+E, D and D+E – which are valid for only 5 years at a time). After reaching their 70th birthday ordinary entitlement holders must make a new application and, if this is granted, each subsequent licence will be valid for 3 years. These 'after 70' licences cost £6 on each 3-yearly renewal.

A provisional hgv licence holder or a provisional (category C or C+E) lgv entitlement holder who passes the lgv driving test will be issued with a test pass certificate which is valid for 2 years during which time the holder may continue to drive, although the DVLA advice is to convert this to a full entitlement as soon as possible – there is no additional fee for this change. Failure to apply for a full licence within 2 years of passing the driving test will result in the need to take and pass the test again to obtain a full licence. This time limit applies irrespective of the category of test taken (ie motorcycle, car, lgv or pcv).

A fee of £6 is payable for the issue of a new licence to drivers who have been disqualified unless this was for drink-driving convictions, in which case a new licence will cost £20. Duplicate and exchange licences cost £6 each, as does the issue of a British licence in exchange for one issued in Northern Ireland, the Isle of Man, the Channel Islands or the EU.

Additional fees are payable for lgv/pcv vocational driving entitlements. Both provisional vocational entitlements and the renewal of existing entitlements cost £21. Replacement of a vocational entitlement following disqualification costs £12.

An lgv/pcv driving entitlement is normally valid for 5 years or until the holder reaches the age of 45 years, whichever is the longer. After the age of 45 years, 5 year entitlements are granted subject to medical fitness, but may be for lesser periods where the holder suffers from a relevant or prospective relevant disability. From the age of 65 years, vocational entitlements are granted on an annual basis.

Tax Deductions
The cost of renewing lgv driving entitlements and for undergoing medical examinations in connection with licence renewals is income tax deductible against Schedule E earnings – but not the costs of first obtaining such an entitlement.

Lost/Mislaid Licences

Drivers who have mislaid their driving licence may obtain a temporary 'Certificate of Entitlement' (commonly referred to as a cover note), valid for 1 month, either from the DVLC at Swansea (free of charge) or from local Vehicle Registration Offices (VROs) at a cost of £3.50, subject to proof of their identity. This document is valid for proving entitlement to drive (eg for the purposes of satisfying the police that a licence is held, hiring a vehicle, applying for an International Driving Permit, or to satisfy an employer). Applicants for these certificates will be able to complete an application for a duplicate licence at the VRO and this will be sent on to the DVLC in Swansea for processing.

Notional Gross Weights

For the purposes of determining driving licence requirements in cases where a goods vehicle or trailer does not have a manufacturer's or DoT specified gross weight or gross train weight (ie for vehicle and trailer combinations) marked on it, a system of 'multipliers' is used to calculate the notional maximum gross weight. The unladen weight of the vehicle (or trailer) is multiplied by the number given in the tables below for that class of vehicle, and the resulting figure is taken to be the gross or gross train weight, but only for the purposes of deciding what driving licence is required and not for any other purpose.

Motor Vehicles

Class of Vehicle	_	*Number*
1. Dual-purpose vehicles not constructed or adapted to form part of an articulated goods vehicle combination		1.5
2. Breakdown vehicles		2
3. Works trucks and straddle carriers used solely as works trucks		2
4. Electrically-propelled motor vehicles		2

5. Vehicles constructed or adapted for, and used solely for, spreading   2
    material on roads to deal with frost, ice or snow
6. Motor vehicles used for no other purpose than the haulage of   2
    lifeboats and the conveyance of the necessary gear of the lifeboats
    which are being hauled
7. Living vans   1.5
8. Vehicles constructed or adapted for, and used primarily for the   1.5
    purpose of, carrying equipment permanently fixed to the vehicle, in a
    case where the equipment is used for medical, dental, veterinary,
    health, educational, display or clerical purposes and such use does
    not directly involve the sale, hire or loan of goods from the vehicle
9. Three-wheeled motor vehicles designed for the purpose of street   2
    cleansing, the collection or disposal of refuse or the collection
    or disposal of the contents of gullies
10. Steam-propelled vehicles   2
11. Vehicles designed and used for the purpose of servicing, controlling,   2
    loading or unloading aircraft on an aerodrome
12. Motor vehicles of a class not mentioned above where equipment,   1
    apparatus or other burden is permanently attached to and forms part
    of the vehicle and where the vehicle is only used on a road for carrying,
    or in connection with the use of, such equipment, apparatus or other
    burden
13. Motor vehicles of a class not mentioned above which are either   2
    (a) heavy motor cars or motor cars first used before 1 January 1968, or
    (b) locomotives or motor tractors first used before 1 April 1973
14. Any motor vehicles not mentioned above.   4

Trailers

Class of Vehicle	_	Number
1. Engineering plant		1
2. Trailers which consist of drying or mixing plant designed for the production of asphalt or of bituminous or tar macadam		1
3. Agricultural trailers		1
4. Works trailers		1
5. Living vans		1.5
6. Any trailers not mentioned above.		3

Articulated Vehicles

Class of Combination	Number
1. Articulated goods vehicle combinations where the semi-trailer is a trailer of a kind mentioned in paragraphs 1, 2, 3, 4 or 5 of 'Trailers' above	1.5
2. Any other articulated goods vehicle combination.	3

Production of Driving Licences

Both the police and enforcement officers of the DoT's Vehicle Inspectorate (VI) Agency can request a driver – or the person accompanying a provisional entitlement holder – to produce his licence showing his ordinary and vocational entitlements to drive. If he is unable to do so at the time, he may produce them without penalty:

    (a) if the request was by a police officer, at a police station of his choice within 7 days; or

    (b) if the request was by an enforcement officer, at the Traffic Area Office within 10 days.

In either case, if the licence cannot be produced within the 7 or 10 days it can be produced as soon as reasonably practicable thereafter. An LA can also require the holder of an lgv/pcv driving entitlement to produce his licence at a Traffic Area Office for examination within 10 days.

Failure to produce a licence on request is an offence.

A police officer can ask a driver to state his or her date of birth – British ordinary driving licences carry a coded number which indicates the holder's surname and their date of birth (but not Northern Ireland licences which carry a photograph of the holder). The name and address of the owner of the vehicle owner can also be requested.

When required by a DoT examiner to produce his licence, an lgv entitlement holder may be required to give his date of birth and to sign the examiner's record sheet to verify the fact of the licence examination. This should not be refused.

Licence holders apprehended for endorsable fixed penalty (ie yellow ticket) offences are required to produce their driving licence to the police officer at that time or later (ie within 7 days) to a police station and surrender the licence for which they will be given a receipt. Failure to produce a licence in these circumstances means that the fixed penalty procedure will not be followed and a summons for the offence will be issued requiring a court appearance. Drivers summoned to appear in court for driving and road traffic offences must produce their driving licence to the court on, at least, the day before the hearing.

Driving Licence Penalty Points and Disqualification

Driving licence holders may be penalised following conviction by a court for offences committed on the road with a motor vehicle. These penalties range from the issue of fixed penalty notices for non-endorsable offences, which do not require a court appearance unless the charge is to be contested and incur no driving licence penalty points although the relevant fixed penalty has to be paid; to those for endorsable offences when penalty points are added on the licence counterpart and the fixed penalty is incurred or a heavy fine imposed on conviction if a court appearance is made.

In other cases, licence disqualification for a period (extending to a number of years in serious cases – especially for drink-driving related offences) and, in very serious instances, imprisonment of the offender may follow conviction in a magistrates' court or indictment for the offence in a higher court. Holders of vocational driving entitlements may be separately penalised for relevant offences which could result in such entitlements being suspended or revoked and in serious circumstances the holder being disqualified from holding a vocational entitlement – see below.

The Penalty Points System on Conviction and Disqualification

Driving licence endorsement of penalty points following conviction for motoring offences is prescribed by the Road Traffic Offenders Act 1988. Further provisions relating to driving offences are contained in the Road Traffic Act 1991 and came into effect on 1 July 1992 – see pp 110–11.

The penalty points system grades road traffic offences according to their seriousness by a number or range of penalty points, between 2 and 10 points, imposed on the driving licence of the convicted offender. Once a maximum of 12 penalty points has been accumulated within a 3 year period counting from the date of the first offence to the current offence (not from the date of conviction), disqualification for at least 6 months will follow automatically.

Most offences rate a fixed number of penalty points to ensure consistency and to simplify the administration; but a discretionary range applies to a few offences where the gravity may vary considerably from one case to another. For example, failing to stop after an accident which only involved minor vehicle damage is obviously less serious than a case where an accident results in injury.

When a driver is convicted of more than one offence at the same court hearing, only the points relative to the most serious of the offences will be endorsed on the licence. The points relative to each individual offence will not be aggregated. Once sufficient points (ie 12) have been endorsed on the driving licence and a period of disqualification has been imposed (6 months for the first totting-up of points), the driver will have his 'slate' wiped clean and those points will not be counted again. Twelve more points would have to be accumulated before a further disqualification would follow, but to discourage repeated offences the courts will impose progressively longer disqualification periods in further instances (minimum 12 months for subsequent disqualifications within 3 years and 24 months for a third disqualification within 3 years).

Licence Endorsement Codes and Penalty Points

Following conviction for an offence, the driver's licence (ie the green counterpart) will be endorsed by the convicting court with both a code (to which employers and prospective employers should refer so they can assess the offences which drivers have committed) and the number of penalty points imposed as follows.

Code	Accident offences	Penalty points
AC 10	Failing to stop after an accident	8–10
AC 20	Failing to report an accident within 24 hours	8–10

AC 30	Undefined accident offence	4–9
	Disqualified driver	
BA 10	Driving while disqualified	6
BA 20	Driving while disqualified on age grounds	2
BA 30	Attempting to drive while disqualified	6
	Careless driving	
CD 10	Driving without due care and attention	3–9
CD 20	Driving without reasonable consideration for other road users	3–9
CD 30	Driving without due care and attention or without reasonable consideration for other road users	3–9
	Construction and use offences	
CU 10	Using a vehicle with defective brakes	3
CU 20	Causing or likely to cause danger by reason of unsuitable vehicle or using a vehicle with parts or accessories (excluding brakes, steering or tyres) in a dangerous condition	3
CU 30	Using a vehicle with defective tyres	3
CU 40	Using a vehicle with defective steering	3
CU 50	Causing or likely to cause danger by reason of load or passengers	3
CU 60	Undefined failure to comply with C&U Regulations	3
	Reckless driving	
DD 30	Reckless driving	10
DD 60	Manslaughter or culpable homicide while driving a vehicle	+
DD 70	Causing death by reckless driving	+
	Drink or drugs	
DR 10	Driving or attempting to drive with alcohol level above limit	+
DR 20	Driving or attempting to drive while unfit through drink	+
DR 30	Driving or attempting to drive then failing to supply a specimen for analysis	+
DR 40	In charge of a vehicle while alcohol level above limit	10
DR 50	In charge of a vehicle while unfit through drink	10
DR 60	Failure to provide a specimen for analysis in circumstances other than driving or attempting to drive	10
DR 70	Failing to provide a specimen for breath test	4
DR 80	Driving or attempting to drive when unfit through drugs	+
DR 90	In charge of a vehicle when unfit through drugs	10
	Insurance offences	
IN 10	Using a vehicle uninsured against third-party risks	6–8
	Licence offences	
LC 10	Driving without a licence	2
LC 20	Driving otherwise than in accordance with a licence	3–6

LC 30	Driving after making a false declaration about fitness when applying for a licence	3–6
LC 40	Driving a vehicle having failed to notify a disability	3–6
LC 50	Driving after a licence has been revoked or refused on medical grounds	2

Miscellaneous offences

MS 10	Leaving a vehicle in a dangerous position	3
MS 20	Unlawful pillion riding	1
MS 30	Play street offences	2
MS 40	Driving with uncorrected defective eyesight or refusing to submit to a test	3
MS 50	Motor racing on the highway	+
MS 60	Offences not covered by other codes	as appropriate
MS 70	Driving with uncorrected defective eyesight	3
MS 80	Refusing to submit to an eyesight test	3
MS 90	Failure to give information as to identity of driver etc	3

Motorway offences

MW 10	Contravention of Special Roads Regulations (excl speed limits)	3

Pedestrian crossings

PC 10	Undefined contravention of Pedestrian Crossing Regulations	3
PC 20	Contravention of Pedestrian Crossing Regulations with moving vehicle	3
PC 30	Contravention of Pedestrian Crossing Regulations with stationary vehicle	3

Provisional licence offences

PL 10	Driving without 'L' plates	3–6
PL 20	Not accompanied by a qualified person	3–6
PL 30	Carrying a person not qualified	3–6
PL 40	Drawing an unauthorised trailer	3–6
PL 50	Undefined failure to comply with conditions of a provisional licence	3–6

Speed limits

SP 10	Exceeding goods vehicle speed limits	3–6
SP 20	Exceeding speed limit for type of vehicle (excluding goods or passenger vehicles)	3–6
SP 30	Exceeding statutory speed limit on a public road	3–6
SP 40	Exceeding passenger vehicle speed limit	3–6
SP 50	Exceeding speed limit on a motorway	3–6
SP 60	Undefined speed limit offence	3–6

NB: In all of the speed limit cases above, disqualification is obligatory where the relevant speed is in excess of 30mph over the statutory limit.

Traffic directions and signs

TS 10	Failing to comply with traffic light signals	3
TS 20	Failing to comply with double white lines	3

TS 30	Failing to comply with a 'Stop' sign	3
TS 40	Failing to comply with direction of a constable or traffic warden	3
TS 50	Failing to comply with a traffic sign (excluding 'Stop' signs, traffic lights or double white lines)	3
TS 60	Failing to comply with a school crossing patrol sign	3
TS 70	Undefined failure to comply with a traffic direction or sign	3
	Theft or unauthorised taking	
UT 10	Taking and driving away a vehicle without consent or an attempt thereat	8
UT 20	Stealing or attempting to steal a vehicle	8
UT 30	Going equipped for stealing or taking a vehicle	8
UT 40	Taking or attempting to take a vehicle without consent; driving or attempting to drive a vehicle knowing it to have been taken without consent; allowing oneself to be carried in or on a vehicle knowing it to have been taken without consent	8
UT 50	Aggravated taking of a vehicle	3–11
	Special code	
TT 99	To signify a disqualification under 'totting-up' procedure	–

Where the offence is one of aiding or abetting, causing or permitting or inciting, the codes are modified as follows.

Aiding, abetting, counselling or procuring
Offences as coded, but with zero changed to 2, eg UT 10 becomes UT 12.

Causing or permitting
Offences as coded, but with zero changed to 4, eg LC 20 becomes LC 24.

Inciting
Offences as coded, but with zero changed to 6, eg DD 30 becomes DD 36.

The length of time for periods of disqualification are shown by use of the letters D = days, M = months and Y = years. Consecutive periods of disqualification are signified by an asterisk (*) against the time period. The symbol + means that 4 points are added to a licence if for exceptional reasons disqualification is not imposed.

Disqualification Offences

The endorsing of penalty points will also arise on conviction for offences where disqualification is discretionary and where the court has decided that immediate disqualification is not appropriate (for example, if acceptable 'exceptional' reasons are put forward – see also below). In this case the offender's driving licence will be endorsed with 4 points. The courts are still free to disqualify immediately if the circumstances justify this.

Offences carrying obligatory disqualification, are shown in the following list.
1.   Causing death by dangerous driving and manslaughter.
2.   Dangerous driving within 3 years of a similar conviction.
3.   Driving while unfit through drink or drugs.
4.   Driving or attempting to drive with more than the permitted breath-alcohol level.
5.   Failure to provide a breath, blood or urine specimen.
6.   Racing on the highway.

Driving while disqualified is a serious offence which can result in a fine at level 5 on the standard scale (see below), currently £5000 maximum, or 6 months' imprisonment, or both.

New Procedures and Offences

Since 1 January 1992, short period disqualifications (SPDs) of not more than 56 days have been treated differently from longer period disqualifications. In such cases the disqualification will be marked on the licence by the court and the licence returned to the holder (it will not be held by the DVLA in these circumstances), who must not drive any vehicle, including lgvs, while the ban is in force. Once the terminal date for the disqualification is passed the holder may drive vehicles in categories B and C1 again, but not lgvs in category C and C+E until clearance to do so is received from the LA.

Certain new driving offences were created by the Road Traffic Act 1991. These came into force from 1 July 1992, after which date the previous charge of reckless driving is replaced by a new charge of 'dangerous' driving (which had previously existed until 1977) for which a more specific definition is given in the Act. This states that a person is to be regarded as driving dangerously for these purposes if the way he drives 'falls far short of what would be expected of a competent and careful driver, and it would be obvious to a competent and careful driver that driving in that way would be dangerous'. The Act goes on to say that driving would be regarded as dangerous 'if it was obvious to a competent and careful driver that driving the vehicle in its current state would be dangerous' – this obviously applies to the vehicle's mechanical condition or the way it is loaded. It further states that 'dangerous' refers to danger either of injury to any person or of serious damage to property.

The principal offences to which this relates are dangerous driving and causing death by dangerous driving. Further new provisions are added which make it an offence for any person to cause danger to road users by way of intentionally and without lawful authority placing objects on a road, interfering with motor vehicles, or directly or indirectly interfering with traffic equipment (eg road signs etc).

Penalties under these new provisions are heavily increased in certain cases. For example, for the new offence of causing death by careless driving while under the influence of drink or drugs carries a maximum penalty of up to 5 years in prison and/or a fine. For causing a danger to road users the maximum penalty is up to 7 years' imprisonment and/or a fine.

Other new provisions to be introduced are extended driving tests for those convicted under the new dangerous driving provisions, rehabilitation courses for drink-drivers (see below) and, from 1 October 1992, much heavier maximum penalties for the most serious of driving offences – £5000 maximum fines for example.

Drink Driving and Breath Tests

It is an offence to drive or attempt to drive a motor vehicle when the level of alcohol in the breath is more than 35 microgrammes per 100 millilitres of breath. This is determined by means of an initial breath test, conducted on the spot when the driver is stopped, and later substantiated by a test on a breath-testing machine (ie Lion Intoximeter) at a police station. The breath/alcohol limit mentioned above equates to the blood/alcohol limit of 80 milligrammes of alcohol in 100 millilitres of blood or the urine/alcohol limit of 107 milligrammes of alcohol in 100 millilitres of urine.

If the person suspected of an alcohol-related offence cannot, due to health reasons, produce a breath sample, or if a breath test shows a reading of not more than 50 microgrammes of alcohol per 100 millilitres of breath they are given the opportunity of an alternative test, either blood or urine, for laboratory analysis. This test can only be carried out at a police station or a hospital and the decision as to which alternative is chosen rests with the police (unless a doctor present determines that for medical reasons a blood test cannot or should not be taken). Similarly, if a breath test of a driver shows the proportion of alcohol to be no more than 50 microgrammes in 100 millilitres of breath, the driver can request an alternative test (ie blood or urine) as described above for those who cannot provide a breath sample for analysis.

Prosecution will follow a failure to pass the test which will result in a fine or imprisonment and automatic disqualification from driving. Failure to submit to a breath test and to a blood or urine test are serious offences, and drivers will find themselves liable to heavy penalties on conviction and potentially long-term disqualification or driving licence endorsement (endorsements for such offences remain on a driving licence for 11 years – see p 115).

The police *do not* have powers to carry out breath tests at random but they *do* have powers to enter premises to require a breath test from a person suspected of driving while impaired through drink or drugs, or who has been driving, or been in charge of a vehicle which has been involved in an accident in which another person has been injured.

In a highly publicised case in 1991, charges against a lorry driver breathalysed by police having been woken from sleep on the sleeper berth of his vehicle cab (and with his tachograph set to record rest) were dismissed by the court. Although presumably over the statutory limit (otherwise no charge would have been brought), the driver was clearly making no attempt to drive his vehicle and the vehicle was parked in a proper (ie non-public) place. Since the court dismissed the case there was no opportunity to obtain a ruling from a higher court as to the precise legal position of drivers in such circumstances. It is considered that drinking and then sleeping in a vehicle cab does present risk of prosecution for drivers.

Drink-Driving Disqualification

Conviction for a first drink-driving offence will result in a minimum 1 year period of disqualification and for a second or subsequent offence of driving or attempting to drive under the influence of drink or drugs longer periods of disqualification will be imposed by the court. If the previous such conviction took place within 10 years of the current offence the disqualification must be for at least 3 years.

Drivers convicted twice for drink-driving offences may have their driving licence revoked altogether. Offenders who are disqualified twice within a 10 year period for any drink-driving offences and those found to have an exceptionally high level of alcohol in the breath (ie more than $2\frac{1}{2}$ times over the limit) or those who twice refuse to provide a specimen will be classified as high risk offenders (HROs) by the Driver and Vehicle Licensing Agency. They will be required to show that they no longer have an 'alcohol problem' and be subject to an additional medical examination by a DoT approved doctor (fee £33.40) before their licence will be restored to them. A higher than normal fee (ie £20) will be charged for the renewal of such licences.

Special Reasons for Non-Disqualification

The courts have discretion in exceptional mitigating circumstances (ie when there are 'special reasons') not to impose a disqualification. The mitigating circumstance must not be one which attempts to make the offence appear less serious and no account will be taken of hardship other than exceptional hardship. Pleading that you have a wife and children to support or that you will lose your job is not generally considered to be exceptional hardship for the purposes of determining whether or not disqualification should be imposed.

If account has previously been taken of circumstances in mitigation of a disqualification, the same circumstances cannot be considered again within 3 years.

Where a court decides, in exceptional circumstances as described above, not to disqualify a convicted driver, 4 penalty points will be added to the driver's licence in lieu of the disqualification.

Drink-Driving Courses

The Road Traffic Act 1991 contains provisions whereby certain (but not all) drink-driving offenders may have the period of their disqualification reduced if they agree to undertake an approved rehabilitation course and satisfactorily complete it. The provision applies only where the court orders the individual (who must be over 17 years of age) to be disqualified for at least 12 months following conviction under the Road Traffic Act 1988 for:

- causing death by careless driving when under the influence of drink or drugs;
- driving or being in charge of a motor vehicle when under the influence of drink or drugs;
- driving or being in charge of a motor vehicle with excess alcohol in the body; or
- failing to provide a specimen (of breath, blood or urine) as required.

The court has a duty to ensure that a place on an approved course is available for the offender. It must explain to the offender 'in ordinary language' the effect of the order (to reduce the disqualification period), the amount of the fees and that these must be paid in advance, and it must seek the offender's agreement that the order should be made – in other words, it is a completely voluntary scheme and there is no question of force.

Given these provisos, the court can order the period of disqualification to be reduced by not less than 3 months and not more than one-quarter of the unreduced period (for example, with a 12 month disqualification, the reduction will be 3 months, leaving a 9 month disqualification period to be served).

The latest date for completion of one of these rehabilitation courses is 2 months prior to the last day of the reduced disqualification period (so, in the 9 month example given above, the course would have to be completed by the last day of the seventh month). On completion of the course the offender is given a certificate which must be returned to the clerk of the supervising court (named in the order) in order to secure the reduction. Failure to complete the course or to produce the certificate on time will result in the loss of this facility.

A number of approved courses have now been established around the country (with fees of between £50 and £200) and these will require attendance for between 16 and 30 hours, made up of a number of separate sessions, during which the offender will learn about the effects of alcohol on the body and on driving performance and behaviour, about drink-driving offences, and about the alternatives to drinking and driving. A range of relevant advice will also be given.

This scheme is being conducted by the Department of Transport as a 3-year experiment (successful schemes of this nature are run in the USA and Germany). The DoT says these courses will provide a real opportunity for offenders to change their attitude to drinking and driving before they drive again – but it is not a soft option.

Other Penalties

In addition to the penalties of disqualification and the endorsement of penalty points on driving licences, courts may impose fines and, for certain offences, imprisonment. The maximum fine for most offences is determined by reference to a scale set out in the Criminal Justice Act 1991 as follows:

Level 1	£200
Level 2	£500
Level 3	£1000
Level 4	£2500
Level 5	£5000

Offences such as dangerous driving, failing to stop after an accident or failure to report an accident and drink-driving offences, carry the current maximum fine of £5000, as do certain vehicle construction and use offences (eg overloading, insecure loads, using a vehicle in a dangerous conditon etc) and using a vehicle without insurance.

Unit Fines

The Criminal Justice Act mentioned above also brought into force a system of Unit fines whereby convicted offenders are fined according to their income. In determining the fine to be paid, the courts refer to a laid-down formula of units based on the seriousness of an offence.

The Act specified the maximum number of units as follows:

Level 1 offence	2 units
Level 2 offence	5 units
Level 3 offence	10 units
Level 4 offence	25 units
Level 5 offence	50 units

The number of units is then multiplied by the offender's weekly disposable income – thus the higher paid are penalised more than those on lower incomes. Currently, the minimum fine per unit is £4 for a person earning £20 per week, and the maximum is £100 for a person earning £18,000 per year or more. On this basis, for example, a conviction for not wearing a seat belt (a two unit offence) could result in a fine of £8 only for a low-income offender and £200 for those with a high income: conviction for driving without insurance (a 50 unit offence) could see a fine varying between £200 and £5000 for convicted offenders at opposite ends of the income scale.

Penalties against Vocational Entitlements

Where a licence holder is disqualified from driving following conviction for offences committed with cars or other light vehicles, or as a result of penalty point totting-up, any vocational entitlement which that person holds is automatically lost until the licence is reinstated. Additionally, the holder of an lgv/pcv vocational entitlement may have this revoked or suspended by the DVLA (without reference to the LA – see below) and be disqualified from holding such entitlement, for a fixed or an indefinite period, at any time on the grounds of misconduct or physical disability. Furthermore, a person can be refused a new lgv/pcv driving entitlement following licence revocation, again either indefinitely or for some other period of time which the Secretary of State (ie via the DVLA) specifies. A new vocational test may be ordered before the entitlement is restored – see further below.

Disqualification from holding an lgv vocational entitlement as described above does not prevent a licence holder from continuing to drive vehicles within the category B and C1 entitlements that he holds.

The LAs continue to play a disciplinary role under the new licensing scheme in regard to driver conduct, but only at the request of the DVLA. They have powers under the new provisions to call drivers to public inquiry (PI) to give information and answer questions as to their conduct. Their duty is to report back to the DVLA if they consider that an lgv/pcv entitlement should be revoked or the holder disqualified from holding an entitlement – the DVLA must follow the LA's recommendation in these matters.

Failure to attend a PI when requested to do so (unless a reasonable excuse is given) means that the DVLA will automatically refuse a new vocational entitlement or suspend or revoke an existing entitlement.

Large goods vehicle drivers who have been off the road for a period of time after being disqualified are having to prove themselves capable of driving small goods vehicles legally and safely for a period of time before their lgv driving entitlement may be restored by the LAs. Mr John Mervyn Pugh, LA for the West Midlands Traffic Area has said that banned lgv drivers must prove themselves on 7.5 tonners before being allowed to drive vehicles up to 38 tonnes again.

A recent change in the government's rules on disciplining lgv entitlement holders means that the LAs now follow a set of recommended guidelines in imposing penalties against such entitlements. Under these rules, and where there are no aggravating circumstances, a driver being disqualified for 12 months or less should be sent a warning letter with no further disqualification of the lgv entitlement. When a driving disqualification is for more than 1 year, the offender should be called to appear before the LA and he should incur an additional suspension of his lgv entitlement, amounting to between 1 month and 3 months. The intention here is to allow the person to regain his driving skills and road sense in a car before driving a heavy vehicle again. Where two or more driving disqualifications of more than 8 weeks have been incurred within the past 5 years, and the combined total of disqualification exceeds 12 months, the driver should be called to public enquiry and a further period of lgv driving disqualification imposed, amounting to between 3 and 6 months.

In the case of new lgv entitlement, for applicants who already have 9 or more penalty points on their ordinary licence the guidelines recommend that the LA should issue a warning as to future conduct or suggest that the applicant tries again when the penalty points total on his licence has been reduced.

Removal of Penalty Points and Disqualifications

Penalty points endorsed on driving licences can be removed (by application to the DVLA Swansea on form D1 and on payment of a fee of £6).

The waiting period before which no such application would be accepted is 4 years from the date of the offence, except in the case of reckless/dangerous driving convictions when the 4 years is taken from the date of conviction. Endorsements for alcohol-related offences must remain on a licence for 11 years.

Licences returned after disqualification will show no penalty points but previous disqualifications (within 4 years) will remain and if a previous alcohol/drugs driving offence disqualification has been incurred, this will remain on the licence for 11 years.

Application may be made by disqualified drivers for reinstatement of their licence after varying periods of time depending on the duration of the disqualifying period as follows:
1. Less than 2 years – no prior application time.
2. Less than 4 years – after 2 years have elapsed.
3. Between 4 years and 10 years – after half the time has elapsed.
4. In other cases – after 5 years have elapsed.

The courts are empowered to require a disqualified driver to retake the driving test before restoring a driving licence, and following the introduction of provisions contained in the Road Traffic Act 1991, it is now mandatory for them to impose 'extended' re-tests following disqualification for the most serious of driving offences, namely, dangerous driving, causing death by dangerous driving and manslaughter by the driver of a motor vehicle (in Scotland, the charge is culpable homicide).

The fee charged for replacement licences following disqualification is £6, but where the disqualification was for drink-driving-type offences the fee for a replacement is £20.

Removal of LGV Driving Licence Disqualification

Drivers disqualified from holding an lgv/pcv entitlement, as described above, may apply to have the disqualification removed after 2 years if it was for less than 4 years, or after half the period if the disqualification was for more than 4 years but less than 10 years. In any other case including disqualification for an indefinite period an application for its removal cannot be made until 5 years have elapsed. If an application for the removal of a disqualification fails, another application cannot be made for 3 months.

The DVLA will not necessarily readily restore lgv/pcv driving entitlements on application following disqualification of a driving licence. Applicants may be called by an LA to public inquiry when he will inquire into the events which led to the disqualification and at which he may also decide that the applicant must wait a further period before applying again or must take a new lgv/pcv driving test in order to regain the vocational entitlement.

Appeals

If the DVLA refuses to grant an application for an lgv/pcv driving entitlement or revokes, suspends or limits an existing entitlement, the applicant or entitlement holder may appeal against the decision under the Road Traffic Act 1988. The first step is for them to notify the DVLA, and any LA involved in consideration of the applicant's conduct, of their intention to appeal. The appeal can then be made to a magistrates' court acting for the petty sessions in England or Wales, or in Scotland to the local sheriff.

International Driving Permits

Certain foreign countries do not accept British ordinary driving licences – eg Bulgaria, Hungary and the Commonwealth of Independent Soviet States (CIS) – in which case an international driving permit (IDP) will be required by British licence holders wishing to drive in those countries. These permits are obtainable from the RAC, RSAC, AA or the National Breakdown Recovery Club (NBRC). The fee is £3 and a passport-type photograph is required for attachment to the permit. Applicants must be UK residents, over 18 years of age and hold full driving entitlements for the category of vehicle which the IDP is required to cover.

Exchange of Driving Qualifications

Holders of valid licences in the following countries can exchange these for full British driving licences: Australia, Austria, Barbados, British Virgin Islands,

Cyprus, Finland, Gibraltar, Hong Kong, Japan, Kenya, Malta, New Zealand, Norway, Singapore, Sweden, Switzerland and Zimbabwe. Also, holders of British Forces Germany (BFG) driving licences may exchange these for a British ordinary driving licence. The fee for exchange licences is £6.

British issued EU-model driving licences showing lgv/pcv vocational entitlements are valid for driving the particular categories of vehicle to which they apply in all Community countries (and Northern Ireland). Community driving licence holders visiting or intending to stay and work in Britain may apply to exchange these for British licences showing vocational entitlements without taking the appropriate test provided they have qualified to drive such vehicles in their own country and can prove recent driving experience with such vehicles over a specified period – 6 months within the previous 18 months or 12 months within the previous 3 years.

Visitors Driving in the UK

Visitors to the UK may drive vehicles in the UK provided they hold a domestic driving licence issued in their own country (ie outside the UK and the EU) or a convention driving permit (issued under the Geneva Convention on Road Traffic by a country outside the UK). Holders of such permits are entitled to drive vehicles of a class which their own national or international licence covers for a period of 12 months from the date of their entry into the UK. From 16 July 1994, new UK residents from other EU member states and certain 'designated' countries have up to 5 years in which to exchange their domestic driving licences for a British licence – previously this was required within 1 year. The 'designated' countries, besides other EU member states, are as follows: Australia, Austria, Barbados, British Virgin Islands, Finland, Hong Kong, Japan, Kenya, Malta, New Zealand, Norway, Republic of Cyprus, Singapore, Sweden, Switzerland and Zimbabwe.

The Second Driver Licensing Directive

The so-called 'second' driver licensing directive of the EU (ie EC Directive 439/91) is due to come into effect for the UK on 1 July 1996. Under the provisions of this Directive, drivers who pass the category B car test will be restricted to driving cars and light goods vehicles up to 3.5 tonnes in weight (and with no more than eight passenger seats) and towing a trailer up to a maximum of 750kg gross weight (making a combined maximum weight for the vehicle and trailer of 4.25 tonnes). New drivers wishing to drive vehicles of gross weights over 3.5 tonnes and up to 7.5 tonnes will have to take a new form of category C1 driving test on a goods vehicle of 4 tonnes minimum weight and capable of a speed of up to 80kph. (*It is stressed here that existing holders of a driving entitlement for 7.5 tonners are not affected.*) New drivers wishing to drive vehicles in category C1 (ie 3.5–7.5 tonnes) with a trailer attached and for which a category C1+E entitlement is required will have to first pass a test on a category C1 vehicle (see above) which is towing a trailer of at least 2 tonnes gross weight and making a total combination length of at least 8 metres. Once a C1+E entitlement is gained (ie after 1 July 1996), and providing the person is at least 21 years of age, this would allow him to drive a vehicle of up to 12 tonnes gross weight (ie maximum authorised mass – see p 94 for a definition). The existing 18-year age limit for driving vehicles of up to 7.5 tonnes gross weight in category C1+E will be maintained.

New drivers seeking category C1 and C1+E driving entitlements from 1 July 1996 will be required to meet the medical standards currently applicable for full category C and C+E entitlements, as described earlier in this chapter.

Additionally, the DoT is currently undertaking development work on the establishment of a more stringent system of theory testing ready for 1 July 1996 when the second directive comes into force. This would apply to all new drivers, of cars as well as goods and passenger vehicles.

Unit Fine Scheme

The system of 'Unit Fines' described on p 114 whereby a convicted road traffic offender is penalised in accordance with his or her ability to pay has been dropped in favour of the traditional system where the amount of any fine imposed on conviction is related solely to the seriousness of the offence.

Photographs on Driving Licences

In August 1994 the government announced proposals for a new form of 'easy-to-carry' UK driving licence to incorporate both the holder's photograph and additional relevant information. The proposed licences, to be introduced from 1996 for Britain's 32 million drivers, will be of the shape and size of a credit card and will contain a built-in microchip capable of storing additional data about the holder, or be of the 'smart card' variety with magnetic strips that could include the holder's address and National Insurance number for example.

Driving and Epilepsy

Car and light vehicle drivers who suffer from epilepsy can, since 5 August 1994, obtain a licence to drive such vehicles (but not lgvs and pcvs) if they have been free of an asleep attack for 2 years and an awake attack for 1 year (previously it was 2 years in each case).

7: Driving Tests and Driver Training

As a result of the harmonisation of driver licensing within the European Union, as detailed in the previous chapter, certain aspects of the long-standing British large goods vehicle (lgv) driving test are to be amended from 1 July 1996. However, even after 1996, the main objective of the existing test will remain, which is to ensure that goods vehicle drivers are competent to drive large vehicles on the roads in safety. It also provides a measure of professionalism among commercial vehicle drivers. The test is more comprehensive and more complex than the ordinary driving test and, consequently, demands greater skill and knowledge from the driver who wishes to pass.

For many years the advanced driving test organised by the Institute of Advanced Motorists (IAM) has been looked upon as a severe test of driving skills requiring a high degree of knowledge of the 'rules of the road' for private car drivers. Commercial vehicle drivers who want to show that they too have attained an exceptional level of proficiency can take the advanced driving test designed specially for commercial vehicles and on passing they may display the coveted IAM badge.

This chapter contains details of the ordinary driving test (ie for category B and C1 vehicles), the present lgv driving test (and the proposed changes) which must be passed to gain a licence to drive large goods vehicles, and the advanced driving test which drivers may take voluntarily. The chapter also covers certain aspects of heavy goods vehicle driver training provisions, but the Young HGV Driver Training Scheme detailed in previous editions of this *Handbook* has been 'mothballed' due to lack of continuing support.

Candidates for ordinary and vocational (ie lgv and pcv) driving tests are reminded that they must produce satisfactory evidence of identity when arriving for a test. Acceptable identity documents for this purpose would include existing driving licences (ordinary, lgv or pcv or an overseas driving licence), a passport or an employer-issued identity card bearing the holder's name, signature and photograph. At the examiner's discretion valid bank cheque cards or credit cards could be accepted. If a test candidate cannot produce satisfactory means of identification the test will not be conducted and the fee will be forfeited. This measure is intended to combat a rising incidence of persons with false identities taking multiple lgv tests on behalf of others who cannot themselves pass the test but who are prepared to pay large sums to acquire a test pass certificate.

Ordinary Driving Test

Before people are granted a licence to drive a motor vehicle on the road they must pass a driving test on the class of vehicle for which they require the licence.

The test is carried out by an examiner from the Driving Standards Agency (DSA – Executive Agency of the DoT). The test candidate has to meet the following requirements:

1. They must show that they are fully conversant with the contents of the *Highway Code.*
2. They must prove that they are able to read in good daylight (with the aid of spectacles, if worn) a motor vehicle's registration number in accordance with the vision requirements (see p 101).
3. They must show that they are competent to drive without danger to and with due consideration of other users of the road, including being able to:
   (a) start the engine of the vehicle;
   (b) move away straight ahead or at an angle;
   (c) overtake, meet or cross the path of other vehicles and take an appropriate course;
   (d) turn right-hand and left-hand corners correctly;
   (e) stop the vehicle in an emergency and in a normal situation, and in the latter case bring it to rest at an appropriate part of the road;
   (f) drive the vehicle backwards and while doing so enter a limited opening either to the left or to the right;
   (g) cause the vehicle to face the opposite direction by the use of forward and reverse gears;
   (h) indicate their intended actions at appropriate times by giving appropriate signals in a clear and unmistakable manner (in the case of a left-hand drive vehicle or a disabled driver for whom it is impracticable or undesirable to give hand signals there is no requirement to provide any signals other than mechanical ones);
   (i) act correctly and promptly on all signals given by traffic signs and traffic controllers and take appropriate action on signs given by other road users.

Since 1 April 1991 a reverse parking manoeuvre has been incorporated in the driving test. This requires test candidate to stop their vehicle next to and parallel with a parked vehicle, then reverse to position and park their vehicle

in front of or behind the other vehicle, level with and reasonably close to the kerb.

Candidates who fail the driving test are given an oral explanation of the reasons for their failure.

Driving Test Fee

The fee for an ordinary driving test currently is £27.50 or £37.50 if conducted on a Saturday. Higher test fees of £55 on weekdays and £77.50 on Saturdays are payable where an extended re-test is ordered following an obligatory disqualification of a driving licence.

Driving Instruction

Only DoT-approved instructors (ADIs – approved by the Driving Standards Agency) are permitted to give driving instruction for payment on vehicles legally defined as 'motor cars' which basically means the following classes of vehicle:
1. Private cars.
2. Goods vehicles not exceeding 3050kg unladen weight.

More stringent standards have been introduced for driving instructors, in particular extended training periods are necessary before instructors can become qualified.

Tuition given for payment on heavy goods vehicles does not come within the scope of this legislation despite the opinion of a certain number within the transport industry that it should do so. Consequently, registration of instructors with the DSA for this purpose is not necessary.

The LGV Driving Test

The large goods vehicle driving test is also conducted by DSA examiners (see p 120) and booking has to be made through the local offices of the Driving Standards Agency (see Appendix V for addresses).

Certain test centres now offer Saturday morning lgv and pcv driver testing.

Application and Fees

Applications for the test have to be made on form DLG 26 obtainable from local DSA offices (see Appendix V). The current test fee of £62 (£80 for Saturday and weekday evening tests) must be sent with the application. Applicants are warned to apply for a test in good time. This is important if an lgv driving licence is required – subject to passing the test – from a particular date; and applicants must also ensure that their driving is of a sufficiently high standard to be able to pass the test. Should a candidate need to cancel a test appointment this should be done at least five clear working days in advance, otherwise the fee will be forfeited. Test candidates must be able to produce satisfactory identification on arrival at the test centre otherwise the examiner may refuse to conduct the test and the fee will be forfeited.

Block Bookings

Previous arrangements which allowed driver training establishments to make block bookings in advance without specifying individual candidate names have been put in jeopardy by changes introduced by the DSA. Block bookings could be made for lgv driving tests and in these cases trainees could be named up to 1 week (instead of the normal 28 days) before the test date. Training organisations could make provisional bookings with the local DSA office but the appointment had to be confirmed more than 1 month ahead on forms DLG 26X, Y and Z, together with the fee.

Vehicles for the LGV Driving Test

The candidate has to provide the vehicle (or arrange for the loan of a suitable vehicle) on which he wishes to be tested and it must comply with the following (pre-1 July 1996) requirements:

1. it must be unladen and of the category for which an lgv entitlement is required*, ie
    (a) for a category C entitlement it must be of at least 7.5 tonnes gross weight,
    (b) for a category C+E entitlement it must be either an articulated vehicle or a drawbar combination (in the latter case the trailer must have at least two axles and the combination must have a permissible maximum weight of at least 15 tonnes);
2. it must display either hgv or ordinary 'L' plates at the same time at the front and rear**;
3. it must be in a thoroughly roadworthy condition;
4. seating accommodation in the cab must be provided for the examiner; and
5. it must have sufficient fuel for a test lasting up to 2 hours.

NB: From 1 July 1996, when the second EU Directive on driver licensing comes into force, a more extensive list of minimum test vehicles will become available, as described on p 132.

**NB: The old-type orange-background hgv 'L' plate is to be phased out from 1996 but in the meantime either the normal red 'L' plate can be used or the hgv 'L' plate.*

The LGV Driving Test Syllabus

The DSA has published a recommended syllabus to be followed by candidates preparing for the lgv driving test. Copies of the syllabus are contained within the Agency's recently published book *Your Large Goods Vehicle Driving Test* (available from HMSO price £8.99).

The syllabus comprises ten main sections containing advice and detailing specific requirements for the test as well as an introductory section outlining the prior 'thorough' knowledge that a candidate should acquire before attempting the test.

The Syllabus is as follows:

Knowledge

Candidates must have a thorough prior knowledge of:

- the latest edition of the *Highway Code* – new edition 1993 – (especially those sections concerning lorries)
- the regulations governing goods vehicle driver's permitted hours of work and rest requirements

Candidates must have a thorough understanding of:

- general motoring regulations, especially
  - road traffic offences
  - drivers' and operators' licences (where applicable)
  - insurance requirements
  - vehicle road tax (VED) relating to lgvs and trailers
  - plating of lgvs and trailers
  - annual testing of lgvs.

1. Legal requirements

Lgv driving test applicants must:

- be at least 21 years old
- meet the stringent eyesight requirements
- be medically fit to drive lorries
- hold a full licence for driving motor cars (ie category B)
- hold and comply with the conditions for holding either
  - a provisional lgv driving licence, or
  - a full lgv driving entitlement for a lesser category of vehicle
- ensure that the vehicle being driven
  - is legally roadworthy
  - is correctly plated
  - has a current goods vehicle test certificate
  - is properly (VED) licensed with the correct tax disc displayed
- make sure the vehicle being driven is properly insured for its use (especially if on contract hire)
- display L-plates to the front and rear of the vehicle
- be accompanied by a qualified driver who holds a valid, full, driving entitlement for the category of vehicle being driven
- wear a seat belt, if fitted, unless an exemption applies
  - it is important to ensure that all seatbelts fitted in the vehicle, and their anchorages and fittings, are secure and free from obvious defect
  - the syllabus warns that children should not normally be carried in lgvs, but where they are carried, with permission, the driver must comply with the regulations relating to the wearing of seat belts by children or the use of child restraints
- be aware that it is a legal requirement to notify the DVLC Swansea of any medical condition which could affect safe driving if its duration is likely to be 3 months or more
- ensure that any adaptations for disability purposes are suitable to allow the vehicle to be controlled safely.

2. *Vehicle controls, equipment and components*

Test candidates must:

- understand the function of the
  - accelerator
  - clutch
  - gears
  - footbrake
  - handbrake
  - secondary brake
  - steering

  and be able to use these controls competently
- know the function of all other controls and switches on the vehicle and be able to use them competently
- understand the meanings of
  - gauges
  - warning lights
  - warning buzzers
  - other displays on the instrument panel
- be familiar with the operation of tachographs and their charts
- know the legal requirements which apply to the vehicle
  - speed limits
  - weight limits
  - braking system (ABS)
  - fire extinguishers to be carried
- be able to carry out routine safety checks, and identify defects, especially with
  - power steering
  - brakes (tractive unit + semi-trailer on articulated, or rigid vehicle and trailer combinations)
  - tyres on all wheels
  - seatbelts
  - lights
  - reflectors/reflective plates
  - direction indicators
  - marker lights
  - windscreen wipers and washers
  - horn
  - rear view mirrors
  - speedometer
  - tachograph
  - exhaust system
  - brake line and electric connections on rigid vehicles + trailers or articulated vehicles
  - coupling gear
  - hydraulic and lubricating systems
  - self-loading or tail-lift equipment
  - dropside hinges and tailgate fastenings
  - curtainside fittings and fastenings
  - winches and auxiliary gear where these items are fitted

- know the safety factors relating to
  - stowage
  - loading
  - stability
  - restraint of any load carried on the vehicle
- know the effects that speed limiters will have on the control of the vehicle – especially when intending to overtake
- know the principles of the various systems of retarders which may be fitted to lgvs
  - electric
  - engine driven
  - exhaust brakes and when they should be brought into operation.

3. Road user behaviour
Test candidates must:
- know the most common causes of road traffic accidents
- know which road users are more vulnerable and how to reduce the risks to them
- know the rules, risks and effects of drinking and driving
- know the effects that
  - illness (even minor ones)
  - drugs or cold remedies
  - fatigue

  can have on a driver's performance
- recognise the importance of complying with rest period regulations
- be aware of the age-dependent problems among other road users
  - children
  - young cyclists
  - young drivers
  - more elderly drivers
  - elderly or infirm pedestrians
- concentrate and plan ahead in order to anticipate the likely actions of other road users and be able to select the safest course of action.

4. Vehicle characteristics
Test candidates must:
- know the most important principles concerning braking distances under various
  - road
  - weather
  - load

  conditions
- know the different handling characteristics of other vehicles with regard to
  - speed
  - stability
  - braking

- manoeuvrability
- know that some other vehicles such as bicycles and motorcycles are less easily seen than others
- be aware of the difficulties caused by the characteristics of their own and other vehicles, and be able to take the appropriate action to reduce any risks which may arise, for example
  - lgvs and buses moving to the right before making a sharp left turn
  - articulated vehicles taking what appears to be an incorrect line before negotiating
    - corners
    - roundabouts
    - entrances
  - blind spots which occur with many large vehicles
  - bicycles, motorcycles and high-sided vehicles being buffeted in strong winds, especially on exposed sections of road
  - turbulence created by large goods vehicles travelling at speed affecting pedestrians, cyclists, motorcyclists, vehicles towing caravans, and drivers of smaller vehicles.

At all times they must remember that other road users may not understand the techniques required to manoeuvre a large goods vehicle safely.

5. *Road and weather conditions*

Test candidates must:

- know the various hazards which can arise when driving
  - in strong sunlight
  - at dusk or dawn
  - during the hours of darkness
  - on various types of road such as
    - narrow lanes in rural areas
    - one-way streets
    - two-way roads in built-up areas
    - three lane roads
    - dual carriageways with various speed limits
    - trunk roads with two-way traffic
    - motorways
- gain experience in driving on urban roads with 20mph or 30mph speed limits, and also on roads carrying denser traffic volumes at higher speed limits in both daylight and during the hours of darkness
- gain experience in driving on both urban and rural motorways
- know which road surfaces will provide better or poorer grip when braking
- know all the associated hazards caused by bad weather such as
  - rain
  - snow
  - ice
  - fog
- be able to assess the difficulties caused by
  - road
  - traffic
  - weather conditions

- drive defensively and anticipate how the prevailing conditions may affect the standard of driving shown by other road users.

6. Traffic signs, rules and regulations
Test candidates must:
- have a thorough knowledge and understanding of the meanings of traffic signs and road markings
- be able to recognise and comply with traffic signs such as
  - weight limits
  - height limits
  - signs prohibiting lgvs
  - loading/unloading restrictions
  - traffic calming measures
    - 20mph zones
    - road width restrictions
    - speed reduction humps
  - roads designated as 'red routes'
  - night-time and weekend lorry bans such as those in the London boroughs.

7. Vehicle control and road procedure
Test candidates must have the knowledge and skill to carry out the following list of tasks when appropriate:
  - safely and expertly
  - in daylight and, if necessary, during the hours of darkness.

Where the tasks involve other road users they must:
  - make proper use of the mirrors
  - take effective observation
  - give signals where necessary
- take the following necessary precautions, where they are applicable, before getting into the vehicle
  - ensure number plates are correct and securely fitted
  - check all round for obstructions
  - ensure that any load is secure
  - check air lines are correctly fitted and free from leaks
  - check all couplings to drawing vehicle and trailer
  - check landing gear is raised
  - check trailer brake is released
  - check all bulbs, lenses and reflectors are fitted
  - make sure all lights, indicators and stop lights are working
  - ensure all reflective plates are visible, clean and secure
  - examine tyres for defects
  - examine all load restraints for tension etc
  - ensure any unused ropes are safely stowed
- before leaving the vehicle cab, make sure that
  - the vehicle is stopped in a safe, legal and secure place
  - the handbrake is on
  - the engine is stopped
  - the electrical system is switched off

- – the gear lever/selector is in 'neutral'
- – all windows are closed
- – the passenger door is secure
- – the keys have been removed from the starter switch
- – nobody will be endangered when the cab door is opened
- before starting the engine, carry out the following safety checks
  - – the handbrake is applied
  - – the gear level is in 'neutral'
  - – the doors are properly closed
  - – the driving seat is properly adjusted for
    - height
    - distance from the driving controls
    - back-rest support and comfort
  - – the mirrors are correctly adjusted
  - – their seat belt is fastened and adjusted
- start the engine, but before moving off check
  - – the vehicle (and any trailer) lights are on – if required
  - – gauges indicate correct pressure for the braking system
  - – no warning lights are showing
  - – no warning buzzer is sounding
  - – no ABS fault indicator is lit (where fitted)
  - – all fuel and temperature gauges are operating normally
  - – engine pre-heater (glow plug) lamp is operating (where fitted)
  - – it is safe to move off by looking all round – especially the blind spots
- move off
  - – straight ahead
  - – at an angle
  - – on the level
  - – uphill
  - – downhill
- select the correct road position for normal driving
- take effective observation in all traffic conditions
- drive at a speed appropriate to the road, traffic and weather conditions
- anticipate changes in traffic conditions, adopt the correct action at all times and exercise vehicle sympathy
- move into the appropriate traffic lane correctly and in good time
- pass stationary vehicles safely
- meet, overtake and cross the path of other vehicles safely
- turn right or left at
  - – junctions
  - – crossroads
  - – roundabouts
- drive ahead at
  - – crossroads
  - – roundabouts
- keep a safe separation gap when following other vehicles
- act correctly at all types of pedestrian crossing

- show proper regard for the safety of all other road users, with particular respect for those most vulnerable
- drive on
  - urban roads
  - rural roads
  - dual carriageways

  keeping up with the traffic flow (but still observing seed limits) where it is safe and appropriate to do so
- comply with
  - traffic regulations
  - traffic signs
  - signals given by authorised persons
  - police officers
  - traffic wardens
  - school crossing patrols
- take the correct action on signals given by other road users
- stop the vehicle safely at all times
- select safe and suitable places to stop the vehicle reasonably close to the nearside kerb when requested
  - on the level
  - facing uphill
  - facing downhill
  - before reaching a parked vehicle, but leaving sufficient room to move away again
- stop the vehicle on the braking exercise manoeuvring area
  - safely
  - as quickly as possible
  - under full control
  - within a reasonable distance from a designated point
- reverse the vehicle on the manoeuvring area
  - under control
  - with effective observation
  - on a pre-determined course
  - to enter a restricted opening
  - to stop with the extreme rear of the vehicle within a clearly defined area
- cross all types of level crossing
  - railway
  - rapid transit system (trams) where appropriate.

8. *Additional knowledge*
The test candidate must know:
- the importance of inspecting all tyres on the vehicle for
  - correct pressure
  - signs of wear
  - evidence of damage
  - safe tread depth
  - objects between twin tyres
  - indications of overheating

- safe driving principles in order to prevent skidding
- how to drive when the road is
  - icy
  - snow-covered
  - flooded
  - covered by excess surface water
- what to do if involved in
  - a damage-only traffic accident
  - a road traffic accident involving
    - injury
    - fire
    - spillage of hazardous material
    - danger to other road users due to obstruction by fallen loads etc
  - any type of accident on a motorway
- steps to take if the vehicle breaks down
  - in the daytime
  - at night

  on
  - a bend on a road with two-way traffic
  - a busy dual carriageway
  - a clearway
  - a motorway
- the differences between toughened and laminated glass used in lgv windscreens
- how to use the hammer or similar tool, if fitted, to exit from the vehicle in an emergency
- basic first aid for use on the road
- the precautions to take to prevent theft of
  - the vehicle
  - the load
  - the trailer
  - equipment on the vehicle
- factors to consider when selecting a safe place to leave an unattended trailer or semi-trailer
  - legal (ie not on a public road)
  - safe (so as not to endanger any member of the public)
  - convenient (so as not to block any access or exit)
  - suitable (ie level and firm enough to support the trailer or semi-trailer)
  - secure (ie where the trailer and/or its load is not liable to be stolen)
  - will not create a hazard for other road users
  - whether any anti-theft device can be fitted (eg coupling locking device).

9. *Motorway driving*

Test candidates must have a thorough *practical* knowledge of the special
  - rules
  - regulations
  - driving techniques

which apply on motorways

Candidates will not be asked to drive on the motorway on the lgv driving test, but will be expected to show a thorough understanding and knowledge of all aspects of motorway driving, particularly
- overtaking
- exercising lane discipline
- the effects of speed limiters
- joining and leaving motorways
- breakdown and emergency procedures
- driving in adverse weather conditions
- principal causes of accidents on motorways.

10. *Safe working practices*
Test candidates must:
- adopt the correct method of climbing into the vehicle cab
- avoid the risks involved in jumping down from the cab
- ensure that any tilt-cab locking mechanism is secure (especially after routine maintenance)
- follow safety guidelines when operating
  - under raised tipper bodies
  - near inspection pits
  - tail-lift controls
  - on-board hoists
  - on any walkway
  - under overhead cables
  - refuelling points
  - between parked vehicles
  - between tractive unit and trailer of any kind
  - underneath any vehicle
  - under any vehicle supported by jacks
  or before
  - carrying out roadside repairs
  - removing road wheels
  - inflating tyres.

Preparing for the LGV Driving Test

Detailed technical questions will be asked by the examiner in addition to random questions on the *Highway Code* (in accordance with the syllabus above) to test the candidate's knowledge of the correct action to take in the event of a fault developing in a component affecting the safe operation of the vehicle. 'Workshop-type' answers will not be expected but only sufficient information to satisfy the examiner that the driver knows when a heavy goods vehicle is in a safe condition to be driven. The candidate will also be expected to have a knowledge of safe loading and correct load distribution on his vehicle and the reasons why this is important. In addition he may be expected to give details of the checks that should be carried out before starting on a journey. Drivers of articulated vehicles will be questioned to test their knowledge of the correct procedures for uncoupling and re-coupling a semi-trailer.

Before undertaking the lgv driving test a driver should study carefully the *Highway Code* and the Driving Standards Agency book *Your Large Goods Vehicle Driving Test* (available from HMSO, price £8.99), and put the advice given into practice when driving on the road before taking the lgv test to eliminate any bad habits acquired since passing the ordinary test. The manoeuvring exercises should also be practised as much as possible before taking the test, if facilities are available, to ensure a good chance of passing.

LGV Test Passes

A driver who passes the lgv driving test will be issued with a certificate to that effect. The lgv driving test pass certificate will remain valid for a period of 2 years during which time the holder can apply for an lgv driving entitlement of the appropriate category.

NB: The dual test which previously could give a learner ordinary and learner hgv driver a combined pass in both is now abolished. It is now not possible for a candidate to take the lgv test for category C or C+E vehicles without first holding a full entitlement to drive vehicles in category B.

LGV Test Failures

A driver who fails the lgv driving test will be given a written statement of failure and an oral explanation of the reasons for his failure. He may apply for an immediate re-test.

Minimum Test Vehicles (from 1 July 1996)

From 1 July 1996 changes will be implemented to the current lgv driving test in regard to the minimum weight and speed capability of vehicles which candidates must supply for the test (known as minimum test vehicles – MTVs). The following such vehicles will be required:

1. Vehicles for category B tests must have at least four wheels and be capable of a speed of at least 100kph.
2. For category B+E tests the vehicle itself must comply with category B requirements (see above) and must be drawing a trailer of at least 1 tonne gross weight (ie 1 tonne maximum authorised mass (mam) – see p 94 for a definition).
3. For category C1 tests the vehicle must be of at least 4 tonnes mam and capable of a speed of at least 80kph
4. For category C1+E tests the vehicle must comply with the requirements for category C tests (see above) and must be drawing a trailer of at least 2 tonnes mam – the combined length of the combination must be at least 8 metres.
5. For category C tests (ie rigid goods vehicles exceeding 3.5 tonnes pmw) the vehicle must be of at least 10,000kg permissible maximum weight (or mam – see above), at least 7 metres long and be capable of at least 80kph.
6. For a category C+E test (articulated vehicles) the vehicle must be articulated and have a pmw (or mam – see above) of at least 18,000kg, at least 12 metres long and be capable of at least 80kph.
7. For a category C+E test (drawbar vehicle combinations only) the rigid towing

vehicle should meet the requirements for a category C test and the trailer should be at least 4 metres long. A minimum total weight for the combination of at least 18,000kg is required and the 80kph minimum speed capability also applies.

The Proposed Euro Theory Test

From 1 July 1996, under the terms of the EU 'second' Directive, driving test candidates (both for ordinary and vocational entitlements) will have to face a separate tough, written, theory examination of 20 or more questions (see also p 118). The introduction of a test of this level would bring Britain more into line with other EU member states such as Germany, the Netherlands, Spain and Portugal, and Norway and Sweden (which have a written examination lasting up to 3 hours and which must be passed before a practical driving test is conducted), or possibly with France, Belgium and Luxembourg which use interactive video (as in arcade games) as a means of theory testing for driving test candidates.

Regulations to introduce theory testing will have to be made in time for their introduction by 1996. Member states will be left to determine how they achieve the desired result of improved driving standards across the Community, but the intention is that certain subjects must be included, such as traffic law, mechanical knowledge, safety, motorway driving, environmentally friendly driving and how to deal with accidents.

Advanced Commercial Vehicle Driving Test

The advanced commercial vehicle driving test is organised by the Institute of Advanced Motorists (IAM House, 359–365 Chiswick High Road, London W4 4HS; telephone 081-994 4403; fax 081-994 9249) and is open to any heavy goods vehicle driver, subject to certain conditions as follows:
1. Loads, if carried on test vehicles, must be properly secured.
2. A safe seat at the front of the vehicle must be available for the examiner.
3. The driver must not, by taking the test, contravene the drivers' hours and record-keeping rules.

On passing the test, the applicant becomes eligible for membership of the IAM.

Fees

A fee has to be paid to the Institute of Advanced Motorists for the commercial vehicle test (currently £25) together with the annual subscription to the Institute (currently £12). Applicants for the test must send both the fee and the first year's subscription (ie a total of £37). The annual subscription portion (£12) is refunded if the applicant is unsuccessful in the test.

Exemption from the Test

Certain specially qualified drivers may apply to become Members of the Institute without taking the advanced driving test.

1. Royal Navy, Army and Royal Air Force lgv instructors and qualified testing officers who have passed an lgv instructor's course and whose application is supported by the recommendation of the applicant's commanding officer.
2. Holders of the Road Transport Industry Training Board Instructor's Certificate.
3. Fire Service lgv instructors who have completed an lgv instructor's course, and whose application is supported by the senior instructor.
4. Lgv driving examiners.

The Advanced Test

To pass the test the driver should show 'skill with responsibility' and any driver of reasonable experience and skill should be able to pass without difficulty. The Institute examiners are all ex-police drivers holding a Class 1 Police Driving Certificate, and they test candidates on routes located all over Britain.

The test lasts about 2 hours, during which the test route of some 35–40 miles is covered. The route incorporates road conditions of all kinds including congested urban areas, main roads, narrow country lanes and residential streets. Candidates are not expected to display elaborate driving techniques. Examiners prefer to see the vehicle handled in a steady workmanlike manner without exaggeratedly slow speeds or excessive signalling. Speed limits must be observed (driving in excess of any limit results in test failure) and the driving manner must take into consideration road, traffic and weather conditions. However, the examiners expect candidates to drive briskly within the limits and to cruise at the legal limit (on the road or the vehicle, whichever is lower) whenever circumstances permit.

Drivers will be asked to reverse around a corner and to make a hill start. There will be spot checks on the driver's power of observation (ie the examiner will ask questions about road signs or markings recently passed or about other significant landmarks).

Examiners ask a number of questions of candidates but these are not trick questions. The previous requirement for the driver to give a running commentary during a portion of the route no longer exists – although the test regulations do state that candidates are free to give a commentary if they wish to make extra clear their ability to 'read the road'.

Test Requirements

The examiner will consider the following aspects of driving:

Acceleration: must be smooth and progressive, not excessive or insufficient and must be used at the right time and place.

Braking: must be smooth and progressive, not fierce. Brakes should be used in conjunction with the driving mirror and signals. Road, traffic and weather conditions must be taken into account.

Clutch control: engine and road speeds should be properly co-ordinated when changing gear. The clutch should not be 'ridden' or slipped and the vehicle should not be coasted with the clutch disengaged.

Gear changing: should be smooth and carried out without jerking.

Use of gears: the gears should be correctly selected and used and the right gear engaged before reaching a hazard.

Steering: the wheel should be correctly held with the hands and the 'crossed arm' technique should not be used except when manoeuvring in confined spaces.

Driving position: the driver should be alert and should not slump at the wheel. Resting an arm on the door while driving should be avoided.

Observation: the driver should 'read' the road ahead and show a good sense of anticipation and the ability to judge speed and distance.

Concentration: the driver should keep his attention on the road and should not be easily distracted.

Maintaining progress: taking account of the road, traffic and weather conditions, the driver must keep up a reasonable pace and maintain good progress.

Obstruction: the candidate must be careful not to obstruct other vehicles by driving too slowly, taking up the wrong position on the road or failing to anticipate and react correctly to the traffic situation ahead.

Positioning: the driver must keep to the correct part of the road especially when approaching and negotiating hazards.

Lane discipline: the driver must keep to the appropriate lane and be careful not to straddle white lines.

Observations of road surfaces: the driver must keep an eye on the road surface especially in bad weather and should watch out for slippery conditions.

Traffic signals: signals, signs and road markings must be observed, obeyed and approached correctly and the driver should show courtesy at pedestrian crossings.

Speed limits and other legal requirements: these should be observed. The examiner cannot condone breaches of the law.

Overtaking: must be carried out safely and decisively maintaining the right distance from other vehicles and using the mirror, signals and gears correctly.

Hazard procedure and cornering: road and traffic hazards must be coped with properly, and bends and corners taken in the right manner.

Mirror: the mirror must be used frequently especially in conjunction with signals and before changing speed or course.

Signals: direction indicator, and hand signals when needed, must be given at the right place and in good time. The horn and headlamp flasher should be used as per the *Highway Code.*

Restraint: the driver should show reasonable restraint, but not indecision, at the wheel.

Consideration: sufficient consideration and courtesy should be shown to other road users.

Vehicle sympathy: the driver should treat the vehicle with care, without overstressing it by needless revving of the engine and by fierce braking.

Manoeuvring: manoeuvres (reversing) should be performed smoothly and competently.

Young Driver Training Scheme

Following measures introduced in the 1974 Road Traffic Act, a scheme was established for the training of lgv drivers from the age of 18 years. A National Joint Training Committee was set up through the RTITB (see note following) and it evolved a scheme for youngsters of 16 and upwards to be given progressive training for a category C driving entitlement which they are able to hold at 18, working up to a category C+E entitlement – for articulated vehicles. The course included periods of further education on a day-release basis at colleges.

However, due to a lack of continuing support, the scheme was mothballed from late on in 1993. For this reason, no further details will be included here until such time as it may, if ever, be revived.

The RTITB

In 1992, the Road Transport Industry Training Board (RTITB) was privatised by the government and became a completely independent commercial training organisation called RTITB Services (now renamed Centrex, ie Centres of Training Excellence). Haulage firms which were registered with the Board no longer have to pay an annual levy.

The Board's role in establishing training standards is taken over by the Road Haulage Industry Training Standards Council (RHITSC) which is to establish a career structure in the industry enabling employees to follow a recognised series of qualifications agreed with the National Council for Vocational Qualifications in progressing to positions at supervisory level.

Driver Training for Lorry Loaders

No mandatory requirements exist at the present time for goods vehicle drivers to hold certificates of competence to operate lorry-mounted cranes. However, under the Health and Safety at Work etc Act 1974 employers have a statutory duty to provide adequate instruction and safety training for all employees.

A voluntary certification scheme is run by the Construction Industry Training Board (CITB) to improve safety on construction sites. This scheme is strongly supported by the construction industry which may refuse entry to their own sites to non-certified lorry drivers. Mainly, the Board's scheme is concerned with ensuring a sound understanding of safety procedures for the use of a wide range of equipment including lorry-mounted cranes and skip loaders. Under the scheme, the Board will provide certification of existing lorry-loader and skip-loader operators who can show by means of employer confirmation that they are experienced in the use of such equipment.

Existing operatives who can produce the employer declaration can obtain the Board's safety certificate for lorry-mounted crane operation under a Grandfather Rights arrangement extending until June 1993. A similar arrangement will apply to skip-loader operatives until the end of 1994. Drivers who qualify by

Grandfather Rights have only to attend a short 'safety awareness' course to obtain their certificate. Newcomers seeking certification after the qualifying date will have to undergo more extensive training and testing to show that they can operate such equipment with complete safety.

Another scheme is run by the Contractors' Mechanical Plant Engineers (CMPE), although this requires no specific training, certification being based solely on employers' references. Training is also provided by most member firms of the Association of Lorry Loader Manufacturers and Importers (ALLMI). Additionally, the Association itself publishes a Code of Practice for safe application and use of loaders, and is currently preparing a training programme. Details are available from the Association at 38 Shaw Crescent, Sanderstead, South Croydon, CR2 9JA. Other organisations providing suitable lorry-loader training include the Freight Transport Association, the Road Transport Industry Training Board and a number of commercial training firms.

Driver Training for Dangerous Goods Carrying

Regulations which came into force on 1 July 1992 require drivers of dangerous goods carrying tanker and tank container vehicles (and later, those driving vehicles carrying dangerous goods in packages – see below) to hold ADR-style* training certificates issued by the Driver and Vehicle Licensing Agency, Swansea and gained by attending an approved course and passing a written examination set by the City and Guilds of London Institute (C&G). The Road Traffic (Training of Drivers of Vehicles Carrying Dangerous Goods) Regulations 1992, implement in the UK the requirements of EC Directive 89/684, and replace the training requirements set out in existing dangerous goods legislation (see Chapter 23).

*NB: ADR is the recognised abbreviation for the European Agreement Concerning the International Carriage of Dangerous Goods by Road to which the UK is party.

Relevant Vehicles

Specifically, the 1 July 1992 implementation date related to drivers of road tanker and tank container vehicles with a capacity exceeding 3000 litres, or those exceeding 3.5 tonnes permissible maximum weight operating under the Road Traffic (Carriage of Dangerous Substances in Road Tankers and Tank Containers) Regulations 1992 and to all vehicles carrying explosives under the provisions of the Road Traffic (Carriage of Explosives) Regulations 1989.

Drivers of vehicles carrying dangerous goods in packages under the Road Traffic (Carriage of Dangerous Substances in Packages) Regulations 1992 and radioactive substances under the Radioactive Substances (Carriage by Road) (Great Britain) Regulations 1974, had until 1 January 1995 to comply.

Transitional and Provisional Certification

Existing drivers of tanker vehicles who held Hazchem training certificates, or those who received dangerous goods training within the previous 5 years, could apply for a transitional ADR certificate which was valid both in the UK and internationally for a period of 5 years from the date of the training course or until 31 December 1994, whichever was the earlier.

Drivers who had not received such training but who had been employed continuously in dangerous goods driving during the previous 5 years could obtain a provisional certificate lasting until 31 December 1994, but this was valid only in the UK and did not permit the driver to undertake international journeys with dangerous goods.

Full Certification

Full certification of all drivers of all types of vehicles engaged in carrying dangerous goods by road was required under EC Directive 684/89 by 31 December 1994. This means that from 1 January 1995 it will be illegal for any persons to drive such a vehicle unless they have completed the necessary training, passed the relevant examinations (see below) and hold a vocational training certificate to this end.

NB: At the time of preparing this edition of the Handbook *the EU is proposing to extend the deadline for dangerous goods driver certification as described above from 31 December 1994 to the end of 1997 for tanker vehicle drivers and to the year 2000 for drivers of vehicles carrying packaged dangerous goods. This is because many EU member states are having difficulty in implementing the necessary training and certification arrangements (the UK included).*

Driver Responsibilities

The new-style ADR certificates will remain valid for a period of 5 years and will be renewable, subject to the holder attending an approved refresher course and taking a further examination. Drivers must carry the certificate with them when driving relevant vehicles and must produce it on request by police or a goods vehicle examiner. It is an offence to drive a dangerous goods vehicle without being the holder of a full or transitional certificate, or to fail to produce such a certificate on request.

Employer Responsibilities

It is the responsibility of the employer to ensure that dangerous goods drivers receive training so they understand the dangers arising from the products they are carrying and what to do in an emergency situation, and that they hold relevant full or transitional ADR certificates covering the vehicle being driven and the products carried. The employer has a duty to provide (and pay for) necessary training leading to the certificate and the official C&G examination.

Approved Training and the City and Guilds Examination

The Department of Transport has approved suitable establishments where training can be obtained and the C&G examination taken.

A syllabus has been established for the examination which involves both theoretical sessions and practical exercises. The examination itself comprises a core element designed to assess the candidate's practical and legal knowledge plus a specialist element for either road tanker and tank container drivers or packaged goods drivers (or both if required). Additionally, candidates will have to pass individual 'dangerous substance' examination papers which cover each of nine classes of dangerous goods, to test their specialist knowledge of the products they carry in their work.

Candidates who pass the examination (by achieving a pass mark of at least 75 per cent in each element – ie core, tanker/package and substance) will receive their certificate direct from the DVLA, Swansea. Those who fail can apply to re-sit the examination without further training within a period of 16 weeks from receipt of the notification of failure.

Further information on the training and examinations may be obtained from Area Offices of the Heath and Safety Executive (see local telephone directory).

8: Vehicle Excise Duty

Vehicle Excise Licences

All mechanically propelled vehicles, whether used for private or business purposes, which are used or parked on public roads in Great Britain must have, and display, excise licences (often called road fund licences or road tax) indicating that the appropriate amount of vehicle excise duty (VED) has been paid, unless they are being driven, by previous appointment, to a place where they are to have their annual test. It is an offence under the Vehicles (Excise) Act 1971 to use or keep an unlicensed vehicle on the road for any period, however short. In the event of a conviction for such an offence heavy fines may be imposed and back duty may be claimed. Operators' licence holders also jeopardise their 'O' licences when committing excise duty offences.

The present system of excise licences for most goods vehicles is based on gross weights and axle configurations while goods vehicles not exceeding 3500kg gross weight are in the same class as private vehicles which is called private/light goods (PLG) (see p 401 for full details).

Exemptions

Exemption from vehicle excise duty applies to:
1. Vehicles owned or operated by the Crown.
2. Vehicles used for fire brigade or ambulance services.
3. Road rollers.
4. Vehicles used for the haulage of lifeboats and lifeboat gear.
5. Vehicles travelling to or from a place where they are to have an annual roadworthiness test by prior appointment.
6. Vehicles carrying built-in road construction machinery.
7. Vehicles designed and built specially for spreading grit on snow and ice-covered roads.
8. Tower vehicles used for street lighting maintenance.
9. Local authority road watering and gulley cleaning vehicles.
10. Electric vehicles.
11. Pedestrian-controlled vehicles.

Six-Mile Exemption
In addition to the exemptions mentioned above there is one other important exemption from vehicle excise duty. This applies to vehicles intended to be used on a public road *only* when travelling from one place to another, both of which places must be occupied by the registered owner of the vehicle, and provided the distance travelled on the road is not more than 6 miles in a week.

In order to take advantage of this exemption, application has to be made to a local Vehicle Registration Office (VRO) on form V325/1 (see local telephone directory for address). Details must be given of the roads used, the location of the two points at either end of the journey, the actual distance and the number of trips per week. A special windscreen disc is issued showing the vehicle is exempt from VED.

Payment of Duty

Excise duty is payable once annually, or every 6 months. To assist the private motorist in saving towards vehicle excise duty licence stamps valued at £5 each can be purchased at certain post offices.

Application for initial registration or renewal of duty is made on the appropriate form as follows:

First registration	– Form VE 55/1 for new vehicles
	– Form VE 55/5 for re-imported vehicles
Renewal of duty	– Form V10 for all vehicles up to 3500kg
(12 or 6 months)	– Form V11 renewal reminder which enables vehicles to be re-licensed at a post office
	– Form V85 for heavy goods vehicles which can only be re-licensd at VROs.

A circular licence disc is issued by the Driver and Vehicle Licensing Centre (DVLC) – or a post office when renewing licences here – when the duty is paid and this must be displayed in the vehicle windscreen where it can be clearly seen from the near side. It is an offence to fail to display a current and valid excise licence disc.

Reduced, Extended Periods and Refunds
A vehicle may be licensed for only 6 months at a time if preferred but this method of licensing means that a higher amount of duty is paid annually. All licences are valid only from the first day of the month in which they come into force. Alternatively, new vehicles may be licensed part way through a month (but only at local VROs) with the duty period commencing on the 10th, 17th or 24th day of the month and continuing through to the end of the 12 month period from the first day of the following month.

In cases where a vehicle is only to be used for a shorter period than either 6 or 12 months or is taken out of service during the currency of the licence, a licence has to be taken out for one or other of these periods and then surrendered to the VRO (see local telephone directory for addresses of local VROs) when it is no longer required; the VRO will refund the duty for each complete month remaining on the licence. To gain a full month refund the licence must be surrendered at least by the last day of the previous month. Application for a refund should be made on form V14.

Rates of Duty

The rate of duty payable varies depending on the way in which the vehicle is constructed, the way that it is used and, in the case of light vehicles, its unladen weight, and for heavy goods vehicles, their gross weight and the number of axles.

Duty is categorised in tables as follows:

1. Private/light goods vehicles (ie not exceeding 3500kg gross weight)
   - rates for ordinary vehicles plus those registered by farmers and showmen.
2. Bicycles and tricycles (not over 450kg).
3. Hackney carriages.
4. General haulage vehicles.
5. Showmen's haulage vehicles.
6. Special machines (including agricultural machines, ploughing engines etc).
7. Trade licences.
8. Plateable rigid and articulated vehicles (not exceeding 12,000kg gross weight)
   - rates for ordinary vehicles plus those registered by farmers and showmen
   - up to 7500, and 7500 to 12,000kg.
9. Plateable articulated goods vehicles (exceeding 12,000kg gross train weight)
   - Table A, 2-axle tractive unit used with any semi-trailer (1, 2, 3 or more axles)
   - Table B*, 2-axle tractive unit used with 2- or more axled semi-trailers only
   - Table C*, 2-axle tractive unit used with 3- or more axled semi-trailers only
10. - Table D, 3-axle (or more) tractive unit used with any semi-trailer
    - Table E*, 3-axle (or more) tractive unit used with 2- or more axled semi-trailers only
    - Table F*, 3-axle (or more) tractive unit used with 3- or more axled semi-trailers only.

    (All tables contain rates for goods vehicles and for those registered by farmers and showmen.)

    *NB: Licences taken out at the concessionary rates for these vehicle combinations do not permit the use of semi-trailers with fewer axles – it would be an offence to do so.
11. Recovery vehicles.
12. Plateable rigid goods vehicles (exceeding 12,000kg gross weight)
    - rates for ordinary vehicles plus those registered by farmers and showmen
    - rates for vehicles with 2 axles, 12,000kg to 17,000kg
    - rates for vehicles with 3 axles, 12,000kg to 27,000kg
    - rates for vehicles with 4 or more axles, 12,000kg to 32,000kg.
13. Trailers
    - rates for trailers 4000kg to 12,000kg gross weight (ie towed by vehicles over 12 tonnes gvw).
    - rates for trailers over 12,000kg gross weight (ie towed by vehicles over 12 tonnes gvw).
    - rates for ordinary trailers plus those towed by vehicles registered by farmers and showmen

14. Non-plateable vehicles and vehicles not subject to testing, and 'restricted' vehicles
    – rates for ordinary vehicles and those registered by farmers and showmen
    – single rate only for vehicles falling in this category.
15. Special Types vehicles – for movement of indivisible loads.

Articulated Combinations

In the case of articulated tractive units which are used with a variety of trailers, the determining factor for taxation purposes is the number of axles on the trailer likely to be used with it. This is so that the maximum amount of duty is paid to prevent a vehicle operating on the road at a rate less than that which is applicable. In general terms, the system operates so that the fewer the axles the greater the duty payable and vice versa. This is because the duty relates to the road damage caused and greater damage arises with fewer axles. In consequence of this it is illegal to operate a vehicle on a road for which the incorrect (ie insufficient) excise duty has been paid.

Trailers

Where drawbar trailers are drawn, if the gross weight of the trailer exceeds 4 tonnes and the gross weight of the towing vehicle exceeds 12 tonnes, additional duty is payable in accordance with published duty tables but not otherwise. So, if the towing vehicle or the trailer falls below these weights no trailer duty is payable.

Goods Carrying Vehicles

The goods vehicle rate of duty is payable if the vehicle is built or has been converted to carry goods, and is actually used to carry goods in connection with a trade or business. In general this rate of duty applies to all types of lorries, vans, trucks, estate cars, dual-purpose vehicles and also passenger vehicles if they are converted for carrying goods. If, however, dual-purpose vehicles (see below) and goods vehicles are never used for carrying goods in connection with a trade or business they may be licensed at the private/light goods (PLG) rate of duty which is currently £130 per year.

New Rates for Increased-Weight Goods Vehicles from 1 January 1993

Increases in vehicle weights for certain rigid and articulated configurations were introduced from 1 January 1993 for which three new individual rates of duty are applied as follows:

3-axle rigid vehicle with permissible maximum weight of 26,000 kg – duty rate £2260;

4-axle rigid vehicle with permissible maximum weight of 32,000 kg – duty rate £4250;

4-axle articulated vehicle with permissible maximum weight of 35,000 kg – duty rate £5000.

Driver Training and Non-Goods Carrying Vehicles

Heavy goods vehicles which are used exclusively for driver training purposes may be licensed at the private rate of duty even if they carry ballast (eg concrete blocks) to simulate driving under loaded conditions – *but not other loads*. Similarly, other goods vehicles used for private purposes (ie carriage of goods but not in connection with a trade or business – eg privately used horse boxes) may also be licensed at the private rate of duty – currently £130.

Private Rate Taxation of Heavy Vehicles on International Work
The use of heavy vehicles for goods carrying in the UK or within other EU member states while taxed only at the private/light goods rate of duty (PLG) is illegal both in the UK and in Europe (see also section on penalties p 148). Under the provisions of bilateral agreements and under International Conventions on the Taxation of Road Vehicles, vehicles on which the correct rate of vehicle excise duty has been paid in their country of registration are exempt from payment of further duty on being temporarily imported into the territories of other parties to the agreements. Where the correct duty is not paid in the 'home' country, there is a liability for payment of additional duty in each other country through which the vehicle travels.

Dual-Purpose Vehicles
For the purposes of the regulations a dual-purpose vehicle (as referred to above) is defined as a vehicle, built or converted to carry both passengers and goods of any description, which has an unladen weight of not more than 2040kg and has either four-wheel drive or:
(a)   has a permanently fitted roof
(b)   is permanently fitted with one row of transverse seats (fitted across the vehicle) behind the driver's seat (the seats must be cushioned or sprung and have upholstered back-rests)
(c)   has a window on either side to the rear of the driver's seat and one at the rear.

The majority of so-called estate cars, shooting-brakes, station wagons, hatchbacks, certain Land Rovers and Range Rovers are dual-purpose vehicles under this definition. It should be noted, however, that vehicles used for *dual operations* are not dual-purpose vehicles in terms of the legal requirements.

Special and Reduced Rates of Duty

Restricted HGVs
This class is intended to cover goods vehicles exceeding 3500kg gross weight which are exempt from the plating requirements and those which are purpose-built for specialist uses other than being specific goods vehicles. When such vehicles are first registered, form E100 must be completed and sent to the Goods Vehicle Centre at Swansea together with supporting documentation as evidence of their 'special' purpose. The Vehicle Inspectorate will then decide if the vehicle may be registered in this class so as to gain exemption from the normal goods vehicle rates of duty.

Recently the DVLA has been clamping down on operators claiming exemption within this class to ensure that only those vehicles genuinely falling within the category gain the goods vehicle duty exemption.

Farmers' Vehicles

Goods vehicles registered in the name of a person engaged in agriculture and used on public roads only for carrying the produce of, or materials for, the agricultural land he occupies and for no other purpose, can be licensed at a reduced rate of duty.

These reduced rate licences, known as 'Farmers' or 'F' licences, are issued conditional on the vehicle being used for no purposes other than those stated, except that there is a concession that allows the holder of an 'F' licence to carry the produce of, or articles for, another farmer provided that this is only done occasionally, that the goods carried represent only a small proportion of the total goods carried on the vehicle, and that no payment is asked and no reward is given.

Travelling Showmen

Goods vehicles registered in the name of a travelling showman, and used only for his business, and vehicles fitted with a living van type of body or some special type of body or superstructure, are subject to a special rate of duty. It has been stressed by the DoT that heavy vehicles taxed at the showman's rate of duty must not be used for the carriage of general haulage loads; the goods carried must be directly related to the occupation of a travelling showman.

Works Trucks and Mobile Cranes

Works trucks and trailers, agricultural machines, small dumpers, mobile plant, fork-lift trucks and mobile cranes are all subject to special rates of duty. If a works truck or fork-lift truck is used on a public road only when travelling from one set of premises to another set of premises occupied by the same person or company and it does not travel on the road more than 6 miles in a week, then no licence duty has to be paid. If, however, such vehicles are used on public roads for more than 6 miles in a week or if they are used on a road for loading or unloading, the rate of duty payable is currently £35 per year. Mobile cranes may be licensed at the £35 rate provided they do not carry any load on the road except equipment used in connection with the crane assembly. The heavier types of dumper can be licensed at the £130 private rate only if they are used on roads when travelling empty between sites.

Mobile Engineering Plant

Other types of mobile engineering plant such as mowing and digging machines, excavators, agricultural machines and mobile compressors may be licensed at the £35 annual rate.

Electric Vehicles

Electric vehicles have been exempted from vehicle excise duty altogether since the 1980 Budget.

Registration and Renewal of Duty

A new vehicle has to be registered with local VROs or the DVLC at Swansea and a registration number has to be obtained for the vehicle. The number has to be displayed on a plate at both the front and rear of the vehicle.

Form VE 55, which is used for motor vehicle registrations, consists of a single sheet which will be used for first licensing and registration purposes. Attached to it is a two-sheet section with carbon paper inserts, so that the official details of the vehicle and the dealer's name and town (but nothing more) will copy through on to those sheets. These will be sent to an agency acting for the motor industry which will also act for the DoT in the assembly of official vehicle registration statistics.

A feature on these additional sheets is a voluntary statistical section, in which applicants are invited to supply, if they wish to do so, information which will be of value for government and motor industry statistical purposes. It includes questions about the purchaser's occupation and previous vehicle and about the main use to which he intends to put the new vehicle. The information given will help the industry to obtain a better knowledge of the market and therefore to give the best possible service to its customers. Also, if the purchaser chooses to give his name and address, this will enable the manufacturer to get in touch with him direct over any safety matters which may arise. Any information given in this section will be treated as confidential.

Most of the main vehicle manufacturers and importers will have entered the details of each vehicle on the form before it reaches the dealer. This will have two advantages, both for motor dealers and for the purchaser himself. In the first place, they will no longer have to fill in the details themselves, and, in the second place, the association of one form with one vehicle right from the start is intended to make it more difficult to register a stolen vehicle as new. If the manufacturer or importer has not completed the form in advance, copies are available on demand at licensing offices.

Registration of Trailers
The Vehicle Inspectorate is reported to be considering a plan to formally register trailers in an effort to combat vehicle and load theft.

Documents
Besides the completed form V55, a certificate of insurance and a copy of the supplier's invoice, a Type Approval Certificate (TAC) is required for goods vehicles which are subject to the type approval regulations, or alternatively a Minister's Approval Certificate (MAC); these documents must be produced when first registering a goods vehicle with form V55 and with form V205 if the vehicle is exempt from plating and testing, if a concessionary rate of duty is applicable or if it is to be used with a trailer when the additional trailer duty is payable (ie over 12 tonnes and 4 tonnes gross weight respectively).

Issue of Registration Document/Licence Disc
The DVLC, on receipt of the above mentioned items, will issue a licence disc and a registration document, form V5, for the vehicle showing its registration number and details of the vehicle such as make, type, type of body (van or tipper, for example), engine number, chassis number, colour, method of propulsion (petrol, heavy oil or battery) and the gross weight.

Police officers and certain officers of the DoT may request production of the registration document for inspection at any reasonable time.

Renewal of Licences

When the licence expires after 6 months or 12 months it has to be renewed by completing renewal form V10 or form V85. Renewal reminders are sent out from the DVLC on form V11 and this may be used as an alternative to form V10 to renew the licence at certain post offices (provided the vehicle is not over 3500kg gross weight where form V10 is used, there is no change in ownership or address which has not been recorded and no change to the vehicle or its use) or at local VROs. Form V85 is used for vehicles over 3500kg gross weight. The vehicle registration document, the appropriate fee, a valid certificate of insurance and, if the vehicle is subject to annual testing, a current test certificate, must accompany the application.

Where to Apply

When registering a vehicle for the first time, application must be made to the local VRO, or to the DVLC at Swansea SA99 1AR.

Applications for renewal of duty may be made as follows:

1. If you have the official licence renewal reminder, form V1 (providing there is no change in the details on the form)

   In person to main post offices or VROs, or by post to VROs

2. If you have form V11 but there are changes in the details

   In person or by post to VROs

3. If the vehicle is a heavy goods vehicle for which no form V11 has been received

   In person or by post to VROs.

Replacement Licences and Discs

Duplicates or replacements for lost or defaced vehicle registration documents or windscreen discs are available on application, using form V20 at VROs. They cost £5.50 each (except in certain exceptional cases – eg if stolen with the vehicle, if for a VED exempt vehicle, if lost in the post). As a security measure, when future application is made for a replacement registration document, the DoT may contact the previous vehicle owner to ensure that the applicant is entitled to have possession of the vehicle and the registration document.

Alteration of Vehicles

If a vehicle is altered during its life by adding or removing equipment, by changing the type of body or even the colour of the vehicle, by changing the plated weight or increasing the unladen weight by fitting a heavier body or heavier components, the DVLC must be advised of the changes at once. If the changes mean that the vehicle goes into a higher weight range, then the additional licence duty becomes payable from the date of the changes.

If a van is converted to carry passengers or has side windows fitted to the body behind the driver's seat the owner becomes liable to pay car tax to HM Customs and Excise. The amount of tax payable is based on the current wholesale value of the vehicle. Any person making such a conversion must report the fact to the Customs and Excise authorities immediately.

Sale of Vehicles

When a vehicle is sold, the seller must notify the DVLC by completing the bottom tear-off portion of the registration document (Form V5), which is perforated for this purpose, with the name and address of the new owner of the vehicle. It is an offence to fail to notify a change of vehicle ownership (maximum fine £1000) and it can lead to the original owner being prosecuted for offences committed with the vehicle by the new owner and leaving the previous owner to pay any fixed penalty fines incurred. The new owner has to fill in his name and address in the changes section of the registration document and send the document to the DVLC for registration in his name. If a vehicle is sold for scrap or is broken up, the registration document has to be sent to the DVLC with a note advising them of this.

Production of Test Certificates

A valid goods vehicle test certificate has to be produced at the time of re-licensing any goods vehicle over 1 year old (ie vehicles over 3500kg gross weight and articulated vehicles which are subject to the goods vehicle annual test), and a light vehicle or private car type (ie MOT) test certificate when re-licensing any vehicle not over 3500kg gross weight which is over 3 years old and is subject to the light vehicle annual test scheme.

Vehicles Exempt from Plating and Testing

When applying to license or relicense goods vehicles which are exempt from plating and testing, a declaration has to be made on form V112G (available from VROs) to cover the non-production of a valid test certificate.

Penalties and Payment of Back Duty

Heavy fines may be imposed for vehicle excise duty offences (see below), particularly non-payment of duty, payment at an incorrect (ie too low) rate of duty or payment by means of a cheque which defaults (see also section on p 144). Additionally, the offender may be ordered to pay an amount of back duty. Previously the amount of back duty payable could be reduced if the offender could prove non-use of the vehicle during any particular month, but this provision has been rescinded.

New Scheme for Tax Dodgers

The DVLA has introduced a new national scheme to crack down on vehicle tax dodgers. Conspicuous red warning notices are to be placed on the windscreens of vehicles not displaying a current and valid excise licence disc and details of the vehicle sent to the DVLC. Persistent tax offenders are warned by the DoT that they face fines of up to £1000 for each offence or *five times the annual rate of duty, whichever is the greater*. For maximum capacity articulated vehicles this could mean a penalty of up to £25,000.

Data Protection

In accordance with the provisions of the Data Protection Act the DVLC at

Swansea is registered as a 'Data User' and as such must make available, on request, details held on file concerning individual persons. Such information is normally shown on driving licences and vehicle registration documents but an enquiry to establish details held on file can be made (on payment of a fee) to the Vehicle Enquiry Unit (Data Protection Queries), DVLC, Swansea SA99 1AN.

Trade Licences

Trade licences (trade plates) are available for use by motor traders and vehicle testers to save them the inconvenience of having to license individually every vehicle which passes through their hands.

Fees and Validity

A trade licence can be obtained from the local VRO in whose area the applicant has his business. It is valid either for 1 year or for 6 months (licences are issued on 1 January and 1 July) and the licence fee is £130 for 1 year and £71.50 for 6 months. Replacements for lost or defaced trade licences cost £11 per set or £6.50 for the plate containing the licence and £4.50 for other plates.

Display of Trade Licences

The VRO will, when they have approved the application, issue a pair of special number plates with the registration number in red letters on a white background. One of the plates has a licence holder attached to it and the licence affixed (it does not have to be displayed on the windscreen); this plate must always be carried at the front of the vehicle on which the plates are being used. The other plate is carried at the rear.

Issue of Licences

There are considerable restrictions on the issue and use of trade licences – as already mentioned they are issued only to:

(a) motor traders, defined as 'manufacturers or repairers of, or dealers in mechanically propelled vehicles' (this also includes dealers who are in business consisting mainly of collecting and delivering mechanically propelled vehicles), plus those who modify vehicles (eg by fitting accessories) and those who provide valet services for vehicles who may use the licence for all vehicles which are from time to time temporarily in their possession in the course of their business as motor traders;

(b) vehicle testers for all vehicles which are from time to time submitted to them for testing in the course of their business as vehicle testers.

Vehicles such as service vans or general run-about vehicles owned by motor traders cannot be used under trade licences; the full rate of duty has to be paid for such vehicles.

Use of Trade Licences

The following are the purposes for which vehicles operated under a trade licence may be used by a motor trader or vehicle tester:

1. For test or trial in the course of construction or repair of the vehicle or its accessories or equipment and after completing construction or repair.
2. Travelling to or from a weighbridge to check the unladen weight or travelling to a place for registration or inspection by the Council.
3. For demonstration to a prospective customer and for travelling to or from a place of demonstration.
4. For test or trial of the vehicle for the benefit of a person interested in promoting publicity for the vehicle.
5. For delivering the vehicle to a purchaser.
6. For demonstrating the accessories or equipment to a prospective purchaser.
7. For delivering a vehicle to, or collecting it from, other premises belonging to the trade licence holder or another trader's premises.
8. For going to or coming from a workshop in which a body or equipment or accessories are to be, or have been, fitted or where the vehicle is to be or has been valeted, painted or repaired.
9. For delivering the vehicle from the premises of a manufacturer or repairer to a place where it is to be transported by train, ship or aircraft or for returning it from a place to which it has been transported by these means.
10. Travelling to or returning from any garage, auction room or other place where vehicles are stored or offered for sale and where the vehicle has been stored or offered for sale.
11. Travelling to a place to be tested (and return), dismantled or broken up.

It should be noted that the use of a vehicle on trade plates does not exempt the driver or operator from the need to ensure that it is in sound mechanical condition when on the road, even if being driven for the purposes of road testing or fault finding prior to repair or after repair. The police will prosecute if they find trade licensed vehicles on the road in an unsafe or otherwise illegal condition.

Carriage of Goods on a Trade Licence

Goods may only be carried on a vehicle operating under a trade licence:

1. When a load is necessary to demonstrate or test the vehicle, its accessories or its equipment – the load must be returned to the place of loading after the demonstration or test unless it comprised water, fertiliser or refuse.
2. When a load consists of parts or equipment designed to be fitted to the vehicle being taken to the place where they are to be fitted.
3. When a load is built in or permanently attached to the vehicle.
4. When a trailer is being carried for delivery or being taken to a place for work to be done on it.
5. If the goods are another fully licensed vehicle being carried for the purpose of travel from or to the place of collection or delivery (ie the driver's own transport to get him out or back home).

Carriage of Passengers on a Trade Licence

The only passengers who are permitted to travel on a trade licensed vehicle are:

1. The driver of the vehicle, who must be the licence holder or his employee
    - other persons may drive the vehicle with the permission of the licence holder but they must be accompanied by the licence holder or his employee (this latter proviso does not apply if the vehicle is only constructed to carry one person).
2. Persons required to be on the vehicle by law, a statutory attendant for example.
3. Any person carried for the purpose of carrying out his statutory duties of inspecting the vehicle or trailer.
4. Any person in a disabled vehicle being towed including persons from the disabled vehicle being carried provided this is not for hire or reward.
5. A prospective purchaser or his servant or agent.
6. A person interested in promoting publicity for the vehicle.

NB: It is illegal for transport fleet operators to road test their own vehicles on trade plates. This has been established on the grounds that the vehicles are not 'temporarily' in their possession and are therefore outside the permitted terms of a trade licence use.

Recovery Vehicles

Since 1 January 1988 a recovery vehicle is no longer permitted to undertake recovery operations on trade plates. A separate class of VED at an annual rate of duty of £85 and £46.75 for 6 months (currently) applies to these vehicles. Any vehicle used for recovery work which does not conform to the definition given below must be licensed at the normal goods vehicle rate according to its class and gross weight.

Definition of Recovery Vehicle

For the purpose of this new taxation class, a recovery vehicle is one which is 'either constructed or permanently adapted primarily for the purpose of lifting, towing and transporting a disabled vehicle or for any one or more of those purposes'.

A vehicle will no longer be a recovery vehicle under the regulations (ie Vehicles (Excise) Act 1971, Schedule 3, as amended by the Finance Act 1987) if at any time it is used for a purpose other than:

(a)   the recovery of a disabled vehicle (this has been extended to two vehicles);
(b)   the removal of a disabled vehicle from the place where it became disabled to premises at which it is to be repaired or scrapped;
(c)   the removal of a disabled vehicle from premises to which it was taken for repair to other premises at which it is to be repaired or scrapped;
(d)   carrying any load other than fuel and any liquids required for its propulsion and tools and other articles required for the operation of, or in connection with, apparatus designed to lift, tow or transport a disabled vehicle.

NB: It has been held in the High Court (DPP v Yates 1988) that a recovery vehicle towing a partly suspended broken down vehicle is not carrying a load.

Plating and Testing

Recovery vehicles licensed under the recovery vehicle taxation class are not exempt from goods vehicle plating and testing unless they satisfy the definition of a 'breakdown vehicle'.

Operation of Recovery vehicles

Licensed recovery vehicles are exempt from 'O' licensing, the EU drivers' hours rules and the tachograph requirements but those persons who drive them must comply with the British Domestic driving hours rules (see pp 59–60 for details).

Rebated Heavy Oil

Commercial vehicles powered by diesel (heavy oil) engines must use diesel fuel on which the full rate of duty has been paid. A lower rate of duty is payable on fuel used for purposes other than driving road vehicles, such as driving auxiliary equipment, for contractors' plant which does not use public roads, bench testing of engines and space heating. Fuel on which the lower rate of duty has been paid is known as rebated heavy oil but is more commonly called gas oil or red diesel.

Rebated heavy oil must be marked, when delivered from bonded oil warehouses, with a red dye so that its use can easily be detected, and the supplier must deliver to the recipient a delivery note bearing a statement that the oil is 'not to be used as road fuel'. If both rebated and unrebated oils are stored in the same place a notice bearing the same wording must be placed at the outlet of the rebated oil supply.

Road fuel testing units staffed by officers of Customs and Excise operate throughout the UK to test fuel in vehicles and in storage tanks. Under the Hydrocarbon Oil Regulations 1973, Customs and Excise officers are empowered to examine any vehicle and any oil carried in it or on it and may also enter and inspect any premises and inspect, test or sample any oil on the premises whether the oil is in a vehicle or not. Vehicle owners and drivers must give the officers facilities for inspecting oils in vehicles or on premises.

The following vehicles may use rebated heavy oil as fuel. All other vehicles must use unrebated (full-duty paid) oil at all times:
- Vehicles not used on public roads and not licensed for road use
- Road rollers
- Road construction machinery (vehicles used or kept on a road solely for carrying built-in road construction machinery)
- Vehicles exempted from excise licence duty which use public roads for not more than 6 miles in a week
- Agricultural machines
- Trench digging and excavating machines

- Mobile cranes
- Mowing machines
- Works trucks.

Heavy penalties, including fines and repayment of duty, are imposed on offenders convicted of using illegal diesel fuel.

It should be noted particularly by international hauliers that use of 'red diesel' (ie untaxed diesel fuel) carried in reserve or 'belly' tanks and used once outside the UK is an illegal practice (ie not the *carrying* of,* but the *use* of, such fuel). Although Customs checks abroad on vehicle fuel tanks are limited, any operator found running on such fuel within the EU is likely to face heavy penalties.

NB: It is being reported that the French authorities are clamping down on the use of belly tanks on international road haulage vehicles – in fact what the French want is for the additional fuel to be carried as hazardous goods subject to the full requirements of ADR (see p 137). A number of UK operators have been heavily fined in France for carrying such tanks and the DoT's current advice is to avoid France if possible when running with these tanks.

Similarly, supplies of so-called 'green' diesel have become available from the Irish Republic. This fuel is to be regarded in the same way as 'red' diesel. In other words, it is illegal for normal goods vehicle use on UK roads and in Europe.

9: Insurance – Vehicle, Premises and Business

Owners and operators of motor vehicles using the public highway must insure against third-party* injury and passenger claims. Further, an essential part of any investment in property (buildings, vehicles, plant, etc) is to obtain protection by insurance against loss or damage by theft, fire or any other eventuality. It is also wise to be protected against claims made by third parties for compensation following injury to themselves or damage to their property as a result of some occurrence involving you, your employees, your property, or taking place on your premises.

The insurance company is the first party; the insured person(s) is the second party, and anybody else involved (particularly if they make a claim for compensation) is termed the third party.

Motor Vehicle Insurance

Third-Party Cover

The Road Traffic Act 1988 (s143) requires that all motor vehicles, except invalid carriages and vehicles owned by local authorities or the police, used on a road must be covered against third-party risks. This can be achieved by means of an insurance policy or, alternatively, by a deposit of £500,000 in cash or securities to the Accountant-General of the Supreme Court but this is only applicable if authorisation is granted by the Secretary of State for Transport. Normally, such authorisation is only granted to public bodies and authorities and to major organisations with access to the substantial funds which may be needed to meet major accident claims.

Where the cover is obtained by conventional insurance means the Road Traffic Act 1988 stipulates that cover is only valid if taken out with insurers who are members of the Motor Insurers' Bureau (MIB) which is a body established to meet claims for compensation (in respect of death or personal injuries only) by third parties involved in accidents with motor vehicles which subsequently prove to be uninsured against third-party risks.

Sections 145(3a) and (3c) of the 1988 Act state that the insurance policy:

> . . . must insure such person, persons or classes of persons as may be specified in the policy in respect of any liability which may be incurred by him or them, in respect of the death or bodily injury to any person caused by, or arising out of, the use of the vehicle on a road . . . (and) . . . must also insure him or them in respect of any liability which may be incurred by him or them . . . relating to payment for emergency treatment.

The emergency treatment referred to is that provided at the scene of an accident by a doctor or hospital authority and that provided by and charged for by a hospital for in-patient or out-patient care.

Passenger Liability

Passenger liability insurance cover for motor vehicles is compulsory. This requirement applies to all vehicles which are required by the Road Traffic Act to have third-party insurance and the cover must extend to authorised passengers (other than employees of the insured who are covered separately by the compulsory employers' liability insurance), other non-fare paying passengers and also to what may be termed 'unauthorised passengers' such as hitch-hikers and other people who are given lifts.

Unauthorised Passengers
The display in a vehicle of a sign which says 'No passengers . . .' or 'No liability . . .' does not indemnify a vehicle operator or driver from claims by so-called 'unauthorised' passengers who may claim for injury or damage received when travelling in or otherwise in connection with the vehicle resulting from the driver's or vehicle operator's negligence. The law ensures that such liabilities are covered within the vehicle's policy on insurance.

Property Cover

In accordance with an EU Directive (EC 5/84), from 1 January 1989 all UK motor insurance policies have been required to cover liability for damage to property (up to a maximum liability of £250,000 arising from one accident or a series of accidents from one cause). Damage to property in this context includes that caused by the weight of the vehicle (eg to road surfaces, paving slabs etc) and by vibration which may damage services (eg gas and water mains, gullies and sewers, telephone cables etc) below the road surface and third parties whose property is damaged in a vehicle accident have the right to request details of the vehicle insurance.

Certificate of Insurance

A policy of insurance does not provide the cover required by the Act until the insured person or organisation has in their possession a Certificate of Insurance. Possession meaning, in this context, exactly what it says: 'promised' or 'in the post' is not sufficient to satisfy the law. The policy itself is not *proof* of insurance cover it only sets out the terms and conditions for the cover and the exclusion and invalidation clauses.

The Certificate (or a temporary cover note proving cover until the Certificate is issued) which is *proof* (or evidence) of cover must show the dates between which the cover is valid, give particulars of any conditions subject to which the policy is issued (eg the permitted purposes for which the vehicle may be used and those which are not permitted) and must relate to the vehicles covered, either individually by registration number or by specification and to the persons who are authorised to drive them.

Production of Insurance Certificate

It is necessary to produce a current Certificate of Insurance when making application for a excise licence (road tax) for a vehicle. Alternatively, a temporary cover note may be produced and this will be accepted, but the insurance policy itself is not acceptable.

The owner (ie registered keeper) of a motor vehicle must produce a Certificate of Insurance relating to the vehicle if required to do so by a police officer. If he is not able to produce the Certificate on the spot, or if an employed driver is required to produce a Certificate of Insurance for the vehicle he is driving, it may be produced for inspection, no later than *7* days from the date of the request by the police officer, at any police station which the owner or driver, chooses. The person to whom the request is made does not have to produce the Certificate personally, but may have somebody else take it to the nominated police station for him. A valid temporary cover note would suffice instead of the Certificate if this has not yet been issued.

Duty to Give Information

If requested to do so, the owner of a vehicle must give the police any information they ask for to help determine whether on any particular occasion a vehicle was driven without third-party insurance cover in force. The owner must also give information about the identity of a driver who may at any time have been driving a vehicle which is registered in his name, or information which may lead to identification of a driver if he is asked to do so by the police.

When the vehicle or vehicles concerned in such a request are the subject of a hiring agreement, the term 'owner' for the purposes of these insurance provisions includes each and every party to the hiring agreement.

Invalidation of Cover

Insurance cover may be invalidated and claims refused if policy conditions are not strictly adhered to. In particular these circumstances may arise if the vehicle is operated illegally (for example, in excess of its permissible weight, without a valid test certificate, in an unsound mechanical condition, outside the terms of an 'O' licence, with an incorrectly or unlicensed driver or one who is disqualified, or if replacement components fitted to the vehicle are not to manufacturer's specification).

By way of example, a case was reported where liability was rejected by an insurance company when a fast sports saloon motor car was fitted with tyres not approved for the top speed of which the car was capable (well in excess of 100mph), although the claim arose out of an accident at less than 30mph.

It is important to stress the need to examine carefully all the clauses contained in a motor insurance policy and to take steps to avoid any action which may invalidate the policy. The employment of unlicensed or incorrectly licensed drivers or the use of unroadworthy vehicles (ie vehicles which do not comply with legal requirements or are found to be on the road in a dangerous condition) are two examples of the most likely ways of invalidating a motor insurance policy. Similarly, the policy should cover *all persons* who may be required (or may need in an emergency) to drive vehicles, not just employees.

Use of Unfit Drivers

It is important not to use drivers who are, or who are believed to be, medically unfit to drive. Insurance companies have a duty to notify the Secretary of State for Transport of the names and addresses of people refused insurance cover on medical grounds so their driving licences can be withdrawn.

Payment to Travel

Previously, motor insurance cover could be invalidated if passengers paid towards the cost of car-running expenses. As a result of provisions in the Transport Act 1980, the receipt of travel expenses contributions from passengers in private cars does not invalidate insurance policies provided no profit is made.

Cancellation of Insurance

When an insurance policy is cancelled, the Certificate of Insurance – there may be one or more depending on the number of vehicles covered by the policy – relating to that policy must be surrendered to the insurer within 7 days of the cancellation date.

Cover in EU Countries

It is a requirement that every motor insurance policy issued in an EU country, including Britain, must include cover against those liabilities which are compulsorily insurable under the laws of every other EU member state (Article 7 (2) of the EEC Directive on Insurance of Civil Liabilities arising from the use of Motor Vehicles (No 72/166/CEE)) and some non-EU states. Motor policies issued in the UK contain provisions for such cover but this only provides very limited legal minimum cover and, while an international motor insurance 'green card' is no longer essential to enable EU member state boundaries to be crossed by vehicles (private or commercial), it is wise to obtain a green card when travelling or sending goods vehicles abroad, in order to obtain the much wider cover provided by the policy. Possession of a green card provides adequate evidence of insurance when abroad and it also eliminates problems of language and different procedures in foreign countries.

Cover in Non-EU Countries

The EU insurance arrangements described above have been extended to five non-EU countries in Europe – Austria, Finland, Norway, Sweden (*NB: These four countries are due to become EU member states from 1995*) and Switzerland. Vehicles from these countries are exempt from checks on their insurance documents at UK ports of entry and drivers of British vehicles will enjoy reciprocal arrangements in these five countries as well as in the other 11 EU member countries (ie France, Belgium, Holland, Germany, Luxembourg, Italy, Denmark, Eire, Greece, Spain, Portugal).

British motor policies have been extended to cover vehicles in all 16 countries so that they meet the national law of each country on motor insurance. In their own interests, drivers should continue to carry either a green card or their British Insurance Certificate when travelling abroad because, although they will not be subjected to routine border crossing checks for insurance, they may be required to produce evidence of insurance in the event of an accident.

International Accident Report Form

A special accident report form has been devised by the European Insurance Committee (CPA) for use as an agreed statement and accident report to be completed by drivers at the time of an accident when travelling in a foreign country (ie a country other than that in which the vehicle is insured). The use of such a form (available from insurers), particularly when dealing with persons from other countries who cannot speak your language, can help to resolve matters later.

Fleet Insurance

Most large fleet operators obtain insurance cover on a 'blanket' basis. Under this arrangement vehicles are not specified on the Certificate of Insurance by registration number but there is a statement on the Certificate to the effect that cover is provided for any vehicle owned, hired or temporarily in the possession of the insured person or company. With blanket insurance it is normal to advise the insurance company by means of a quarterly return of the registration numbers of all vehicles added to or deleted from the fleet strength during that period.

The basic insurance premium is calculated on the total fleet at the beginning of the policy year and adjustments are made by the insurance company issuing debit or credit notes as necessary following receipt of the quarterly returns. This system saves the insurance companies having continually to issue and cancel cover notes and Certificates for vehicles in fleets where there may be many changes during a year because of staggered replacement programmes. The insured company also benefits by not having to get in touch with their insurers every time a vehicle is obtained or disposed of and, further, by being able to obtain an excise licence for a new vehicle without having to wait to receive a cover note from the insurance company.

Additional Vehicle Insurance Cover

Extended Cover

The minimum cover against third-party risks mentioned above is not sufficient protection for the owner of a vehicle in the event of it being involved in an accident, damaged in any way (eg by vandals or by another vehicle when the driver was not present) or stolen. To obtain extra protection against such contingencies it is necessary to extend the insurance cover beyond the third-party legal minimum. This can be done in varying stages depending on what the vehicle owner considers necessary for his purpose. The basic policy can be extended to cover loss of the vehicle or damage to it as a result of fire or theft. The insurance can be further extended to give comprehensive cover which provides protection against third-party claims, fire and theft risks and accidental damage to the vehicle itself.

Loading and Unloading Risks

Goods vehicle insurance policies should include clauses which give protection against claims arising from the loading or unloading of vehicles or the activities of employees engaged on such work.

Loss of Use

Most motor vehicle policies do not include cover for the loss of use of a vehicle or for the hire of a replacement vehicle following an accident. If the accident proves to be the fault of the third party, a claim has to be made against him for the loss of use or the hiring charges incurred but such claims are often difficult to substantiate (particularly the value of loss of use of the vehicle) and may result in only meagre awards. An extension to the policy covering such eventualities is the most satisfactory means of protection against this type of loss.

Mechanical Failure

Mechanical failure is another item which is not normally included in motor insurance policies and generally it is not possible to obtain this type of cover for motor vehicles (although it is for some items of heavy engineering plant). Damage to engines caused by frost is covered in the majority of commercial vehicle insurance policies although there are certain qualifications. It is necessary, for example, if a claim is to be met, for the vehicle to have been sheltered in a properly constructed garage between specified hours of the night. It is a condition of all policies that all reasonable steps should be taken to safeguard the vehicle from such loss or damage, and this clause particularly is one which the insurance company can use to escape a claim if it feels the policy conditions have not been complied with.

Windscreen Breakage

Insurance companies normally provide cover for windscreen breakage within the standard motor insurance policy. Claims made for broken windscreens are generally limited to a fixed amount but are paid to the policy holder without detriment to any existing no-claims bonus and irrespective of whether or not an 'excess' clause is in force on the policy. Similar cover applies on most goods vehicle policies providing for the cost of replacement of the broken windscreen and for repairs to paintwork damaged by the broken glass.

Towing

Insurance cover for towing a vehicle which has broken down is normally provided under a goods vehicle policy, but the cover does not extend to damage caused to the vehicle while it is being towed, or to loss or damage of any goods being carried by the broken-down vehicle.

Damage by Weight

A goods vehicle insurance policy should provide cover against claims for damage caused to roads, bridges, manhole covers and such like by the weight of the vehicle passing over them. Some policies have a limit on the maximum liability acceptable for damage to property and this amount should be checked to ensure that it is adequate to meet likely claims in this respect.

Defence Costs

A motor insurance policy can be extended to cover legal costs incurred in defending a driver faced with manslaughter or causing death by reckless driving charges.

Goods in Transit Insurance

A motor vehicle insurance policy does not provide cover for claims made for damage or loss to goods carried on or in the vehicle. Goods in Transit (GIT) insurance cover is needed to provide protection for this eventuality.

Most GIT insurance policies provide cover in accordance with the limits included in the Road Haulage Association Conditions of Carriage which is normally a maximum liability of £1300 per tonne for goods carried within the UK. If goods are of relatively low value (bulk traffics, such as coal, gravel and other excavated materials, for example) a lower limit of liability and consequently a lower premium can be considered. However, in many cases the £1300 per tonne limit can be totally inadequate. Many loads these days are valued at tens of thousands of pounds with an equivalent value per tonne way in excess of the RHA level and it is necessary to ensure that the amount of insurance cover is adequate to cover the value of such loads.

When goods of this level of value are carried regularly the insurance company will provide suitable annual cover but in some instances goods vehicle operators may find that a lower level of cover is suitable for most of their activities since they only occasionally carry high-value loads. It is important when this happens that the haulier makes himself aware of the load value and that the insurance company is advised of such loads and the appropriate cover obtained. Failure to do so could leave a haulier facing expensive loss or damage claims from his own pocket.

Some GIT policies specifically exclude certain high-risk loads such as cigarettes, tobacco, spirits, livestock, computers, etc, so the operator faced with a request to carry such a load should check that his policy covers the value and consult his insurers before accepting an order to move the goods.

The GIT policy can be on an 'All-Risks' basis but it is usual for the policy to meet the particular requirements of the operator to give him protection against the liabilities he assumes when he accepts goods for carriage. Such liabilities may be accepted under Conditions of Carriage (see later in this chapter) or under a contract or agreement or, in the absence of any specific contract or conditions, at common law. If the operator is carrying his own goods in addition to other people's he should make sure that these are also covered under the policy.

Liability for Goods

Most transport managers are aware that it is essential that they should effect a GIT insurance policy in respect of the goods carried, but it is most important to consider very carefully the liabilities which are assumed for the goods handled on behalf of customers and that there is full understanding of the GIT insurance contract which has been arranged.

If transport contractors for commercial reasons assume total responsibility for very high-value loads and do not in any way limit their liability by contract or by the application of conditions of carriage (see p 164), they will soon realise that the claims which are being handled by their GIT insurers become so expensive that the premium subsequently demanded will be far too high to bear.

It is for this reason that the majority of transport contractors find it sensible to limit their liability in accordance with Conditions of Carriage such as those published by the RHA (it has drawn up new Conditions, dated 1991, which are the copyright of the RHA and may not be used by non-members) where the liability is based on a value of £1300 per tonne on the actual weight of the goods carried or on the computed weight if the volume of the goods exceeds 80 cubic feet per tonne. This limit can be varied on the insurance policy to suit individual demands but otherwise retaining for the operator the legal liability limitations.

New conditions have also been drawn up by the FTA in conjunction with the Institute of Purchasing and Supply, and these specify a liability limit of £2000 per tonne. These are not copyright and may be used by hauliers although legal difficulties may arise where RHA conditions and the new FTA/IPS conditions are applied to the same movement contract (ie one set by the customer and one by the haulier).

High-Value Loads

There may be a temptation for haulage contractors to accept high-value loads because the freight rate being offered by the consignor is generous in comparison to normal haulage rates. Such loads, however, are notoriously attractive to thieves so it is important before accepting them to examine the GIT insurance policy to make sure that such high-value goods are not specifically excluded and that any special requirements which the insurance company may have imposed regarding overnight parking and general vehicle security and protection are complied with.

Night Risk and Immobiliser Clauses

Insurance policies frequently contain clauses requiring vehicles carrying high-value goods to be securely parked in locked or guarded premises overnight (known as the 'Night Risk' clause) or to be fitted with approved vehicle protection devices such as steering column locks, engine immobilisers and alarm systems (known as the 'Immobiliser' clause). It is a condition of the insurance that such devices must be maintained in good working order and must be put into effect when the vehicle is left unattended. Failure to comply with such conditions can render the cover invalid.

Sub-Contracting

Before valuable loads are sub-contracted to other hauliers, operators should take considerable pains to satisfy themselves as to the genuineness of any driver calling at their premises for a load (telephoning the driver's employer is one suggested method of checking). Experience shows that drivers with criminal intent will state that they are employed by a certain firm and that they require a return load; documents are frequently handed to the driver in such cases; he picks up the load and disappears. A few days later, when investigations are made, it is discovered that the driver obtained the load by false pretences. GIT insurers may not accept responsibility for such losses, or indeed any losses involving sub-contracted loads unless they have had prior notification of the loads and the circumstances.

GIT on Hired Vehicles

The increasing use of vehicles on contract hire raises an important issue regarding liabilities for goods carried. Normally under the terms of the hire contract it is made quite clear by the hire company supplying the vehicle that it assumes no responsibility for loss or damage to the goods which are carried on the vehicle.

Vehicles Hired with Drivers

If, however, under the terms of the contract the hire company offers to provide a driver, the driver acts under its instructions. Should he act in a way which would be considered contrary to normal reasonable action (for example, leaving a fully laden vehicle overnight in the open when he had been specifically instructed to empty the vehicle or to place it in a locked garage), the hire company may find itself held liable at law for a 'fundamental breach of contract' and be faced with having to pay the full amount of any loss incurred. A method of overcoming this difficulty is for the hire company to arrange with the owner of the goods for a GIT insurance policy to be effected in their joint names and for the owners of the goods to pay the premium in the contract hire agreement.

Conditions of Carriage

An operator carrying goods for hire or reward is advised to set out conditions of carriage under which he contracts to carry goods. In these conditions the carrier can define his liabilities by stipulating limits on the value of goods for which he will normally accept responsibility, with goods of higher value being carried only on special terms, and stating circumstances and provisions under which no compensation is payable. For example, an operator could make it a condition that he accepts no liability under the following circumstances:

- Act of God.
- Act of war or civil war.
- Seizure under legal process.
- Act or omission of the trader, his employees or agents.
- Inherent liability to wastage in bulk, or weight, latent defect inherent defect vice or natural deterioration of the merchandise.
- Insufficient or improper packing.
- Insufficient labelling or addressing.
- Riots, civil commotions, strikes, lock-outs, stoppage or restraint of labour from whatever cause.
- Consignee not taking or accepting delivery within a reasonable time.
- Loss of a particular market whether held daily or at intervals.
- Indirect or consequential damages.
- Fraud on the part of the trader. In this context trader means either consignor or consignee.
- If non-delivery of a consignment, whether in part or whole, is not notified in writing within a specified number of days of despatch and a claim made in writing within a further specified number of days of despatch.
- If pilferage or damage is not notified in writing within a specified number of days of delivery, and a claim made in writing within a further specified number of days of delivery.

A note to the effect that goods are carried only under the Conditions of Carriage should be made on all relevant business documents, particularly consignment and delivery notes, invoices and quotations. Conditions of Carriage should always be drawn to the customer's attention and copies made available for customers to examine before they give orders for movements to commence.

RHA Conditions of Carriage

The Conditions of Carriage used by members of the Road Haulage Association are an excellent example of the sort of conditions which could be used by a haulage contractor to define his responsibilities. The Association published new Conditions in 1991 which include an increase in the limit of liability to £1300. Further sets of conditions for livestock carrying and sub-contracting are also prepared for members.

NB: The RHA conditions of carriage are the copyright of the Association and their use by a non-member would be a breach of copyright and therefore illegal – the RHA has taken legal action in cases of unauthorised use of its conditions.

Cover for International Haulage Journeys

The GIT cover described above is not sufficient or even legally acceptable where vehicles are engaged on international haulage work. In most cases, such operations are governed by the provisions of the *Convention on the Contract for the International Carriage of Goods by Road*, commonly known and referred to as the CMR Convention. This Convention automatically applies where an international haulage journey takes place between different countries at least one of which is party to the Convention (with the exception of UK–Eire and UK mainland–Channel Islands journeys which are ruled not to be international journeys for this purpose).

Road hauliers who carry goods on any part of an international journey, whether they know it or whether they choose to or not, fall within the legal confines of the CMR Convention under which compensation levels for loss or damage to goods are much higher than the standard Conditions of Carriage GIT cover applicable in National transport operations. CMR levels of cover vary according to a set standard which is published daily in the financial press. For this reason it is important to obtain adequate cover when involved in international transport.

Additionally, where hauliers undertake cabotage operations, they should discuss the levels of cover required with their insurers – indeed they should see if cover is available to cover certain liabilities such as losses from unattended vehicles. Difficulties may arise where local conditions of carriage are imposed and claims and legal wrangling arise under law other than English law.

Unfair Contract Terms

The Unfair Contract Terms Act 1977 affects such contracts as Conditions of Carriage. The effect of this legislation is to increase the liability of transport operators, particularly in respect of instances where liability for negligence is disclaimed by contract or by notice. Further, it prevents a business from excluding its liabilities for breach of contract when dealing with the general public. Consequently, any term in a contract purporting to exclude liability for personal injury by negligence is void. Any term excluding liability for damage to property by negligence is also void unless the term used is reasonable as between the parties to the contract.

Security

Insurance claims relating to vehicle thefts have increased in recent years and this is a serious problem of concern to insurance companies as well as to vehicle operators and the police. It is important that when away from base drivers should be encouraged to park their vehicles, especially if they are loaded with valuable goods, in guarded security parks. While the number of suitable security parks is limited and they are not conveniently located, it is nevertheless in everybody's interest that vehicles should not be left parked overnight on the roadside or on pieces of wasteland.

Security Warning

The RHA issues a security warning to members as follows:

1. Make every effort to ensure you are employing honest staff. Take up references over at least the previous 5 years and be suspicious of unexplained gaps. When checking references by telephone be sure to look in the telephone directory yourself for the number. A number supplied by a dishonest applicant could connect you to his accomplices. A staff enrolment form is available on application to RHA Area Offices and this form or a similar one should be completed.
2. Until you have seen his driving licence and have in your possession his P45 tax form and photograph, do not allow a newly-engaged driver to take out a vehicle.
3. Fit a vehicle immobiliser and/or alarm in as inaccessible a position as possible. Choose one which provides protection without the driver having to perform any operation which he normally would not have to do to stop his vehicle. Inspect the device frequently.
4. Drivers of vehicles carrying valuable loads should not get out of their cab if stopped. Even if a policeman requests them to do so, they should offer to go to the nearest police station. Bolts on the inside of the cab doors give added protection against hijackers.
5. A trouble-free cash bonus, from which a driver can be fined if he does not observe your security drill, is helpful.
6. Vehicles should not be left unattended for long periods, especially at night. At no time should keys be left in an unattended vehicle. Remember, a stationary vehicle with its windscreen wipers or indicators operating gives a clear signal to any watching criminal that it is his for the taking.
7. Discourage drivers from using the same cafes at the same time each day, particularly where their vehicles are not parked within sight.
8. Starter or ignition switches, security lock keys: remove numbers and keep the keys for each vehicle on a ring which is welded so that they cannot be separated.
9. If a vehicle's keys are lost, change switches and locks. It is much cheaper than losing a load.
10. Invite drivers to report to the police any suspicious circumstances, such as transfer of goods from one vehicle to another without apparent reason, which they might see on their travels, or the registration number of any vehicle which is persistently following them.
11. When disposing of a vehicle, remove the name of your firm so that a thief cannot use it to secure a load by false pretences.

Other Insurance – Business and Premises

While vehicle operators may be particularly concerned about obtaining appropriate insurance cover for their vehicles and the loads carried, they also need adequate insurance protection for other business contingencies in exactly the same way as any other employer or firm. Some of these insurances are listed and briefly described below.

Employers' Liability

It is a legal requirement for employers to cover their liabilities for any bodily injury incurred or disease contracted by their employees during or arising from their employment in the employer's business activities. In the case of a haulage contractor the business activities for which cover is required would be both as a haulage contractor and as an owner or occupier of property (ie the business premises).

The minimum cover required by law is £2 million, but usually policies are issued with unlimited liability. Premiums are based on the total payroll of the firm divided into categories (eg clerical staff, drivers, maintenance staff). Extensions to the basic policy can provide cover for employees engaged on private work for directors or senior management against liability incurred in work/employment related, sporting, social, first aid or welfare activities and against liabilities arising through sub-contracting work. Legal costs are recoverable in addition to any compensation awards.

Public Liability

Public liability policies provide cover against legal liability to third parties for bodily injury or illness or loss of or damage to their property arising out of the insured's business activities because of the insured's negligence. The policy can be extended to cover contractual liability which is essential for hauliers operating under printed Conditions of Carriage.

The policy conditions should be such that they cover contingencies not covered by the motor vehicle policies for damage caused by the vehicle or driver. It should also provide cover for liability arising from goods or vehicles being sold, supplied, altered, serviced or repaired.

Indemnity is usually set at a fixed figure per occurrence. A minimum of £500,000 or preferably £1 million is recommended as court awards have reached this figure for injuries to just one person.

Money Cover

Insurance cover can be obtained for the loss of cash, bank notes, currency notes, cheques, postal orders, postage or revenue stamps, national insurance stamps (now available only for limited use by self-employed people), 'holiday with pay' stamps, luncheon vouchers, trading tokens, credit vouchers, travellers' cheques, VAT vouchers and other negotiable instruments.

Extensions to the policy should include assault benefits to employees and other persons lawfully carrying these items and provide cover in respect of damage to and loss from slot machines, franking machines and safes. Premiums are usually based on estimated carryings of money to and from banks in a year.

Credit Insurance

Cover is available for financial loss because of customers defaulting on payment or their insolvency.

Indirect or Consequential Loss

When operating under RHA Conditions of Carriage which exclude liability for indirect or consequential loss and delay, the owner of the goods is not covered under a GIT policy. Cover is available under indirect or consequential loss insurance at a premium amounting to approximately 20 per cent of the GIT insurance premium.

Theft

Theft cover provides protection in the event of loss or damage resulting from entry to or exit from property by violent or forcible means. Usually, no cover is provided in such policies against larceny if there is no damage or visible sign of entry.

Petrol Installations and Oil Storage Tanks

Cover for accidental damage to fuel pumps, surface tanks and piping (but not for the loss of the contents, which is covered under the normal theft or fire insurance) is provided by such insurance. It also covers collapse, rupture or weld failure of these items.

Storm Damage

These policies compensate for losses resulting from the entry of water into petrol and diesel tanks as a result of heavy rain storms.

Fire and Special Perils

Cover can be obtained under this form of insurance for damage caused by fire, aircraft falling on the premises, explosion, riot and civil commotion, lightning, impact (including by own vehicles which is important because you cannot claim for this under your own vehicle insurance), and burst pipes.

Particular attention has been drawn to the need for adequate cover in respect of riot and civil commotion following well-publicised disturbances in some of Britain's inner city areas. Unless a fire insurance policy is extended to provide such cover, fire damage by rioters is not covered. It is possible to claim on the local authority under the Riot Damages Act 1886 but this would hardly provide adequate compensation for full reinstatement of premises.

Glass

This cover provides compensation for broken windows and for the cost of temporary boarding up.

Pressure Vessels and Boilers

This type of policy includes provision for the regular and statutory inspection of compressors and air receivers, together with compensation in the event of their explosion or collapse.

Lifting Equipment

The law requires such equipment to be regularly inspected and certificated. Insurance cover can be obtained which provides the inspection and protection against claims arising from the use of hoists, lifts, cranes, fork-lift trucks, pulley blocks, chains, and so on.

Personal Accident

A firm can cover its principals, directors, staff, drivers and maintenance staff against personal accident while driving or while away from the premises (not on the premises). This will provide set levels of regular income for various contingencies or lump-sum damages for loss of limbs. Where the firm is large enough there are special schemes available which have significant tax advantages to the employer.

Medical Expenses

When drivers are required to travel abroad they should be covered for medical expenses incurred in foreign countries, for compensation for taking relatives out to visit them if they are detained in hospital and for bringing the patient back to the UK for further treatment if necessary. Compensation for the loss of drivers' personal effects and baggage can usually be included in this type of policy.

Legal Expenses

Cover may be obtained to protect against the legal costs incurred in contesting unfair dismissal claims and other breach of employment contract issues such as pension rights and matters arising from legislation on equal pay, sex discrimination and race relations. Such policies also usually provide cover for legal expenses incurred in other disputes over liability or responsibility for the action of individuals or firms and for legal expenses following criminal prosecution under any statute. Other schemes are available to cover actual compensation awards.

Loss of Profit

This type of insurance would apply if a transport operator lost his vehicles and warehouse as a result of, for example, a fire. While the fire insurance would cover the cost of damage incurred, a loss of profit policy would compensate the operator for his business losses to competitors during the period of disruption until he got his business back on its feet again.

Usually the policy would provide this cover for at least 12 months and possibly even longer in view of the delays experienced in obtaining new vehicles and in getting premises replanned and rebuilt.

Book Debts

This insurance provides cover for debts caused by loss of records as a result of fire or other physical causes which leave no trace of amounts owed.

Fidelity Guarantee

Fidelity guarantee insurance provides cover against fraud or dishonesty by employees in connection with their employment during or within 18 months of the period of the fraud occurring or termination of the employment, whichever comes first.

Hired-In Plant

This type of policy covers all the liabilities imposed by hiring agreements including damage to the plant (vehicles, fork-lift trucks, cranes, etc), losses in hiring revenue incurred by the owner and claims made by third parties. 'The Contractors' Plant Association Conditions of Hire' also impose such liabilities while hired plant is on the highway. A motor insurance certificate is also required in these instances.

Motor Contingencies

Operators will find that their existing motor policies and other insurances exclude such things as liabilities incurred when hiring-in vehicles and drivers. A motor contingency policy would protect the insured by covering the liabilities of the person or firm owning the vehicle or employing the driver if its insurance was not current or was invalidated for some reason (eg premiums not paid). Similarly, a motor contingency policy would cover employees' use of their own cars on company business. The cover would be for any liability of the employer resulting from the use of the car on his business (eg if the employee had not paid his premium or, for example, if the vehicle was not taxed or not in roadworthy condition thus causing his cover to be invalid), but it would not provide any cover for the liability of the employee himself.

All Risks

This is a policy which could be used to cover any eventuality which is not specifically covered in any other policy held by a business. For example, it would provide compensation for loss or damage to any valuable paintings or antiques on company premises or in directors' offices, the firm's sporting trophies, awards of merit, and so on.

Computers

Policies are available to provide cover for material damage to computer systems resulting from accidental causes, electrical or mechanical breakdown. Such policies also cover consequential loss. Further policies cover software against loss or damage – even coffee spillage – and accidental or malicious erasure of data and provide for the costs of re-establishing information.

Obtaining the Best Cover

When negotiating with an insurance company for cover the premiums and policy conditions should be carefully considered and it is usually advisable to compare the terms offered with those available from other insurers. The services of an insurance broker can be helpful in finding the most satisfactory terms and competitive premium rates. Brokers retained for this purpose will give advice on the terms and conditions, handle claims and generally ensure that you are getting the best possible cover at economic rates. No payment is made to insurance brokers; they obtain their payment by way of commissions or discounts on premiums paid to the insurance company.

Insurance Claims

When making claims on insurance policies there are a number of points which deserve particular attention if a broker has not been engaged to handle these problems. The first and most important point in regard to claims following motor vehicle accidents is that insurance companies must be given immediate notice of an accident to any vehicle for which they supply cover, followed by a properly and fully completed accident report and claim form. A time limit is specified for this, normally 7 days.

Completion of Claim Form

If the claim is being made as a result of a vehicle accident the driver, if possible, should complete a claim form giving as much detail of the accident as he can: time, place, conditions of the weather and road, his position on the road and his speed, his direction of travel and the location of identifying objects, and the names and addresses of other parties involved and of any witnesses. A description of events leading up to the occurrence and a sketch of the position of the vehicles involved, both before and after the collision, should also be made on the report form, together with an indication of the damage to vehicles and property. The driver should not make any statement at the scene of the accident admitting or indicating liability (eg by apologising for his mistake).

Processing Claims

Once the insurance company has received the claim form they will get on with the business of deciding where the responsibility for the accident lies and how it should be apportioned. They will arrange for their motor vehicle assessor to examine the damaged vehicle and give permission for the repairs to be carried out if the estimate which the repairer has submitted is acceptable and, of course, if the vehicle is repairable. If the vehicle is beyond economical repair the assessor will authorise a 'write-off'.

Recovery of Uninsured Loss

In the event of the third party being at fault in an accident and the insured having an excess on the policy (by which the insured person volunteers to pay part of the cost of the repairs to their own vehicle, usually the first £25 to £250, for which there is usually a reduction in premium), he will need to make a claim against the third party for recovery of the excess (usually termed the 'uninsured loss'). The insurance company (or brokers) will deal with this matter if the damage is more than the excess and they are meeting the difference, but if the damage is slight and it is not intended to make a claim on the insurance company, a claim must be made direct to the third party for the 'uninsured loss'.

Action on Truck Thefts

Significant increases in truck and load thefts has prompted yet another measure to combat this rising menace, which is causing disruption to the industry and can also spell absolute disaster to the individual haulier whose vehicle is stolen.

A joint campaign by *Commercial Motor*, the RHA and Iveco Ford Truck has led to the production of the *Trucktheft Action Pack* (available from *Commercial Motor* [UK telephone 0371 810433, price £9.95 plus £2.35 p&p] or Iveco Ford Truck dealers). The pack contains useful advice to hauliers to help them fight crime and lists producers and installers of security equipment. It also provides them with some useful information on what to do if they are unfortunate enough to become the victim of such crime.

10: Road Traffic Law

Road traffic regulations are very complex and are to be found in a number of Acts and statutory instruments relating to all aspects of road use by pedestrians, cyclists and motorcyclists, motorists and, of course, heavy goods and passenger vehicle drivers and operators. Following a recent major review of road traffic law, a number of new offences were incorporated into the Road Traffic Act 1991 and mainly brought into effect from 1 July 1992.

The new provisions include a 'dangerous driving' offence, tougher measures to deal with drink drivers (with 5 years' imprisonment for drink drivers who cause a death), a new 'endangerment' offence to deal with vandals who place road users' lives at risk by placing dangerous objects on a road or interference with traffic signs or signals (with up to 7 years' imprisonment for convicted offenders), a new tougher driving test to be taken by bad drivers before their licences are restored following disqualification, the introduction of new technology to improve the detection of speeding and traffic light offences.

Besides offences and penalties, road traffic law deals with a number of other issues which affect the vehicle driver and other road users. Many of the particular legal requirements relating to the use of vehicles on the road are identified in the *Highway Code* (a new edition of the *Code* was published in 1993) along with much useful advice on driving and road usage although it should be remembered that the *Code* is intended for guidance and is not, in itself, a book of traffic law. Among more recent provisions included in the *Code* are such topics as child safety in cars, alcohol and the road user and the use of in-car telephones.

In this chapter some of the more important aspects of road traffic regulations affecting goods vehicle operators are covered.

Speed Limits

Three levels of speed limit are imposed on road users:
1. Limits applying to particular roads.
2. Limits applying to particular classes of vehicle (including limits imposed by the mandatory fitment of speed limiter devices).
3. Temporary speed limits introduced for special reasons such as in potentially hazardous situations and in times of fuel shortages.

Speed Limits on Roads

On roads where street lights are positioned at intervals of not more than 200 yards (defined as a 'restricted' road), an overall speed limit of 30mph applies to all classes of vehicle unless alternatively lower speeds are indicated by signs or unless the vehicle itself is subject to a lower limit by reason of its construction or its use. In some instances, speeds in excess of 30mph are permitted on such roads and this is indicated by appropriate signs showing the higher maximum limits.

The present maximum speed limits on roads outside built-up areas are 60mph on single-carriageway roads and 70mph on dual-carriageway roads and motorways, except where specified temporary or permanent lower limits are in force. In certain high-accident risk areas (eg housing estates) 20mph speed limit zones are being introduced in conjunction with road humps. The first was started in Norwich in January 1991 and others have followed. Other so-called 'traffic calming' measures being increasingly used to improve road safety include road-narrowing chicanes.

Advisory speed limits on motorways should be observed. These are shown by illuminated signs and indicate hazardous situations and road works ahead and by temporary speed limits signs at road works. The amber flashing warning lights positioned on the nearside of motorways (two lights, one above the other) indicate an advisory slowing down until the danger, and the next non-flashing light, is passed. Mandatory speed limits may also been seen at roadworks sites on motorways (indicated by white signs with black letters and a red border). Failure to comply with these mandatory motorway speed warning signs can result in prosecution.

Speed Limits on Vehicles

Vehicles are restricted to certain maximum speeds according to their construction, weight or use but when travelling on roads which themselves are subject to speed restrictions it is the lowest permitted speed (ie of the vehicle or of the section of road) which must be observed.

Private cars
Motor cars and dual-purpose vehicles must observe the 70mph limit on motorways and dual-carriageway roads and 60mph on single-carriageways and the appropriate lower limits on all other occasions.

Car-Derived Vans
Light vans derived from private-type motor cars (ie car-derived vans which also includes car-derived open-back, pick-up trucks) may travel at the same speeds as private cars on these roads (ie 70mph on motorways and dual-carriageways and 60mph on other roads unless lower limits are in force).

Private Cars and Car-Derived Vans Towing Trailers
The speed limit for motor cars (including dual-purpose vehicles), car-derived vans and motor caravans towing trailers and caravans is 50mph on single-carriageway roads and 60mph on dual-carriageways and motorways. There is no longer any legal requirement for such vehicles towing trailers and caravans to display a '50'mph plate at the rear.

Light Goods Vehicles
Light goods vehicles (apart from car-derived vans mentioned above) for the

purposes of speed limits are vehicles up to and including 7.5 tonnes maximum laden weight (mlw). Speed limits for rigid vehicles in this category are 50mph on single-carriageway roads, 60mph on dual-carriageway roads and 70mph on motorways.

The maximum speed for rigid vehicles up to 7.5 tonnes maximum laden weight drawing trailers and articulated vehicles up to 7.5 tonnes maximum laden weight on single-carriageway roads is 50mph and on dual-carriageways and motorways the limit is 60mph.

Large Goods Vehicles
There is no distinction in terms of maximum speed between rigid and articulated large goods vehicles and those with drawbar trailers. Large goods vehicles are those over 7.5 tonnes maximum laden weight (mlw) which, under other regulations, are required to display rear reflective markers so making them readily identifiable for speed limit enforcement purposes. Speed limits for all vehicles in this category are 40mph on single-carriageway roads, 50mph on dual-carriageway roads and 60mph on motorways.

Many new goods vehicles are now required under UK legislation to be fitted with speed limiters set to restrict their top speed to 60mph, while under EU legislation certain other heavy vehicles are restricted to 90kph (approx 56mph) (see Chapter 13 for full details).

Passenger Vehicles
The maximum speed limit for buses and coaches over 3.05 tonnes unladen weight and with more than eight passenger seats is dependent upon the overall length. For those not exceeding 12 metres in length the limits are 50mph on single-carriageway roads, 60mph on dual-carriageway roads and 70mph on motorways. For those over 12 metres in length the limits are 50mph on single-carriageway roads and 60mph on dual-carriageways and motorways.

NB: In all cases mentioned above, it must be stressed that these limits only apply where no lower limit is in force.

Special Types Vehicles
Vehicles operating outside the Construction and Use (C&U) Regulations for the purposes of carrying abnormal indivisible loads come within scope of the Special Types General Order (STGO) as described in Chapter 22 and must conform to specified speed limits depending on their category. These speed limits are stated on p 369 but are repeated here with other speed limits for ease of reference:

Vehicle category	Motorways	Dual-carriageways	Single-carriageways
Category 1	60mph	50mph	40mph
Category 2	40mph	35mph	30mph
Category 3	30mph	25mph	20mph

NB: The category 3 speed limits also apply when wide loads between 4.3 metres and 6.1 metres are being carried on Category 1 and 2 Special Types vehicles.

Works Trucks and Industrial Tractors
The maximum speed limit for works trucks and industrial tractors is 18mph but the latter are not permitted on motorways.

Agricultural Vehicles
Agricultural vehicles are limited to 40mph. They are not permitted on motorways.

Motor Tractors/Locomotives
Where such vehicles (including their trailers) are fitted with springs and wings their maximum permitted speeds are 40mph on motorways and 30mph on other roads. When they do not have springs and wings the maximum speed limit is 20mph on all roads.

Track-Laying Vehicles
These vehicles are limited to maximum speeds of 20mph or 5mph depending on their construction (ie whether they have springs and wheels fitted with pneumatic or resilient tyres – see below).

Vehicles with Non-Pneumatic Tyres
Vehicles with resilient, non-pneumatic (ie solid) tyres are restricted to a maximum speed of 20mph on all roads. Those with non-resilient tyres (eg traction engines) are restricted to 5mph – see also above: track-laying vehicles.

Emergency Vehicles
Fire, police and ambulance service vehicles are exempt from all speed limits if, by observing the speed limit, they would be hampered in carrying out their duties. However, drivers of such vehicles have a duty to take particular care when exceeding statutory limits and could face proceedings if an accident results while exceeding the limits.

Table of Vehicle Speed Limits

	Motorway (mph)	Dual-carriageways (mph)	Other roads (mph)
Private cars			
– solo	70	70	60
– towing caravan or trailer	60	60	50
Buses and coaches			
– not over 12 metres length	70	60	50
– over 12 metres length	60	60	50

Goods vehicles
Car-derived vans

– solo	70	70	60
– towing caravan/trailer	60	60	50

Not Exceeding 7.5 tonnes mlw

– solo	70	60	50
– articulated	60	60*	50
– draw-bar	60	60*	50

Over 7.5 tonnes mlw

– solo	60	50	40
– articulated	60	50	40
– draw-bar	60	50	40

Note : mlw means maximum laden weight (ie maximum gross weight for a vehicle as specified in construction and use regulations).
* In Northern Ireland the speed limit for vehicles in these two categories is 50mph only.
NB: See above for speed limits for Special Types and other vehicles.

Speed-enforcement Cameras
The Road Traffic Act 1991 legally authorised the use of photographs as evidence by courts in cases of alleged speeding (and traffic-light jumping). Progressively, cameras have been installed at key sites since 1 July 1992 to catch speeding drivers and traffic light offenders. The so-called Gatso cameras (named after their inventor, Dutch ex-racing driver Maurice Gatsonides) produce film showing the vehicle and its number plate (which can be enhanced by scientific means for purposes of clarity – and can even decipher the registration number of vehicles fitted with plates which have been photo-reflective sprayed), date, time and the vehicle speed.

Lighting-Up Time

All mechanically propelled vehicles must display front and rear position lights between sunset and sunrise and headlamps (where required by regulations) during the hours of darkness, which is from half-an-hour *after* sunset to half-an-hour *before* sunrise (see Chapter 15 for lighting details) and during daytime hours when visibility is seriously reduced.

Night Parking

Goods vehicles not exceeding 1525kg unladen do not require lights at night when standing on restricted roads (ie on which a 30mph speed limit – or lower limit – is in force), if they are parked either in a recognised parking place (ie outlined by lamps or traffic signs) or on the nearside, close to and parallel with the kerb, facing the direction of travel and with no part of the vehicle within 10 metres of a junction (ie on the same side as the vehicle or on the other side of the road). On any road where these conditions are not met lights must be shown (ie front and rear position lights).

All goods vehicles exceeding 1525kg unladen weight must display lights at all times when parked on roads between sunset and sunrise. Trailers and vehicles with projecting loads must not be left standing on roads at night without lights.

Vehicles should be parked on the nearside of the road when left standing overnight except when parked in a one-way street or in a recognised parking place and they must not cause obstruction.

Increasing attention is being given by the police and local authorities to drivers sleeping overnight in heavy vehicles with sleeper cabs while parked in lay-bys. Drivers should be warned against this practice which is usually considered illegal on the grounds that the vehicle is causing an obstruction. A similar situation applies when draw-bar trailers and semi-trailers are left in lay-bys.

Stopping, Loading and Unloading

Leaving Engine Running

Whenever a driver leaves his vehicle on a road, the engine must be stopped (except in the case of fire, police or ambulance service vehicles or when the engine is used to drive auxiliary equipment or for providing power for batteries to drive such equipment).

Obstruction

A vehicle must not be left in a position where it is likely to cause obstruction or danger to other road users, eg near an entrance to premises, near a school, a zebra crossing or a road junction (see p 183 for full list).

Trailers must not be left on a road when detached from the towing vehicle.

Loading and Unloading Restrictions

Vehicles must not stop or park on clearways to load or unload. In some areas loading and unloading restrictions are indicated by yellow lines painted on the kerb at right angles to it (Figure 10.1) as follows:
1. A single yellow line at intervals indicates restrictions during certain hours, for example, peak hours.
2. Double yellow lines at intervals indicate a complete ban on loading and unloading during the whole of the working day, for example, 8.00 am to 6.30 pm.
3. Three yellow lines at intervals indicate a ban at all times.

The times quoted vary from place to place but where this type of road marking is used there is always a sign stating the hours during which loading and unloading must not take place.

Parking Meter Zones
Loading and unloading in parking meter zones during the working day (the times are indicated on signs) is not allowed unless a gap between meter

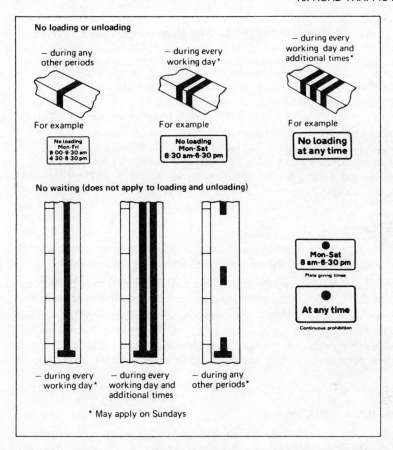

Figure 10.1 *Waiting and loading restriction road markings and plates giving times*

areas or a vacant meter space can be found. A vehicle using a meter space for loading or unloading can stop for up to 20 minutes without having to pay the meter fee (this does not apply when parking for any purpose other than loading or unloading the vehicle).

Waiting and Parking Restrictions
Single, double or broken yellow lines painted on the road parallel to the kerb apply to waiting and parking at various times, but they do not indicate a ban on loading or unloading and the same applies to 'no waiting' prohibitions indicated by 'no waiting' signs (see *Highway Code* for full details of waiting and parking restrictions).

Motorway Driving

Motorway driving requires special care and observance of the motorway regulations. In particular, vehicles not capable of exceeding 25mph on the level are prohibited from using motorways, including vehicles operating under the

Special Types General Order (see also Chapter 22) if they cannot exceed this speed.

Vehicles must not stop on motorways except through mechanical defect or lack of fuel, water or oil, due to an accident, the illness of a person in the vehicle or for other emergency situation (including giving assistance to other persons in an emergency), to permit a person from the vehicle to recover or remove objects from the carriageway. It is illegal to drive on the hard-shoulder or the central reservation, to reverse or to make a 'U-turn' on a motorway. Vehicles which must use the hard-shoulder for emergency reasons as described above must remain there only for so long as is necessary to deal with the situation.

Use of Lanes

Goods vehicles with operating weights in excess of 7.5 tonnes and vehicles drawing trailers (and certain other heavy motor cars not included in the categories mentioned) must not use the outer or offside lane of three- and four-lane motorways.

On some steep slopes of two-lane motorway sections, heavy goods vehicles (ie over 7.5 tonnes) are banned from using the outside lane; these bans are clearly signposted on the approaches to the appropriate section indicating the extent of the banned section and the vehicles prohibited from using the outer lane.

Temporary Speed Limits

Where carriageway repairs take place on motorways or where contraflow traffic systems are used an *advisory* 50mph speed limit is usually imposed. This is considered by the police to be a maximum speed, and they may prosecute drivers found speeding in these sections for a 'driving without due consideration . . .' type of offence. However, it is becoming more common for a *mandatory* temporary speed limit to be imposed in such cases and where drivers are detected speeding in these sections they will be prosecuted for this offence.

Speed Limits for Recovery Vehicles on Motorways

For the purposes of motorway speed limits, recovery vehicles may travel at up to 60mph. This follows a High Court ruling (on an appeal by the Director of Public Prosecutions) that such vehicles are constructed to carry a load and therefore may travel at the same maximum speed as other heavy goods vehicles. Previously such vehicles were required to observe the 40mph maximum limit applicable to vehicles classified as motor tractors, light and heavy locomotives.

Other Vehicles on Motorways

Light and heavy locomotives which do not comply with C&U regulations, dump trucks, engineering plant and vehicles for export which do not comply with the

C&U regulations may be driven on motorways provided they are capable of attaining a speed of 25mph on the flat when unladen and so long as they are not drawing a trailer.

Learner Drivers on Motorways

Learner drivers are not allowed to drive on motorways, but holders of provisional lgv driving entitlements may drive heavy goods vehicles on motorways provided they hold a full ordinary entitlement (ie category B) and are accompanied by a qualified driver.

Lights on Motorways

Hazard Warning
Motorways are equipped with amber hazard warning lights located on the nearside verge and placed at 1 mile intervals. When these lights flash, vehicles must slow down until the danger which the lights are indicating, and a non-flashing light, has been passed.

Rural Motorways
Rural motorways have amber lights, placed at not more than 2 mile intervals and usually located in the central reservation, which flash and indicate either a maximum speed limit or, by means of red flashing lights, that one or more lanes ahead are closed. The speed limit indicated applies to *all* lanes of the motorway and should not be exceeded.

Urban Motorways
Urban motorways have overhead warning lights placed at 1000-yard intervals. Amber lights flash in the event of danger ahead and indicate a maximum speed limit or an arrow indicating that drivers should change to another lane. If red lights flash above any or all of the lanes, vehicles in those lanes must stop at the signal. It is as much of an offence to fail to stop at these red lights as it is to ignore automatic traffic signals.

Motorway Road Markings
Experimental road markings designed to reduce the risk of nose-to-tail collisions are being tried out on some sections of motorway (eg the M1 near Leicester). The chevron-shaped markings are painted on the road surface at 10 metre intervals for 5 kilometres (ie 3 miles). Drivers are advised to keep at least two chevrons (ie 2 seconds) between themselves and the vehicle in front.

Local Radio Station Frequency Signs
Under arrangements with the DoT, local radio stations can have their broadcasting frequencies indicated on motorway signs. It is a condition of such signposting that the station in question provides traffic news relevant to the location of the signs and of benefit to long-distance travellers 24 hours a day, 7 days a week with at least four broadcasts per hour at peak times and two per hour during off-peak times with programme interruptions for important announcements. The signs do not carry the station name or logo, only the

broadcasting frequency enabling drivers quickly to tune into the appropriate wavelength.

Motorway Fog Code

To help drivers avoid the grave hazards of fog on motorways and to meet the special dangers of mixed traffic, an eight-point drivers' Code applies as follows:

Fog Code
1. Slow down; keep a safe distance. You should always be able to pull up within your range of vision.
2. Don't hang on to someone else's tail lights; it gives you a false sense of security.
3. Watch your speed; you may be going much faster than you think.
4. Remember that if you are in a heavy vehicle you need a good deal longer to pull up.
5. Warning signals are there to help and protect. Do observe them.
6. See and be seen – use headlights or fog lamps.
7. Check and clean windscreen, lights, reflectors and windows whenever you can.
8. If you must drive in fog, allow more time for your journey.

Segregation
1. Drivers of cars, light goods vehicles and coaches should move out of the left-hand lane when it is safe to do so but not if they will soon be turning off the motorway. When they want to leave the motorway, they should start their move to the left well before the exit. They should be prepared to miss the exit if they cannot reach it safely.
2. Drivers of heavy lorries should keep to the left-hand lane, but be ready to let other drivers into the lane at entry points and well before exit points.

The Code and the Segregation advice, which can be found in the *Highway Code*, apply on motorways throughout Great Britain whenever there is fog.

M25 Automatic Fog Warning System
A new automatic fog warning system has been established on the M25 motorway. Detectors installed alongside the motorway identify when visibility falls below 300 metres and automatically switch on the existing matrix signals to display the message 'FOG'. The signals are located at strategic points where unexpected pockets of fog may occur as identified by the Meteorological Office (30 such zones were identified and 54 danger spots within those zones). When the fog signs are on drivers should slow down and proceed at a speed where they can safely stop within their range of vision.

Hazard Warning

Four-way direction-indicator flasher systems fitted to vehicles may be legally used to indicate that a vehicle is temporarily obstructing the road or any part of the carriageway either while loading or unloading, when broken down or for emergency reasons (previously their use was only permitted in emergencies). Further details of vehicle lighting requirements are to be found in Chapter 15.

Lights During Daytime

If visibility during the daytime is poor, because of adverse weather conditions, drivers of all moving vehicles must switch on both front position lights and headlamps or front position lights and matched fog and spot lights. This applies in the case of heavy rain, mist, spray, fog or snow or similar conditions. When vehicles are equipped with rear fog lights (see pp 270–1) these should be used when the other vehicle lights are switched on in poor daytime visibility conditions. Further details of vehicle lighting requirements are to be found in Chapter 15.

Parking

Drivers who park their vehicles in a position which causes danger or obstruction to other road users can be prosecuted and their driving licence endorsed with penalty points on conviction (usually 3) or they can be disqualified from driving.

Danger or obstruction may be caused by a parked vehicle:
   (a)   in a 'no-parking' area
   (b)   on a clearway
   (c)   alongside yellow lines
   (d)   where there are double white lines
   (e)   near a road junction
   (f)   near a bend
   (g)   near the brow of a hill
   (h)   near a humpback bridge
   (i)   near a level crossing
   (j)   near a bus stop
   (k)   near a school entrance
   (l)   near a pedestrian crossing
   (m)   on the right-hand side of the road at night
   (n)   where the vehicle would obscure a traffic sign
   (o)   on a narrow road
   (p)   on fast main roads and motorways
   (q)   near entrances and exits used by emergency service vehicles
   (r)   near road works
   (s)   alongside or opposite another parked vehicle.

This list does not leave many alternative places for parking for the goods vehicle driver who has collections or deliveries to make, particularly in town, and for this reason drivers should be instructed to take reasonable care when parking in congested areas to avoid causing obvious obstruction or danger. For example, drivers should not double park, block entrances and exits of business or private premises, park near dangerous junctions or near pedestrian crossings, as well as avoid the areas mentioned above.

Parking on Verges

The Road Traffic Act 1988 (sections 19 and 20) makes it an offence to park a heavy commercial vehicle (ie a vehicle over 7.5 tonnes maximum laden weight including the weight of any trailer) on the verge of a road, on any land between two carriageways or on a footway whether the vehicle is totally parked on those areas or only partially so.

There are exemptions to this: when a vehicle is parked on such areas with the permission of a police officer in uniform, or in the event of an emergency, such as for the purposes of saving life or extinguishing fire, or for loading and unloading, provided that the loading or unloading could not have been properly performed if the vehicle had not been so parked and that the vehicle was not left unattended while it was parked.

It is an offence (under the Road Traffic Act 1988 s34) for any person to drive a motor vehicle on to common land, moorland or other land which does not form part of a road or on any footpath or bridleway, beyond a distance of 15 yards except where legal permission exists to do so but then only for the purpose of parking or to meet an emergency such as saving life or extinguishing fire.

Lorry Routes and Bans

Local authorities identify preferred routes for heavy vehicles passing through their areas and display on them appropriate signs. The type of sign shown in Figures 10.2 and 10.3 are used to:
1. Mark the most suitable route between dock areas and the nearest convenient connection with the primary route/motorway network.
2. Mark a suitable alternative route at any place on the primary route network where drivers of goods vehicles might be advised to avoid a particular part of that route, but where it is not appropriate to direct all traffic on to the alternative route, or to the primary route itself.
3. Mark routes from the primary/motorway route system to local inland centres which generate a high level of goods vehicle traffic (industrial estates, for example).

Certain areas and especially London impose bans on the movement of goods vehicles and the parking of goods vehicles. These bans and restrictions are always marked with appropriate signs and operators are advised to ensure that their drivers observe them.

London Bans

Vehicles more than 12.2 metres long are banned from Central London unless they are delivering to or collecting from specific addresses within the Central London area.

Vehicles over 16.5 tonnes pmw are prohibited from travelling along many routes through Greater London at certain times unless the operator holds an exemption permit (issued under the London Boroughs Transport Scheme) which must be carried on the vehicle and the vehicle must display exemption plates at the front and rear in a conspicuous position. Vehicles are also required to be fitted with air-brake silencers (hush kits).

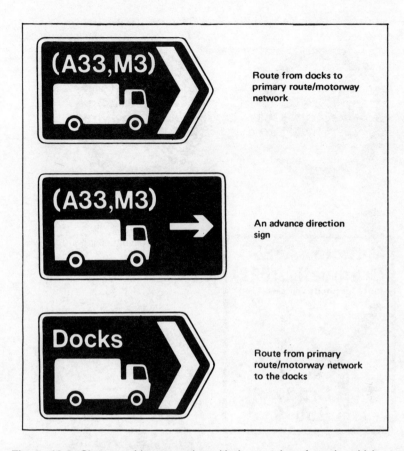

Figure 10.2 *Signs used in connection with the routeing of goods vehicles to avoid congestion and unsuitable areas*

The routes on which the ban applies are well signposted and the times at which it applies are also given on the signs. It is an offence for a driver of a goods vehicle over 16.5 tonnes pmw to travel on the banned routes at the relevant times unless a valid exemption permit has been issued and is carried on the vehicle. Application for exemption permits and vehicle plates should be made to the London Boroughs Transport Committee, Rooms 301–305, Hampton House, 20 Albert Embankment, London SE1 7TJ, Tel 071-582 6220.

The London ban applies at the following times:

Sunday	at all times
Monday to Friday	midnight to 7.00 am and 9.00 pm to midnight
Saturday	midnight to 7.00 am and 1.00 pm to midnight

The government has plans to simplify the operation of this night-time and weekend lorry ban via the Deregulation Bill which is before parliament at the time of preparing this 1995 edition of the *Handbook*.

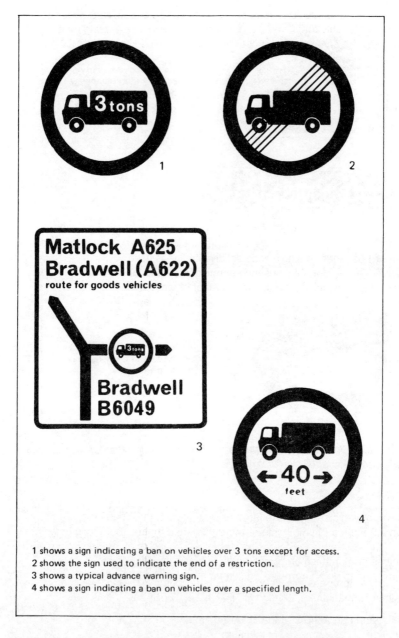

1 shows a sign indicating a ban on vehicles over 3 tons except for access.
2 shows the sign used to indicate the end of a restriction.
3 shows a typical advance warning sign.
4 shows a sign indicating a ban on vehicles over a specified length.

Figure 10.3 *Signs indicating route bans on heavy goods vehicles*

London 'Red Route' Scheme
A new scheme to prevent traffic congestion on certain primary routes in London
has been introduced. Currently about 6.5 miles of roadway between Archway
Road and the Angel, Islington has been designated as a 'Red Route' and on
this section, identified by single or dual red road markings and accompanying

'Red Route' signs, vehicles may only stop to collect or deliver at specified times (or not at all), only for limited periods, and only in marked loading bays. Eventually the scheme is intended to cover about 500 kilometres (300 miles) of London's roads.

Failure to comply with the restrictions can lead to severe penalties. Initially a new-type red parking ticket (penalty currently £40) will be issued by police or traffic wardens, but if the driver argues or refuses to move, an impoundment order may be issued and the vehicle taken away by police. This will cost the driver or operator £95 to get the vehicle back, plus the £40 penalty, a total of £135.

Bus-Only Lanes

Traffic lanes on urban roads reserved solely for use by buses are a common feature in many towns and cities. Uniform traffic signs and road markings indicate bus lanes. A single wide solid white line is used to mark the edge of the reserved lanes. Upright signs incorporating international symbols combined with arrows will show to other traffic the number of lanes available for their use. When the signs and restrictions are in operation on a road, all other vehicles, except for pedal cycles (and taxis if signed to this effect), are prohibited from using the bus lane.

Level Crossings

Most railway level crossings are now fitted with automatic half-barrier crossing gates and appropriate warning signs are given in advance. When a train is approaching such crossings, red lights flash and a bell rings to warn drivers and pedestrians. Once these warnings start the barrier comes down immediately, and drivers should not zig-zag around the barriers. When the train has passed, the barriers will rise unless another train is following, in which case the warnings will continue.

Drivers of vehicles which are large or slow (ie with their loads that are more than 2.9 metres (9ft 6in) wide or more than 16.8 metres (55ft) long or weighing more than 38 tonnes gross or incapable of a speed of more than 5mph) wishing to cross one of these crossings must, before attempting to cross, obtain permission to do so from the signalman by using the special telephone which is provided at the crossing. Failure to do this is an offence. In the event of a vehicle becoming stuck on the crossing the driver should advise the signalman immediately by using the telephone.

Weight-Restricted Roads and Bridges

Where signs indicate that a particular section of road or a bridge is restricted to vehicles, not exceeding a specified weight limit or axle weight limit, unless otherwise expressly stated, the weight limit shown relates to the actual weight of the vehicle or to an individual axle of the vehicle, not the relevant plated weights. Where doubt exists about any particular sign it is advisable to consult the local authority responsible for its erection and to determine the precise wording of the Traffic Management Order under which authority the sign would have been erected.

Owner Liability

Under the Road Traffic Offenders Act 1988 responsibility for payment of fixed penalty fines or excess parking charges rests with the registered vehicle owner (ie the keeper of the vehicle, not necessarily the legal owner), if the driver who committed the offence cannot be identified or found. The registered owner of the vehicle is sent details of the alleged offence and is obliged to pay the fine or submit a 'Statutory Statement' of ownership in which he states whether he was the vehicle owner at the time of the alleged offence (in which case he should name the driver), had ceased to be the owner at that time or had not yet become the owner at that time. Where the person was not the owner he must give the name of the previous owner or the new owner to whom he transferred the vehicle if he knows it.

When a vehicle is hired out for less than 6 months, and such an incident arises, the hiring company can declare that the vehicle was on hire and send a copy of the hiring agreement together with a signed statement of liability from the hirer accepting responsibility for the fine or excess parking charge. Such a clause is normally included in the hiring agreements which the hirer signs. Failure to pay a fixed penalty or excess charge, or to give information as required by the police in such matters, can result in a fine of up to £1000 on conviction (or even £5000 in certain circumstances).

Fixed Penalties

In order to reduce the pressure on the courts, a system of fixed penalties exists by which both traffic wardens and the police can issue fixed penalty notices requiring the vehicle driver or owner to pay the fixed penalty or to elect to have the case dealt with in court in the normal way. The fixed penalty system operates on two levels; non-endorsable offences (mainly dealt with by traffic wardens), and driving licence endorsable offences which only the police can deal with since traffic wardens have no general authority to request the production of driving licences (see below). Additionally, two London area parking offences are included in the fixed penalty system.

Non-Endorsable Offences

For non-endorsable offences a white ticket/notice (penalty £20) is issued either to the driver if present or is fixed to the vehicle windscreen. Since no driving licence penalty points are involved for such offences there is no requirement to examine the licence. Traffic wardens have authority to issue fixed penalty tickets for the following non-endorsable offences:
1. leaving a vehicle parked at night without lights or reflectors;
2. waiting, loading, unloading or parking in prohibited areas;

3. unauthorised parking in controlled parking zone areas;
4. contravention of the Vehicles (Excise) Act 1971 by not displaying a current licence disc;
5. making 'U' turns in unauthorised places;
6. lighting offences with moving vehicles;
7. driving the wrong way in a one-way street;
8. overstaying on parking meters, returning to parking places before the expiry of the statutory period, or feeding meters to obtain longer parking facilities than those permitted in a meter zone;
9. parking on pavements or verges by commercial vehicles exceeding 3050kg unladen weight (see pp 183–4).

Endorsable Offences

The extended fixed penalty system covers driving licence endorsable offences which can be dealt with only by the police (ie not traffic wardens who have no general authority to examine driving licences) – this includes some 250 driving and vehicle use offences. For endorsable offences a yellow ticket/notice with a different level of penalty applies as described below.

For driving licence endorsable offences the police issue a yellow ticket/notice for which a penalty of £40 is payable. These tickets are only issued after the police officer has seen the offender's driving licence and has established that the addition of penalty points appropriate to the current offence, when added to any points already on the licence, will not result in automatic disqualification under the 12-point totting-up procedure. If this is the case, the ticket will be issued and the driving licence will be confiscated (an official receipt covering the holder for non-possession or production of his licence will be given, valid for 2 months) being returned to the holder with the appropriate penalty points added when the penalty has been paid.

If the offender does not have his driving licence with him at the time, the penalty notice will not be issued on the spot but will be issued at the police station if the driving licence is produced there within 7 days – subject again to the number of penalty points already on the licence.

Where the addition of further points in respect of the current offence would take the total of penalty points on the licence to 12 or more thus leading to automatic disqualification, the ticket will not be issued and the offence will be dealt with by the offender being summoned to appear in court in the normal manner.

London Area Parking Offences

The fixed penalty system includes two specific London area parking offences, namely parking on a 'Red Route' (see pp 186–7) for which the penalty is currently £40 and parking in other prohibited places for which the penalty is £30.

Payment or Election to Court

Fixed penalty notices must be paid in accordance with the instructions on the notice and within the specified time limit of 28 days. Alternatively, the offender can elect to have the charge dealt with by a court so he has the opportunity of defending himself against the charge or, even if he accepts that he is guilty of the offence, of putting forward mitigating circumstances which he feels may lessen any penalty which may be imposed.

The address of the fixed penalty office to which the penalty payment should be sent is given in the notice together with instructions for making application for a court hearing if this course of action is chosen.

Failure to Pay

With both the white and yellow ticket systems, failure to pay the statutory penalty within the requisite period of 28 days will result in the offender being automatically considered guilty and the penalties being increased by 50 per cent (ie to £30 and £60 respectively). These increased amounts become fines and continued non-payment will lead to the arrest of the offender and appearance before a court in the district where the offence was committed. This could be many miles from where the offender lives and may necessitate him being transported there under arrest and possibly held overnight.

Summary of Offences

Among the offences covered by the extended scheme are the following:

White Ticket (non-endorsable)
1. Not wearing seat belt.
2. Driving and stopping offences (reversing, parking, towing, etc).
3. Contravention of traffic signs, box junctions, bus lanes, etc.
4. Contravening driving prohibitions.
5. Vehicle defects (speedometer, wipers, horn etc).
6. Contravening exhaust and noise regulations.
7. Exceeding weight limits (overloading, etc).
8. Contravention of vehicle lighting requirements.
9. Contravention of vehicle excise requirements.

Yellow Ticket (endorsable)
1. Speeding.
2. Contravention of motorway regulations.
3. Defective vehicle components (brakes, steering, tyres, etc) and vehicle in dangerous condition.
4. Contravention of traffic signs.
5. Insecure and dangerous loads.
6. Leaving vehicles in dangerous positions.
7. Contravention of pedestrian rights.

This is only a summary of a very extensive list of offences included within the scheme.

Traffic Wardens

In addition to the powers of traffic wardens to issue fixed penalty tickets as described in the previous section, they also have powers to act as parking attendants at street parking places, to carry out special traffic control duties, to inquire into the identity of drivers of vehicles, to act in connection with the custody of vehicles at car pounds and to act as school crossing patrols.

It is *only* when on duty at 'car pounds' that traffic wardens have the power to request production of a driving licence.

Pedestrian Crossings

There are two types of pedestrian crossing. Zebra crossings are bounded on either side by areas indicated by zig-zag road markings in which overtaking, parking and waiting are prohibited. The marked areas extend to about 60ft on either side of the crossing. A 'give-way' line 3ft from the crossing is the point at which vehicles must stop to allow pedestrians to cross. Pelican crossings are controlled by traffic lights which vehicle drivers must observe and pedestrians should cross only when the green light signal indicates that they should do so (at many Pelican crossings an audible bleeper is provided to assist blind people to cross with safety). With this type of crossing, if there is a central refuge for pedestrians, each side of the refuge is still considered to be part of a single crossing. Only if the two parts of the crossing are offset does it become two separate crossings.

Builders' Skips

Provisions are contained in the Highways Act 1980 to control the placing of builders' skips on the road. Before such a skip is placed on the road, permission must be obtained from the local authority. This will be given subject to conditions relating to the size of the skip, its siting, the manner in which it is made visible to oncoming traffic, the care and disposal of its contents, the manner in which it is lit or guarded and its removal when the period of permission ends.

Owners must ensure that skips carry proper reflective markers (see p 280), are properly lit at night, are clearly marked with their name and telephone number or address, and that they are moved as soon as is practical after they have been filled.

The police and highway authorities have powers to re-position or remove a skip from the road and recover the cost of doing so from the owner, and a fine may be imposed.

For the purposes of the Act the definition of a builder's skip is 'a container designed to be carried on a road vehicle and to be placed on a highway or other land for the storage of builders' materials, or for the removal and disposal of builders' rubble, waste, household and other rubbish or earth'.

Abandoned Motor Vehicles

It is an offence under the Refuse Disposal (Amenity) Act 1978 to abandon a motor vehicle or any part of, or part removed from, a motor vehicle in the open air or on any other open land forming part of a highway. Such offences, on conviction, can lead to fines of up to £200 for a first offence and up to £400 and a term of up to 3 months imprisonment or both for a subsequent offence.

Vehicles which are illegally or obstructively parked can be removed and a statutory charge of £105 imposed. An additional charge for storage of a removed vehicle is £12 per day and £50 is charged for its disposal. Removed and impounded vehicles are not released until all relevant charges have been paid.

Wheel Clamps

Vehicles which are illegally parked or which cause obstruction, in a wide area of central London, will be immobilised by the Metropolitan Police or by contractors on their behalf. A wheel clamp, known as the 'Denver Boot', will be fixed to one wheel of the vehicle thereby preventing it being driven away. A notice will be stuck to the vehicle giving the driver notice of the offence committed and instructions for securing release from the clamp.

It is an offence to try to remove a wheel clamp or to attempt to drive off with one fitted. Vehicle drivers finding a clamp fixed to their vehicle must go to the Metropolitan Police pound in Hyde Park underground car park and request removal of the clamp. Both a removal charge (currently £105) and a fixed penalty (currently £20) have to be paid before the clamp is removed.

Overloaded Vehicles

It is an offence to drive an overloaded vehicle on a road. Under the Road Traffic Act 1988 (sections 70 and 71) an authorised examiner or police officer may prohibit the use of an overloaded vehicle on the road until the weight is reduced to within legal limits and may direct, in writing, the person in charge of an overloaded vehicle to remove the vehicle to a specified place. See also p 224.

The government is currently committed to increasing enforcement of vehicle weight regulations and proposes to weigh at least 100,000 heavy vehicles annually. It intends that penalties for serious or persistent offenders should be such as to deter others from overloading their vehicles. In the mean time efforts are still being made to get the government to accept a 'due diligence' clause to relieve operators of the liability for overloading where they can show they took all possible steps to avoid an offence and at the same time to make the consignor liable where he knowingly causes an operator's vehicle to overload (eg through not declaring or properly calculating consignment weights).

A driver may be instructed to drive for a distance of up to 5 miles to a weighbridge for the weight of his vehicle and load to be checked. If he is directed to drive more than 5 miles to the weighbridge and his vehicle is found to be within the maximum permitted weights then a claim may be made against the appropriate highway authority for the costs incurred.

The maximum fine for an overloading offence is £5000 but any one instance of an overloaded vehicle could result in conviction for more than one offence, each of which carries this maximum penalty. Subsequent convictions for such offences could lead to higher fines. Convictions for overloading offences also jeopardise the operator's licence (see also Chapter 12 on vehicle weights and overloading).

Dynamic Weighing

The use of dynamic axle weighing machines has increased for weight enforcement purposes. These are permitted under regulations (the Weighing of Motor Vehicles (Use of Dynamic Axle Weighing Machines) Regulations 1978) which specify that an enforcement officer can require a vehicle to be driven across the weighing platform of a machine for this purpose. The permitted weights for the vehicle are measured to within plus or minus 150kg for each axle and within plus or minus 150kg multiplied by the total number of axles to determine the tolerance on the total vehicle weights.

The government is stepping up its enforcement activities on vehicle overloading and warns operators that they must allow a margin for safety if they cannot keep within the limits.

Accident Reporting

Any driver involved in a road accident in which a personal injury is caused to any person other than himself, or damage is caused to any vehicle other than his own vehicle, or damage is caused to any animal* other than animals carried on his own vehicle or to any roadside property (see below for definition) MUST STOP. Failure to stop after an accident is an offence and fines of up to £5000 can be imposed on conviction.

The driver of a vehicle involved in an accident must give to anybody having reasonable grounds for requiring it his own name and address, the name and address of the vehicle owner and the registration number of the vehicle.

If the accident results in injury or damage to any person other than the driver himself or to any other vehicle or to any reportable animal,* or to roadside property, then the details of the accident must be reported to the police *as soon as reasonably practicable afterwards, but in any case no later than 24 hours after the event*. This obviously does not apply if police at the scene of the accident take all the necessary details. Failure to report an accident is an offence which also carries a maximum fine of £5000.

* NB: For these purposes an animal means any horse, ass, mule, cattle, sheep, pig, goat, dog (in Northern Ireland only, a 'hinnie' is added to this list).

Under the Road Traffic Act 1988 the need to stop following accidents extends to cover any damage caused to any property 'constructed on, fixed to, growing on, or otherwise forming part of the land in which the road is situated or land adjacent there to'. This means that if a vehicle runs off the road and no other vehicles or persons are involved the driver still has to report damage to fences, hedges, gate-posts, street bollards, lamp-posts, and so on.

Third parties injured in accidents who find when making claims for damages that the vehicle was uninsured at the time can make a claim for personal injuries only to the Motor Insurers' Bureau (MIB) (see also Chapter 9). More recently the MIB compensation scheme has been extended to cover claims for damage to property by uninsured vehicles. Such claims for property damage will only be accepted if the vehicle driver is traced – not otherwise – and provided no claim for the damage can be made elsewhere. There is a limit of £250,000 on claims and they are subject to a £175 excess clause.

Road Humps

Regulations permit the construction of road humps on sections of the highway where a 30mph speed limit is in force. The road hump will be treated as part of the highway provided it complies with the regulations. New regulations were introduced in 1989 to enable local authorities to reduce pedestrian casualties by the introduction of more road humps (see also p 174 regarding 20mph speed limit zones).

Sale of Unroadworthy Vehicles

It is an offence to sell, supply, offer to sell or expose for sale or supply a vehicle which is in such a condition that it does not comply with the construction and use regulations, and is therefore legally unroadworthy. Under the Road Traffic Act 1988 offenders are liable to a fine of up to £5000.

This means that to display a vehicle for sale which needs attention to bring it up to the required standard is an offence, even though the intention would have been to remedy any defects before a purchaser paid for or took the vehicle away. However, it would not be an offence if the buyer was made aware of the defects and he intended to remedy them or have them remedied before using the vehicle on the road.

It is also an offence to fit any part to a vehicle which, by its fitting, makes the vehicle unsafe and causes it to contravene the regulations. For example, fitting a tyre which is below the limits regulating tread depth would be to commit such an offence.

Under the Road Traffic Act 1988 (effective from 1 July 1992) the law relating to the sale of unroadworthy vehicles has been tightened: in particular, the seller of such a vehicle now has a statutory duty to take steps to ensure that the buyer is aware that the vehicle is in an unroadworthy condition. Previously the seller could rely on the defence that he believed that the vehicle was not to be used on the road until made roadworthy.

Seat Belts

Legislation specifying the compulsory wearing of seat belts in motor vehicles came into force on:

- *31 January 1983*
  for drivers and front-seat passengers of motor cars, light vehicles not exceeding 1525kg unladen weight registered on or after 1 April 1967 and vehicles not exceeding 3500kg gross weight registered since 1 April 1980.
- *1 July 1991*
  for adults (ie persons aged 14 years or over) travelling in the rear of motor cars to wear seat belts where they are fitted – irrespective of the age of the car.
- *2 February 1993*
  for the driver and any person sitting in a seat which is equipped with a seat belt (even on a voluntary basis) to wear the belt provided – this includes drivers of goods vehicles over 35 tonnes gross weight which are fitted with seat belts on a voluntary basis since there is no legal requirement for such vehicles to be so fitted.
  Also, from this date:
  - it is illegal to carry any unrestrained child in the front seat of a vehicle,
  - children under 3 years of age travelling in the front seat of a vehicle must be restrained by a suitable child restraint,
  - children under 12 years of age and under 150 centimetres (4ft 11ins) in height travelling in vehicles must use a restraint if there is a suitable one anywhere in the vehicle.

NB: Seat belts have been fitted to all new cars and taxis since 1987.

Responsibility for Seat Belt Wearing
It is the individual responsibility of the vehicle driver and any adult passengers (ie persons 14 years of age and over) to wear the seat belt provided for the seat in which they are sitting (ie front or rear). The driver is *not* liable where adult passengers fail or refuse to comply with the law in this regard. However, in the case of children under 14 years of age it is the driver's responsibility to ensure that the law is complied with, irrespective of whether the child's parents, guardians or other responsible person in whose charge they are, are in the vehicle.

Bench-Type Seats
Where a light goods vehicle is fitted with bench-type or double front passenger seat, the requirement for seat belt wearing applies only to the person riding in the outer part of the seat (ie furthest from the driver) and not to a person in the middle part of the seat provided that the nominated passenger seat is occupied. It is illegal to occupy the centre part of the seat (ie next to the driver) where a belt may not be provided if the outer part of the seat with the belt provided is unoccupied.

Failure to Wear Seat Belts
Failure by a person to wear a seat belt as required by law could result in a fixed penalty (currently £20) or a fine of up to £500 if convicted by a court. In a case

relating to illegally carrying an unrestrained child a fine of up to £200 could be imposed.

A court has ruled that if, as a result of an accident, injuries were sustained which may have been prevented or lessened had the injured person been wearing a seat belt, then the damage awarded to that person in any claim should be reduced by an appropriate amount. Subsequently, other cases involving motor accident claims have followed the same lines.

Exemptions
Exemption from seat belt wearing applies:
1. When holding a valid medical certificate giving exemption (see further details below).
2. When driving a vehicle contructed or adapted for the delivery or collection of goods or mail to consumers or addresses, while engaged in making local rounds of deliveries or collections.
3. When driving a vehicle at the time of carrying out a manoeuvre which includes reversing.
4. When accompanying a learner driver as a qualified driver and supervising the provisional entitlement holder while that person is performing a manoeuvre which includes reversing.
5. In the case of a driving test examiner (but *not* an instructor) who is conducting a test of competence to drive and who finds that wearing a seat belt would endanger himself or any other person.
6. In the case of a person who is driving or riding in a vehicle being used for fire brigade or police purposes, or for carrying a person in lawful custody, including a person being so carried.
7. In the case of a driver of a licensed taxi who is seeking hire, answering a call for hire, or carrying a passenger for hire; or of a driver of a private hire vehicle which is being used to carry a passenger for hire.
8. When *riding* in a vehicle being used under a trade licence for the purposes of investigating or remedying a mechanical fault in the vehicle. *Note: this particular exemption refers specifically to 'riding' in a vehicle and does not include 'driving' a vehicle for the same or similar purposes – therefore it must be concluded that the driver of a vehicle using it for the purpose described would not be exempt from wearing a seat belt whereas a passenger riding in the vehicle for the same purpose would be exempt.*
9. In the case of a disabled person, wearing a disabled person's seat belt.
10. In the case of a person *riding* (see note above) in a vehicle while it is taking part in a procession organised by or on behalf of the Crown. This exemption also applies to a person riding in a vehicle which is taking part in a procession held to mark or commemorate an event which is commonly or customarily held in the police area in which it is being held, or for which a notice has been given under the Public Order Act 1986.

The regulations also do not apply to a person who is:
1. driving a vehicle if the driver's seat if not provided with an adult seat belt;
2. riding in the front of a vehicle in which no adult belt is available to him/her;
3. riding in the rear of a vehicle in which no adult belt is available to him/her.

It should be noted that these exemptions relate to the non-wearing of seat belts where they are not provided, but this circumstance may involve other infringements of the law relating to the non-fitment of seat belts.

Use of Radios and Telephones in Vehicles

The *Highway Code* contains advice against using a hand-held microphone or telephone handset while the vehicle is moving except in an emergency. Drivers should not stop on the hard shoulder of a motorway to answer or make a call no matter how urgent. The *Code* recommends that a driver should only speak into a fixed, neckslung, or clipped-on microphone when it would not distract his attention from the road.

Autoguide in London

The government is proceeding with its plans for the Autoguide scheme in London. Licences were issued from 1 June 1990 for the operation of a pilot scheme, with approximately 1000 vehicles involved in the trials. The scheme covers a 400 sq mile area within the M25 London Orbital motorway ring controlled by 200 roadside beacons. It is anticipated that motorists will be able to have the Autoguide equipment in their cars within the 1990s.

Autoguide is a system that gives drivers recommended routes to their destination using a display fitted to the dashboard of their vehicle. Roadside beacons transmit the advice data to vehicles and also keep the central Autoguide computer abreast of current traffic conditions. The system is self-adjusting so it will not shift traffic congestion from one location to another.

A second system, Trafficmaster, which is less complex than Autoguide, has also been licensed and introduced on the M25 to help drivers avoid congestion and to monitor speeds on the motorway. Trafficmaster operates from a system of infra-red beacons mounted on motorway bridges which send signals to a central control when traffic flow speeds fall below 25mph. The information is then relayed to vehicle dashboard-mounted screens and updated every 15 seconds. The cost to motorists is currently estimated at £300 per vehicle installation and £222 annual subscription.

Bridge Bashing

A government campaign to reduce the number of so-called 'bridge-bashing' incidents may result in legislation requiring all large goods vehicles (probably over 3 metres – 9ft 10in – high) to have 'in-cab' displays of overall vehicle travelling height and load dimensions (not confirmed at the time of preparing this 1995 edition of the *Handbook*) – see also p 216.

In 1990, 800 such accidents were recorded, of which 11 resulted in train derailments. Now some 20 or so key railway bridges have been identified as providing risk of potential disaster, should they be struck by a large vehicle. Infra-red warning systems are to be installed at these key sites which will trigger alarms when over-height vehicles approach.

11: Northern Ireland Operations

Vehicles based and registered in England, Wales and Scotland must comply with all the normal legal requirements (eg vehicle condition, excise duty, insurance and observance of traffic rules) set out in this *Handbook* when operating in Northern Ireland, but particularly so in regard to 'O' licensing (Chapter 1), professional competence (Chapter 2), drivers' hours and record-keeping regulations (Chapters 3 and 4), tachographs (Chapter 5), driver licensing and testing (Chapters 6 and 7) and plating and testing (Chapter 16). It should be noted that in regard to road traffic and road traffic offences there are differences between the Northern Ireland requirements and those on the British mainland. A separate edition of the *Highway Code* is published for Northern Ireland.

Vehicles based and operated in Northern Ireland must comply with the law as it applies in the Province which, although substantially similar to that applicable in the rest of the United Kingdom, does vary in some respects as shown in this chapter.

Road Freight Operator Licensing

In order to carry goods by road for reward (but not for purely own-account operations) with vehicles over 3.5 tonnes permissible maximum weight in Northern Ireland it is necessary to hold a road freight operator's licence (RFOL) issued under the provisions of the Transport Act (Northern Ireland) 1967.

NB: The Department of the Environment in Northern Ireland is proposing to introduce 'O' licensing for own-account operators, and environmental controls for all licence holders and applicants.

Since 1 January 1978 regulations have been in effect to ensure that road freight operators are better qualified. These regulations, The Road Transport (Qualification of Operators) Regulations (Northern Ireland) 1977, implement EC Directive 561/74 *On admission to the occupation of road haulage operator*. Thus the requirements for obtaining a road freight operator's licence in Northern Ireland take account of this requirement and depending on the qualifications of the operator the licence may or may not be restricted to operations within the United Kingdom. The type and scope of the licence will be indicated clearly in the licence – covering either national operations or both national and international operations.

Conditions for Grant of 'O' Licence
Under the statutory requirements for the grant of a road freight operator's licence, an operator has to satisfy the issuing authority that he is:
1. Of good repute

2. Of appropriate financial standing
3. Professionally competent or that he employs a full-time manager who is professionally competent and of good repute.

The requirements of good repute and appropriate financial standing are as stated in detail in Chapter 1 of the *Handbook*. While the professional competence requirement in Northern Ireland is largely as explained in Chapter 2 there are some differences as shown below.

Professional Competence by Examination

A separate syllabus for the Royal Society of Arts examination is applicable to Northern Ireland. It differs mainly in section C, *Access to the Northern Ireland Market*, which covers vehicle and operators' licences and professional competence requirements. Copies of the syllabus *Examinations for the Certificates of Professional Competence in Road Transport (Northern Ireland)* may be obtained from: Royal Society of Arts Examinations Board, Westwood Way, Westwood Business Park, Coventry CV4 8HS. The current examination syllabus applicable to Northern Ireland national goods and national passenger operations was published in 1988 and both NI national examinations have taken the modular form, with a core examination common to both passenger and goods sectors.

Issuing Authority

For operations in Northern Ireland, the issuing authority for operators' and vehicle licences is: Department of the Environment (DoE) for Northern Ireland, Road Transport Licensing Branch, Upper Galwally, Belfast BT8 4FY.

Objections to Road Freight 'O' Licences

Objections to the grant of a road freight 'O' licence may be made to the DoE by the Road Haulage Association (RHA) and the Freight Transport Association (FTA) on the grounds that the statutory conditions for the grant of such a licence are not or will not be met.

Appeals

Appeals against refusal by the DoE to grant a road freight operator's licence or where such a licence has been suspended or revoked may be made within 28 days to the County Court.

Application for Road Freight Operator's Licence

Application for a road freight operator's licence must be made to: Department of the Environment (NI), Road Transport Licensing Branch, Upper Galwally, Belfast BT8 4FY.

Form RFL 1 is used for the application. This requires details to be supplied as follows:
1. Name and place of business.
2. Type of licence required (ie national or national and international operations).

3. Address of operating centre (ie where vehicles are to be parked when not in use). Where premises are not owned the 'original' rental agreement must be sent with a first application.
4. Number of vehicles to be operated (original receipts or hire purchase agreements for vehicles and trailers to be operated must be sent with the application).
5. Whether professional competence requirements are met.
6. If applicant is not professionally competent, the name and address of the full-time transport manager who is qualified must be given.
7. Evidence of professional competence (the original certificate must be sent).
8. Whether the applicant or his named transport manager has any convictions offences in the past 5 years which are not spent.
9. Details of convictions.
10. Names and addresses (and past addresses) of the applicant, the named transport manager, and all directors and/or partners.
11. Name and address of bank from which a reference can be obtained.
12. Declaration (signature, status and date).

Licences are usually valid for 3 years.

Further information may be required in support of a road freight operator's licence regarding the applicant's financial status and a statement of estimated operating costs for vehicles specified in the application. The former requires the applicant to send the following items:
1. Bank status report
2. Details of any credit facilities from bank or any other financial resources available (eg deposit accounts)
3. Copy of latest bank statement
4. Rental agreement for operating premises
5. Receipts or hire purchase agreements in respect of vehicles
6. Original certificate of competence
7. A statement showing a breakdown of estimated annual costs and receipts of running the proposed road transport business.

The requirement to supply statements of estimated operating costs for vehicles involves the need to complete a form designed for this purpose. The following information must be provided:
1. Details of vehicles
    (a) type and model
    (b) registration number and date of first registration
    (c) type of trailer
    (d) gross weight
    (e) unladen weight
    (f) carrying capacity
    (g) number of tyres fitted and size (including trailer)
    (h) estimated fuel consumption (miles per gallon)
    (i) estimated annual mileage
    (j) vehicle length (in feet).
2. Estimated running costs and standing costs for 1 year (one vehicle only):
    (a) wages (including National Insurance)
    (b) motor tax

(c) insurance for vehicle

(d) insurance for goods in transit

(e) fuel

(f) maintenance

(g) tyres

(h) shipping charges if applicable

(i) hire purchase payments on vehicle and trailer (if applicable)

(j) rent and rates for business premises

(k) depreciation on vehicle

(l) miscellaneous expenses (eg telephone, stationery, etc)

Total costs

3. Estimated income from 1 year's operation (one vehicle only).

It is generally believed (although never officially stated) that the DOE is looking for 'O' licence applicants to have at least 25 per cent of first year operating costs in the bank at the start of the operation or assets equivalent to £7500 for each rigid vehicle and £15,000 for each articulated vehicle either in the form of bank credits or agreed overdraft facilities.

Penalties for Improper Use

Anyone operating internationally on a road freight 'O' licence that is restricted to national operations only will be liable, on conviction, to a fine and his 'O' licence and vehicle licences may be revoked or suspended. There are also penalties laid down for using unlicensed vehicles on national operations.

Changing Licences

The holder of a road freight 'O' licence which is restricted to national transport operations may have the restriction removed to enable him to engage in both national and international operations if he can meet the additional requirements (namely professional competence in both national and international transport operations) for engaging in international operations.

Vehicle Licences

Northern Ireland motor vehicles over 3.5 tonnes gross weight may not be used to carry goods for hire or reward except under a road freight vehicle licence granted under Section 17 of the Transport Act (NI) 1967 (as amended). Vehicle licences may only be granted to a person who holds a road freight 'O' licence and will cover either:

1. National transport operations only (ie covering Northern Ireland and the remainder of the UK) – Blue disc.
2. National and international transport operations – Green disc.

Vehicle licences are valid for 1 year and are issued in the form of a disc for display in the vehicle windscreen. They are serial numbered and show the vehicle registration number, the name of the owner and the date of expiry. Application is made on form RFL 3 to the DoE Northern Ireland (see address below) enclosing both the vehicle excise licence disc and the goods vehicle certificate as evidence that the vehicle is taxed and tested, if applicable (see p 205), on the date of issue of the vehicle licence.

Application for Road Freight Vehicle Licences

Applications for road freight vehicle licences have to be made on form RFL3 to the DoE at the address given below. Details which must be provided by the person or firm using the vehicles are as follows:

1. Name, address and telephone number of applicant
2. Road freight 'O' licence number
3. Whether the RFOL is valid for international operations
4. (a) Registration numbers of vehicles which are owned or in possession under an agreement for hire, hire purchase, credit sale or loan for which road freight licences are required
   (b) Type of vehicle
   (c) Type of body
   (d) Gross weight of vehicle
5. Address where vehicles will be parked overnight
6. Declaration that statements given are true.

Penalties for Illegal Use

Making a false statement to obtain the grant of a road freight operator's licence or a road freight vehicle licence is an offence punishable on conviction by a fine or imprisonment for up to 6 months or both. The 'O' licence could also be suspended or revoked. Use of a motor vehicle on a road for the carriage of goods for reward without a road freight vehicle licence can result in a fine which increases for subsequent convictions.

Own-Account Transport Operations

Under Northern Ireland operations, own-account operators who do not engage in hire or reward haulage operations are exempt from the requirement to hold road freight operators' licences. (*NB: See note at beginning of this chapter regarding proposed changes to NI operator licensing.*) However, if an own-account operator wishes to send vehicles across to Great Britain, they must carry an 'own-account document' indicating details of the user of the goods vehicle, his trade or business, the goods being carried, their loading and unloading points, the vehicle and the route.

This information may be set out on a firm's letter-headed notepaper but generally form GV243 should be obtained from the Road Freight Licence Office to formalise presentation of the information. The purpose of this document is to cover vehicles travelling in Great Britain without displaying a British 'O' licence disc.

Period Permits for Northern Ireland Operators

Hauliers who wish to engage in cross-border haulage of goods for reward into Eire must obtain a Period Permit for the territory of the Republic of Ireland. In order to obtain a Period Permit an applicant must hold a current road freight operator's licence valid for international operations and current road freight vehicle licences. Permits do not permit the operator to engage in internal transport in the Republic of Ireland. It is reported that there is no longer any check on these permits when vehicles cross the border and that NI operators are no longer applying for them.

Validity and Duration of Permits

Period permits are valid for any number of cross-border journeys within the period stipulated on the Permit but may only be used in respect of one vehicle at any one time. Permits are normally issued for 12 months but they may be issued for shorter periods in certain circumstances.

A Period Permit is not transferable to any other person and must be returned to the issuing office within 15 days of the expiry date.

Statistical Returns

Within 10 days of the end of each month, the holder of a Period Permit has to make a statistical return to the issuing office giving detailed information of all loads carried across the border in each direction. If no cross-border haulage has been undertaken during a month a 'nil' return must be made.

Application

Application has to be made on form RFL 58 to: Department of the Environment (NI), Road Transport Licensing Branch, Upper Galwally, Belfast BT8 4FY.

Exemptions from Period Permits

The following transport activities are exempt from the need for Period Permits for cross-border haulage into the Republic of Ireland:
1. Types of carriage listed in Annex 1 of the First Directive of the Council of the EC of 23 July 1952 as amended, that is:
   (a) Frontier traffic in an area extending for a distance of 25 kilometres as the crow flies, on each side of the frontier, provided that the total distance covered does not exceed 100 kilometres as the crow flies
   (b) Occasional carriage of goods to or from airports, in the event of air services being diverted
   (c) Carriage of luggage in trailers coupled to passenger-carrying vehicles and the carriage of luggage in all types of vehicle to and from airports
   (d) Carriage of mails
   (e) Carriage of vehicles which have suffered damage or breakdown
   (f) Carriage of refuse and sewage
   (g) Carriage of animal carcasses for disposal
   (h) Carriage of bees and fish fry
   (i) Funeral transport
   (j) Carriage of goods in motor vehicles the permissible laden weight of which, including that of trailers, does not exceed 6 tonnes or the

permissible payload of which, including that of trailers, does not exceed 3.5 metric tons

(k) Own-account transport as provided for in EC Council Directive 80/49/EC of 20 December 1979

(l) Carriage of articles required for medical care in emergency relief, in particular for natural disasters

(m) Carriage of valuable goods (eg, precious metals) effected by special vehicles accompanied by the police or other security guards.

2. The carriage of goods under an authorisation granted pursuant to Council for the carriage of goods by road between member states as amended, or by virtue of a licence issued pursuant to the scheme adopted by Resolution of the Council of Ministers of the European Conference of Ministers of Transport (ECMT) on 14 June 1973.

3. Transport on own account, provided that an appropriate own-account document is carried on the vehicle.

4. Transport of spare parts and provisions for ocean-going ships.

5. Transport of works and objects of art for fairs and exhibitions or for commercial purposes.

6. Transport of articles and equipment intended exclusively for advertising and information purposes.

7. Unladen runs by goods vehicles, but if the vehicle is entering to collect goods, the permit applying to the transport operation which is to follow should be carried on the vehicle on entry.

8. Transport of properties, accessories and animals to or from theatrical, musical, film, sports or circus performances, fairs or fêtes and those intended for radio, recording, or for film or television production.

Cabotage

Northern Ireland licensed operators must not collect and deliver goods within the British mainland. Similarly, British 'O' licensed operators must not collect and deliver goods within the province of Northern Ireland. These prohibited activities are known as cabotage.

Now that cabotage is legalised throughout the EU countries, Northern Ireland operators can take advantage of the opportunity to undertake such work within European states provided that they hold the necessary cabotage permit issued by the IRFO at Newcastle upon Tyne. Only a very limited number of such permits have been made available initially to NI road hauliers.

Drivers' Hours and Records

The EU requirements regarding drivers' hours and record keeping as described in Chapters 3 and 4 of the *Handbook* apply in Northern Ireland on substantially the same basis.

Tachographs

Legal requirements for the fitment and use of tachographs in Northern Ireland operations are the same as those described for the United Kingdom in Chapter 5.

Goods Vehicle Drivers' Licensing

Drivers of large goods vehicles in Northern Ireland must comply with the regulations for lgv driving licences and lgv driving tests as set out in Chapters 6 and 7.

Plating & Testing of Vehicles (Northern Ireland Certification)

The United Kingdom system of goods vehicle plating and annual testing does not apply in Northern Ireland. The Province has its own Goods Vehicle Certification scheme (commonly known and referred to in the Province as the PSV test) which requires heavy goods vehicles to be submitted to the Department of the Environment (NI) for an annual mechanical examination. Under The Goods Vehicles (Certification) Regulations (Northern Ireland) 1982, owners of goods vehicles (other than those specifically exempted – see p 208) must obtain a test certificate for each vehicle no later than 1 year from the date of first registration and annually thereafter (see exemption 10 on p 208).

Application for a certificate must be made to the following address at least 1 month before the date on which the certificate is to take effect: Department of the Environment, Vehicle and Driving Test Centre, Balmoral Road, Belfast BT12 6QL.

Re-test fees apply where application is made within 21 days from the date of service of the notice and the vehicle is presented for re-examination within 28 days.

Applications by Non-NI Based Bodies

Where an application is made by a corporate body with its principal or registered office outside Northern Ireland or by a person residing outside Northern Ireland the following conditions must be observed:
1. During the currency of the certificate a place of business must be retained in NI.
2. They must be prepared to accept, at such a place of business, any summons or other document relating to any matter or offence arising in NI in connection with the vehicle for which the certificate is applied for.
3. They must undertake to appear at any court as required by such a summons or by any other document.
4. They must admit and submit to the jurisdiction of the court relative to the subject matter of such summons or other document.

Failure to comply with any of the above-mentioned requirements will involve immediate revocation of the certificate.

Examination of Vehicle

When notified by the Department the applicant must present the vehicle for examination, in a reasonably clean condition, together with the registration book and previous certificate, if any, at the time and at the centre specified in the notice.

Issue of Certificate
If, after examining the vehicle, the Department is satisfied that it complies in all respects with the regulations in respect of the construction, use, lighting and rear marking of vehicles, a certificate will be issued.

Refusal of Certificate
If the vehicle does not meet the requirements of the regulations a certificate will be refused and the applicant will be notified of the reasons why.

Re-Examination of Vehicles
When a certificate has been refused and the defects specified in the notice have been put right, an application may be made for a further examination of the vehicle. A reduced re-test fee will be payable if the re-test is conducted within 21 days of the original test.

Refund of Fees

Prepaid test fees may be refunded in the following circumstances:
1. If an appointment for an examination of a vehicle is cancelled by the Department;
2. If the applicant cancels the appointment by giving the Department (at the centre where the appointment is made) 3 clear working days' notice;
3. If the vehicle is presented to meet the appointment but the examination does not take place for reasons not attributable to the applicant or the vehicle;
4. If the applicant satisfies the Department that the vehicle could not be presented for examination on the day of the appointment because of exceptional circumstances which occurred no more than 7 days before the day of the appointment, and provided that notice is given to the centre where the examination was to take place within 3 days of the occurrence.

Duplicate Certificates

Duplicate certificates may be issued in replacement of those which have been accidentally lost, defaced or destroyed. A fee is payable for replacement certificates. If subsequently the original certificate is found, it must be returned to the nearest examining centre or to any police station.

Display of Certificates

The certificate issued on satisfying the examiners must be attached to the vehicle in a secure, weather-proof holder and must be displayed on the nearside windscreen or on the nearside of the vehicle not less than 610mm and not more than 1830mm above the road surface so that the particulars of the certificate are clearly visible (ie at eye level) to a person standing at the nearside of the vehicle.

Conditions of Certificate

It is a condition of the certificate that the vehicle owner:
1. Must not permit the vehicle to be used for any illegal purpose;
2. Must not deface or mutilate the certificate or permit anybody else to do so;
3. Must, at all reasonable times, for the purpose of inspection, examination or testing of the vehicle to which the certificate relates:
   (a) produce the vehicle at such a time and place as may be specified by any Inspector of Vehicles;
   (b) afford to any Inspector of Vehicles full facilities for such inspection, examination or testing including access to his premises for that purpose;
   (c) must ensure that the vehicle and all its fittings are maintained and kept in good order and repair and must take all practical steps to ensure that all parts of the mechanism, including the brakes, are free from defects and are in efficient working order;
   (d) must immediately notify the nearest examination centre of any alteration in design or construction of the vehicle since a certificate was issued.

Transfer of Certificates

If a vehicle owner sells or changes the ownership of a vehicle, he must return the certificate for the vehicle to the nearest examination centre and notify the Department of the name and address of the transferee. The Department may then transfer the certificate on request by the new owner.

If a vehicle owner dies or becomes infirm of mind or body, on application of any person the Department may transfer the certificate to such a person.

Change of Address

If a certificate holder changes his address during the currency of a certificate, he must notify details of such changes to the nearest examination centre.

Markings on Vehicles

When certificates have been issued for vehicles, those vehicles must be marked with:
1. The name and address of the owner in legible writing and in a conspicuous position on the nearside of the vehicle; and
2. Where the unladen weight of the vehicle exceeds 1020kg, the unladen weight should be painted, or otherwise clearly marked, in a conspicuous position on the offside of the vehicle. In the case of articulated vehicles the unladen weight of the tractive unit and the trailer must be marked on the respective unit and trailer.

Offences

It is an offence to operate when a certificate has expired, to alter, deface, mutilate or fail to display a certificate. Failure to observe such rules will result in the certificate being declared invalid. It is also an offence to assign or to transfer a certificate to another person with the same resultant penalty. A fine or 6 months' imprisonment may be imposed on summary conviction for such offences or up to 2 years' imprisonment upon any further conviction or indictment.

Renewal of Certificates

At least 1 month before the expiry date of a certificate the holder should apply for a new one using an application form obtainable from any examination centre or Local Vehicle Licensing Office of the Department.

Exemptions from Certification

The following vehicles are exempt from the requirements of NI certification:
1. Vehicles constructed or adapted for the sole purpose of spreading material on roads or used to deal with frost, ice or snow;
2. A land tractor, land locomotive or land implement;
3. An agricultural trailer drawn on a road only by a land tractor;
4. A vehicle exempted from duty under section 7(i) of the Vehicles (Excise) Act (Northern Ireland) 1972 and any trailer drawn by such a vehicle;
5. A motor vehicle for the time being licensed under the Vehicles (Excise) Act 1971, paragraph (a);
6. A trailer brought into NI from a base outside NI if a period of 12 months has elapsed since it was last brought into NI;
7. A pedestrian-controlled vehicle;
8. A track-laying vehicle;
9. A steam-propelled vehicle;
10. A vehicle used within a period of 12 months prior to the date of it being registered for the first time in NI or the UK; or, where a vehicle has been used on roads in NI or elsewhere before being registered, the exemption applies for the period of 12 months from the date of manufacture rather than from the date of registration – for this purpose any use before the vehicle is sold or supplied retail is disregarded.

Builders' Skips

Under regulations applicable in Northern Ireland, builders' or rubbish skips must not be left on any footpath, cycle track, central reservation or verge on roads where a vehicle is prohibited from waiting or which are subject to a 30 mph speed limit without the written consent of the Northern Ireland Department of the Environment.

Skips must be clearly visible, both ends must be painted yellow and a red and yellow reflective chevron strip must be shown along the top. The skip owner's name and address or telephone number must be displayed. Traffic cones must be placed on the approaches to a skip and yellow lights must be attached to each corner at night. No danger must be caused through the siting of the skip.

12: Vehicle Weights and Dimensions

European and British law relating to the weights and dimensions of goods vehicles and trailers is extremely complex. Much of it is difficult for operators to comprehend and apply, and, even worse, for their drivers who may, when loading, have to make on-the-spot decisions which could later prove to be wrong, thereby breaching the law and bringing possible prosecution for themselves and their employers. The major weight and dimensional changes in recent years are as follows:

May 1983, the maximum permissible weight for certain articulated vehicles was increased from 32.5 tonnes to 38 tonnes and other dimensional changes were introduced.

March 1990, the maximum length for certain articulated vehicles was increased to 16.5 metres with corresponding increases in semi-trailer length under the provisions of the Road Vehicles (Construction and Use) (Amendment) Regulations 1990. But these changes were combined with a more technically complex method of measurement and with very restrictive turning circle limitations which have caused many headaches in their interpretation and application. Other such vehicles not complying with the specification set out in regulations remain restricted to 15.5 metres.

October 1991, the maximum length for drawbar vehicle combinations was increased to 18.35 metres with various restrictive intermediate dimensions also being imposed which effectively takes away any benefits of additional load platform length which might otherwise have been expected.

1 January 1993, three-axle rigid vehicles permitted up to 25,000kg gross weight on a 4.9 metre outer-axle spread (with conventional suspension) and up to 26,000kg gross weight provided that they have twin-wheel drive axles, an outer-axle spread of at least 5.2 metres and are equipped with 'road-friendly' suspensions. Four-axle rigid vehicles are permitted up to 32,000kg gross weight, provided that they have twin-wheel drive axles and 'road-friendly' suspensions, with a stipulation that the gross weight does not exceed 5 tonnes per metre of outer-axle spread (effectively meaning a maximum outer-axle spread of 6.4 metres), and four-axle articulated vehicles and four- and five-axle drawbar combinations are permitted at 35,000kg provided that they have 'road-friendly' suspensions.

March 1994, six-axle articulated vehicles and drawbar combinations with 'road-friendly' suspensions which are used exclusively in combined road-rail transport operations are permitted to operate at 44 tonnes gross weight. The maximum weight for drawbar combinations is increased from 35 tonnes to 38 tonnes.

NB: A 'road-friendly' suspension is an air suspension system, or one which is equivalent under EC Directive 92/7 (Annex III). An air suspension system qualifies if at least 75 per cent of the spring effect is caused by an air spring or one which is air equivalent.

The UK regulations in which these vehicle weights and dimension provisions are to be found are the Road Vehicles (Construction and Use) Regulations 1986 (as amended) as well as in specific EU Directives. These UK regulations also cover many other aspects of vehicle construction and use, but for ease of understanding the subject has been split into two separate chapters in this *Handbook*. This chapter deals with vehicle weights and dimensions and Chapter 13 deals with the other important topics under the construction and use regulations.

Length

Rigid and Articulated Vehicles

The maximum overall length permitted for rigid vehicles is 12 metres. For certain articulated vehicles the maximum permitted length is 16.5 metres (see also below), provided the combination can turn within minimum and maximum swept inner and outer concentric circles of 5.3 metres radius and 12.5 metres radius respectively (see Figure 12.1), otherwise the maximum permitted length remains at the old limit of 15.5 metres. The swept circle requirements do not apply to low loader or step-frame low-loader combinations, car transporters, articulated vehicles constructed to carry indivisible loads of exceptional length, articulated vehicles with semi-trailers built or converted to increase their length prior to 1 April 1990 or to articulated vehicles not exceeding 15.5 metres overall length.

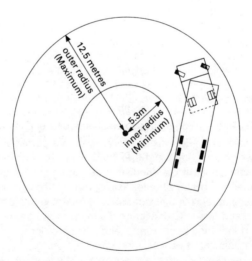

Figure 12.1 *The maximum and minimum outer and inner swept circles within which a new 16.5 metre long articulated vehicle must be able to turn*

For the purposes of enforcement of the turning circle requirements, the DoT has notified vehicle manufacturers that it will take a notional measurement from the king-pin to the centre line of the semi-trailer bogie. Where such a dimension does not exceed 7.8 metres the combination will be assumed to comply. Where this dimension exceeds 7.8 metres the DoT reserves the right to demand a turning circle demonstration on a steering pad.

The maximum overall length for an articulated vehicle incorporating a low-loader semi-trailer (but not a step-frame semi-trailer) is increased to 18 metres. Such vehicles do not have to meet the turning circle requirements described above.

Where an articulated vehicle is designed to carry indivisible loads of exceptional length there is no length restriction (an indivisible load means 'a load which cannot without undue expense or risk of damage be divided into two or more loads for the purpose of conveyance on a road').

Trailers

The maximum length for any drawbar trailer which has four or more wheels and is drawn by a vehicle which has a maximum gross weight exceeding 3500kg, is 12 metres. The same 12 metre maximum length limit also applies to agricultural trailers. The maximum permitted length for all other drawbar trailers is 7 metres.

In the case of composite trailers (see p 260) having at least four wheels and drawn by a goods vehicle over 3500kg permissible maximum weight or by an agricultural vehicle, these may be up to 14.04 metres long.

Vehicle and Trailer Combinations (Drawbars)

When a rigid motor vehicle is drawing a trailer the maximum overall length for the combination is 18.35 metres. Subject to certain other minimum dimensional requirements being met (see below).

NB: Although the maximum individual lengths for both drawing vehicle and trailer are 12 metres as stated, two such maximum length units obviously cannot be combined within the 18.35 metre limit.

The EC Directive (EC 91/60) permitting the increase in maximum length for drawbar combinations to 18.35 metres combines that dimension with limits on:

- maximum loadspace – 15.65 metres to be shared between the two bodies (previous close-coupled drawbars operating within the 18 metre length limit could achieve up to 16.2 metres loadspace);
- a minimum coupling dimension of 0.35 metres (to provide a 16 metre 'envelope' of load and coupling space); and
- a minimum cab length of 2.35 metres (see Figure 12.2).

Vehicles to these dimensions are permitted in UK domestic operations as well as on international journeys.

Existing high-volume drawbar combinations that do not conform to the new requirements are permitted to continue in operations until 31 December 1998, provided they were first registered before 1 October 1991.

When a trailer is designed for carrying indivisible loads of exceptional length the length of the drawing vehicle must not exceed 9.2 metres and the whole combination must not exceed 25.9 metres.

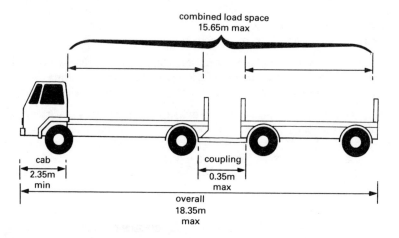

Figure 12.2 *The maximum dimensions for drawbar vehicle combinations from 1 October 1991 for domestic and international operations*

When two or more trailers are drawn the overall length of the combination must not exceed 25.9 metres unless an attendant is carried and 2 days' notice is given to the police. When two trailers are drawn within the 25.9 metre limit mentioned here (ie only legally permissible with a vehicle classed as a motor tractor or locomotive), only one of the trailers may exceed an overall length of 7 metres. When three trailers are drawn (ie only legally possible with a vehicle classed as a locomotive) none of the trailers may exceed a length of 7 metres.

The limits do not apply when a broken-down vehicle (which is then legally classed as a trailer) is being towed.

Articulated Semi-Trailers

The maximum permitted length for certain articulated vehicles is 16.5 metres. This applies where such vehicles include a semi-trailer with a distance from the centre-line of the king-pin to the rear of the trailer which does not exceed 12 metres and where the distance from the king-pin (or rearmost king-pin if more than one) to the furthest point on the front corner of the trailer does not exceed 2.04 metres (4.19 metres for car transporters) – see Figure 12.3.

In practical terms this provides a loadspace length of up to 13.61 metres (including front and rear wall thicknesses/headboards, etc) with flat platform or dry-freight trailers (at 2.5 metres wide) or a maximum of only 13.57 metres in the case of refrigerated semi-trailers (at 2.6 metres wide) – see Figure 12.4. Dry freight semi-trailers built to this length can accommodate 26 metric pallets (ie 1000 x 1200mm) or 33 Europallets (ie 800 x 1200mm).

Articulated semi-trailers built since 1 May 1983 are limited to a maximum length of 12.2 metres. There is no specified length limit for semi-trailers built

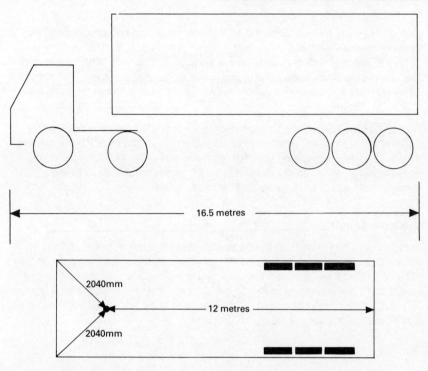

Figure 12.3 *The maximum dimensions for new articulated vehicles and semi-trailers*

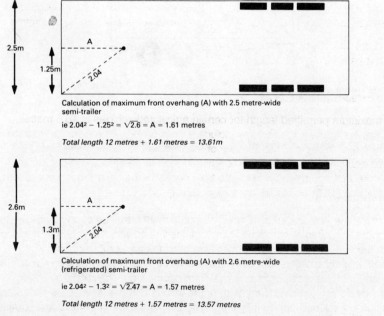

Calculation of maximum front overhang (A) with 2.5 metre-wide semi-trailer

ie $2.04^2 - 1.25^2 = \sqrt{2.6} = A = 1.61$ metres

Total length 12 metres + 1.61 metres = 13.61m

Calculation of maximum front overhang (A) with 2.6 metre-wide (refrigerated) semi-trailer

ie $2.04^2 - 1.3^2 = \sqrt{2.47} = A = 1.57$ metres

Total length 12 metres + 1.57 metres = 13.57 metres

Figure 12.4 *Diagram showing the calculation of semi-trailer lengths for operation at 16.5 metres overall length*

prior to this date. In measuring the 12.2 metre dimension no account need be taken of the thickness of front or rear walls or any parts in front of the front wall or behind the rear wall or closing device (ie door, shutter, etc). The thickness of any internal partitions must be included in the length measurement. This means effectively that the dimension relates only to load space between the front and rear walls (Figure 12.5).

The 12.2 metre length limit for semi-trailers as described does not apply to a trailer which is normally used on international journeys part of which are outside the UK. Similarly, articulated vehicles operating within the 15.5 metre limit on international journeys do not have to meet the turning circle requirements described above for 16.5 metre long vehicles.

Minimum Length

In the case of articulated combinations with maximum permissible weights above 32.5 tonnes, specified minimum overall length dimensions have to be observed. These range from 10 metres minimum length for 33 tonnes maximum weight to 12 metres minimum length for 38 tonnes maximum weight (see table on p 220). There is no minimum length requirement for articulated vehicles up to and including 32.5 tonnes maximum gross weight.

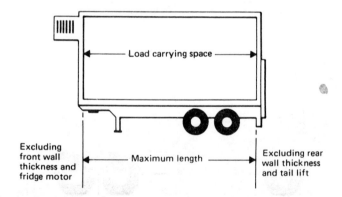

Figure 12.5 *Measurement of the maximum length dimension for pre-1990 and other semi-trailers not conforming to the 1990 regulations*

Measurement of Length

In measuring vehicle or trailer length account must be taken of any load-carrying receptacle (eg demountable body or container) used with the vehicle. Excluded from the overall length measurement are such things as rubber or resilient buffers and receptacles for customs seals. In the case of drawbar combinations the length of the drawbar is excluded from overall length calculations. With dropside-bodied vehicles the length of the tailboard in the lowered (ie horizontal) position is excluded unless it is supporting part of the load in which case it must be included in the overall length measurement and for the purposes of establishing overhang limits (see below).

Overhang

Overhang is the distance by which the body and other parts of a vehicle extend beyond the rear axle. The maximum overhang permitted for rigid goods vehicles (ie motor cars and heavy motor cars) is 60 per cent of the distance between the centre of the front axle and the point from which the overhang is to be measured. The point from which overhang is measured is, in the case of two-axled vehicles, the centre line through the rear axle, and in the case of vehicles with three or more axles two of which are rear axles, 110mm to the rear of the centre line between the two rear axles (Figure 12.6).

This regulation does not apply to vehicles used solely in connection with street cleansing; the collection or disposal of refuse; the collection or disposal of the contents of gullies or cesspools; works trucks; or tipping vehicles, provided the total overhang does not exceed 1.15 metres (3ft 9in approx). There is no specified overhang limit on trailers.

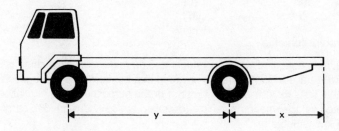

Two-axled vehicles: Overhang 'x' must not exceed
60 per cent of length 'y'

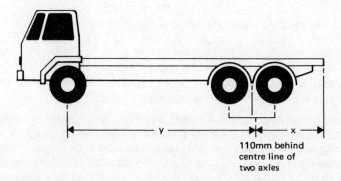

110mm behind
centre line of
two axles

Vehicles with three axles or more: Overhang 'x' measure from 110mm behind the centre line of the two rear axles must not exceed 60 per cent of the length of 'y' which is the distance between the centre line of the front wheel to the centre line of the two rear axles plus 110mm.

Figure 12.6 *How to measure overhang on vehicles with two axles, three axles and more. This measurement applies equally to two- and three-axle tractive units*

Width

Motor Vehicles

The overall width of motor tractors, motor cars and heavy motor cars (most goods vehicles are included in these classifications) must not be more than 2.5 metres and the maximum width of locomotives must not be more than 2.75 metres.

NB: In due course an increase in width to 2.55 metres will be permitted to allow easier stacking of standard pallets. No date for implementation was available at the time of preparing this edition of the Handbook.

Trailers

The maximum permissible width for trailers is 2.5 metres provided the drawing vehicle has a maximum permissible weight exceeding 3500kg. If the towing vehicle is below this weight, the width of the trailer must not exceed 2.3 metres.

Refrigerated Vehicles

The maximum permitted width for refrigerated (ie reefer) vehicles, semi-trailers and drawbar trailers is 2.60 metres provided that the thickness of the side walls (inclusive of insulation) is at least 45mm. For the purposes of this regulation 'refrigerated vehicle' means a vehicle (or trailer) specially designed to carry goods at low temperature.

Height

Apart from the particular case mentioned below, there are currently no legal maximum height limits for other goods vehicles or for loads in Britain but these are, obviously, governed by the height of bridges on the routes on which the vehicles are operated. For general information, the minimum height of bridges on motorways is normally 16ft to 16ft 6in and the maximum heights for buses is 4.57 metres (see also item regarding high loads on p 373).

The operator must bear in mind, however, that if he loads vehicles to a height which could cause danger he would be liable to prosecution under the construction and use regulations. Also, if a vehicle were to be loaded to a height whereby the load hit a bridge on the route being used, the operator could be accused under these regulations of using a vehicle on a road for a purpose for which it was so unsuitable as to cause, or to be likely to cause, danger.

EU regulations specify a height limit of 4 metres but this does not apply in the UK.

Height Limit on Vehicles Over 35 tonnes

A maximum height limit of 4.2 metres is imposed on both articulated vehicles and drawbar combinations operating at (ie laden to) weights in excess of 35 tonnes (previously 32.5 tonnes).

Measurement of the height is taken when the vehicle is standing on level ground and it extends to the top of any part of the structure of the vehicle, including the top of any detachable structure attached to the vehicle for containing a load (eg, demountable bodies and containers).

The height limit does not include loads themselves or sheets, ropes, webbing straps, nets, chains or any other flexible securing device. It does not apply at the time when vehicles are being loaded or unloaded nor does it apply when their laden weight is reduced to 35 tonnes or less.

Height Marking

When vehicles or trailers are used to carry containers, engineering equipment or skip loaders, the overall height of the vehicle and its load must be indicated to the driver in the vehicle cab if the height exceeds 3.66 metres. The height marking, in letters and figures at least 40mm tall, must show the 'travelling height' to within 25mm. For the purpose of this regulation 'containers' means containers of a type which are not in themselves a vehicle or trailer (ie it excludes box vans and box van trailers) and which have a volume of at least 8 cubic metres. This clearly means that demountable bodies are included. Travelling height is the maximum height of the vehicle or its load measured from a level road surface.

NB: Proposals are currently under discussion to extend the requirement for height marking described above to all vehicles over 3 metres (9ft 10in) high to avoid further instances of 'bridge bashing' – see also item in Chapter 10.

Weight

Maximum permitted weights (the total weight of the vehicle and load, including the weight of fuel, and the driver and passenger if carried) for goods vehicles and trailers depend on their wheelbase, the number of axles, the outer axle spread (the distance between the centre of the wheels on the front and rearmost axles) or the relevant axle spacing in the case of articulated vehicles (see p 220).

All rigid goods vehicles over 3500kg gross weight, articulated vehicles and trailers over 1020kg unladen weight should be fitted with a DoT plate (see p 291) on which is shown, for that vehicle, the maximum permissible axle weights and gross weight (or in the case of articulated vehicles, the combined weight of the tractive unit and trailer) for that vehicle in Great Britain.

The maximum permitted laden weights for different types of vehicle are shown in the following tables.

Rigid Vehicles and Trailers (not fitted with road-friendly suspensions)

	Description	kg
1.	Trailer with two closely spaced axles and with a distance between the foremost axle of the trailer and the rearmost axle of the drawing vehicle of at least 4.2 metres	18,000
2.	Trailer with three closely spaced axles and with a distance between the foremost axle of the trailer and the rearmost axle of the drawing vehicle of at least 4.2 metres	24,000
3.	Two-axled *vehicle* with a distance between the foremost and rearmost axles of at least 3.0 metres	17,000
4.	Two-axled *trailer* with a distance between the foremost and rearmost axles of at least 3.0 metres	18,000

Rigid Vehicles (not falling within above table)

	No of axles	Distance between foremost and rearmost axles	kg
1.	2	less than 2.65 metres	14,230
2.	2	at least 2.65 metres	16,260
3.	3 or more	less than 3.0 metres	16,260
4.	3 or more	at least 3.0 but less than 3.2 metres	18,290
5.	3 or more	at least 3.2 but less than 3.9 metres	20,330
6.	3 or more	at least 3.9 but less than 4.9 metres	22,360
7.	3	at least 4.9 metres	25,000
8.	4 or more	at least 4.9 but less than 5.6 metres	25,000
9.	4 or more	at least 5.6 but less than 5.9 metres	26,420
10.	4 or more	at least 5.9 but less than 6.3 metres	28,450
11.	4 or more	at least 6.3 metres	30,000

Maximum permitted gross weights for rigid vehicles (not including articulated tractive units) with drive axles fitted with twin tyres and road-friendly suspensions are as follows.

	No of axles	Distance between the foremost and rearmost axles	kg
1.	2	less than 2.65 metres	14,230
2.	2	at least 2.65 metres	16,260
3.	3 or more	less than 3.0 metres	16,260
4.	3 or more	at least 3.0 but less than 3.2 metres	18,290
5.	3 or more	at least 3.2 but less than 3.9 metres	20,330
6.	3 or more	at least 3.9 but less than 4.9 metres	22,360
7.	3 or more	at least 4.9 but less than 5.2 metres	25,000
8.	3	at least 5.2 metres	26,000

9.	4 or more	at least 5.2 but less than 6.4 metres	the distance in metres between the foremost and rearmost axles multiplied by 5000 and rounded up to the next 10kg
10.	4 or more	at least 6.4 metres	32,000

Articulated Vehicle Weights

The maximum weights for different classes of articulated vehicles, except those used in combined transport operations (see pp 223–4), based on the number of axles (subject to plated weights and axle spacings) are as shown in the table below.

	Type of articulated vehicle	Maximum weight (kg)
1.	Motor vehicle first used on or after 1 April 1973 and semi-trailer having a total of five or more axles	38,000
2.	Motor vehicle with two axles first used on or after 1 April 1973 and semi-trailer with two axles being used on international transport	35,000
3.	Motor vehicle with two axles first used on or after 1 April 1973 with driving axles having twin tyres and and road-friendly suspension, and two-axle semi-trailer	35,000
4.	Motor vehicle and semi-trailer with a total of four axles (not covered in items 2 and 3 above)	32,520
5.	Motor vehicle with two axles first used on or after 1 April 1973 with driving axles having twin tyres and road-friendly suspension, and single-axle semi-trailer	26,000
6.	Motor vehicle with two axles and single-axle semi-trailer (not covered in item 5 above)	25,000

Maximum Permitted Laden Weights for Articulated Vehicles
The maximum permitted laden weight for *complete articulated vehicles* (ie tractive unit and semi-trailer) depending on the number of axles of the tractive unit, the relevant axle spacing (ie rearmost axle of tractive unit to rearmost axle of semi-trailer – see Figure 12.7) and the minimum overall length (in some cases only) is as given in the following table.

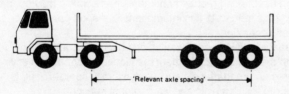

'Relevant axle spacing'

Figure 12. 7 *Measurement of axle spacing for articulated vehicles*

Relevant axle spacing (m)

Where motor vehicle has 2 axles	Where motor vehicle has at least 3 axles	Maximum weight (kg)	Minimum overall length (m)
at least 2.0	at least 2.0	20,330	–
at least 2.2	at least 2.2	22,360	–
at least 2.6	at least 2.6	23,370	–
at least 2.9	at least 2.9	24,390	–
at least 3.2	at least 3.2	25,410	–
at least 3.5	at least 3.5	26,420	–
at least 3.8	at least 3.8	27,440	–
at least 4.1	at least 4.1	28,450	–
at least 4.4	at least 4.4	29,470	–
at least 4.7	at least 4.7	30,490	–
at least 5.0	at least 5.0	31,500	–
at least 5.3	at least 5.3	32,520	–
at least 5.5	at least 5.4	33,000	10.0
at least 5.8	at least 5.6	34,000	10.3
at least 6.2	at least 5.8	35,000	10.5
at least 6.5	at least 6.0	36,000	11.0
at least 6.7	at least 6.2	37,000	11.5
at least 6.9	at least 6.3	38,000	12.0

Axle Spacing for Articulated Vehicles
For the purposes of determining articulated vehicle weights a dimension referred to as the 'relevant axle spacing' is used. This is defined as 'the distance between the rearmost axle of the drawing vehicle (ie the tractive unit) and the rearmost axle of the semi-trailer' (Figure 12.7).

Lorry and Trailer Combination Weights

Lorry and trailer (ie drawbar) combinations must not exceed 24,390kg gross weight unless the trailer is fitted with power-assisted brakes which remain operative even when the drawing vehicle's engine is not running and a brake pressure warning device is provided in the driver's cab in which case the permitted maximum train weight for the combination is 32,520kg or more, depending on specification – see below.

1. Where a drawbar combination was first used on or after 1 April 1973 and complies with the braking requirement mentioned above it may operate at up to 35,000kg gross train weight in international transport operations – but not in UK domestic operations.
2. Where a four-axle drawbar combination first used on or after 1 April 1973 complies with the braking requirement mentioned above and its driving axles have twin tyres and road-friendly suspension it may operate at up to 35,000kg in both UK domestic and international transport.
3. Where a drawbar combination with five or more axles first used on or after

1 April 1973 complies with the braking requirements mentioned above and all its driving axles have twin tyres and road-friendly suspension suspension (or do not exceed 8.5 tonnes) it may operate at up to 38,000kg.

4. Where a drawbar combination with at least six axles complies with the braking requirements mentioned above and all its driving axles have twin tyres and road-friendly suspension (or do not exceed 8.5 tonnes) it may operate at up to 44,000kg in combined transport operations only – see pp 223–4.

Trailer Weights

The maximum laden weight permitted for unbraked trailers is not more than half the unladen weight of the towing vehicle.

Trailers with overrun brakes are limited to a maximum laden weight of 3500kg unless they were first manufactured before 27 February 1977 in which case their maximum laden weight is 3560kg (these limits do not apply to agricultural trailers).

The maximum permissible weight for trailers used in draw-bar combinations is 18 tonnes provided the trailer has an axle spacing of at least 3 metres and is plated for 18 tonnes.

Axle and Wheel Weights

Maximum permitted axle and wheel weights (ie the total weight which may be imposed by an axle or a wheel on the road surface when the vehicle is fully loaded) for vehicles is as follows.

Axle/wheel type	Maximum permitted weight transmitted to road (kg)
1. Two wheels in line transversely each of which is fitted with a wide tyre or with two pneumatic tyres having the centres of their areas of contact with the road, not less than 300mm apart, measured at right angles to the longitudinal axis of the vehicle	
(a) if the wheels are on the sole driving axle of a motor vehicle (ie not a bus)	10,500
(b) if the vehicle is a bus which has 2 axles and of which the weight transmitted to the road surface by its wheels is calculated in accordance with C&U regulations	10,500
(c) in any other case	10,170
2. Two wheels in line transversely otherwise than as mentioned in item 1	9,200
3. More than two wheels in line transversely	
(a) in the case of a vehicle manufactured before 1 May 1983 if the wheels are on one axle of a group of two closely spaced axles or on one of three adjacent axles as mentioned in the C&U regulations	10,170

	(b) in the case of a vehicle manufactured on or after 1 May 1983	10,170
	(c) in any other case	11,180
4.	One wheel not transversely in line with any other wheel	
	(a) if the wheel is fitted as described in item 1	5,090
	(b) in any other case	4,600

Wheel and axle weights for motor vehicles and trailers not included above:

5.	More than two wheels transmitting weight on to a strip of the road surface on which the vehicle rests contained between two parallel lines at right angles to the longitudinal axis of the vehicle	
	(a) less than 1.02m apart	11,180
	(b) 1.02m or more apart but less than 1.22m apart	16,260
	(c) 1.22m or more apart but less than 2.13m apart	18,300
6.	Two wheels in line transversely	9,200
7.	One wheel, where no other wheel is in the same line transversely	4,600

Maximum permitted weights for vehicles with two closely spaced axles

Description of vehicle	Maximum permitted weight for two closely spaced axles (kg)
1. Motor vehicle or trailer with a distance between the two closely spaced axles in either case being less than 1.3 metres	16,000
2. Motor vehicle with a distance between the two closely spaced axles of at least 1.3 metres, or a trailer with a distance between the two closely spaced axles of at least 1.3 but less than 1.5 metres (and not being a vehicle described in item 3 or 4 below)	18,000
3. Motor vehicle with a distance between the two closely spaced axles of at least 1.3 metres and with every driving axle fitted with twin tyres and either road-friendly suspension or neither of them with an axle weight above 9500kg	19,000
4. Trailer with two closely spaced axles driven from the motor vehicle, and fitted with twin tyres and either road-friendly suspension or neither of them with an axle weight over 9500kg	19,000
5. Trailer with a distance between the two closely spaced axles of at least 1.5 but less than 1.8 metres	19,320
6. Trailer with a distance between the two closely spaced axles of at least 1.8 metres	20,000

Maximum permitted weights for three closely spaced axles

Description of vehicle	Maximum permitted weight for three closely spaced axles (kg)
1. Vehicle where smallest distance between any two of the three closely spaced axles is less than 1.3 metres	21,000
2. Vehicle where smallest distance between any two of the three closely spaced axles is at least 1.3 metres and at least one of the axles does not have air suspension	22,500
3. Vehicle where the smallest distance between any two of the three closely spaced axles is at least 1.3 metres and all three axles have air suspension	24,000

Vehicles for Combined Transport

New rules were introduced in March 1994, as part of the government's plans to boost rail freight in the run-up to privatisation, permitting swap-body and container-carrying lorries running to and from rail terminals to operate at up to 44 tonnes gross weight. The new limit applies only to articulated vehicles and drawbar combinations equipped with at least six axles and road-friendly suspensions (or having no axle exceeding 8.5 tonnes), and to articulated vehicles comprising specially built bi-modal semi-trailers (ie capable of running on road or rail), used in such combined road-rail transport operations.

For articulated vehicles the maximum laden weight permitted is dependent upon specified minimum axle spacings as follows:

Minimum axle spacing (metres)	Maximum laden weight (kg)
6.7	39,000
7.1	40,000
7.4	41,000
7.6	42,000
7.8	43,000
8.0	44,000

For the purposes of the legislation, containers and swap-bodies are defined as being receptacles at least 6.1 metres long designed for repeated carriage of goods and for transfer between road and rail vehicles.

In order to take advantage of the weight increase, the driver must carry with him documentary evidence to show that the swap-body or container load is on its way to a rail terminal (the document must show the name of the rail terminal, the date of the contract and the parties to it), or is on its way back from a rail terminal (in which case the document must show the terminal and the date and time that the unit load was collected). There is no restriction on the distance that may be travelled to or from a rail terminal for the purposes of complying with this legislation.

Overall Weight Limits

The total weight of the load on a vehicle, together with the weight of the vehicle itself, must not exceed the maximum permitted weight for each individual axle or for the vehicle.

Notional Weights (Multipliers)

Regulations enable a notional gross weight to be determined for unplated vehicles and trailers from an unladen weight for driver licensing purposes. The regulations specify a wide variety of vehicle types but the most important are:

1. Heavy motor cars or motor cars first used before 1 January 1968 or locomotives or motor tractors first used before 1 April 1973 – *multiply unladen weight by factor of 2.*
2. Articulated vehicles – *multiply the combined unladen weights of the tractive unit and the semi-trailer by a factor of 2.5.*

See pp 103–4 for full list of multipliers.

Weight Offences

It is an offence on the part of both the driver and the vehicle operator (ie the driver's employer) to operate a goods vehicle on a road laden to a weight above that at which it has been plated by the DoT (ie above the maximum permitted gross and individual axle weights) and both are liable to prosecution. Such offences are 'absolute' in that once the actual overweight has been established the fact that it was a deliberate action to gain extra revenue or purely accidental, unintentional, outside the driver or vehicle operator's control or loading was in a place where no weighing facilities existed, is of no consequence in defending against the charge.

Currently, the policy adopted by the Vehicle Inspectorate is that if a vehicle exceeds its weight limits by more than 5 per cent (up to a maximum of 1 tonne) it will be prohibited from proceeding on its journey until the weight is reduced to within legal limits – see below. Where an overload exceeds 10 per cent of the vehicle's legal limits or a maximum of 1 tonne, a prosecution will follow with heavier penalties being imposed on conviction. Trading standards officers may not take such a lenient view in their dealing with overloaded vehicles.

Defence

There is no defence of 'due diligence' against charges of overloading, something for which the Road Haulage Association has been campaigning, but it is a defence under the Road Traffic Act 1988 (section 42) to prove that at the time the contravention was detected the vehicle was proceeding to the nearest available weighbridge or was returning from such weighbridge to the nearest point at which it was reasonably practicable to remove the excess load. There have also been instances reported where successful defences have been made against conviction where the operator was able to show that he had no way of knowing or controlling the weight placed on a vehicle.

It is a further defence, where the weight exceeds maximum limits by not more than 5 per cent, to prove that the weight was within legal limits at the time of loading the vehicle and that no person had since added anything to the load.

Penalties
Overloading offences are looked upon very seriously by the enforcement authorities and by the courts and very heavy penalties are imposed on offenders. In addition to punitive fines (currently a maximum of £5000 per offence – an overloaded vehicle could result in a number of individual offences related to gross and axle weights), the operator risks losing his 'O' licence and the driver could put his lgv driving entitlement in jeopardy.

Prohibition of Overweight Vehicles
Any vehicle on a road found by a vehicle examiner to be overloaded to the extent that it could endanger public safety will be ordered off the road immediately. The necessary powers to enable this step to be taken are included in section 70(2) and (3) of the Road Traffic Act 1988 and they empower an authorised officer to prohibit the driving of a goods vehicle on a road if after having it weighed it appears to him that the vehicle exceeds the relevant weight limits imposed by the construction and use regulations, and as a result would be an immediate risk to public safety if it were used on a road. The officer may be one of the examiners or a specially authorised weights and measures inspector or a police constable.

*NB: A new endangerment offence in the Road Traffic Act 1991 (effective from 1 July 1992) could be applied to vehicle overload offenders in appropriate circumstances and on conviction or indictment they could be faced with heavy fines and even imprisonment for up to 7 years.

A prohibition notice, form GV160, will be issued to the driver of a vehicle found to be overweight and it is the driver's responsibility to remove the excess weight to his own satisfaction and clear the GV160 before proceeding on his journey. The penalty for ignoring a prohibition notice is a fine of up to £5000.

Official Weighing of Vehicles
Under the Road Traffic Act 1988 an authorised officer can request the person in charge of a vehicle to drive to a weighbridge to be weighed. If the journey is more than five miles and on arrival the vehicle is found to be within the legal limit then the vehicle owner can claim for the loss involved in making the journey. It is an offence to refuse to go to a weighbridge if requested (maximum fine currently £5000). Once a vehicle has been weighed the official in charge should give the driver a 'certificate of weight' (whether it is overloaded or not) and the vehicle will be exempt from further requests for weighing while carrying the same load on that journey.

Dynamic Weighers
Provisions regarding the weighing of vehicles are contained in the 1974 and 1988 Road Traffic Acts which give the Secretary of State for Transport powers to make regulations regarding the method of weighing vehicles and to specify

the limits within which a weight determined by a weighbridge is presumed to be accurate. Regulations are in force to permit the dynamic weighing of vehicles at the roadside (see also p 192).

Portable Weighers
There have been reports of Trading Standards officers in some areas using portable axle weigh-pads as a means of checking vehicle/axle weights and have brought overloading prosecutions against operators based on such weighings. There is considerable doubt about the accuracy of such machines and the Road Haulage Association has expressed its concern about their use.

Compliance with EU Weight and Dimension Increases

Membership of the EU imposes a clear mandate, as part of a Common Transport Policy, to harmonise goods vehicle weights and dimensions in order to establish a regime of fair competition among transport operators of the Community. Due to our commitment to European unity, Britain has agreed to conform to EU weights and dimensional requirements and has acceded to 40-tonne lorries – but the latter only from from 1 January 1999.

The EU requirements for harmonised vehicle and trailer weights and dimensions are to be found in EC Directive 85/3 and further in EC Directives 89/338 and 89/461, which amend the 1985 version.

13: Construction and Use of Vehicles

In constructing goods vehicles and trailers, manufacturers and bodybuilders must observe requirements regarding the specification and standards of construction of components and the equipment used in the manufacture. While some of these items are covered by the Type Approval scheme (see Chapter 14), the majority are included in the Road Vehicles (Construction and Use) Regulations 1986, Statutory Instrument (SI) 1078/1986 (available from HMSO) and subsequent amendments to these regulations which, collectively, are commonly referred to as the C&U regulations. Certain EU legislation also applies, as indicated in the text.

Once a goods vehicle or trailer has been built and put into service, it is the operator as the vehicle user (see pp 4–5) who must then ensure that it complies fully with the law regarding its construction and use when on the road. It is worth pointing out that where a vehicle on the road is found to contravene the constructional aspects of the regulations, it is the operator who will be prosecuted and if convicted he will be liable to meet the penalty and may find that it jeopardises his 'O' licence. It is no defence or excuse to say that the fault with the vehicle rests with the manufacturer, bodybuilder or even the supplying dealer.

Only the main items from the regulations and amendments which concern the goods vehicle operator and transport manager are included in this chapter. Those aspects of the regulations dealing with the limitations on vehicle weights and dimensions are covered in Chapter 12. The provisions of the C&U regulations dealing with safe operation and safety of vehicles and loads are explained in Chapter 20. Many other points relating to a wide variety of other types of vehicle (motor bicycles and invalid carriages, for example) are not included because they are not generally thought to be relevant to the reader of this *Handbook*.

Definitions of Vehicles

For the purpose of these regulations the following definitions, as given in the Road Traffic Act 1988, apply:

- A **goods vehicle** is a vehicle or a trailer adapted or constructed to carry a load.
- A **motor car** is a vehicle which, if adapted for the carriage of goods, has an unladen weight not exceeding 3050kg but otherwise has an unladen weight not exceeding 2540kg.
- A **heavy motor car** is a vehicle constructed to carry goods or passengers with an unladen weight exceeding 2540kg.
- A **motor tractor** is a vehicle which is not constructed to carry a load and has an unladen weight not exceeding 7370kg.

- A **light locomotive** is a vehicle which is not constructed to carry a load and which has an unladen weight of more than 7370kg but not exceeding 11,690kg.
- A **heavy locomotive** is a vehicle which is not constructed to carry a load and which has an unladen weight exceeding 11,690kg.
- An **articulated vehicle** as defined in the C&U regulations is a motor car or heavy motor car with a trailer so attached that when the trailer is uniformly loaded, at least 20 per cent of the weight of the load is imposed on the drawing vehicle.
- A **composite trailer** is a combination of a converter dolly and a semi-trailer, and is treated as one trailer only when considering the number of trailers which may be drawn.
- **Engineering plant** means movable plant or equipment in the form of a motor vehicle or trailer which is specially designed and constructed for the purposes of engineering operations and which cannot, for this reason, comply with the C&U Regulations. Also it is constructed to carry only materials which it had excavated from the ground and which it is specially designed to treat while being carried. It also means mobile cranes which do not conform in all respects with the C&U Regulations.
- A **land tractor** means a tractor with an unladen weight not exceeding 7370kg which is designed and primarily used for work on the land in connection with agriculture, grass cutting, forestry, land levelling, dredging or similar operations. To comply with this definition the tractor must be owned by a person engaged in agriculture or forestry or a contractor engaged in this business; it must not be constructed or adapted to carry any load other than its own loose equipment, or a load consisting of agricultural or woodland produce within 15 miles of a farm or forestry estate owned by the person who owns the tractor, or implements fitted to the tractor, for work on land or for forestry.
- A **pedestrian controlled vehicle** means a motor vehicle which is controlled by a pedestrian and which is not constructed or adapted to carry a driver or passenger.
- A **works truck** means a motor vehicle (other than a straddle carrier) designed for use in private premises and used on a road only in delivering goods from or to such premises, or from a vehicle on a road in the immediate neighbourhood, or in passing from one part of the premises to another or to other private premises in the immediate neighbourhood, or in connection with road works or in the immediate vicinity of the site of such works.
- A **works trailer** means a trailer used for the same purposes as a works truck.

Constructional Requirements

Brakes

All goods vehicles must meet specified braking efficiencies. The regulations state minimum efficiencies for the service brake, for the secondary brake, and for the parking brake or handbrake. On pre-1968 vehicles the secondary brake can be the handbrake and on post-1968 vehicles it can be a split or dual system operated by the footbrake. If it is the latter it must be capable of meeting the secondary requirement if part of the dual system fails. The parking brake must achieve the required efficiency by direct mechanical action or by the energy of a spring without the assistance of stored energy.

Every vehicle must have a parking brake system to prevent at least two wheels from turning when it is not being driven. All vehicles first used after 1 January 1968 must have an independent parking brake.

Anti-lock (ie anti-skid) braking systems are required on certain new articulated vehicles and drawbar trailer combinations under EU legislation (for which Category 1 – wheel-by-wheel systems – are necessary) and under British C&U regulations. The vehicles affected are articulated tractive units and rigid goods vehicles over 16 tonnes equipped to draw trailers and first used from 1 April 1992, and trailers over 10 tonnes built on or after 1 October 1991.

Specified Braking Efficiencies
Vehicles used before 1 January 1968:

Two-axle rigid vehicles	Service brake	45 per cent
	Secondary brake	20 per cent
Multi-axled rigid vehicles, trailer combinations and articulated vehicles	Service brake	40 per cent
	Secondary brake	15 per cent

Vehicles first used on or after 1 January 1968:

All vehicles	Service brake	50 per cent
	Secondary brake	25 per cent
	Parking brake – must be capable of holding the vehicle on a gradient of at least 1 in 6.25 without the assistance of stored energy (1 in 8.33 with a trailer attached).	

Maintenance of Brakes
The braking system on a vehicle, including all of its components and means of operation must be maintained in good and efficient working order and must be properly adjusted at all times.

Braking Standards
The regulations reflect the requirements for braking standards laid down in EC Directives 320/1973, 524/1975 and 489/1979 (as amended). These call for the fitting of load sensing valves or anti-lock braking on drive axles – most modern tractive units already comply with these requirements – and the overall emphasis is on stability, eliminating jack-knifing and trailer swing. EU rules permit the use of two-line air braking systems instead of the traditional British three-line systems. Most older tractive units in the UK have three-line braking systems fitted with yellow, blue and red couplings.

Vehicles built to the European standard have only two lines (red and yellow or two black). Coupling three-line tractive units to three-line trailers, two-line tractive units to two-line trailers and two-line tractive units to three-line trailers presents no difficulties. Problems arise when coupling three-line tractive units to two-line trailers. Such combinations must not be used unless they are specially designed or modified by fitting a fourth coupling or internal valves and connecting pipework.

On tractive units first registered before 1 April 1983 a notice should be displayed stating that the system is suitable for coupling. If no such notice is displayed the driver should check the system to see if it is suitable for coupling.

Endurance Braking Systems
From 1 January 1995 it is likely that UK braking requirements will be harmonised with those of the EU, thereby requiring new vehicles from this date carrying dangerous goods to be fitted with endurance braking systems. Older vehicles are likely to have to be retro-fitted by 1 January 2000. EU vehicles over 16 tonnes gross weight on cross-border ADR work have required such braking systems from 1 July 1993.

The purpose of endurance braking systems is to relieve excess loadings on the normal vehicle service brakes, particularly, for example, on long downhill gradients where they tend to fade due to overheating. A variety of proprietary retarders are available to meet this requirement.

Brakes on Trailers
Trailers constructed before 1 January 1968 must have an efficient braking system on half the number of wheels. Trailers constructed after this date must be fitted with brakes operating on all wheels which are capable of being applied by the driver of the drawing vehicle and having maximum efficiencies matching the braking requirement for the drawing vehicle, emergency brakes operating on at least two wheels and a parking brake capable of holding the trailer on a gradient of at least 1 in 6.25.

Overrun Brakes
Overrun brakes may be fitted to trailers not exceeding 3500kg gross weight (or 3560kg if made before 27 February 1977). Overrun brake couplings must be damped and matched with the brake linkage. Normally, to ensure that these standards are met, the coupling design needs to be type approved. Trailer braking efficiency must be at least 45 per cent and the parking brake must be capable of holding the laden trailer on a gradient of 1 in 6.25 (ie 16 per cent). Modern braked trailers must also be fitted with an emergency device which automatically applies the brakes if the trailer becomes uncoupled from the towing vehicle. This does not apply to single axle trailers up to 1500kg gross weight provided they are fitted with a safety chain or cable to stop the coupling head touching the road if the trailer becomes detached.

Light Trailer Brakes
Light trailers must be fitted with brakes if:
1. Their maximum gross weight exceeds 750kg and their unladen weight exceeds 102kg; or

2. Their maximum gross weight exceeds 750kg and they were built on or after 1 October 1982; or

3. Their laden weight on the road exceeds half the towing vehicle's kerbside weight (this does not apply to agricultural trailers or to trailers whose unladen weight does not exceed 102kg and which were built before 1 October 1982).

Unbraked trailers must have their maximum gross weight marked in kilograms in a conspicuous position on the nearside.

Parked Trailers
When trailers are detached from the towing vehicle they must be prevented from rolling by means of a brake, chain or chock applied to at least one of their wheels.

Lifting Axles

Draft amendments to the construction and use regulations previously circulated which were intended to introduce controls on the use of manually and automatically retractable axles have been delayed pending the introduction of Euro-wide legislation from the EU Commission.

The original object of the regulations was to ensure that the use, or incorrect use, of such axles does not result in the other axles of the vehicle becoming overloaded. The regulations would have banned in-cab control of lift-axles except for a facility to provide additional traction for brief periods only (60 seconds or 90 seconds with trailer lift-axles). When one axle of a tandem bogie is retracted, it proposed that the remaining axle be limited to 50 per cent of the permitted bogie weight and with tri-axle bogies the remaining axles should bear not more than 60 per cent of the maximum permitted bogie weight.

The definition of overhang was to be amended to disregard rear-fitted retractable axles (measurement of the wheelbase with a tandem-axle bogie would then be to the drive axle, not from a point behind the midway point between the axles – see Figure 12.6) and the regulations would have classified self-tracking axles as steering axles for the purposes of definitions.

Tyres
It is an offence to use, or cause or permit to be used on a road, a vehicle or a trailer with a pneumatic tyre which is unsuitable for the use to which the vehicle is being put. It is also an offence to have different types of tyres fitted to opposite wheels of the vehicle or trailer; for example, radial-ply tyres must not be fitted to a wheel on the same axle as wheels already fitted with cross-ply tyres and vice versa. Tyres must be inflated to the vehicle or tyre manufacturers' recommended pressures so as to be fit for the use to which the vehicle is being put (for example, motorway work or cross-country work). No tyre must have a break in its fabric or a cut deep enough to reach the body cords, more than 25mm or 10 per cent of its section width in length, whichever is the greater; also there must be no lump, bulge or tear caused by separation or partial fracture of

its structure, neither must there be any portion of the ply or cord structure exposed.

Approval Marks on Tyres

It is an offence to sell motor car tyres unless they carry an 'E' mark to show compliance with EU load and speed requirements. It is also an offence to sell retreaded car or lorry tyres unless they are manufactured and marked in accordance with British Standard BS AU 144b 1977.

Since 1 October 1990 it has been a requirement for tyres on heavy goods vehicles to show load and speed markings in accordance with UN ECE Regulation 30 or 54. It is a legal requirement that these limits of both loading and speed performance are strictly observed. Failure to do so can result in prosecution and could invalidate insurance claims in the event of an accident to a vehicle loaded above the weight limit of the tyres or travelling at a speed in excess of the tyre limit.

Tread Depth

All tyres on goods vehicles of over 3500kg gross weight must have a tread depth of at least 1mm across three-quarters of the breadth of the tread and around the entire circumference of the tyre. This 1mm tread depth must be in a continuous band around the entire circumference of the tyre. Further, on the remaining one-quarter of the width of the tyre where there is no requirement for the tread to be 1mm deep, the base of the original grooves must be clearly visible.

The minimum tread depth for cars, light vans (not exceeding 3500kg gross weight) and light trailers was increased to 1.6mm from 1 January 1992, and this applies across the central three-quarters of the width of the tyre and in a continuous band around the entire circumference. The 1mm limit stated above remains in force for heavy goods vehicles.

Recut Tyres

Recut tyres may be fitted to goods vehicles of over 2540kg unladen weight which have wheels of at least 405mm rim diameter and to trailers weighing more than 1020kg unladen weight and electric vehicles. They must not be used on private cars, dual-purpose vehicles, goods vehicles or trailers of less than the weight or wheel size specified.

Run-Flat and Temporary Use Spare Tyres

The regulations permit the legal use of 'run-flat' tyres in a partially inflated or flat condition and what are described as temporary use spare tyres. This is of more consequence to motor car users because the only tyres of the former (ie run-flat) type currently available are designed for a limited range of private cars (eg Dunlop Denovo tyres). Where a temporary use spare tyre is being used the vehicle speed must not exceed 50mph otherwise the legal provision which permits their use ceases to apply. The temporary use spare tyre or the wheel to which it is fitted must be of a different colour to the other wheels on the vehicle and a label must be attached to the wheel giving clear information about the precautions to be observed when using the wheel.

Lightweight Trailer Tyres
Tyres fitted to lightweight trailers since 1 April 1987 must be designed and maintained to support the maximum axle weight at its maximum permitted speed (ie 60mph).

Windscreen Wipers and Washers

All vehicles must be fitted with one or more efficient automatic windscreen wipers capable of clearing the windscreen to provide the driver with an adequate view to the front and sides of the vehicle. They must be maintained in good and efficient working order and must be adjusted properly. This provision does not apply if the driver can get an adequate view of the road without looking through the windscreen.

A vehicle should not be driven on the road with defective windscreen wipers or washers. This is an offence which could result in a fine of up to £1000 (as with many other C&U regulation offences).

Washers
Vehicles required to be fitted with windscreen wipers must be fitted with a windscreen washer which is capable, in conjunction with the wipers, of clearing the area of the windscreen swept by the wipers of mud or dirt. Washers are not required on land tractors, track-laying vehicles and vehicles which cannot travel at more than 20mph.

View to the Front

Drivers must have a full view of the road and traffic ahead at all times when driving. Obstructing the windscreen with mascots, stickers, stone guards and other such equipment could result in prosecution and/or failure of the vehicle when it is presented for annual test.

Mirrors

Goods vehicles and dual-purpose vehicles must be fitted with at least two mirrors. One of these must be fitted externally on the offside and the other must, in the case of vehicles first used since 1 June 1978, be fitted in the driver's cab or driving compartment. When an interior mirror does not provide an adequate view to the rear, a mirror must be fitted externally on the near side. The mirrors must show traffic to the rear or on both sides rearwards. Mirrors fitted to vehicles over 3500kg must conform to Class II and those fitted to other vehicles with Class II or III as described in EC Directive 127/71.

Fitment of Mirrors
External mirrors with a bottom edge less than 2 metres from the ground (when the vehicle is loaded) must not project more than 20cm beyond the overall width of the vehicle or vehicle and trailer. Mirrors fitted on the offside must be adjustable from the driving seat unless they are of the spring-back type. Internal mirrors fitted to vehicles first registered on or after 1 April 1969 must be framed with some material (usually plastic beading) which will reduce the risk of cuts to passengers who may be thrown against the mirror.

Type Approval for Mirrors
Since 1 June 1978 vehicles must be fitted with rear view mirrors bearing the EU type approval 'E' mark. The relevant date for the implementation of this requirement to Ford Transit vehicles was 10 July 1978 instead of 1 June 1978 which was the date for all other vehicles.

Wide Angle Mirrors
Goods vehicles over 12 tonnes maximum permissible weight first used since 1 October 1988 must be fitted with additional mirrors which provide close proximity and wide angle vision for the driver in accordance with EC Directives 205/85 and 562/86.

Horn

All vehicles with a maximum speed exceeding 20mph, except works trucks and passenger-controlled vehicles, must be equipped with an audible warning instrument. The sound emitted by a horn must be continuous and uniform and not strident. Gongs, bells, sirens and two-tone horns are only permitted on emergency vehicles but a concession allows similar instruments, except two-tone horns, to be used on vehicles from which goods are sold to announce the presence of the vehicle to the public. Any vehicle first used since 1 August 1973 must not be fitted with multi-toned or musical horns.

Restriction on Sounding Horns
Audible warning instruments must not be sounded at any time while the vehicle is stationary or in built-up areas (ie where 30mph speed restriction is in force) between 11.30 pm and 7.00 am (see also item about reversing alarms on p 261). The use of the horn on a stationary vehicle in an emergency situation (ie 'at times of danger due to another moving vehicle on or near the road') is allowed.

Horns Used as Anti-Theft Devices
Audible warning instruments which are gongs, bells or sirens may be used to prevent theft or attempted theft of a vehicle provided a device is fitted which will stop the warning sounding continuously for more than 30 seconds (from 1 August 1992). Hazard lights and interior lights may be set to operate continuously for five minutes if the vehicle is tampered with.

Speedometer

Speedometers (and/or tachographs, as appropriate – see Chapter 5) must be fitted to all vehicles registered since 1 October 1937 except those which cannot or are not permitted to travel at more than 25mph, agricultural vehicles which are not driven at more than 20mph and works trucks first used before 1 April 1984. In the case of vehicles first used since 1 April 1984 the speedometer must indicate speed in both miles per hour and kilometres per hour. The instrument must be maintained in good working order at all material times and kept free from any obstruction which might prevent it being easily read.

Defence

It is a defence to be able to show that a defect to a speedometer or a tachograph occurred during the journey the vehicle was on when the offence was detected or that, at that time, steps had been taken to get the defect repaired as soon as reasonably practicable.

Seat Belts

Seat belts for the driver and one front-seat passenger must be fitted to goods vehicles not exceeding 1525kg unladen registered since 1 April 1967 and goods vehicles not exceeding 3500kg gross weight first used since 1 April 1980. From 1 October 1988 goods vehicles over 3500kg must be fitted with seat belt anchorage points for each forward facing seat to which lap-strap type seat belts can be fixed (See Chapter 10 for details of seat belt fitment requirements in fleet cars).

Vehicles to which this regulation applies, first used since 1 April 1973, must be fitted with belts which can be secured and released with one hand only and also with a device to enable the belts to be stowed in a position where they do not touch the floor. Vehicles to which this requirement applies will fail the annual test if seat belts are not fitted, are permanently obstructed or are not in good condition.

The legal requirement for drivers and passengers to wear seat belts came into effect in January 1983 with certain exemptions (see p 196).

Silencer

An adequate means of silencing exhaust noise and of preventing exhaust gases escaping into the atmosphere without first passing through a silencer must be fitted to all vehicles. Silencers must be maintained in good and efficient working order and must not be altered so as to increase the noise made by the escape of exhaust gases.

Safety Glass

Goods vehicles must be fitted with safety glass (ie toughened or laminated, when fractured, does not fly into fragments likely to cause severe cuts) for windscreens and windows in front of and on either side of the driver's seat. The windscreen and all windows of dual-purpose vehicles must be fitted with safety glass. Glass bearing an approval mark under EC Directive 92/22 is regarded as meeting the requirements stated above for safety glass.

The glass must be maintained so as not to obscure the vision of the driver while the vehicle is being driven on the road. This means that a driver could be prosecuted for having a severely misted up, iced up or otherwise dirty windscreen.

Wings

Goods vehicles and trailers must be fitted with wings to catch, as far as practicable, mud and water thrown up by the wheels unless adequate protection is provided by the bodywork.

Articulated vehicles and trailers used for carrying round timber are exempt from these requirements in respect of all except the front wheels of the tractive unit. Vehicles and trailers in an unfinished condition which are proceeding to a body builder for work to be completed and works trucks are also exempt from the need to have wings.

Fuel Tanks

Vehicle fuel tanks must be constructed so as to prevent any leakage or spillage of fuel. In particular, there is concern about leakage from heavy vehicle fuel tanks on to the road where it causes exceptional danger to cyclists and motorcyclists. Failure to maintain diesel fuel tanks in good condition (and especially filler caps) is now an offence in its own right and gives the police opportunity to prosecute without having to actually observe spillage from the tank. Fines of up to £2000 could be imposed on conviction for such offences.

Vehicles first used since 1 July 1973 and manufactured since 1 February 1973 which are propelled by petrol engines must have metal fuel tanks fitted in a position to avoid damage and prevent leakage. This provision does not apply where the vehicle complies with relevant EU regulations (EC 221/70) and is marked accordingly.

Power-to-Weight Ratio

All goods vehicles powered by diesel engines first used on or since 1 April 1973 must comply with a power-to-weight ratio of 4.4kW per 1000kg (due to increase to 5kW per tonne under proposed EU regulations). The specific requirement in the regulations is that the power of the engine shown on the manufacturer's plate must be at least 4.4 times the gross weight shown on the plate (in kg).

When a vehicle is fitted with engine-driven ancillary equipment designed for use when the vehicle is in motion at a speed in excess of 5mph, the power from the vehicle's engine must be sufficient to provide power to drive such equipment and leave at least 4.4kW/1000kg available to drive the vehicle.

Exemptions to this requirement apply to vehicles first manufactured before 1 April 1973 and those first manufactured before 1 April 1973 and propelled by a Perkins 6.354 engine.

Underrun Bumpers

Rear underrun bumpers (referred to in legislation as rear underrun protection) must be fitted to most rigid goods vehicles over 3.5 tonnes gross weight manufactured since 1 October 1983 and first used since 1 April 1984. Trailers, including semi-trailers, over 1020kg unladen weight manufactured since 1 May 1983 must also be fitted with bumpers. Certain vehicles and trailers are exempt (see below) from the fitting requirements and there were no retrospective fitting requirements for existing vehicles.

Strength of Bumpers
Rear underrun bumpers must be constructed so they are capable of

withstanding a force equivalent to half the gross weight of the vehicle or trailer or a maximum of 10 tonnes, whichever is the *lesser,* without deflecting more than 400mm measured from the rearmost point of the vehicle or trailer – not from the original vertical position of the bumper.

Fitment of Bumpers
Bumpers must be fitted as near as possible to the rear of the vehicle and the lower edge must not be more than 550mm from the ground (see Figure 13.1). Normally, only one bumper would be fitted but where a tail-lift is fitted or the bodywork or other parts of the vehicle make this impracticable, two or more bumpers may be fitted. When a single full-width bumper is fitted it must extend on each side of the centre to within at least 100mm from the outermost width of the rear axle, but must not in any case extend beyond the width of the rear axle measured across the outermost face of the tyres. When two or more bumpers are fitted, for the reasons mentioned above, the space between each part of the bumper must not exceed 500mm and the outermost edge of the bumpers must extend to within at least 350mm from the outermost width of the rear axle (see Figure 13.2). Bumpers must not protrude beyond the width of the vehicle or trailer and the outside ends of the bumper must not be bent backwards.

Maintenance of Bumpers
Rear bumpers must be maintained free from any obvious defect which would adversely affect their performance in giving resistance to impact from the rear. It is also important to ensure that the dimensional requirements are met – particularly if the bumper is damaged (for example by fork-lift truck impact or reversing on to loading bays).

Exemptions
Rear underrun bumpers do not have to be fitted to vehicles and trailers in the following list:
1. Vehicles incapable of a speed exceeding 15mph on the level under their own power.
2. Tractive units of articulated vehicles.
3. Agricultural trailers, trailed appliances and agricultural motor vehicles.
4. Engineering plant.
5. Fire engines.
6. Road spreading vehicles (ie for salt and grit).
7. Rear tipping vehicles.
8. Naval, military or airforce vehicles.
9. Vehicles being taken to have bodywork fitted, or being taken for quality or safety checks by the manufacturer, distributor or dealer in such vehicles.
10. Vehicles being driven or towed to a place to have a rear underrun bumper fitted by prior arrangement.
11. Vehicles designed to carry other vehicles which are loaded from the rear (eg car transporters).
12. Trailers designed and constructed (not just adapted) to carry round timber, beams or girders of exceptional length.
13. Vehicles fitted with tail-lifts where the tail-lift forms part of the floor of the vehicle and extends to a length of at least one metre.

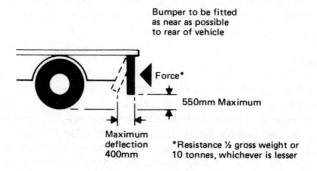

Figure 13.1 *Illustration of the rear underrun bumper force resistance requirements and ground clearance dimension*

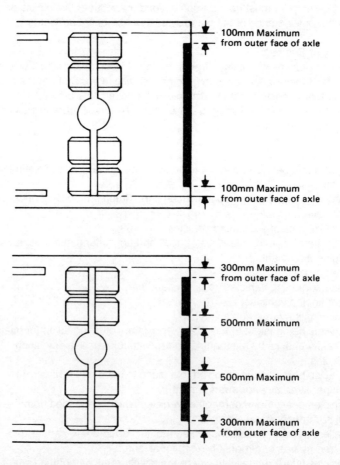

Figure 13.2 *Illustrations of the fitting dimensions for single and multiple rear underrun bumpers*

14. Temporarily imported foreign vehicles and semi-trailers.
15. Vehicles specially designed (not just adapted) for carrying and mixing liquid concrete.
16. Vehicles designed and used solely for the delivery of coal by means of a conveyor fixed to the vehicle so as to make the fitment of a rear underrun bumper impracticable.

Sideguards

Most heavy vehicles and trailers must be fitted with sideguards to comply with legal requirements except certain vehicles and trailers which are exempt from the fitting requirement as listed at the end of this section.

Sideguards must be fitted to the following vehicles and trailers:
1. Goods vehicles exceeding 3.5 tonnes maximum gross weight manufactured since 1 October 1983 and first used since 1 April 1984.
2. Trailers exceeding 1020kg unladen weight manufactured since 1 May 1983 and which, in the case of semi-trailers, have a distance between the foremost axle and the centre line of the kingpin (or rearmost kingpin if there is more than one) exceeding 4.5 metres (Figure 13.3).
3. Semi-trailers made before 1 May 1983 with a gross weight exceeding 26,000kg and used in an articulated combination with a gross train weight exceeding 32,520kg.

Sideguards are not required on vehicles and trailers, other than semi-trailers, where the distance between any two consecutive (ie front and rear) axles is less than 3 metres (Figure 13.4).

Strength of Sideguards
Sideguards must be constructed so they are capable of withstanding a force of 200kg (2 kilonewtons) over their length, apart from the rear 250mm, without deflecting more than 150mm. Over the last 250mm the deflection must not be more than 30mm under such force (Figure 13.5). These force resistance requirements *do not* apply where sideguards were fitted to existing semi-trailers (ie those built before 1 May 1985, and which were used at weights above 32,520kg).

Fitment of Sideguards
The fitting position for sideguards depends on the type of vehicle or trailer as follows:
1. Rigid vehicles: at front – not more than 300mm behind the edge of the nearest tyre and the foremost edge of the sideguard; at rear – not more than 300mm behind the rearmost edge of the sideguard and the edge of the nearest tyre (Figure 13.6).
2. Trailers: at front – not more than 500mm behind the edge of the nearest tyre and the foremost edge of the sideguard; at rear – not more than 300mm behind the rearmost edge of the sideguard and the edge of the nearest tyre (Figure 13.7).
3. Semi-trailers with landing legs: at front – not more than 250mm behind the centre line of the landing legs and the foremost edge of the sideguard; at rear – not more than 300mm behind the rearmost edge of the sideguard and the edge of the nearest tyre (Figure 13.8).

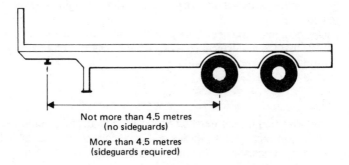

Not more than 4.5 metres
(no sideguards)

More than 4.5 metres
(sideguards required)

Figure 13.3 *Measurements of relevant distance between foremost axle and centre of kingpin for semi-trailers to determine if sideguards must be fitted*

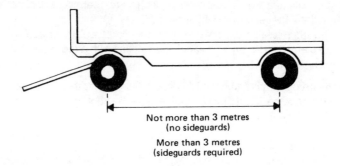

Not more than 3 metres
(no sideguards)

More than 3 metres
(sideguards required)

Figure 13.4 *Measurement of two consecutive axles on draw-bar trailers – the same dimension applies to rigid vehicles – to determine if sideguards must be fitted*

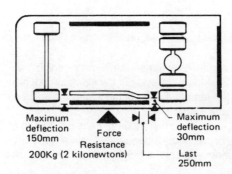

Maximum
deflection
150mm

Force
Resistance
200Kg (2 kilonewtons)

Maximum
deflection
30mm

Last
250mm

Figure 13.5 *How force resistance applies to sideguard on new vehicles and trailers (It does not apply to sideguard fitted to existing semi-trailers)*

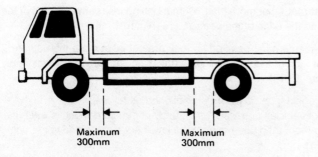

Figure 13.6 *Fitting position for sideguard on rigid vehicles*

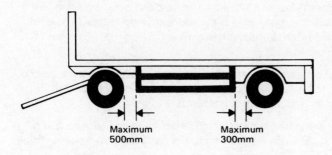

Figure 13.7 *Fitting position for sideguard on trailers*

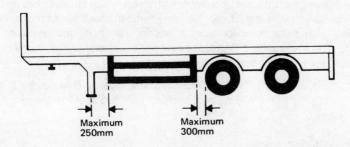

Figure 13.8 *Fitting position for sideguards on semi-trailer with landing legs*

4.   Semi-trailers without landing legs: at front – not more than 3 metres behind the centre line of the rearmost kingpin and the foremost edge of the sideguard; at rear – not more than 300mm behind the rearmost edge of the sideguard and the edge of the nearest tyre (Figure 13.9).

In all cases sideguards must be fitted so they are not inset more than 30mm from the external face of the tyre excluding any distortion due to the weight of the vehicle (Figure 13.10).

The upper edge of sideguards must be positioned as follows:
1.   In the case of vehicles or trailers with a body or structure which is wider than the tyres, no more than 350mm from the lower edge of the body or structure (Figure 13.11).
2.   In the case of vehicles or trailers with a body or structure which is narrower than the tyres or which does not extend outwards immediately above the wheels, a vertical plane taken from the outer face of the tyre must be measured upwards for 1.85 metres above the ground. If this plane is dissected by the vehicle structure within 1.85 metres from the ground the sideguard must extend up to within 350mm of the structure where it is cut by the vertical plane (Figure 13.12); if the vertical plane is not dissected by the vehicle structure the upper edge of the sideguard must extend to be level with the top of the vehicle structure to a minimum height of 1.5 metres from the ground (Figure 13.13).

The lower edge of sideguards must not be more than 550mm from the ground. This dimension is to be measured on level ground and, in the case of a semi-trailer, when its load platform is horizontal.

When sideguards are to be fitted to extendible trailers and to vehicles and trailers designed to carry demountable bodies or containers, the following fitting provisions apply:
1.   Sideguards must be fitted to extendible trailers in compliance with the original fitting specifications in regard to spacings from the nearest wheel, kingpin or landing leg when the trailer is at its shortest length. When the trailer is extended beyond its minimum length the spacings between the front edge of the sideguard and the semi-trailer landing legs or kingpin (if it has no landing legs) and the rear edge of the sideguard and the foremost edge of the tyre nearest to it are no longer applicable.
2.   Sideguards must be fitted to vehicles and trailers which are designed and constructed (not merely adapted) to carry demountable bodies or containers so that when the body or container is removed the sideguards remain in place. This means that if the vehicle runs without a body or container it must still comply with the sideguard requirements.

Vehicles which are fitted with sideguards complying with EC Directive 89/279 do not have to comply with the fitting requirements specified in UK regulations as detailed above.

Construction of Sideguards
All parts of the sideguard which face outwards must be 'smooth, essentially rigid and either flat or horizontally corrugated'. Each face of the guard must be a minimum of 100mm wide (including the inward face at the forward edge) and the vertical gaps between the bars must not be more than 300mm wide.

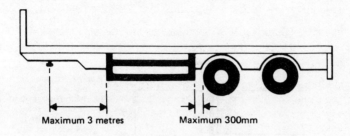

Maximum 3 metres   Maximum 300mm

Figure 13.9 *Fitting position for sideguards on semi-trailer without landing legs*

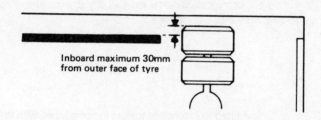

Inboard maximum 30mm
from outer face of tyre

Figure 13.10 *Inboard mounting position for sideguards - all vehicles and trailers*

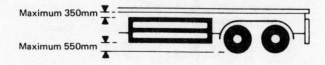

Maximum 350mm

Maximum 550mm

Figure 13.11 *Fitting position for sideguards where body or structure is wider than the tyres*

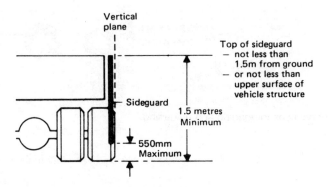

Figure 13.12 *Fitting position for sideguards where body or structure is narrower than the tyres up to a height of 1.85 metres*

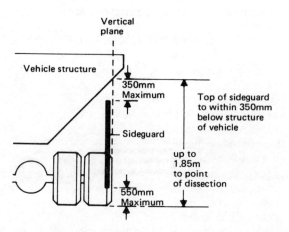

Figure 13.13 *Fitting position for sideguards where body or structure is narrower than the tyres*

Maintenance of Sideguards
Sideguards must be maintained free of any obvious defect which would impair their effectiveness. It is important to ensure that the fitting dimensions are observed particularly when the sideguards are damaged (eg by fork-lift truck impact).

Exemption to Fitting Dimensions
The specific requirements relating to the fitting positions for sideguards as previously described only apply so far as is practicable in the case of the following vehicles and trailers:

1. Those designed solely for the carriage of a fluid substance in closed tanks permanently fitted to the vehicle and provided with valves and hose or pipe connections for loading and unloading.
2. Those vehicles which require additional stability during loading and unloading or while working and which are fitted with extendible stabilisers on either side (eg lorry mounted cranes, tower wagons, inspection platforms).

Exemptions
Sideguards do not have to be fitted to vehicles and trailers in the following list:

1. Vehicles incapable of a speed of more than 15mph on the level under their own power.
2. Agricultural trailers.
3. Engineering plant.
4. Fire engines.
5. Land tractors.
6. Side and end tipping vehicles and trailers.
7. Vehicles with no bodywork fitted and being driven or towed for the purposes of a quality or safety check by the manufacturer, distributor or dealer in such vehicles, or being driven by prior arrangement to have bodywork fitted.
8. Vehicles being driven or towed to a place by prior arrangement to have sideguards fitted.
9. Vehicles designed solely for use in connection with street cleansing, the collection or disposal of refuse or the collection or disposal of the contents of gullies or cesspools.
10. Trailers specially designed and constructed to carry round timber, beams or girders of exceptional length.
11. Articulated tractive units.
12. Naval, military or airforce vehicles.
13. Trailers specially designed and constructed (not merely adapted) to carry other vehicles loaded from the front or rear (eg car transporters).
14. Temporarily imported foreign semi-trailers.
15. Low-loader trailers where:
    (a) the upper surface of the load platform is not more than 750mm from the ground, and
    (b) no part of the edge of the load platform is more than 60mm inboard from the external face of the tyre (discounting the distortion caused by the weight of the vehicle).

Ground Clearance for Trailers

Minimum ground clearances are specified for goods carrying trailers manufactured since 1 April 1984. Such trailers must have a minimum ground clearance of 160mm if they have an axle interspace of more than 6 metres and not more than 11.5 metres. If the interspace is more than 11.5 metres the minimum clearance is 190mm (Figures 13.14 and 13.15).

Measurement of the axle interspace is taken from the point of support on the tractive unit in the case of semi-trailers or the centre line of the front axle in other cases to the centre line of the rear axle or the centre point between rear axles if there is more than one (Figure 13.15).

In determining the minimum ground clearance no account should be taken of any part of the suspension, steering or braking system attached to any axle, any wheel and any air skirt. Measurement of the ground clearance is taken in the area formed by the width of the trailer and the middle 70 per cent of the axle interspace (Figure 13.16).

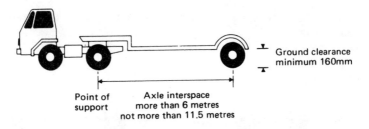

Point of support

Axle interspace
more than 6 metres
not more than 11.5 metres

Ground clearance
minimum 160mm

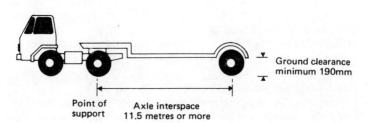

Point of support

Axle interspace
11.5 metres or more

Ground clearance
minimum 190mm

Figure 13.14 *Ground clearance for trailers - measure from the point of support to the centre line of the axle*

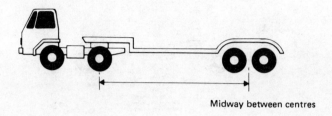

Midway between centres

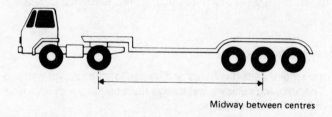

Midway between centres

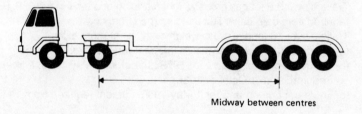

Midway between centres

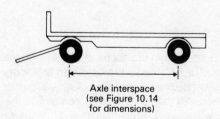

Axle interspace
(see Figure 10.14
for dimensions)

Figure 13.15 *The point for measuring axle interspace on multi-axle semi-trailers and on other trailers*

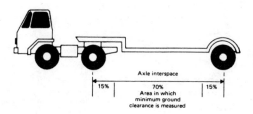

Figure 13.16 *The area in which minimum ground clearance is measured*

Anti-Spray Equipment

Regulations require certain vehicles and trailers to be equipped with anti-spray devices. The following vehicles and trailers must be fitted with approved equipment from the dates shown:
1. Motor vehicles over 12 tonnes gross weight made on or after 1 October 1985 and first used on or after 1 April 1986. Fitment required from date when vehicle first used on road from 1 April 1986.
2. Trailers over 3.5 tonnes gross weight made on or after 1 May 1985. Fitment required from date trailer first used on road from new.
3. Trailers over 16 tonnes gross weight with two or more axles
    (a) made before 1 January 1975. Fitment required from 1 October 1987;
    (b) made on or after 1 January 1975 but before 1 May 1985. Fitment required from 1 October 1986;
    (c) trailers made on or after 1 May 1985. Fitment required from 1 May 1985.

NB: The EU proposes reducing the weight threshold at which the fitment of anti-spray devices becomes mandatory from the present 12 tonnes to 3.5 tonnes.

Retrospective Fitting
When trailers which come within the scope of 3(a) and 3(b) above were in service before the end of 1984 and were fitted with anti-spray equipment or other spray suppression devices at that time, such equipment or devices will remain legally acceptable if they 'substantially conform' to the requirements of the regulations. Legal precedence has established that the word 'substantially' can mean as little as 60 per cent. The operator would also have to be in a position to prove fitment before the end of 1984.

Exemptions
Anti-spray requirements do not apply to those vehicles and trailers which are fitted with such devices in accordance with EC Directive 91/226 (and marked

accordingly) and to those which are exempt under the C&U regulations from the need for wings. Further exemptions are as follows:

1. Four-wheel and multi-wheel drive vehicles.
2. Vehicles with a minimum of 400mm (approximately 16ins) ground clearance in the middle 80 per cent of the width and the overall length of the vehicle.
3. Works trucks.
4. Works trailers.
5. Broken down vehicles.
6. Vehicles which cannot due to their construction exceed 30mph on the level under their own power.
7. Vehicles specified in regulations as exempt from sideguards
   (a) agricultural trailers and implements
   (b) engineering plant
   (c) fire engines
   (d) side and end tippers
   (e) military vehicles used for naval, military or airforce purposes
   (f) vehicles with no bodywork fitted being driven on road test or being driven by prior appointment to a place where bodywork is to be fitted or for delivery
   (g) vehicles used for street cleansing, the collection or disposal of the contents of gullies or cesspools
   (h) trailers designed and constructed to carry round timber, beams or girders of exceptional length
   (i) temporarily imported foreign semi-trailers.
8. Concrete mixers.
9. Vehicles being driven to a place by prior arrangement to have anti-spray equipment fitted.
10. Land locomotives, land tractors and land implement conveyors.
11. Trailers forming part of an articulated vehicle or part of a combination of vehicles, having in either case, a total laden weight exceeding 46,000kg.

British Standard

The British Standard on spray suppression was originally contained in two documents, namely BS AU 200 (parts 1 and 2) 1984, which applied to vehicles fitted before 1 May 1987, but this has now been replaced by new British Standards BS AU 200 (part 1a and 2a) 1986 which apply to fitment since this date. The regulations require anti-spray devices to conform to the standard set out in these documents. However, there is no requirement for existing equipment fitted to the requirements of the original BS AU standard to be changed to equipment which meets the new BS AU standard.

In order to comply with the law, relevant vehicles and trailers must be fitted with anti-spray systems which fall into one of two main categories:

1. A straight valance across the top of the wheel and a flap hanging vertically behind the wheel all made from approved spray suppressant material; or
2. A semi-circular valance following the curvature of the wheel with either:
   (a) air/water separator material round the edge; or
   (b) a flap of spray suppressant material hanging from the rear edge.

Spray Suppressant Material

Two types of material are referred to in the Standard. These are generally identifiable as follows:

1. Spray suppressant material – designed to absorb or dissipate the energy of water thrown from the tyre in order to reduce the degree to which water shatters into fine droplets on hitting a surface.
2. Air/water separator – 'a device forming part of the valance and/or wheel flap which permits air to flow through while reducing the emission of spray'.

Maintenance of Anti-Spray Equipment and Devices

The regulations stipulate that all devices fitted to comply with the legal requirement (and every part of such device) must be maintained, when the vehicle is on the road, so that they are free from 'any obvious defect which would be likely to affect adversely the effectiveness of the device'. It is also important that fitting dimensions are maintained especially if the flaps are damaged.

Fitment – Valances and Flaps

Where the choice is for spray suppression to be achieved by the use of valances and flaps (particularly on rear vehicle wheels and trailer wheels) the specific requirements for fitment are as follows:

NB: Capital letters in brackets in the following text refer to items on the adjacent diagrams illustrating fitment details.

- Valances of spray suppressant material must extend across the top of the tyre from a line vertical with the front edge of the tyre (A) to a line beyond the rear wheel which will allow the rear flap to be suspended no more than 300mm from the rear edge of the tyre (B). The valance must be at least 100mm deep (C).

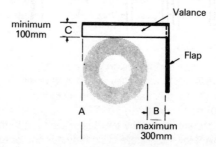

- The valance must extend downwards to be level with the top of the tyre (D) or it may overlap the top of the tyre (E).

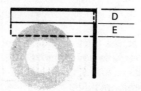

- In the case of multiple axle bogies the relevant dimensions are shown above with the additional requirement that where the gap between the rear edge of the front tyre and the front edge of the rear tyre is greater than 250mm (F) a flap must be fitted between the two.
  Note: no middle flap is required if the distance does not exceed 250mm.
  Note: The top of the valance may be in two separate sections (see shaded part) so long as it otherwise conforms to the dimensions.

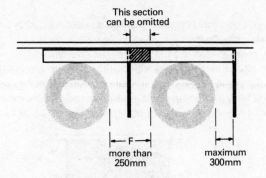

- Valances must extend the full width of the tyre and beyond to a maximum of 75mm (G) in the case of the rear wheels (non-steerable) and 100mm (H) in the case of steered wheels.

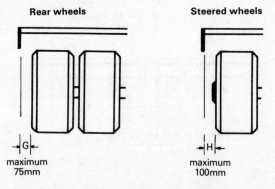

- If the valance extends below the level of the tyre on fixed wheels the gap between the tyre face and the valance can be extended to 100mm (J). There must be no gaps between the valance and the vehicle body.

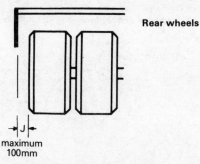

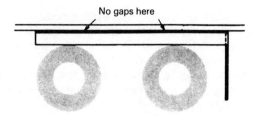

No gaps here

- Flaps used in conjunction with valances as described above must conform to the following dimensions:
  1. They must extend the full width of the tyre/tyres (K).
  2. They must reach down to within 200mm of the ground (L) when the vehicle is unladen (300mm on rearmost axles of trailers used on roll-on/roll-off ferries or on any axle where the radial distance of the lower edge of the valancing does not exceed the radius of the tyres fitted).

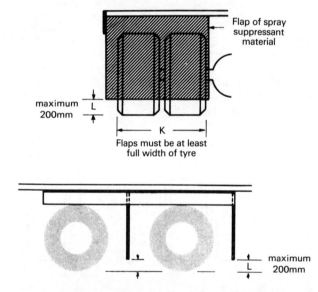

Flap of spray suppressant material

maximum 200mm

Flaps must be at least full width of tyre

maximum 200mm

3. When flaps are used in conjunction with mudguard valances the top of the flap must extend upwards at least to a point 100mm above the centre line of the wheel irrespective of the position of the lower edge of the mudguard.

Flap used in conjunction with mudguard valance

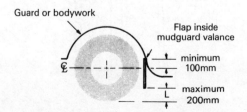

Guard or bodywork

Flap inside mudguard valance

minimum 100mm

maximum 200mm

4. Where the flap extends inside the guard then it must be at least the width of the tyre tread pattern.

- If the flap used is of a type with an air/water separator device (ie bristles) fitted to the bottom edge the following dimensions apply:
  1. Rear edge of tyre to flap – maximum distance 200mm (M).
  2. The edge of the device must come to within 200mm of the ground (N).

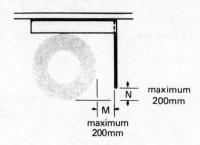

Deflection of Flaps
Wheel flaps must not be capable of being deflected rearwards more than 100m when subjected to a force of 3N (ie 4lbs) applied near the bottom of the flap.

Fitment – Mudguards and Air/Water Separator Devices
Where the choice for compliance with the regulations is by means of conventional mudguarding there are specific dimensions to be observed:

- If the mudguard is covering a steerable wheel (see later note about steerable axles on drawbar trailers) the radius of the edge of the valance must not be more than 1.5 times the radius of the tyre measured at three points (P) and in the case of non-steerable wheels, 1.25 times the same radius.
  1. Vertically above the centre of the tyre.
  2. A point at the front of the tyre 20 degrees above the horizontal centre line of the tyre (non-steerable wheels) or a point 30 degrees above the horizontal centre line of the tyre (steerable wheels).
  3. A point at the rear of the tyre 100mm above the horizontal centre line of the tyre.

Three points of measurement of radial for mudguards (P)

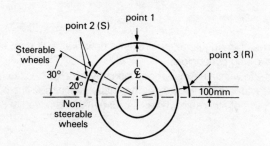

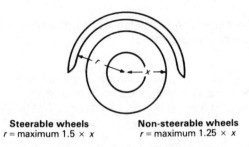

Steerable wheels
r = maximum 1.5 × x

Non-steerable wheels
r = maximum 1.25 × x

- Mudguard valances must be at least 45mm deep behind a point vertically above the wheel centre. They may reduce in depth forward of this point (Q).

Depth of mudguard valances

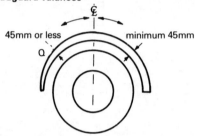

45mm or less   minimum 45mm

Q

- In the case of drawbar trailers, the 1.5 times radius dimension applies as above for the front steerable axle unless the mudguards are fitted to the turntable and thus turn with the wheels in which case the maximum radius for the valance is 1.25 times the tyre radius.
- If the valancing on fixed wheel mudguards is provided by means of air/water separator material (ie bristles) the edge must follow the periphery of the tyre. On steerable wheels the edge must be not more than 1.05 times the tyre radius.

Fixed wheel   **Steerable wheel**

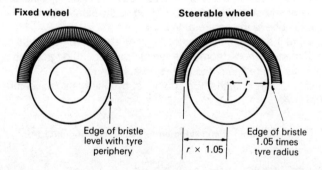

Edge of bristle
level with tyre
periphery

r × 1.05

Edge of bristle
1.05 times
tyre radius

The valances on mudguards must extend downwards at front and rear to at least the following dimensions:

1. at rear – to within 100mm above the centre line of the axle (point 3) (R).
2. at front – to within a line 20 degrees above the centre line of the axle.
   In the case of steerable wheels this dimension is raised to 30 degrees (point 2) (S).

- In the case of multi-mudguarding over tandem axles or bogies, the intersection of the guards between the wheels must conform to one of the two dimensions:
  1. The gap between the guards at the valance edges must not exceed 60mm (T); or
  2. The edges must come down to within 150mm of the horizontal centre line across the wheels (V).

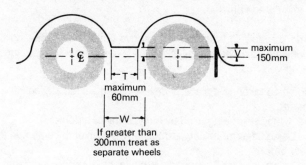

- If the gap between the tyre edges is greater than 250mm a flap must be provided between the wheels. If the gap is more than 300mm the wheels should be treated as though separate for mudguarding purposes (W).

Notes:
1. All dimensions in the regulations are to be taken when the vehicle is unladen, when steerable wheels are straight ahead and when the load platforms of articulated semi-trailers are level.
2. All suppression material or devices and air/water separator material or devices must be permanently and legibly marked with the following mark: BS AU 200/2, plus 'the name, trademark or other means of identification of the responsible manufacturer'.

Speed Limiters

In a move to reduce the number and severity of road accidents, legislation has been introduced requiring certain heavy vehicles to be fitted with speed limiters. The government says that this will also have beneficial effects for operators in terms of fuel economy – possibly saving as much as 150 million litres annually when all relevant vehicles are fitted – and benefit the environment by an annual reduction of an estimated 0.5 million tonnes of carbon monoxide (CO) pumped, by way of exhaust emission, into the atmosphere. At the same time, the EU requires certain vehicles also to be fitted with speed limiters set to a maximum of 90kph (approx 56mph) leading to no end of confusion since the two sets of requirements do not align.

The UK regulations, which came into effect on 1 August 1992, limit relevant vehicles to a maximum speed of 60mph (ie 96.5kph). They apply:

- to all new goods vehicles exceeding 7.5 tonnes permissible maximum weight which are capable of a speed in excess of 60mph on the flat, and first registered on or after 1 August 1992;
- from 1 August 1993, to vehicles exceeding 16,000kg gross weight which are capable of a speed in excess of 60mph on the flat, were first registered on or after 1 January 1988 and which are
  - rigid vehicles constructed to draw trailers, having a difference between the plated gross weight and gross train weight of at least 5000kg,
  - articulated vehicles.

Vehicles not capable of travelling at or above 60mph (eg refuse collection and certain highway maintenance vehicles) are exempt from the regulations, as are the following vehicles:

1. those being taken to a place to have a speed-limited device fitted or calibrated;
2. those owned and being used by navy, army or airforce;
3. those being used for military purposes while driven by a person under military orders;
4. those being used for fire brigade, ambulance or police purposes;
5. those exempt from excise duty under the provisions of the Vehicles (Excise) Act 1971 (ie which do not travel more than 6 miles per week on the road).

Speed limiter equipment must comply with BS AU 217 (or an acceptable equivalent), be calibrated to a set speed not exceeding 60mph and sealed by an 'authorised sealer'. Existing speed limiters, fitted on a voluntary basis prior to 1 August 1992, were permitted and did not have to be sealed by an 'authorised sealer' as stated above.

Speed limiters must be maintained in good working order but it is a defence to show that where a vehicle is driven with a defective limiter, the defect occurred during that journey or that at the time it is being driven to a place for the limiter to be repaired.

Speed Limiter Plates
Vehicles which are required to be fitted with speed limiter equipment under this legislation must also carry a plate (fitted in a conspicuous and readily accessible position in the vehicle cab) on which is shown the words SPEED LIMITER FITTED, the standard with which the installation complies, the speed setting in mph/kph and the name/trademark of the firm which carried out the calibration. Normally such plates will be provided by the authorised sealer.

EU Requirements
The speed limiter requirements of EC Directive 1992/6 apply from 1 January 1994 and require new goods vehicles first registered from this date and which exceed 12 tonnes gross weight to be fitted with limiters set to 90kph (approx 56mph) – ie, 'so adjusted that the stabilised speed of the vehicle does not exceed . . .'. Vehicles above 12 tonnes which were first registered between 1 January 1988 and 1 January 1994 must be fitted by 1 January 1995 (but those used exclusively in national transport operations have 1 year longer, until 1 January 1996, to comply).

Air-Brake Silencers (Hush Kits)

Goods vehicles exceeding 16.5 tonnes gross weight which travel within residential areas of London at nights or weekends must be fitted with air-brake silencers under the terms of Condition 11 of the London Lorry Ban (see also pp 184–5). A House of Lords ruling (July 1991) upheld an an appeal by the London Boroughs Transport Committee (LBTC) that it should be allowed to impose this requirement. The transport trade associations had contested the LBTC's requirement on the grounds that it was unlawful and exceeded the statutory requirements of the Construction and Use regulations. Failure to fit the so-called 'hush kits' will result in refusal of an exemption plate by the LBTC.

Use of Vehicles

In addition to the foregoing constructional requirements which are mainly the responsibility of the vehicle manufacturer or the person building the bodywork, there are requirements regarding the use of vehicles which are the responsibility of the operator. However, as mentioned earlier, the vehicle user carries full legal responsibility for the mechanical condition of a vehicle on the road and its compliance with the constructional requirements even if the fault which led to an offence could be laid at the door of the chassis manufacturer, the bodybuilder, an ancillary equipment supplier or the dealer.

These requirements cover such items as vehicle weights (which were dealt with earlier), towing, fumes, the condition and maintenance of vehicles and their components, noise, smoke, and general safety in the use of vehicles. They also deal with the regulations regarding the number of trailers which a vehicle may draw.

Noise

It is an offence to use, or cause or permit to be used on a road a motor vehicle or trailer which causes an excessive noise because of a defect, lack of repair or faulty adjustment of components or load. Also no motor vehicle must be used on a road in such a manner as to cause any excessive noise which could have been reasonably avoided by the driver. Noise for these purposes is the combined noise emitted by the exhaust plus that from the tyres, engine, bodywork and equipment and the load. Noise levels for goods vehicles are measured by special meters either by the police or by DoT examiners at goods vehicle testing stations and occasionally on roadside tests.

Noise Limits
The permissible noise levels specified in the regulations are based on a 'constructional' standard and an 'in use' standard. The 'in use' standards are as follows:

Motor vehicles first used from 1 April 1970
1.   Plated goods vehicles with a maximum gross weight   89dB(A)
     exceeding 3560kg

2. Other goods vehicles   85dB(A)
3. Motor tractors, locomotives, land tractors, works trucks and   89dB(A)
   engineering plant

Motor vehicles first used from 1 October 1983
1. Goods vehicles with a maximum gross weight not   81dB(A)
   exceeding 3500kg
2. Plated goods vehicles with a maximum gross weight over   86dB(A)
   3500kg
3. Goods vehicles over 12,000kg maximum gross weight and   88dB(A)
   with engine power of at least 200hp (DIN)

More stringent noise limits for goods vehicles were introduced from 1 October 1990 in accordance with EC Directive 424/84. Broadly this mean a reduction to 81dB(A) for vehicles over 3.5 tonnes with engine power less than 75kW per 1000kg, 83dB(A) for those with power outputs between 75 and 150kW and 84dB(A) for vehicles with engine power at or above this figure.

Exemption from Noise Limits
These limits do not apply in the case of a vehicle going by appointment to have the noise measured or for mechanical adjustments to reduce the noise level, or returning from such an appointment. It also does not apply to a stationary vehicle using a power take-off or a vehicle first used before 1 November 1970 if using an exhaust brake.

Smoke

Vehicles must not emit smoke, visible vapour, grit, sparks, ashes, cinders or oily substances which may cause damage to property or injury or danger to any person.

Excess Fuel Devices
Excess fuel devices must not be used on diesel vehicles while the vehicle is in motion. Such devices are incorporated in the vehicle fuel pump to enable extra fuel to be fed to the engine to aid cold starting. Their use when the engine is warm slightly increases the power of the engine, but in doing so black smoke is emitted from the exhaust. For this reason their use is forbidden while the vehicle is in motion.

Smoke Opacity Limits
Diesel-engined vehicles first used after 1 April 1973 (but not manufactured before 1 October 1972) must comply with smoke opacity limits specified in BS AU 141a/1971. Engines fitted to such vehicles must be of a type for which a type test certificate in accordance with the British Standard Specification for *The Performance of Diesel Engines for Road Vehicles* has been issued by the Secretary of State. The certificate will indicate that engines of that type do not exceed the emission of smoke limits set out in the BS Specification.

Exemptions
Land tractors, industrial tractors, works trucks and engineering plant propelled by diesel engines with not more than two cylinders are exempt from this

requirement; so too are vehicles fitted with the Perkins 6.354 engine manufactured before 1 April 1973.

Offences
It is an offence to use a vehicle to which this type test applies if the fuel injection equipment, the engine speed governor or others parts of the engine have been altered or adjusted in such a way that the smoke emission of the vehicle is increased. An offence is committed if a vehicle emits black smoke or other substances even without alteration or adjustment of the parts (eg as a result of lack of maintenance). The Vehicle Inspectorare has promised tougher enforcement of goods vehicle smoke emissions, particularly at the time of submitting vehicles for annual test.

Control of Fumes
Petrol-engined vehicles first used after 1 January 1972 must be fitted with a means of preventing crank-case gases escaping into the atmosphere except through the exhaust system.

Exhaust Emissions

Since 1 April 1991 newly registered diesel-engined goods vehicles exceeding 3.5 tonnes gross weight have been required to comply with EC Directive 88/77 which sets gaseous emission (ie exhaust emission) limits (see Chapter 24 for details of emission requirements for fleet cars and light vans). These limits are already applied in Type Approval regulations to vehicles first used from 1 April 1991 (later for cars/vans with engines over 1400cc and those with diesel engines).

The purpose of the EU's three-part legislative programme (under EC Directive 91/542) is to reduce the amount of nitrous oxide (NOx), carbon monoxide (CO) and unburnt hydrocarbons (HC) blown into the atmosphere from vehicle exhausts. The first stage of this programme (ie the so-called Euro-1 standard) applies to all new goods vehicles Type Approved since 1 July 1992 (and to earlier Type Approved vehicles from 1 October 1993). The tougher Euro-2 standards are to be applied to vehicles over 3.5 tonnes pmw from the end of 1995 with further, more stringent (ie Euro-3) emission controls in prospect from 1999.

The vehicle user is required by law to keep the engine of his vehicle and any emission control equipment (ie catalytic converter) in good working order and in tune.

Gas Powered Vehicles

Regulations specify technical standards for fuel tanks or containers, the filling system and valves and general requirements for gas propulsion systems in motor vehicles. The regulations permit the use of LPG only in gas propelled vehicles, although this may be combined with petrol fuel systems, but the use of methane or hydrogen is prohibited. The DoT has published a free guide *Gas Installations in Motor Vehicles and Trailers* which is available from: Department of Transport, B3, Victoria Road, South Ruislip, Middlesex HA4 0NZ.

Towing

Goods vehicles may draw (ie tow) only one trailer. An exception to this is when a rigid goods vehicle tows a broken-down vehicle on a towing ambulance or dolly in which case although this is counted as towing two trailers it is allowed. In a case where an articulated vehicle has broken down, this may be towed by a rigid goods vehicle so long as the articulated vehicle is not loaded. In these circumstances the outfit is treated as one trailer only, but if it is loaded an articulated outfit being towed is considered to be two trailers and it is illegal for a normal goods vehicle (ie a heavy motor car) to tow it. Only a locomotive can tow a broken-down articulated vehicle which is laden (see below).

Motor tractors may draw one laden or two unladen trailers and locomotives may draw three trailers (see pp 227–8 for definitions).

Composite Trailers
The C&U regulations make it permissible for rigid goods vehicles (apart from locomotives and motor tractors) to draw two trailers instead of only one, when one of the trailers is a towing implement (ie a dolly) and the other is an articulated-type semi-trailer secured to and resting on, or suspended from, the dolly. This combination of dolly and semi-trailer is known as a composite trailer (Figure 13.17).

To comply with the regulations, the dolly needs to have two or more wheels and be specifically designed to support a superimposed semi-trailer. Dollies must display a manufacturer's plate and they are subject to the annual heavy goods vehicle test.

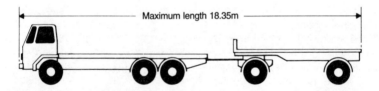

Figure 13.17 *A conventional six-wheeled rigid vehicle drawing a dolly mounted semi-trailer for which the maximum overall length is 18.35 metres*

Towing Distance
The distance between the nearest points of two vehicles joined by a tow rope or chain must not exceed 4.5 metres. When the distance between the two vehicles exceeds 1.5 metres the rope, chain or bar must be made clearly visible from both sides of the vehicles. There is no specified maximum distance limit if a solid tow-bar is used for towing.

Televisions in Vehicles

It is illegal for a vehicle to be fitted with television receiving apparatus where the driver can see the screen either directly or by reflection except where such equipment displays nothing other than information:
1. About the state of the vehicle or its equipment.
2. About the location of the vehicle and the road on which it is located.
3. To assist the driver to see the road adjacent to the vehicle (eg to the rear when reversing).
4. To assist the driver to reach his destination.

Reversing Alarms

It is legally permissible to fit reversing alarms to certain goods and passenger vehicles if desired – *the regulations do not make it mandatory to do so.* Such alarms may be fitted and used on the following vehicles:
1. Commercial vehicles over 2 tonnes gross weight
2. Passenger vehicles with nine or more seats
3. Engineering plant
4. Works trucks.

Time Restriction on Use of Reversing Alarms
The alarms are subject to the same night-time restrictions that apply to the sounding of horns in built-up areas (ie not after 11.30 pm and before 7.00 am – 23.30 to 07.00) and the sound emitted must not be capable of being confused with the Pelican crossing 'safe to cross' signal.

Restriction on Fitment of Reversing Alarms
Such alarms *must not* be fitted to light goods vehicles below 2 tonnes gross weight or to motor cars.

Advice on Use of Alarms
A number of cases have arisen following reversing accidents resulting in death or injury where the Health and Safety Executive have prosecuted the vehicle operators concerned for not voluntarily fitting reversing alarms. In other words the HSE line is that the offender had not taken sufficient steps to ensure safety when his vehicles were reversing by fitting equipment which the law permits, but not mandatorily requires, him to do.

14: Type Approval

Type Approval is a scheme which requires vehicle manufacturers to submit new vehicles (ie new designs, new models and changes of specifications for existing approved models) for approval before they are put on the market. The Department of Transport examines the vehicle submitted to ensure that it meets all legal requirements and also meets minimum standards of construction and performance. When a vehicle has been approved the manufacturer is then required by law to build all vehicles of a similar type to exactly those standards and certify this fact to the customer by the issue of a Certificate of Conformity.

An EU directive lays down the basic procedures for the Type Approval scheme for vehicles and components. Subsidiary directives have also been issued setting out agreed standards on some aspects of vehicle safety or pollution. They cover the same ground as existing national regulations. The directives do not yet cover all vehicle features which need to be regulated so until the programme is complete both EU directives and national regulations apply to relevant items. The UK established a non-compulsory scheme to enable exporting vehicle manufacturers to gain the necessary Type Approval in order to sell their products in EU countries.

Eventually EU Type Approval requirements will supersede UK construction and use regulations but initially vehicles which have a Type Approval Certificate will be exempt from certain construction and use requirements.

Vehicles Covered

Type Approval requirements apply to all passenger cars first licensed for use in the UK from 1 April 1978. The Motor Vehicles (Type Approval for Goods Vehicles) (Great Britain) Regulations 1982 (as amended) make the application of the Type Approval scheme compulsory for goods vehicles, motor caravans, motor ambulances and bi-purpose vehicles constructed for the carriage of both goods and passengers (but not more than eight passengers) which are not included in the car Type Approval scheme and which have been manufactured since 1 October 1982 and first used since 1 April 1983. Applications for Type Approval could be made from 1 October 1981 onwards.

While there is currently no Type Approval scheme for trailers or semi-trailers the EU is proposing that they should be included in Type Approval requirements and is currently endeavouring to harmonise national legislation to provide for the introduction of an EU-wide package covering marking, brakes, lighting, underrun protection and sideguards, spray suppression and bulkhead load strength by 1993.

Exemptions from Type Approval

Type Approval does not apply to the following vehicles:
1. Vehicles manufactured before 1 October 1982 whenever they are first registered.
2. Vehicles manufactured on or after 1 October 1982, providing they are first licensed before 1 April 1983.
3. Temporarily imported vehicles.
4. Vehicles proceeding for export from the UK.
5. Vehicles in the service of visiting forces or headquarters.
6. Certain vehicles which are, or were formerly, in use in the public service of the Crown.
7. Prototypes which are not intended for general use on the roads.
8. Motor tractors, light locomotives and heavy locomotives.
9. Engineering plant, pedestrian-controlled vehicles, straddle carriers, works trucks and track-laying vehicles.
10. Vehicles specially designed and constructed for use in private premises for moving excavated materials, vehicles fitted with movable platforms and vehicles designed and constructed for the carriage of abnormal indivisible loads.
11. Tower wagons.
12. Fire engines.
13. Road rollers.
14. Steam-propelled vehicles.
15. Vehicles constructed for the purpose of preventing or reducing the effect of snow or ice on roads.
16. Two-wheeled motorcycles with or without sidecars.
17. Electrically propelled vehicles.
18. Breakdown vehicles.
19. Any vehicle not exceeding 3500kg unladen weight which is constructed or assembled by a person not ordinarily engaged in the manufacture of goods vehicles of that description.
20. Vehicles not exceeding 3500kg unladen weight providing that:
    (a) the vehicle has been purchased outside the UK for the personal use of the individual importing it or his dependants
    (b) the vehicle has been so used by that individual or his dependants on roads outside the UK before it is imported;
    (c) the vehicle is intended solely for such personal use in the UK; and
    (d) the individual importing the vehicle intends, at the time when the vehicle is imported, to remain in the UK for not less than 12 months from that date.

Responsibility for Compliance

Responsibility for complying with complex construction standards will rest with the manufacturer although users will still be responsible for maintaining vehicles in roadworthy condition. The construction standards applied by the scheme are limited to those which can be approved during the primary stage of manufacture. The standards are identical to those already required under the construction and use regulations but under this scheme vehicles have to be approved before they can be used on the road.

Effects on Plating and Testing

The scheme requires plated weights for heavy goods vehicles to be set during the Type Approval process instead of waiting until the first annual plating and testing examination. This means that heavy vehicle operators need a Type Approval Certificate in order to get a Ministry plate for display in the vehicle cab (see Chapter 16). However, annual testing is retained so as to check the condition of vehicles and to ensure that plated weights are accurate.

Responsibility for Type Approval

All aspects of Type Approval are the responsibility of the DoT Vehicle Inspectorate.

The Standards Checked

To obtain goods vehicle national Type Approval it is first necessary to obtain individual systems approvals for the following items:
1. Power-to-weight ratio (not applicable to petrol-engined vehicles or dual-purpose vehicles).
2. Gaseous exhaust emissions (petrol-engined vehicles only).
3. Particle emission (ie exhaust smoke) (diesel-engined vehicles only).
4. External noise level.
5. Radio-interference suppression (petrol-engined vehicles only).
6. Brakes.

Arrangements for First Licensing of Vehicles

For vehicles which are over 1525kg unladen weight or which form part of an articulated vehicle, application for first licensing on Form V55 must be accompanied by two copies of the Type Approval Certificate, which should have been supplied with the vehicle. On one of these the applicant must complete a declaration saying whether or not the vehicle is exempt from the plating and testing regulations (see Chapter 16 for full details) and whether it has been altered in any way that has to be notified to the DoT under the Type Approval regulations and, if so, whether any action arising from the notification has been satisfactorily completed.

Issue of Plates

When application is made for first licensing a vehicle which is subject to plating and testing, the local Vehicle Registration Office (VRO) will send a copy of the Type Approval Certificate with the applicant's declaration to the Goods Vehicle

Centre (GVC) at Swansea. The second copy will be stamped and returned to the applicant to serve as a temporary Ministry plate. When the GVC receives the copy of the certificate, and if the details compare satisfactorily with those on the copy sent direct by the vehicle manufacturers, the GVC will issue a Ministry plate and laminated plating certificate. These will be sent direct to the person or company in whose name the vehicle is registered. Thus, operators buying new vehicles receive their first plate and plating certificate for the vehicle from the GVC at the time of licensing rather than from the lgv testing station when the vehicle is presented for its first annual test as under previous arrangements. When application is made for licensing a vehicle which is exempt from plating and testing, a copy of the certificate and the declaration will be sent to the GVC so that they are aware that it is exempt.

Refusal to Licence

Since 1 April 1983 no vehicle subject to the Type Approval regulations will be first licensed unless the DoT registration Form V55 has a valid Type Approval number on it or it is an exempt vehicle.

Alteration to Vehicles

If a vehicle, which has been issued with an Approval Certificate and supplied to a dealer or direct to an operator, is modified by them, prior to first licensing, they must notify the VCA at Bristol and send the certificate for the vehicle together with full technical details, drawings of the alterations and details of the weights on the certificate which need, or may need, changing.

The VCA will judge whether the alterations affect the vehicle's compliance with the regulations; if they do not affect compliance the certificate will be returned so the vehicle can be licensed. If they do contravene compliance, the certificate will be cancelled and fresh approval will need to be obtained before the vehicle can be licensed. This is a complex and costly procedure that most operators will want to avoid; they can do so by registering and licensing the vehicle before any alterations are carried out.

15: Vehicle Lighting and Marking

The legal requirements for vehicles to be fitted with and to display lights at night and other times, and for the fitment of reflectors and other markings on vehicles are contained in the Road Vehicles Lighting Regulations 1989 (as amended). These regulations specify in considerable detail all the requirements for the position of lamps and reflectors and the angles from which they must be visible. This chapter can only include a summary of the relevant requirements and dimensions as they apply to goods vehicles which, for most normal purposes, is satisfactory because vehicles and trailers are generally ready-fitted with lamps and reflectors conforming to legal requirements when supplied from new.

However, for a variety of reasons, operators may find it necessary to replace and relocate lamps from time to time when carrying out repairs and conversions, and at this time they are advised to check the regulations carefully to ensure strict compliance with the law. Not only can prosecution follow for incorrectly positioned or non-functioning lights and reflectors, but vehicles could fail their annual test on this account which adds to operating costs and wastes time. There are also the safety considerations with the lives of both vehicle drivers and other road users at risk if vehicles are not showing correct or adequate lights.

The regulations require that lights and reflectors which are fitted to vehicles must be maintained so as to enable them to be driven on a road between sunset and sunrise (times are published in most daily and local newspapers), or in seriously reduced visibility between sunrise and sunset, or to be parked on a road between sunset and sunrise, or between the hours or darkness, without contravening the regulations. All lights must be kept clean and in good working order. It is an offence to cause undue dazzle or discomfort to other road users by the use of lights or through their faulty adjustment, or to have defective or obscured lighting on a vehicle at any time. There is no longer a defence to a charge of having defective lights on a vehicle (see below).

Obligatory Lights

Between sunset and sunrise vehicles used on a public road must display the following obligatory lights, other lights and reflectors:
1. Two front position lamps (ie sidelamps) showing white lights to the front.
2. Two rear position lamps (ie rear lamps) showing red lights to the rear.
3. Two headlamps showing white lights to the front (alternatively, the light may be a yellow light) – required to be shown during the hours of darkness (see below).

4. Illumination for the rear number (ie registration) plate when the other vehicle lights are on.
5. Certain goods vehicles and trailers additionally require side marker lamps plus side-facing reflectors and rear reflective markings – see below.
6. One or two red rear fog lamps on post-1 April 1980 vehicles.
7. Two red reflex retro reflectors at the rear.
8. End-outline marker lamps.
9. Direction indicators on either side showing to the front and rear and capable of giving hazard warning on post-1 April 1986 vehicles.
10. And any other lights or lighting devices with which the vehicle is fitted (eg stop lamps, hazard warning signals, running lamps, dim-dip devices and headlamp levelling devices).

It is illegal except in certain specified cases for a goods vehicle to show a white light to the rear (showing such a light when reversing and indirect illumination of the rear registration plate are permitted for example) or a red light to the front.

No Defence

There is no longer a defence to a charge of having defective lights on a vehicle where it could be proved that the defect occurred during the journey on which the contravention had been detected or that when the contravention had been detected steps had been taken to have the defect remedied 'with all reasonable expedition'. This defence did not, in any case, apply where a vehicle was being used with defective lights during the period when the law requires the lights to be in use (ie between sunset and sunrise).

Headlamps

Motor vehicles must be fitted with two headlamps capable of showing a white or yellow light to the front – both lamps must emit the same colour light. Headlamps must be either permanently dipped or fitted with dipping equipment. Vehicles first used since 1 April 1987 must have dim-dip lighting devices unless their lighting equipment complies with EU requirements (see below).

Headlamps must be mounted so that they are not lower than 500mm from the ground and not higher than 1200mm. They must be placed on either side of the vehicle with their illuminated areas not more than 400mm from the side of the vehicle. They must be equipped with bulbs or sealed-beam units of not less than 30 watts in the case of vehicles first used before 1 April 1986. For vehicles used since this date no minimum wattage requirement is specified.

Headlamp Exemptions

Certain vehicles are exempt from the headlamp requirements. These include vehicles with less than four wheels, pedestrian-controlled vehicles, agricultural implements, land tractors, works trucks, vehicles not capable of travelling at a speed of more than 6mph and military vehicles.

Headlamps on Electric Vehicles

Electrically propelled goods vehicles with four or more wheels registered before October 1969 and electric vehicles with two or three wheels first used before 1 January 1972 and capable of a speed of more than 15mph are required to comply with the headlamp requirements. Those electrically propelled vehicles which are incapable of speeds of more than 15mph are exempt from the headlamp requirements.

Use of Headlamps

Headlamps must be adjusted so that they do not cause dazzle or discomfort to other road users. When vehicles which require headlamps are being driven on unlit roads during the hours of darkness (ie half an hour after sunset to half an hour before sunrise) and in seriously reduced daytime visibility the headlamps must be illuminated. They must be switched off when the vehicle is stationary except at traffic stops.

NB: Unlit roads are roads on which there are no street lamps or on which the street lamps are more than 200yds apart.

Headlamps in Daylight
It is a legal requirement for vehicles to use side position lights (ie sidelights) and dipped headlights when travelling in seriously reduced daytime visibility conditions such as in fog, smoke, heavy rain, spray or snow. If matching fog or fog and spot lights are fitted in pairs these may be used instead of headlights, but sidelights must still be used and the other vehicle lights must be on (eg side marker lights).

There is no specific definition of 'seriously reduced visibility' in the regulations. It is left to the driver to judge whether it is advisable and sensible to use his lights to enable his vehicle to be seen by others.

Dim-Dip Lighting

Since 1 April 1987 newly registered vehicles must be fitted with dim-dip lighting devices which operate automatically when the obligatory lights of the vehicle are switched on and ensure that either 10 per cent (with halogen) or 15 per cent (with grading filament lamps) of the normal dipped beam intensity shows when the vehicle ignition key is switched on or the engine is running. The European Court of Justice has ruled that it is unfair for the British government to legislate for the fitment of dim-dip lighting devices for vehicles which already comply with the EU lighting directive (EC 756/1976).

Front Position Lamps (Sidelamps)

Two front position lamps (ie sidelamps) emitting a white light through a diffused lens must be fitted to all motor vehicles and trailers. If such lamps are incorporated within a headlamp showing a yellow light then the side position lamps may be yellow. No minimum wattage is specified for these lights. The lights must be equal in height from the ground and mounted not more than 1500mm from the ground (in exceptional circumstances this height can be increased to 2100mm) in the case of vehicles first used on or after 1 April 1986 and 2300mm in other cases, and not more than 400mm from the outer edge of the vehicle for vehicles first used since 1 April 1986 and 510mm in other cases. No minimum height above the ground is specified.

Exemptions

Exemptions apply to trailers not more than 1600mm wide and those no longer than 2300mm (excluding the drawbar), built before 1 October 1985.

Optional Lamps

Optional main-beam headlamps (which include spot lamps) may be fitted to a vehicle but they must be capable of being dipped, and if fitted as a matched pair they must also be capable of being switched off together – not individually. They must emit either a white or yellow light, must be adjusted so that they do not cause dazzle to other road users and must not be lit when the vehicle is parked. If optional front fog lamps are fitted and used singly, the headlamps must also be illuminated.

These optional lamps should be positioned not more than 1200mm from the ground and not more than 400mm from the sides of the vehicle. They should be aligned so that the upper edge of the beam is, as near as practicable, 3 per cent below the horizontal when the vehicle is at its kerbside weight and has a weight of 75kg on the driver's seat.

Vehicles first registered since 1 April 1991 may be fitted with only one pair of extra dipped-beam headlamps and then only on vehicles intended to be driven on the right-hand side of the road. They must be wired so that only one pair of dipped-beam headlamps can be used at any one time. Pre-April 1991 registered vehicles may have any number of additional dipped-beam headlamps.

Any number of extra main-beam headlamps (including spot/driving lamps) may be fitted and there is no restriction on their fitment or use except that they must not cause dazzle to other road users.

Reversing Lamps

White reversing lamps (not more than two) may be fitted to vehicles provided that they are only used while the vehicle is reversing and operate automatically only when reverse gear is selected. Alternatively, they may be operated manually by a switch (which serves no other purpose) in the driver's cab provided that a warning device indicates to the driver that the lights are illuminated. The lights must be adjusted so as not to cause dazzle to other road users. Such lamps when bearing an 'e' approval mark do not have to meet minimum wattage requirements but those without approval marks must not exceed 24 watts.

Number Plate Lamp

Rear number plates (ie registration plates) on vehicles must be indirectly illuminated when the other obligatory lamps on the vehicles are lit. The light must be white and must be shielded so that it only illuminates the number plate and does not show to the rear.

Rear Position Lamps (Rear Lamps)

Two red rear position lamps (ie rear lamps) must be fitted to all motor vehicles and trailers. There is no specified wattage for these lights. They must be mounted not less than 350mm and not more than 1500mm (2100mm in exceptional circumstances) from the ground. They must be at least 500mm apart (no specified distance on pre-1 April 1986 registered vehicles) and not more than 400mm (800mm on pre-1 April 1986 registered vehicles) from the outside edge of the vehicle.

Stop Lamps

All goods vehicles (except those not capable of more than 25mph) must be fitted with red stop lamps which are maintained in a clean condition and in good and efficient working order. Vehicles registered before 1 January 1971 need only one such lamp which must be fitted at the centre or to the offside of the vehicle, although a second matching lamp may be fitted on the near side. Vehicles registered since that date need two such lamps (specified wattage 15 to 36 watts except with pre-1 January 1971 registered vehicles) mounted not less than 350mm from the ground and not more than 1500mm (in exceptional circumstances this may be increased to 2100mm) and they must be at least 400mm apart. Such lamps must be visible horizontally from 45 degrees on either side and normally from 15 degrees above and below vertically (from only 5 degrees below where fitted less than 750mm from the ground and only 10 degrees below when fitted not more than 1500mm from the ground).

Rear Fog Lamps

Rear fog lamps (at least one, but two may be fitted) must be fitted to new vehicles and trailers manufactured on or after 1 October 1979 and first used since 1 April 1980. There is no legal requirement to fit such lamps on pre-1 October 1979 registered vehicles but if they are fitted voluntarily they must comply with the regulations in regard to mounting position, method of wiring and use.

Rear Fog Lamps on Articulated and Towing Vehicles

In the case of articulated combinations, the relevant date in this connection is the date of the older of the tractive unit or semi-trailer. Thus if a post-April 1980 registered tractive unit is coupled to a pre-October 1979 built trailer there appears to be no legal requirement for the vehicle to carry rear fog lamps. This means there is no retrospective fitting requirement for such lamps on older trailers. A broken-down vehicle being towed does not need rear fog lamps. However, it is important to remember the dangers which arise if such a combination is used when the tractive unit or towing vehicle itself has rear fog lamps which would, in some instances, be visible to following motorists who, in bad visibility, might not be aware of some 40 feet of trailer or another vehicle on tow behind the lights.

Mounting of Rear Fog Lamps

Rear fog lamps must be mounted either singly on the offside of the vehicle or in a matched pair not less than 250mm and not more than 1000mm from the ground (in the case of agricultural vehicles this height limit is increased to 1900mm or in cases where, because of the shape of the vehicle, 1000mm is not practical it may be increased to 2100mm). The lamps must be at least 100mm from existing stop lamps.

Restriction on Wiring of Rear Fog Lamps

The lights must be wired so that they only operate when the other statutory lights on the vehicle are switched on; they must not be wired into the brake/stop light circuit and the driver must be provided with an indicator to show him when the lights are in use.

Use of Rear Fog Lamps

The lights should only be used in conditions affecting the visibility of the driver (ie in fog, smoke, heavy rain or spray, snow, dense cloud, etc), when the vehicle is in motion or during an enforced stoppage (a motorway hold-up, for example). They must not cause dazzle. The *Highway Code* recommends these lights should not be used unless visibility is below 100 metres.

Side Marker Lamps

Side marker lamps must be fitted on long vehicles and trailers as follows:
1. Vehicles first used on or after 1 April 1991 and trailers made from 1 October 1990 and being over 6 metres long:
    (a) one lamp on each side within 4 metres of the front of the vehicle;
    (b) one lamp on each side within 1 metre of the rear of the vehicle;
    (c) additional lamps on each side at 3 metre intervals (or, if impracticable, 4 metres) between front and rear side marker lamps.
2. Vehicles (including a combination of vehicles) over 18.3 metres long (including the length of the load):
    (a) one lamp on each side within 9.15 metres of the front of the vehicle;
    (b) one lamp on each side within 3.05 metres of the rear of the vehicle;
    (c) additional lamps on each side at 3.05 metre intervals between front and rear side marker lamps.
3. Vehicles in combination between 12.2 metres and 18.3 metres long (but not articulated vehicles) carrying a supported load:

(a) one lamp on each side within 1530mm of the rear of the rearmost vehicle in the combination;

(b) one lamp within 1530mm of the centre of the load, if the load extends further than 9.15 metres to the rear of the drawing vehicle.

4. Trailers more than 9.15 metres long (6 metres for post-1 October 1990 trailers):

(a) one lamp on each side within 1530mm of the centre of the trailer length.

Side marker lamps fitted to pre-1 October 1990 built trailers may show white side marker lights to the front and red lights to the rear, in all other cases such lights must be amber. They must be positioned not more than 2300mm from the ground.

End-Outline Marker Lamps

Vehicles (except those less than 2100mm wide and those first used before 1 April 1991) and trailers (except those less than 2100mm wide and those built before 1 October 1990) must be fitted with two end-outline marker lamps visible from the front and two visible from the rear. They must be positioned no more than 400mm in from the outer edges of the vehicle/trailer and mounted at the front at least level with the top of the windscreen. They must show white lights to the front and red lights to the rear.

Lighting Switches

On vehicles first used since 1 April 1991 a single lighting switch only must be used to illuminate all front and rear position lamps, side and end-outline marker lamps and rear number plate lamp, although one or more front or rear position lamps may be capable of being switched on independently.

Visibility of Lights and Reflectors

A part, at least, of each front and rear position light, front and rear-mounted direction indicator lamp and rear retro-reflector required to be fitted to a vehicle/trailer must be capable of being seen from directly in front or behind the lamp or reflector when the vehicle doors, tailgate, boot lid, engine cover or other movable part of the vehicle is in a fixed open position.

Lights on Projecting Loads

Details of the requirements for the display of lights on projecting loads are given fully in Chapter 22 which covers this subject.

Direction Indicators

All goods vehicles must be fitted with amber-coloured direction indicators (on pre-September 1965 registered vehicles, indicators can be white facing to the front and red facing to the rear) at the front and rear which must be fixed to the vehicle not more than 1500mm (2300mm in exceptional cases) and not less than 350mm above the ground, at least 500mm apart and not more than 400mm from the outer edges of the vehicle. Side repeater indicators are required on vehicles first used since 1 April 1986 and these must be fitted within 2600mm of the front of the vehicle. Normally vehicles should have one indicator on each side at the front and rear but may have two on each side at the rear. They must not have more than one on each side at the front.

Indicators bearing approval marks do not have to meet minimum wattage requirements but those without such marks must be between 15 and 36 watts. They must flash at a rate of between 60 and 120 times a minute and a visible or audible warning must indicate to the driver when they are operating. The indicators must be maintained in a clean condition and in good and efficient working order.

Semaphore Arm Indicators

Vehicles first registered before 1 September 1965 are allowed to have either semaphore arm or flashing indicators. The semaphore arm type must be amber in colour but the flashing indicators may show a white light to the front and a red light to the rear.

Hazard Warning

Direction indicators operating on both sides of the vehicle simultaneously as a hazard warning to other road users are required by law on all vehicles first used since 1 April 1986. They must be actuated by a switch solely controlling that device and a warning light must indicate to the driver that the device is being operated. The hazard indicators may be used when the vehicle is stationary on a road, or any part of the road (ie not just the carriageway), because of a breakdown of it or another vehicle, an accident or other emergency situation, or when the vehicle is causing a temporary obstruction on a road when loading or unloading.

Emergency Warning Triangles

As an additional warning of a hazard, drivers *may* (ie it is not compulsory to do so) place a red warning triangle on the road to the rear of a vehicle causing a temporary obstruction (eg through breakdown). The triangle must be made and marked to British Standard Specification BS AU47: 1965. It must be placed upright on the road, 45 metres to the rear of the obstruction and on the same side.

Other safety devices may be used to warn of vehicles broken down on the roadside. These include traffic cones, warning lamps and traffic pyramids, as well as conventional warning triangles mentioned above.

Warning Beacons

Amber warning beacons must be fitted to vehicles with four or more wheels and having a maximum speed no greater than 25mph when using unrestricted dual-carriageway roads except where such use is merely 'for crossing the carriageway in the quickest manner practicable in the circumstances'.

Amber warning beacons may also be fitted to vehicles used at the scene of an emergency, when it is necessary or desirable to warn of the presence of a vehicle on the road (eg Special Types vehicles carrying abnormal loads) and to breakdown vehicles used at the scene of accidents and breakdowns and when towing broken-down vehicles.

Green warning beacons may be used on vehicles by medical practitioners registered with the General Medical Council when travelling to or dealing with an emergency.

Blue warning beacons and other special warning lamps may only be used on emergency vehicles (ie ambulance, fire brigade or police service vehicles; Forestry Commission fire fighting vehicles; military bomb disposal vehicles; RAF Mountain Rescue vehicles; Blood Transfusion Service vehicles; British Coal mines rescue vehicles; HM Coastguard and Coast Life Saving Corps vehicles; RNLI vehicles and those used primarily for transporting human tissue for transplanting).

In all cases such beacons should be fitted with their centres no less than 1200mm from the ground and visible from from any point at a reasonable distance from the vehicle. The light itself must show not less than 60 and not more than 240 times per minute.

Swivelling Spotlights (Work Lamps)

White swivelling spotlights may be used only at the scene of an accident, breakdown or roadworks to illuminate the working area or work in the vicinity of the vehicle provided it does not cause undue dazzle or discomfort to the driver of any vehicle.

Rear Retro Reflectors

Motor vehicles must be fitted with two red reflex retro reflectors facing squarely to the rear. Reflectors must be fitted not more than 900mm and not less than 350mm from the ground. For normal goods vehicles and trailers they must be within 400mm of the outer edge of the vehicle or trailer and not less than 600mm apart. Reflectors must be capable of being seen from an angle of 30 degrees on either side.

Triangular Rear Reflectors
Triangular rear reflectors if used on a voluntary basis may only be fitted to trailers or broken-down vehicles being towed. There is no longer any specific legal requirement for such reflectors to be fitted. They must not, in any event, be used on other vehicles.

Side Retro Reflectors

Vehicles more than 6 metres long first used since 1 April 1986 (more than 8 metres long if first used before 1 April 1986) and trailers more than 5 metres long must be fitted with two (or more as necessary) amber side retro reflectors on each side. One reflector on each side must be fitted not more than 1 metre from the extreme rear end of the vehicle and another no more than 4 metres from the front of the vehicle with further reflectors at minimum 3-metre intervals (or can be 4-metre intervals) along its length. They must be mounted not more than 1500mm and not less than 350mm from the ground. On pre-April 1986 vehicles one reflector must be positioned in the middle third of the vehicle length and the other within one metre of the rear. Where such reflectors are mounted within one metre of the rear of the vehicle/trailer they may be coloured red instead of amber.

Front Retro Reflectors

Trailers built since 1 October 1990 must be fitted with two obligatory front retro reflectors, white in colour and mounted facing forwards, at least 350mm but not more than 900mm from the ground and no more than 150mm in from the outer edges of the trailer and at least 600mm apart.

VEHICLE MARKINGS

Number (Registration) Plates

All vehicles first registered since 1 January 1973 must be fitted with number plates made of reflecting material complying with BS AU 145a. This requirement does not apply to goods vehicles over 7.5 tonnes gross weight which are required to display rear reflective markers (see p 276) or works trucks, agricultural machines and trailers or pedestrian-controlled vehicles. If a vehicle over 3050kg unladen is exempt from the requirement to fit rear reflective markers then it must be fitted with reflective number plates.

Unladen Weight

Goods vehicles of more than 3050kg unladen weight (ie heavy motor cars), motor tractors and locomotives must have their unladen weight shown in a conspicuous place on the outside of the vehicle where it can easily be seen. This does not apply where the unladen weight is shown on the DoT (ie Ministry) plate attached to the vehicle.

DoT Plates

Goods vehicles over 1525kg and trailers over 1020kg unladen weight must display a DoT plate and/or a manufacturer's plate showing the maximum permissible gross vehicle weight and individual axle weights (see Chapter 16). Since 1 April 1991 the requirement for displaying DoT plates applied only to rigid goods vehicles over 3500kg, articulated vehicles and goods carrying trailers over 1020kg.

Special Types Plates

Since 1 October 1989 it has been a legal requirement for vehicles carrying abnormal loads to display a manufacturer's plate showing the maximum weights at which the vehicle can operate and the relevant speeds for travel at those weights. The weights shown must be those approved by the vehicle or trailer manufacturer and the vehicle must not exceed specified Special Types speed limits when travelling loaded to the weight shown on the plate (see details of speeds for different classes of Special Types vehicles, p 369).

Food Vehicles

Vehicles which are used in connection with a food business or from which food is sold must display in a clearly visible place the name and address of the person carrying on the business and the address at which the vehicle is kept or garaged. If the vehicle bears a fleet number and is kept or garaged on that person's premises the garage address is not required but the local authority must be notified.

Height Marking

The travelling height of vehicles and trailers carrying engineering equipment, containers and skips, where the height of the vehicle and load exceeds 3.66m, must be marked in the cab where the driver can see it. Further new provisions to extend this requirement are currently under discussion (see p 217 for further details).

Hazard Marking

Vehicles which carry hazardous, radioactive or explosive loads must display appropriate hazard warning symbols on the vehicle whether a bulk tanker, a tank container or a normal delivery vehicle used for carrying hazardous consignments and on the individual packages too in the latter case. Further details are given in Chapter 23.

Rear Reflective Markings

All vehicles with a maximum permissible weight exceeding 7500kg and trailers with a maximum permissible weight exceeding 3500kg (see also p 375) must be fitted with rear reflective markers which make them more conspicuous at night and in poor visibility (Figures 15.1 and 15.2). The markers may also be displayed on loads such as builders' skips (see below).

Which Markers to be Fitted

Vehicles not exceeding 13 metres in length and trailers in combinations not exceeding 11 metres must be fitted with the markers shown in diagrams 1 and 2 in Figure 15.1 or if this is not practical the markers shown in diagram 3 may be fitted. Trailers in combinations of more than 11 metres but not more than 13 metres may fit the markers shown in diagrams 1, 2, 3, 4 or 5. Vehicles more than 13 metres long and trailers in combinations more than 13 metres long must be fitted with the markers shown in diagram 4 or 5.

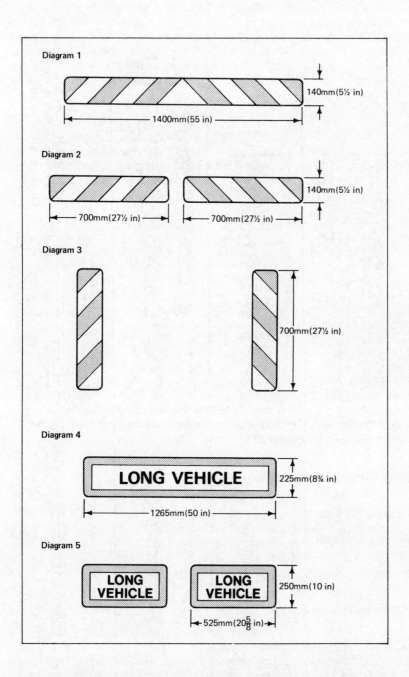

Figure 15.1 *Rear reflective markers required on certain goods vehicles. The plates comprise red fluorescent material background. The lettering is in black on yellow reflex reflecting*

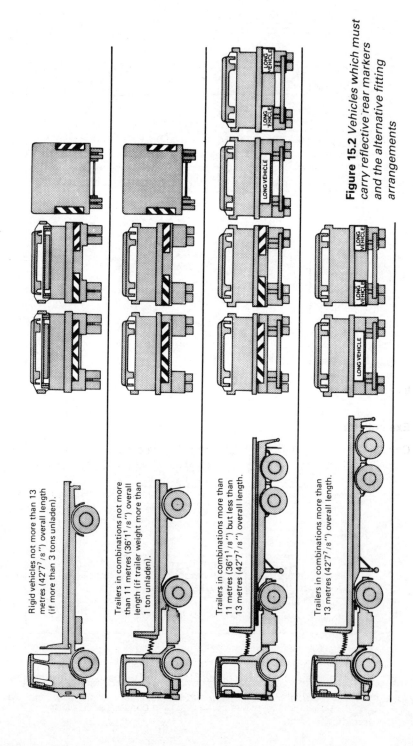

Figure 15.2 *Vehicles which must carry reflective 'rear markers and the alternative fitting arrangements*

Rigid vehicles not more than 13 metres (42'7⁷/₈'') overall length (if more than 3 tons unladen).

Trailers in combinations not more than 11 metres (36'1¹/₈'') overall length (if trailer weight more than 1 ton unladen).

Trailers in combinations more than 11 metres (36'1¹/₈'') but less than 13 metres (42'7⁷/₈'') overall length.

Trailers in combinations more than 13 metres (42'7⁷/₈'') overall length.

Types of Markers

There are two types of markings, each in two sizes:
1. Alternating red fluorescent and yellow reflective diagonal strips – diagrams 1, 2, 3.
2. A central yellow reflective panel overprinted with the words LONG VEHICLE and having a red fluorescent surround – diagrams 4 and 5.

Specification for Markers

Markers fitted to vehicles and trailers must comply with the regulations regarding size and colour and they must be in the form of durable plates stamped with the mark 'BS AU 152'. They must not be simulated by being painted on the vehicle and the plates must not be defaced, cut or modified to aid fitting to the vehicle.

Fitting Position

The height from the ground to the lower edge of the marker when fitted must not exceed 1700mm but must be at least 400mm. It must be fitted parallel to the ground and be facing square to the rear.

Alternative Fitting Position

When a vehicle, which by law requires rear reflective markers to be displayed, is carrying a load which obscures partly or wholly the markers so that they are not clearly visible from the rear, the reflective markers may be fitted to the rear of the load.

Exemptions

Certain vehicles as indicated in the following list are exempt from the requirement to fit these markers. Previously it was illegal to fit the markers to these vehicles, but a change in the regulations permits the fitting of such markers to exempt vehicles on a voluntary basis, provided the vehicles exceed the specified weight limit. It is illegal to display these markers on vehicles which do not require them by law except as mentioned below.

Exempt Vehicles
1. Vehicles with a maximum gross weight not exceeding 7500kg
2. Passenger vehicles other than articulated buses
3. Land tractors, land locomotives, land implements, land implement conveyors, agricultural tractors or industrial tractors
4. Works trucks or works trailers
5. Vehicles in an unfinished condition proceeding to a works for completion or to a place where they are to be stored or displayed for sale
6. Motor vehicles constructed or adapted for the purpose of forming part of articulated vehicles
7. Broken-down vehicles while being drawn in consequence of the breakdown

8. Engineering plant
9. Trailers, not being part of an articulated bus, drawn by public service vehicles
10. Vehicles designed for fire fighting or fire salvage purposes
11. Vehicles designed and used for the purpose of servicing or controlling aircraft
12. Vehicles designed and used for the transportation of two or more motor vehicles carried thereon, or of vehicle bodies or two or more boats
13. Vehicles proceeding to a place for export
14. Vehicles brought temporarily into Great Britain by persons residing abroad
15. Vehicles in the service of a visiting force or of a headquarters
16. Motor vehicles first used before 1 January 1940
17. Vehicles owned or in the service of the Army, Navy or Air Force
18. Vehicles designed for heating or dispensing tar or similar material for road construction or maintenance
19. Trailers being drying or mixing plant designed for the production of asphalt, bitumen or for macadam
20. Trailers made before 1 August 1982 with an unladen weight not exceeding 1020kg
21. Trailers with a gross weight not exceeding 3500kg.

Builders' Skips

Rear reflective markings of the type described above (as shown in diagram 3, Figure 15.1) must be fitted to the ends of builders' skips which are placed on the highway. They must be fitted as a matched pair as near to the outer edge as possible, mounted vertically and no more than 1.5 metres from the ground to the top edge. They must be kept clean, in good order and be visible from a reasonable distance. Such skips are required to be illuminated when standing on roads at night.

16: Goods Vehicle Plating, Annual Testing and Vehicle Inspections

Most goods vehicles are required to be tested annually to ensure they are safe to operate on the road and meet the legal requirements relating to mechanical condition. In particular, this annual inspection is intended to determine whether vehicles and trailers meet the standards specified in the Road Vehicles (Construction and Use) Regulations 1986 (as amended). Additionally, it is necessary for certain goods vehicles and trailers to be 'Ministry plated' to show the maximum permissible gross weight and maximum axle weights at which they may be operated on roads in Great Britain. The requirement for annual plating and testing of goods vehicles is contained in the Road Traffic Act 1988 and is detailed in the Goods Vehicles (Plating and Testing) Regulations 1988 as amended.

The Department of Transport's Vehicle Inspectorate Executive Agency (VI) is responsible for the operation of goods vehicle testing stations.

Annual Testing

Articulated tractive units, rigid goods vehicles over 3500kg gross weight, goods carrying semi-trailers and drawbar trailers over 1020kg unladen weight and converter dollies must be tested annually at a DoT goods vehicle testing station (see Appendix IV for list). Certain specialised vehicles are exempt from the test, as shown on pp 296–7, and the regulations do not apply to vehicles used under a trade licence. Goods vehicle test stations also carry out the Class V test, which is the vehicle MoT-type test, on large passenger vehicles which cannot get into normal MoT test garages (see also Chapter 17).

A Class VII test has been established to bring into the MoT test goods vehicles of more than 3000kg but not more than 3500kg design gross weight which were formerly subject to goods vehicle plating and testing when they had an unladen weight exceeding 1525kg.

Types of Test

There are various types of goods vehicle test as follows:
1. *First test:* the first annual test of the vehicle or trailer (carried out no later than the end of the anniversary month in which it was first registered) at which the information shown on the vehicle's DoT plate (issued from Swansea) is verified.
2. *Part 2 re-test:* examines the vehicle which has failed its first test.
3. *Periodical test:* the annual test which applies to all relevant vehicles after the first test.

4. *Part 3 re-test:* retests a vehicle which has failed its annual or periodical test.
5. *Part 4 test:* a test provided for in the regulations which may be required if a notifiable alteration has been made to the vehicle or if the operator wants the plating certificate amended to show different weights as a result of changes to the vehicle or, for example, if different tyre equipment has been fitted.
6. *Re-test following appeal:* to area mechanical engineer or the Secretary of State for Transport.

Test Dates

Vehicles may be submitted for test at any one of the full-time or part-time goods vehicle test stations selected by the vehicle operator. Vehicles are due for test each year no later than the end of the anniversary month in which they were first registered (eg a vehicle registered on 1 January 1989 would be due for its first test no later than 31 January 1990 and for subsequent tests by 31 January in each following year).

Trailer Test Dates

Articulated semi-trailers and other goods carrying trailers which come within the scheme are due for test during the month indicated by the last two figures of the serial number which the DoT allocates to all trailers when making application for their first test (eg if the last two figures of the serial number are 01 the trailer has to be tested in January each year; 07 means testing in July; 12 means testing in December; and so on). The first test for trailers is due by the end of the first anniversary month from *when they were sold*. At this test the serial number mentioned above is given.

Year of Manufacture/Registration

There is an anomaly with due test dates when a vehicle or trailer is manufactured in one year and is not registered (or sold in the case of a trailer) before 1 July of the following year. In this case they must be tested by the end of December in the year in which they were first registered or sold.

Phased Programmes and Missed Test Dates

The DoT allows vehicle operators the facility of having vehicles voluntarily tested before their due date to accommodate phased programmes of test preparation rather than having a large number of vehicles due for preparation and test in any particular month of the year. This concession is only permitted once during a vehicle's life and when the particular date has been chosen the vehicle will become due for test in the same month in subsequent years, and not in the month of its first registration.

When a vehicle misses its due test date (possibly because it has been off the road for a period) then provided it is tested more than 10 months but less than 1 year after the date it was originally due for its test, a test certificate issued will be valid for up to 14 months from the date of issue. This will save vehicles being tested twice within a short period of time.

Test Applications

Initial application for a first test and for subsequent tests of vehicles and trailers has to be made direct to the DoT, Goods Vehicle Centre, Welcombe House, 91/92 The Strand, Swansea (not to be confused with the DVLC at Swansea). The following forms, which may be obtained from goods vehicle test stations or Traffic Area Offices, are used for making the application:

VTG1L	First test of a vehicle
VTG2L	First test of a trailer
VTG40L	For subsequent tests of both vehicles and trailers

Time for Application

Applications for test should be made during the 2 months prior to the month in which the test is required to take place (ie the last day on which the vehicle may legally operate without a test certificate). Applications should ideally be made at least one month before the date on which the test is preferred by the operator (see note below about the Saturday test facility).

Saturday Testing

A scheme has been introduced for Saturday testing at certain selected test stations (currently 17 stations offer the service) at an additional fee. Operators wishing to take advantage of this facility must mark their applications very clearly 'SATURDAY TEST' and show the appropriate date.

On-Site Testing

There is a suggestion that in the future vehicle tests may be carried out on operator's premises by VI examiners, provided that the facilities and equipment available are brought up to the required standard (at the operator's own expense). New legislation would be required to enable this plan to be put into action and there would be supplementary fees to be paid for on-site testing.

Test Fees

The appropriate test fee, as follows, should be sent to the Goods Vehicle Centre with the application form:

Motor vehicle annual test	
2-axle vehicle	£32.70
3-axle vehicle	£33.70
4-axle vehicle	£34.70
Trailer annual test	
1-axle trailer	£16.70
2-axle trailer	£17.10
3/4-axle trailer	£17.90
Re-tests	
motor vehicles	£16.60
trailers	£8.30

Notifiable alterations* £13.50

Saturday supplement (paid separately)

	(for normal tests)	(for re-tests)
– motor vehicles	£19.50	£9.80
– trailers	£12.30	£6.10

Saturday supplement for notifiable alterations* £8.40

*NB: These are examinations carried out at the operator's request following an alteration to a vehicle or a trailer.

Block Bookings

Large fleet operators may make block bookings for vehicles or trailers of similar type to be tested at a test station so that any available vehicle or trailer of the block may be submitted for test at the appointed time. The test station must, however, be advised 2 or 3 days before the appointed date as to which particular vehicle or trailer will be submitted for test.

Trailer Testing

Many operators have more semi-trailers and trailers than tractive units or drawing vehicles and in order to have these additional trailers tested it may be necessary for them to be submitted for test with a vehicle which has already been tested and has a current valid test certificate. In these cases only the trailer will be examined and the fee payable will be the trailer fee only.

Test Appointments

Following application to the Goods Vehicle Centre for a first or subsequent test (see note above about Saturday testing facility), the test station selected by the operator will confirm the test booking in due course with an appointment card, and all further communications regarding the test must be made with the test station, not with Swansea. If, owing to excessive workload or staff shortage, the chosen test station cannot accommodate the test an appointment will be made at the nearest alternative test station and a card will be sent from that station. The appointment card and the vehicle registration document must be produced on arrival at the test station.

It is essential that vehicles arrive at the test station at the appointed time. Late or non-arrival of a vehicle can mean cancellation of the test and the fee will be forfeited unless an acceptable reason citing 'exceptional circumstances' is put forward. Exceptional circumstances include accident, fire, epidemic, severe weather, failure in the supply of essential services or other unexpected happenings. Breakdown, mechanical defect or non-availability of the vehicle because of shortages in spare parts supply or for operational reasons for example are not looked on as exceptional circumstances.

Cancellations

If it is necessary to cancel a test booking after making application, provided 7 days' notice is given to the test station either a new test date will be arranged or the fee will be refunded after deduction of a £1.50 charge. In exceptional circumstances, such as an accident to the vehicle on the way to the test station, if notification is given to the station within 3 days of the accident the fee will be carried forward or refunded less £1.50. If other exceptional circumstances arise within 7 days of the due date for the test, providing satisfactory evidence is given to the test station, the fee will be similarly carried forward or refunded less £1.50.

Refusal to Test

Test station officials have the right to refuse to test a vehicle or trailer for the following reasons, in which circumstances form VTG12 will be issued:

1. Arrival after the appointed time.
2. Appointment card or vehicle registration document not produced.
3. If it is found that the vehicle brought to the test station does not conform to the details given on the application form.
4. If the vehicle was booked for the test with a trailer but the trailer is not taken to the test station.
5. If the chassis number cannot be found by the examiner or if the serial number given for the trailer by the DoT is not stamped on it.
6. If the vehicle is in a dirty or dangerous condition.
7. If the vehicle does not have sufficient fuel or oil to enable the test to be carried out.
8. If the test appointment card specified that the vehicle should be loaded for the test and it is taken to the test station without a load. Under normal circumstances the decision whether the vehicle is to be tested in a laden or unladen condition is left to the owner to suit his convenience but in some circumstances the test station may request that the vehicle is fully or partially loaded to enable the brakes to be accurately tested on the roller brake tester.
9. In the case of a trailer if the vehicle submitted with it is not suitable to draw it.
10. If the vehicle breaks down during the test.
11. If the vehicle is submitted for its annual test (ie not for the first test) or a re-test and the previous test and plating certificates are not produced.

Test Procedure

Goods vehicle test stations vary in size and in the number of examination staff. Testing normally takes approximately 45 minutes during which the driver must be available to assist and move the vehicle as required. Examination of vehicles is carried out by DoT examiners based at the station in accordance with the *Heavy Goods Vehicle Inspection Manual* published by HMSO (current edition 1993). All items which have to be inspected are listed in the *Manual* together with, where necessary, details of how the inspection of each item should be carried out and the reasons for failing the item. Under the 'reasons for rejection' column in the *Manual* where the item inspected is one that is subject to wear, the maximum tolerance will be indicated.

The examination is conducted in four stages: items 1 to 40 in the *Inspection Manual* are inspected on the hard standing outside the test building. Items 41 to 61 are inspected at the second stage of the examination over the test pit inside the building. The third section of the test, items 62 to 69, are inspected at the next point where a beam setter is used for checking headlamp alignment. The braking tests, under items 70 to 73, which comprise the fourth section of the examination, are the last to be carried out at the end of the test line. A roller brake tester is used for these purposes; the machine indicates on dials on a console the braking force of all the wheels on an axle and each individual wheel (or a pair of twin wheels) when the vehicle's footbrake, handbrake and emergency brake are applied by the driver when directed to do so.

The Smoke Test
Free acceleration smoke emission testing of diesel engines by means of a calibrated smokemeter (to measure the density of smoke in a vehicle exhaust) was introduced as part of the goods vehicle annual test from 1 September 1992. Where it is necessary to carry out a purely visual test, the driver is asked to depress the accelerator pedal firmly from the engine idle position (preferably with the engine already warm) to the maximum fuel delivery position and, immediately the governor operates, release the pedal until the engine slows to a steady idling speed. Smoke emission from the first attempt is ignored but the procedure has to be repeated for a maximum of ten times until the smoke emission is considered to be of equal density for two successive accelerations.

Inspection Card
During the test the examiner has an inspection card on which all items in the *Inspection Manual* are listed and in the case of failure of any item the card is marked accordingly. Four inspection cards are used as follows:

1. Form VTG4A   For tests of rigid vehicles
2. Form VTG4B   For re-tests of rigid vehicles
3. Form VTG4C   For tests of trailers
4. Form VTG4D   For re-tests of trailers

Items to be Inspected
The list of items shown in the *Heavy Goods Vehicle Inspection Manual* to be inspected is as follows:

1. –
2. –
3. Seat belts*
4. –
5. Smoke emission*
6. Road wheels and hubs
7. Size and type of tyres
8. Condition of tyres
9. Sideguards, rear underrun device and bumper bars
10. Spare wheel carrier
11. Tractor to trailer coupling
12. Trailer emergency brake
13. Trailer landing legs
14. Spray suppression, wings and arches
15. Cab security*
16. Cab doors*
17. Cab floor and steps*

18. Driving seat*
19. Security of body
20. Condition of body
21. –
22. Mirrors*
23. View to front*
24. Glass or other transparent material*
25. Windscreen wipers and windscreen washers
26. Speedometer*
27. Audible warning (horn)*
28. Driving controls*
29. Tachograph*
30. Play at steering wheel*
31. Steering wheel*
32. Steering column*
33. –
34. Pressure/vacuum warning*
35. Build-up of pressure/vacuum*
36. Lever operating mechanical brakes*
37. Service brake pedal*
38. Service brake operation
39. Hand-operated brake control valves
40. –
41. Condition of chassis
42. Electrical wiring and equipment
43. Engine and transmission mountings*
44. Oil leaks*
45. Fuel tanks and systems*
46. Exhaust systems*
47. –
48. Suspension pins and bushes
49. Suspension spring units and linkages
50. Attachment of spring units, linkages and sub-frames
51. Shock absorbers (dampers)
52. –
53. Axles, stub axles and wheel bearings
54. Steering linkage
55. Steering gear*
56. Power steering*
57. Transmission*
58. –
59. Mechanical brake components
60. Brake actuators
61. Braking systems and components
62. Rear markings
63. Front position lamps*
64. Rear position lamps and rear fog lamps
65. Reflectors
66. Direction indicators
67. Aim of headlamps*
68. Headlamps*

69. Stop lamps
70. Trailer parking brake
71. Service brake performance
72. Secondary brake performance*
73. Parking brake performance
74. –
75. –

These items do not apply when trailers are being inspected.

– The blank spaces are left to enable the VI to add any further items at a later time.

Test Pass

When a goods vehicle is found to be in satisfactory order a test certificate is issued by the test station. For goods vehicles the certificate is form VTG5 and for goods-carrying trailers it is VTG5A which is issued together with a trailer test disc (VTG5B). The trailer test disc is included with the certificate and this must be removed and fixed on to the trailer in a protective holder in a position where it is conspicuous, readily accessible and clearly visible from the nearside.

The heavy goods vehicle test certificate must be produced when applying for an excise licence for a vehicle which comes within the scope of these regulations and otherwise at the request of a competent authority.

Replacement Documents

Replacement test certificates and trailer test date discs and replacement plates and plating certificates (see later in chapter) may be obtained from the Goods Vehicle Centre at Swansea at a cost of £9.50 each. Application in both cases should be made on form VTG59 obtainable from goods vehicle test stations or Traffic Area Offices. Automatic replacement of lost or defaced documents is not guaranteed. The Secretary of State has powers to order a re-test before issuing such replacements, in which case full test fees become payable.

Test Failure and Retests

When the vehicle is sent for test, it is recommended that a mechanic with a tool-kit and minor spares items (light bulbs, for example) accompanies it so that any minor defects can be rectified on the premises and the test can be completed. The examiner may allow certain minor defects to be repaired during the test if they do not take up too much time and delay the test schedule. In some instances the vehicle may be allowed to be taken out of the test line for minor repairs to be carried out but this again is at the discretion of the examiner.

If it is necessary to take the vehicle away to get the defects rectified and it is submitted again later that day or during the next working day, no additional charge is made. These free re-tests are restricted to those cases where the vehicle failed because of certain prescribed defects in items as follows:

- Legal plate position

- Legal plate details
- Bumper bars
- Spare wheel carrier
- Cab doors
- Mirrors
- View to front
- Speedometer
- Audible warning
- Oil leaks
- Fuel tanks, pipes and system
- Obligatory sidelamps
- Obligatory rear lamps
- Reflectors
- Direction indicators
- Headlamps – vertical aim
- Obligatory headlamps
- Obligatory stop lamps.

If the vehicle which fails the test is submitted to the same test station again within 14 days, a reduced retest fee is charged and only the items on which the test was failed are re-examined. The re-test fee is £16.60 for vehicles and £8.30 for trailers. Arrangements for re-tests have to be made with the manager of the test station concerned.

Appeals against Test Failure
If a vehicle or trailer undergoing test or re-test failed for a reason which the operator believes is not justified he has a right of appeal to the DoT Area Mechanical Engineer, who is based at the Traffic Area Office, provided it is made within 10 days of the test. If the operator is not satisfied with the Area Mechanical Engineer's decision, a further appeal may be made to the Secretary of State for Transport within 14 days of this decision.

Appeals must be made on form VTG8 and the fees are £15 in the case of appeals to the Area Mechanical Engineer and £25 for appeals to the Secretary of State. Appeal fees are refunded if the appeal is upheld.

Plating of Goods Vehicles and Trailers

Manufacturer's Plating

All new goods vehicles and new trailers over 1020kg unladen weight must be fitted with a plate (see Figure 16.1) by the manufacturer which shows specified information as follows:

- The manufacturer's name
- The date of manufacture
- Vehicle type
- Engine type and power rating

- Chassis or serial number
- Number of axles
- Maximum weight allowed on each axle
- Maximum gross weight for the vehicle (including the weight imposed on the tractive unit by a semi-trailer in the case of articulated vehicles)
- Maximum train weight.

The plate containing this information is normally fitted inside the driver's cab on the nearside.

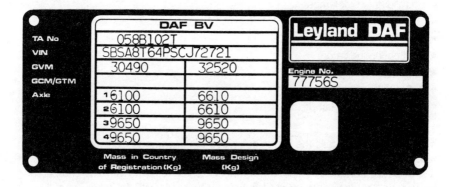

Figure 16.1 *An example of a manufacturer's plate for a rigid 8-wheel chassis*

For trailers, the information shown on the plate is:
- The manufacturer's name
- Date of manufacture
- Chassis or serial number
- Number of axles
- Maximum weight allowed on each axle
- Maximum weight imposed on the drawing vehicle in the case of semi-trailers
- Maximum gross weight for the trailer.

The plate for trailers is usually riveted to the chassis frame on the nearside.

Design Weights
The weights stated are those at which the manufacturer has designed the vehicle to operate. Where these weights exceed those permitted by law (ie in the construction and use regulations) for the type of vehicle in question then, until such time as the vehicle is plated by the Department of Transport, the lower statutory weight limits apply. Conversely, if the manufacturer's design weight is lower than that permitted by law for the type of vehicle then it is the lower limit which applies.

DoT (ie 'Ministry') Plating

When Type Approved goods vehicles over 3500kg gross weight are first registered an official plate is issued showing the maximum permissible gross vehicle weight and individual axle weights at which the vehicle or trailer is allowed to operate within Great Britain.

A plating certificate (Figure 16.2) and a plate (Figure 16.3), giving similar details to those on the manufacturer's plate, are issued by the Goods Vehicle Centre at Swansea after receipt of the necessary documents (including the Type Approval Certificate) when a new vehicle is first registered. The plating certificate must be retained by the vehicle operator but the plate (made of paper and protected in a laminated casing, not metal as is the manufacturer's plate) which is also issued must be fixed to the vehicle in an easily accessible position. Generally it should be fitted inside the cab of vehicles on the nearside (but not affixed to the door) and in a suitable position on the nearside of trailers (it is usually fitted to the chassis frame). In all cases the plates should be protected against the weather, kept clean and legible and secure against accidental loss.

Tractive units and semi-trailers in articulated outfit combinations are plated separately and the individual plates must be fixed separately to the tractive unit and the trailer.

Articulated Vehicle Matching

When planning loads for or when loading articulated outfits, the plated weights of both the tractive unit and the trailer must be taken into consideration. If the outfit normally operates as a matched pair at all times there is little difficulty, but if various trailers covering a range of plated weights are used with the tractive unit, care must be taken by the people concerned (drivers, loaders, weighbridge staff, etc) to ensure that the correct weights are observed so that no offence is committed. For example, if a two-axle tractive unit plated for operation at 38 tonnes gross weight is normally used at this weight with a tri-axle semi-trailer but on a particular occasion a tandem-axle semi-trailer is used, a lower weight, limited by the semi-trailer's gross plated weight (say 32.5 tonnes), must be observed and an offence would be committed if the outfit was still loaded to 38 tonnes gross weight. With 38 tonnes operation permitted for certain articulated vehicles, it is important in this context to ensure that the minimum overall length requirements are also observed (see p 214).

Standard Lists

Vehicles are plated by the DoT in accordance with 'standard lists'. These lists, one for each make of vehicle and trailer, are published by HMSO and they are based on information provided by the manufacturer relating to all the models and types of vehicle or trailer, with their respective serial/chassis numbers, manufactured in recent years.

The designed axle weights and gross vehicle weights are shown in the lists together with differences in design weights which apply when alternative options in suspension systems (such as a differing number of leaves in the

PLATING CERTIFICATE

Department of Transport
Road Traffic Act 1972, Sections 40 and 45
Examination of Goods Vehicles

2057258

DOE REF. No.

A. GENERAL PARTICULARS AND PLATED WEIGHTS

REGISTRATION/ IDENTIFICATION MARK	CHASSIS/ SERIAL No.	YEAR OF ORIGINAL REGISTRATION	YEAR OF MANUFACTURE	MAKE AND MODEL

(1) DESCRIPTION OF WEIGHTS APPLICABLE TO VEHICLE	(2) WEIGHTS NOT TO BE EXCEEDED IN GT. BRITAIN (See note 6 overleaf)	(3) DESIGN WEIGHTS (if higher than shown in col. (2)) (See note 6 overleaf)	(4) TYRES (fitted at time of issue of Certificate)		
	KILOGRAMS	KILOGRAMS	SIZE	PLY RATING	S or T*
AXLE WEIGHT (Axles numbered from front to rear) — AXLE 1.					
AXLE 2.					
AXLE 3.					
AXLE 4.					
GROSS WEIGHT (See note 4 overleaf)			* "S" indicates single wheels and "T" twin wheels		
			Tyre Use Conditions applicable to Vehicle		
TRAIN WEIGHT (See note 5 overleaf)			For meaning of symbol inserted opposite see note 7 overleaf		

B. NOTIFIABLE ALTERATIONS (Section 45 (6) (d) of the Road Traffic Act 1972)

The following alterations in the vehicle or its equipment must be notified to the Secretary of State (See notes 1 to 3 overleaf);

(a)   an alteration made in the structure or fixed equipment of the vehicle which varies the carrying capacity of the vehicle;

(b)   an alteration, otherwise than by way of replacement of a part, adversely affecting any part of a braking system with which the vehicle is equipped or of the means of operation of that system; or

(c)   any other alteration made in the structure or fixed equipment of the vehicle which materially renders the vehicle unsafe to travel on roads at any weight equal to any plated weight shown in column (2) of this plating certificate.

C.   FORM OF CERTIFICATE

I certify that the plated weights and other plated particulars specified in this plating certificate are those determined for the vehicle concerned under Section 45 of the Road Traffic Act 1972 and the regulations made under that Section.

Signature _____

Date of Issue _____   Vehicle Testing Station No. _____

1.	2.	3.	
			VTG7

Figure 16.2 *DoT plating certificate VTG 7*

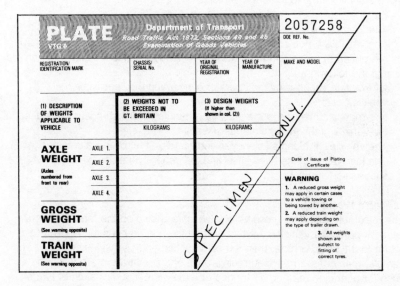

Figure 16.3 *DoT plate (form VTG 6)*

springs), wheels and tyres are selected on any particular model. For example, the use of leaf springs with one or more additional leaves or the use of tyres of a different size or ply rating could possibly mean an increase of a few kilos on the axle or gross vehicle weights. Conversely, by selecting an alternative lower specification, reduced axle or gross vehicle weights may apply (eg to come below the levels at which 'O' licensing or lgv drivers' licensing apply).

Provided vehicles submitted for test comply with the details set out in the standard lists, they will be plated by the DoT at the standard list weights. Plating at lower than standard list weights will only be done if the tyres fitted to the vehicle at the time of the test are not of the ply rating shown in the standard list for that vehicle. In such circumstances the vehicle will be plated at axle and gross weights suitable for the tyres fitted. The weights will be decided by reference to a tyre data sheet, also obtainable from HMSO, which indicates the loadings on tyres of various size and ply ratings.

Down-plating for operational reasons, such as to obtain the benefit of lower rates of VED (eg, in certain circumstances, an extra 1 tonne of plated weight could incur £1000 extra annual duty at current rates), will only be carried out by the DoT if the vehicle is suitably modified by the fitting of weaker parts (eg lower rated tyres or lighter duty propshaft). A campaign has been mounted by the road haulage industry to get the DoT to change this ruling which, according to the Road Haulage Association, results in the needless weakening of vehicles.

In *no* circumstances will a vehicle which has an inefficient braking performance or which is not maintained to the required standard be plated at lower than standard list weights to compensate for these defects. In such cases the examiners will refuse to plate the vehicle.

Non-Standard Vehicles

Vehicles which are non-standard and therefore do not appear in the standard lists are plated at gross and axle weights decided by the test station staff on the basis of the following:

1. Information supplied by the operator concerning the specification of the vehicle.
2. Information contained in the particular manufacturer's standard list relating to the vehicle of the nearest standard type.
3. The vehicle's braking efficiency as indicated on the roller-brake tester.
4. The load rating of the tyres fitted and the general chassis, axle and suspension structure.

Applicable Weights

If the maximum axle and gross vehicle weights shown on the manufacturers' plates for particular vehicles are higher than the current construction and use limit for that type of vehicle, then the construction and use weights prevail. For example, many of the heaviest tractive units and semi-trailers are designed and plated for operation at up to 44 tonnes or even more in anticipation of future increases in permitted vehicle weights or for carrying abnormal loads, but for the present such vehicles, while operating in the UK on normal operations (ie not Special Types applications), are limited to the construction and use maximum gross weight of 38 tonnes which applies in this country.

International Plates

International Proof of Compliance plates are available to and may be voluntarily fitted to vehicles by international hauliers. These plates show compliance with the weights and dimension requirements of EC Directives 85/3 and 86/364. They have been introduced to speed up clearance times through customs for vehicles on international journeys, showing that the vehicles to which they are fitted do meet the regulations on weights, widths and lengths. Operators wishing to fit these plates should apply to the Goods Vehicle Centre at Swansea using application forms obtainable from Swansea direct or from heavy goods vehicle testing stations. Eventually the new style plate will replace the present style of 'Ministry' plates.

Notifiable Alterations

Operators who make any alteration to the structure of their vehicles must notify the DoT before the vehicle is used on the road. Details of the alterations which require notification on form VTG10 are as follows:

(a) *Alterations to the structure or fixed equipment of a vehicle which vary its carrying capacity.*
     These include alterations to any of the following items:
     (i) *Chassis frame or structure*
         Any alteration which increases or decreases the front or rear overhang by more than 1 foot. Any structural alteration (other than

normal adjustment of an extensible structure) which reduces or extends the wheelbase (or in the case of a semi-trailer the equivalent distance). Any other extension, deletion or alteration including cutting, welding, riveting, etc which materially weakens the chassis frame or structure or changes its torsional stiffness.

(ii) *Steering suspension, wheels and axles (including stub axles and wheel hubs)*
The fitting of steering gear, axles, hubs or road springs of a different design or load bearing capacity. The fitting of additional wheels and axles, or the removal of such items. Any addition, deletion or alteration which reduces the inherent strength of the above components.

(iii) *The fitting of an alternative body of different design, construction or type*
Any alteration which reduces materially the strength of the body structure or the means by which it is attached to the chassis. Any alteration which causes the body to extend beyond the rear of the chassis frame.

(b) *Alterations to braking system*
These comprise alterations which adversely affect either the braking system or the braking performance of the vehicle. They include the addition or deletion of components such as reservoirs, servo motors, brake actuators, exhausters and compressors. They would also include the addition of any equipment which it is necessary to connect to any part of the braking system, and the fitting of different brake drums or shoes or liners of a smaller contact area.

(c) *Other alterations to the structure or fixed equipment*
Any other alteration made in the load bearing structure or fixed equipment of the vehicle, eg the coupling gear, which could make the vehicle unsafe to travel on roads at any weight shown on the plate and plating certificate. In the case of a motor vehicle this could include such alterations as changing the type of engine or re-positioning the engine or its mountings (eg petrol to diesel, normal control to forward control, etc).

Exemptions from Plating and Testing

Vehicles which are subject to the Goods Vehicle (Plating and Testing) Regulations 1988 (as amended) are exempt from the need to hold current plating and testing certificates while being taken to a test station, when used on a road during the test, returning to base from the test station after a test and being taken (unladen) to a working or repair centre for work to be carried out on them in connection with the test.

Temporary Exemption
Temporary exemption from the need to hold a test certificate for not more than 3 months can be granted by the test station manager if for some special reason it is not possible for a vehicle or trailer to be submitted for test by the last day of the month in which it was due to be tested.

Temporary exemption may be granted in the event of severe weather, fire, epidemic, a failure in the supply of essential services, an industrial dispute, or other unexpected happenings to either a vehicle or the test station. A normal mechanical breakdown of the vehicle will not be considered a circumstance for temporary exemption. A certificate (form VTG33) confirming the exemption will be issued and this can be produced either to the police or examiners of the Vehicle Inspectorate if they ask for the test certificate, or when applying for an excise licence for the vehicle.

Exempt Vehicles
Vehicles to which the regulations do not apply are as follows:
1. Dual-purpose vehicles not constructed or adapted to form part of an articulated vehicle.
2. Mobile cranes as defined in Schedule 3 of the Vehicles (Excise) Act 1971.
3. Breakdown vehicles.
4. Engineering plant and movable plant and equipment specially designed and constructed for the special purposes of engineering operations.
5. Trailers being drying or mixing plant designed for the production of asphalt or of bituminous or tar macadam.
6. Tower wagons as defined in Schedule 4 of the Vehicles (Excise) Act 1971.
7. Road construction vehicles as defined in section 4(2) of the Vehicles (Excise) Act 1971 and road rollers.
8. Vehicles designed for fire fighting or fire salvage purposes.
9. Works trucks, straddle carriers used solely as works trucks, and works trailers.
10. Electrically propelled motor vehicles.
11. Motor vehicles used solely for clearing frost, ice or snow from roads by means of a snow plough or similar contrivance, whether forming part of the vehicle or not.
12. Vehicles constructed or adapted for, and used solely for, spreading material on roads to deal with frost, ice or snow.
13. Motor vehicles used for no other purpose than the haulage of lifeboats and the conveyance of the necessary gear of the lifeboats which are being hauled.
14. Living vans not exceeding 1525kg unladen weight.
15. Vehicles constructed or adapted for, and used primarily for the purpose of, carrying equipment permanently fixed to the vehicle which equipment is used for medical, dental, veterinary, health, educational, display or clerical purposes, such use not directly involving the sale, hire or loading of goods from the vehicle.
16. Trailers which have no other brakes than a parking brake and brakes which automatically come into operation on the overrun of the trailer.
17. Vehicles exempted from duty because they do not travel on public roads for more than six miles in any week and trailers drawn by such vehicles (Vehicles (Excise) Act 1971, section 7[1]).
18. Agricultural motor vehicles.
19. Agricultural trailers and trailed appliances drawn on roads only by a land tractor.
20. Passenger-carrying vehicles and hackney carriages.

21. Vehicles used solely for the purpose of funerals.
22. Goods vehicles proceeding to a port for export and vehicles in the service of a visiting force.
23. Vehicles equipped with new or improved equipment or types of equipment and used solely by a manufacturer of vehicles or their equipment or by an importer of vehicles, for or in connection with the test or trial of any such equipment.
24. Motor vehicles temporarily in Great Britain.
25. Motor vehicles for the time being licensed in Northern Ireland.
26. Vehicles having a base or centre in any of the following islands, namely Arran, Bute, Great Cumbrae, Islay, Mull or North Uist, from which the use of the vehicle on a journey is normally commenced.
27. Trailers temporarily in Great Britain, a period of 12 months not having elapsed since the vehicle in question was last brought into Great Britain.
28. Track-laying vehicles.
29. Steam-propelled vehicles.
30. Motor vehicles manufactured before 1 January 1960 used unladen and not drawing a laden trailer, and trailers manufactured before 1 January 1960 and used unladen.
31. Three-wheeled vehicles used for street cleansing, the collection or disposal of refuse, and the collection or disposal of the contents of gullies.
32. Vehicles designed and used for the purpose of servicing or controlling aircraft, while so used on an aerodrome within the meaning of the Airports Authority Act 1965 or on roads to such extent as is essential for the purpose of proceeding directly from one part of such an aerodrome to another part thereof or, subject as aforesaid, outside such an aerodrome unladen and not drawing a laden trailer.
33. Vehicles designed for use, and used on an aerodrome mentioned in the last preceding paragraph solely for the purpose of road cleansing, the collection or disposal of refuse or the collection or disposal of the contents of gullies or cesspools.
34. Vehicles provided for police purposes and maintained in workshops approved by the Minister as suitable for such maintenance, being vehicles provided in England and Wales by a police authority or the receiver for the metropolitan police or, in Scotland, by a police authority or a joint police committee.
35. Heavy motor cars or motor cars constructed or adapted for the purpose of forming part of an articulated vehicle which are used for drawing only a trailer falling within a class of vehicle specified in paragraphs 14, 15 or 16 above or a trailer being used for or in connection with any purpose for which it is authorised to be used on roads under the Special Types General Order.
36. Play buses.

NB: The following definitions apply in the above exemptions:

'Breakdown vehicle' means a motor vehicle on which is permanently mounted apparatus designed to lift, wholly or partly, one disabled vehicle and tow the vehicle when so lifted and which is not equipped to carry any load other than articles required for the operation of or in connection with that apparatus, or for repairing disabled vehicles.

'Engineering plant' means movable plant or equipment being a motor vehicle or trailer (not constructed primarily to carry a load) specially designed and constructed for the purposes of engineering operations.

'Works truck' means a motor vehicle designed for use in private premises and used on a road only in delivering goods from or to such premises to or from a vehicle on a road in the immediate neighbourhood, or in passing from one part of any such premises to another or to other private premises in the immediate neighbourhood or in connection with road works while at or in the immediate neighbourhood of the site of such works.

Taxing Exempt Vehicles
When applying for a vehicle excise licence for a vehicle exempt from plating and testing, it is necessary to complete declaration form V112G (available from VROs) in order to obtain the licence without a valid test certificate.

Tachograph Testing
Inspection of tachographs is now included in the annual test, but only in respect of the following items:
1. That a tachograph is fitted (ie where the vehicle requires to have it fitted by law).
2. That it can easily be seen from the driving seat.
3. The condition of the instrument.
4. That the instrument can be illuminated.
5. That all seals are present and intact.

Operators submitting vehicles for test, which are exempt from the requirement for tachograph fitment, must declare the exemption on an appropriate form which lists the exempt categories (see pp 81–3 for list of tachograph exemptions).

Production of Documents

The police and examiners of the VI can request production of both test and plating certificates for goods vehicles when such vehicles have been involved in an accident or if they believe an offence has been committed. If these documents cannot be produced at the time they may be produced within 7 days at a police station convenient to the person to whom the request was made, or as soon as reasonably practicable thereafter.

DoT Checks on Vehicles

Euro-wide Enforcement Link-up

Enforcement agencies from most EU member countries are teaming up to harmonise spot-checks on vehicles and legal documentation. Under the EU's Karolus programme, national civil servants are participating in exchange programmes throughout the Community for the purpose of enforcing a range of single market legislation, including road transport matters such as drivers' hours and tachograph use, mechanical condition of vehicles and overloading. In time, harmonised enforcement documentation will be issued by way of standardised prohibition notices and enforced rest notices.

Roadside Checks

In addition to carrying out annual vehicle tests at the goods vehicle testing stations, examiners of the VI operate roadside checks on commercial vehicles. These roadside checks are carried out at intervals on main roads by examiners who usually take over a lay-by which will accommodate several large vehicles. The VI has stepped up its normal system of roadside checks in recent times and occasionally carries out checks during the night and during weekends. A police officer, in uniform, standing on the road directs vehicles (*NB: only a police officer in uniform can stop a moving vehicle*) which are required for examination into the lay-by where they are inspected mainly for visible wear and defects of the brakes, steering gear, silencers, tyres, lights and reflectors and for the emission of black smoke when the engine is revved up. The examiners only have a limited amount of equipment on these checks so the extent of their examination is correspondingly limited but nevertheless the inspection is undertaken by skilled and observant people and very little escapes their attention.

Vehicle Inspections on Premises

Besides roadside checks, VI examiners and police officers in uniform are, at any reasonable time, free to enter any premises on which goods vehicles are kept and examine the vehicles. The owner's consent is not needed in this case but to examine any vehicle (ie other than goods vehicles) the owner's consent has to be obtained to carry out the examination or he must be given at least 48 hours' notice of such a proposal to carry out an examination. (*If the notice is sent by recorded delivery post the period is increased to 72 hours.*) In the latter case the owner of the premises, if different from the vehicle owner, must also consent. If on these inspections defects are found on the vehicles examined, the same procedure applies regarding the prohibition of their use as explained for roadside checks.

A police officer in uniform or a VI examiner on production of suitable identification can instruct a driver in charge of a stationary goods vehicle on a road (by the issue of form GV3) to take the vehicle to a suitable place to be examined but this must not be for a distance of more than 5 miles.

Prohibition notices can be issued by examiners at a goods vehicle test station if defects of a serious enough nature are found. Also the police may be notified if prosecution is warranted, although it is not usual for vehicles in such a precarious state to be submitted for test.

Powers of Police and VI Examiners and Certifying Officers
The Road Traffic Act 1988 gives authorised examiners (specifically, examiners of VI, London taxi examiners, authorised police officers and persons appointed for the purpose by a chief officer of police and certifying officers appointed under

the Road Traffic Act 1991) powers to test and inspect vehicles (and examine vehicle records), on production of their authority, as follows.

1. They can test any motor or trailer on a road to check that legal requirements regarding brakes, silencer, steering, tyres, lights and reflectors, smoke and fumes are complied with, and may drive the vehicle for this purpose.

2. They can test a vehicle for the same purposes on premises if the owner of the premises consents or has been given at least 48 hours' notice, except where the vehicle has been involved in a notifiable accident, when there is no requirement to give notice. If the notice given is in writing it must be sent by recorded post and the time limit is extended to 72 hours.

3. They may at any time enter and inspect any goods vehicle and goods vehicle records, and may at a reasonable time enter premises on which they believe a goods vehicle or goods vehicle records are kept.

4. They can request a driver of a stationary goods vehicle to take the vehicle to a place for inspection up to 5 miles away.

5. They can at any reasonable time enter premises where used vehicles are sold, supplied or offered for sale or supply or exposed or kept for sale or supply to ensure that such vehicles can be used on a road without contravening the appropriate regulations. They may drive a vehicle on the road for this purpose.

6. They may enter at any reasonable time premises where vehicles or vehicle parts are sold, supplied, offered for sale or supply, exposed or kept for sale or supply.

7. They may require the person in charge of any vehicle to take it to a weighbridge to be weighed. If the vehicle is more than 5 miles from the place where the request is made and the vehicle is found not to be overloaded, the operator can claim against the highway authority for any loss sustained.

8. When a goods vehicle has been weighed and found to exceed its weight limit and its use on a road would be a risk to public safety, they can prohibit its road use by the issue of form GV160 until the weight is reduced to within the legal limit.

9. If they find that a goods vehicle is unfit or likely to become unfit for service they can prohibit the driving of the vehicle on the road either immediately or from a later date and time by the issue of a prohibition notice (form PG9 – see below).
   NB: Police powers in this respect are restricted under the provisions of the Road Traffic Act 1991 *(effective from 1 July 1992) to the issue only of immediate prohibitions in circumstances where they consider that the driving of a defective vehicle would involve a 'danger of injury to any person' as opposed to the VI examiner's right to prohibit a vehicle which is 'unfit for service'.*

10. Where a prohibition order has been placed on a vehicle for various reasons, they are empowered to remove the prohibition (by the issue of form PG10) when they consider the vehicle is fit for use.
    Since 1 July 1992 clearance of prohibitions has required the vehicle to be subjected to a full roadworthiness test by the VI at a Goods Vehicle

Testing Station – and payment of the full test fee (currently £34.70 for vehicles and £17.90 for trailers).

11. They can ask the driver of a goods vehicle registered in an EU member state, fitted with a tachograph, to produce the tachograph record of the vehicle when it is used in this country and ask to examine the official calibration plaque in the instrument. They can at any reasonable time enter premises where they believe such a vehicle is to be found or that tachograph records are kept, and may inspect the vehicle and records (ie tachograph charts).

Police Constables

1. A police constable *in uniform* can stop a moving vehicle on a road ('constable' includes any rank of uniformed police officer). [*Note: In GB, only a uniformed constable has this power but in NI DoE examiners in plain clothes may also stop a moving vehicle.*] They can test any motor vehicle or trailer on a road to check that legal requirements regarding brakes, silencer, steering, tyres, lights and reflectors, smoke and fumes are complied with. They may drive the vehicle for this purpose.
2. They can test a vehicle for the same purposes on premises if the owner of the premises consents or has been given at least 48 hours' notice (or 72 hours if given by recorded post), except that consent is not necessary where the vehicle has been involved in a notifiable accident.

NB: This authority applies to all constables even if not specially authorised under the Road Traffic Act 1988 *as mentioned above.*

Trading Standards Officers

Trading standards officers (ie employed by local authorities) can request the driver of a vehicle which is carrying goods that need an official conveyance note (ie ballast which includes sand, gravel, shingle, ashes, clinker, chippings, including coated material, hardcore and aggregates) to take that vehicle to a weighbridge to be weighed. Goods may have to be unloaded if necessary.

Defect Notices and Prohibitions

Defect Notice

When vehicles are checked by VI examiners various forms may be issued if defects are found. Form PGDN35 is used when defects are not of a serious nature but are such that it is in the operator's and other road users' interest to have them quickly rectified.

Prohibition Notices

The driver of any vehicle found by VI examiners to have serious defects is given a form PG9 (Figure 16.4) on which these are listed. The form PG9 is the examiner's authority to stop the use of the vehicle on the road for carrying goods. Depending on how serious the defects are, the prohibition will either take effect immediately, in which case the vehicle, if loaded, has to remain where it is until either it is repaired or has been unloaded and then taken

PROHIBITION ON DRIVING A VEHICLE ON A ROAD

Road Traffic Act 1972

Traffic Area

A	B	C	D
E	F	G	H
J	K	L	M

By virtue of the powers vested in me by the above Act, I hereby prohibit all driving on roads of the vehicle specified below :-

Identification No. of Vehicle .... Make ....

Licence No. (if any) ....

DEFECTS OCCASIONING PROHIBITION

This prohibition shall come into force immediately/at .... hrs the ....
day of .... 19 .... and shall continue in force until it is removed by a goods
vehicle examiner appointed by the Secretary of State for the Environment.
Signed .... Goods Vehicle Examiner
At .... hrs the .... day of .... 19 ....
Signature of person to whom the notice was given at the time
of inspection ....
Copy to .... Traffic Area.
IMPORTANT - Please see Notes overleaf.

The terms of this notice were varied at .... hrs the .... day
of .... 19 .... See attached Variation Form GV9A.

Signed .... Goods Vehicle Examiner

P of GV 58-0018

GV9

Figure 16.4 *Prohibition on driving a vehicle on a road PG9*

away for repair for which the examiner will give authority by issuing form PG9B (Figure 16.5), or it may be delayed for 12 to 24 hours or more depending on the seriousness of the fault (see note at item 9 on p 300). In this case the vehicle may continue to operate until the limit of the period of exemption of the prohibition by which time, if it is not repaired and cleared, it must be taken off the road. If defects recorded as requiring immediate attention are repaired quickly on the roadside (either by the driver or by a mechanic or repair garage staff who come out to the vehicle) the examiner may issue a variation to the PG9 notice with form PG9A (Figure 16.6) which then allows the vehicle to be removed and used until the new time specified on the variation notice (the forms are described further individually below).

C&U Offences
If defects are found at a roadside check which make the vehicle unsafe to be on the road (usually brakes, steering or suspension defects) or if the defects are such that an offence under the construction and use regulations is committed (particularly in respect of lights, reflectors, smoke emission or the horn), the VI examiners will report these items to the police in attendance for consideration for prosecution.

Prohibition Notices
A number of official forms as described below are used by VI examiners in the process of inspecting vehicles, recording defects and prohibiting the use of those which are defective:

New Forms (GV/PG designations)
In order to combine prohibition notices for both goods and passenger vehicles, the DoT has re-numbered certain relevant prohibition documents with a PG prefix. Thus, GV9 has become PG9, GV9A has become PG9A etc. The way in which the prohibition system works as described here is not changed in any way as a result of the renumbering of the relevant documents.

Form GV3
This form is authorisation for the VI examiners to direct a vehicle to proceed to a specified place to be inspected (normally not more than 5 miles away).

Form PGDN35
Following an inspection of a vehicle, this form is issued to indicate to the user one or more minor defects which are in his interest and the interests of other road users to have rectified at an early date – it is not actually a prohibition. It also certifies that a vehicle has been examined and found free of defects.

Form PG9 (GV9)
When an inspection by a VI examiner reveals defects of a serious nature form PG9 will be issued, specifying the defects and stating the precise time at which the prohibition preventing further use of the vehicle comes into force (which could be the time when the notice is written out – ie immediate effect – or later).

Exemption from a Prohibition on the Driving of a Vehicle on a Road

Road Traffic Act 1972

West Midland Traffic Area
Cumberland House
200 Broad Street
Birmingham B15 1TD

Identification No. of Vehicle... Make...

Licence No. (if any)...

By virtue of the powers vested in me by the above Act, and subject to the conditions

specified below at Nos..I hereby exempt the vehicle

specified above from the terms of the prohibition issued with respect to it

at..hrs...............................

the...day of.. 19...............

CONDITIONS

That:—1. The vehicle is unladen

2. The vehicle proceeds at a speed not exceeding...................miles per hour

3. The vehicle does not tow a trailer

4. The vehicle is towed by a rigid tow bar

5. The vehicle is towed by a suspended tow

6. The vehicle is not used after lighting up time

7. The vehicle proceeds only between..
....and....................................

This notice of exemption expires at..........................hrs the..

day of................................19...............

Signed..Goods Vehicle Examiner

At.......................hrs the..........................day of..........................19............

Copy to...Traffic Area

GV9B

IMPORTANT—Please see Note overleaf.

CP 56-00

Figure 16.5 *Exemption from a prohibition on the driving of a vehicle on a road*

Road Traffic Act 1972

Variation in the Terms of a Prohibition on Driving a Vehicle on a Road

West Midland Traffic Area
Cumberland House
200 Broad Street
Birmingham B15 1TD

A	B	C	D
E	F	G	H
J	K	L	M

Identification No. of Vehicle.. Make..

Licence No. (if any)..

By the powers vested in me by the above Act I hereby vary the terms of the prohibition notice,
issued at...hrs the..day of

.... 19................in respect of the vehicle specified above and

1. suspend it until.............................hrs the...day of..............................
   19.....................or

2. alter the time at which the prohibition shall come into force to...hrs
   the............................... day of.. 19................. and/or

3. alter the list of defects occasioning the prohibition.

Defects Occasioning Prohibition

No's...as specified on the above-mentioned prohibition notice.

Signed..
   Goods Vehicle Examiner

The prohibition will continue in force until it is
removed by a Goods Vehicle Examiner appointed
by the Secretary of State for the Environment.

At.................hrs the...day of.. 19...........

Signature of person to whom the notice was given at the time of inspection

Copy to..Traffic Area

IMPORTANT—Please see Notes overleaf.

WM 56-ST

GV9A

Figure 16.6 *Variation in the terms of prohibition on driving a vehicle on a road PG9A*

Road Traffic Act 1972
Removal of Prohibition on Driving a Vehicle on a Road

_____ Traffic Area

Identification No. of Vehicle._____ Make._____

Licence No. (if any) _____

By virtue of the powers vested in me by the above Act, I hereby remove the prohibition on

the driving of the vehicle specified above which was issued in._____ Traffic Area

at_____ hrs the. _____ day of_____ ..19.____

Signed_____ Goods Vehicle Examiner

at_____ hrs the._____ day of_____19 ____

Dd. 0580452 1500 Bks. T.L. LTD. 4/78

GV10

Figure 16.7 *Removal of a prohibition on driving a vehicle on a road PG10*

REFUSAL TO REMOVE A PROHIBITION ON DRIVING A
VEHICLE ON A ROAD

Road Traffic Act 1972

West Midland Traffic Area
Cumberland House
200 Broad Street
Birmingham B15 1TD

Identification No. of Vehicle.................................... Make...............................

Licence No. (if any)..

By virtue of the powers vested in me by the above Act, I hereby refuse to remove the
prohibition on driving the vehicle specified above on a road, which was issued at
............hrs.....................day of......................19........on the grounds that the vehicle is not
fit for service because of defects Nos..............................listed on the prohibition.

Signed..(Goods Vehicle Examiner)/(Certifying Officer)

at..............................hrs. the...................................day of................................19...........

Copy to...Traffic Area
IMPORTANT:–See notes overleaf

CP 56-ST

GV9C

Figure 16.8 *Refusal to remove a prohibition on driving a vehicle on a road
PG9C*

When form PG9 (GV9) has been issued, a copy is given to the driver and this must be carried on the vehicle until the prohibition is removed. Further copies of the notice are sent to the vehicle operator and, if the vehicle is specified on an 'O' licence, to the relevant Licensing Authority.

If the PG9 (GV9) has immediate effect, this means that the vehicle cannot be driven or towed away at least until the vehicle has been unloaded (see below).

Form PG9A (GV9A)

This form (Variation in the Terms of a Prohibition . . .) is issued if the VI examiner wishes to vary the terms of a PG9 notice by either suspending the PG9 until a future time (eg midnight on the day of issue), altering the time (which is effectively the same thing as suspending the notice as mentioned above) or altering the list of defects shown on the PG9 notice.

Form PG9B (GV9B)

A VI examiner may, after issuing a PG9, exempt the vehicle (Exemption from a Prohibition . . .) from the terms of the prohibition and permit its movement on certain conditions, as follows:

1. that the vehicle is unladen;
2. that the vehicle proceeds at a speed not in excess of a specified figure;
3. that the vehicle does not tow a trailer;
4. that the vehicle is towed on a rigid tow-bar;
5. that the vehicle is towed on a suspended tow;
6. that the vehicle is not used after lighting up time (if it has lighting defects);
7. that the vehicle proceeds only between two specified points.

Form PG9C (GV9C)

When a vehicle which is subject to a PG9 notice is presented to a VI examiner for clearance of the defect and the examiner is not satisfied that it is fit for service he may issue form PG9C (Refusal to Remove a Prohibition . . .), which means that the original PG9 notice remains in force until the defects are satisfactorily rectified.

Form PG10 (GV10)

If the defects specified in a PG9 (GV9) notice have been repaired to the satisfaction of the VI examiner to whom the vehicle is presented for clearance and the examiner is satisfied that the whole vehicle is in a satisfactory condition for use on the road, he will issue a PG10 notice (Removal of Prohibition . . .) which removes the prohibition. The LA must be notified of the clearance if the vehicle is specified on an 'O' licence.

Form GV160

This notice (*which does not have a PG prefix*) relates to prohibition on the use of overweight vehicles. It effectively requires the driver of the vehicle to take the vehicle to a weighbridge and, if found to be overloaded, reduce the gross weight to legal limits before proceeding on his journey.

NB It is a defence to a charge of overloading that the vehicle was on its way to the nearest practicable weighbridge or that at the time of loading the weight was within legal limits and was subsequently not more than 5 per cent heavier despite not having any additions to the load en route.

Effects of prohibition

A vehicle must not, under any circumstances, be used to carry goods while it is the subject of a PG9 prohibition notice, but despite the prohibition notice a vehicle may be driven unladen to a goods vehicle test station or to a place agreed with a goods vehicle examiner (both by previous appointment only) in order to have the vehicle inspected. The vehicle may also be driven on the road for test purposes, provided it is unladen, within 3 miles of where it has been repaired.

Forms PG9 and PG9A have a panel of boxes identified by letters of the alphabet A to M (see Figures 16.4 and 16.6). These boxes are used by enforcement staff to codify certain aspects of the vehicle check as follows:

A   Whether vehicle laden or unladen

B   Whether examination took place on a road or off the road

C   To indicate if defects appeared to be from neglect in which case a letter 'N' is entered*

D   To indicate whether a prosecution will follow (the letter 'P' in this box indicates prosecution by the police and 'PE' by the enforcement authorities)

E   If vehicle is issuing smoke ('S' equals smoking in service; 'S/A' equals smoking under free acceleration when vehicle is stationary)

L   Indicates number of pages of prohibition notice served at the time (ie page 1 of 1, page 1 of 2). This indicates to the examiner, who is asked to clear the prohibition, that there was another sheet(s)to the prohibition notice issued at the time

F,G, H,   Not currently used.
K and M

The 'N' is no longer used in box C of a PG9 to indicate that the defect noted was due to neglect.

Appeals against the Issue of Prohibition Notices

There is no appeal against the imposition of a PG9 prohibition notice, but there is a right of appeal to the Area Mechanical Engineer (AME) at the Traffic Area Office, against refusal to remove a prohibition after repair.

Clearance of Prohibition

An operator having a vehicle placed under an immediate or a delayed prohibition notice has to get the defects repaired and then submit the vehicle to his local Goods Vehicle Testing Station either for a full roadworthiness examination in the case of serious defects (with payment of the relevant fees – currently up to £34.70 for vehicles and up to £17.90 for trailers, in each case depending on the number of axles [see full list of fees on pp 283–4]) or, in the case of minor defects, a 'partial clearance' or 'mini' test (for which a reduced fee of £16.60 for vehicles and £8.30 for trailers is payable). A clearance certificate, PG10 (Figure 16.7) is issued if the examiner is satisfied with the repair. If the examiner is not satisfied with the repair he will issue a form PG9C (Figure 16.8) 'Refusal to Remove a Prohibition' or form PG9A if some of the defects are cleared and others are not. If because of the better inspection facilities at the test station he finds further defects another form PG9 may be issued.

Light Vehicle Testing

Light goods vehicles up to 3500kg gross weight and dual-purpose vehicles (see p 144 for definition) under 2040kg unladen weight are required to be tested on the third anniversary of the date of their original registration and annually thereafter (commonly known and referred to as the MoT test). The tests are carried out at private garages approved by the Department of Transport and displaying the blue and white triple-triangle sign. For full details of the testing scheme for these vehicles, see Chapter 17 which deals with light vehicle testing.

17: Light Vehicle (MoT) Testing

Private cars, motor caravans (irrespective of weight), dual-purpose vehicles (see p 144 for definition) under 2040kg unladen weight and light goods vehicles not exceeding 3500kg gross weight are subject to annual testing (commonly referred to as the MoT test) at DoT-approved commercial garages, starting on the third anniversary of the date of their first registration and each year thereafter.

Testing is carried out at garages displaying the blue and white MoT triple-triangle symbol, and application is made direct to the garage for a suitable appointment. Some centres claim to provide MoT testing 'while you wait'. There is no application form to be completed.

Fees for the light vehicle test – payable to the garage at the time of the test – are as follows.

1. Vehicles in Class IV (ie motor cars and heavy motor cars   £26.10
    not included in Classes III, V, VI or VII)
2. Vehicles in Class V (ie large passenger-carrying vehicles   £33.04
    and play buses)
3. Vehicles in Class VII (ie goods vehicles with a design gross   £28.84
    weight of more than 3000kg but less than 3500kg)

Duplicate test certificates cost £9.00 each.

Vehicle Classes

Vehicles subject to the MoT test are classified as follows:

Class I   Light motor bicycles not exceeding 200cc cylinder capacity with or without sidecars.

Class II   All motor bicycles (including Class I) with or without sidecars.

Class III   Light motor vehicles with three or more wheels (excluding Classes I and II) not exceeding 450kg unladen weight.

Class IV   Heavy motor cars and motor cars (excluding Classes III and V); ie any vehicle with an unladen weight of more than 450kg which is:
    (a) a passenger vehicle (ie private car, taxi, vehicle licensed as private, with 12 passenger seats or less, or small public service vehicle with less than 8 passenger seats)
    (b) a dual-purpose vehicle not exceeding 2040kg unladen weight
    (c) a goods vehicle not exceeding 3000kg gross weight (from 1 April 1991)
    (d) a motor caravan irrespective of weight.

Class V   Large passenger carrying vehicles; ie motor vehicles which are constructed or adapted to carry more than 12 seated passengers in addition to the driver, and which are not licensed as public service vehicles.

Class VI   Public service vehicles other than those in Class V above.

Class VII   Goods vehicles of 3001kg to 3500kg gross weight (these vehicles may be tested at MoT garages – where the facility to test such vehicles is available – or at goods vehicle testing stations).

Exemptions

Public service vehicles with seats for eight or more passengers excluding the driver, track-laying vehicles, vehicles constructed or adapted to form part of an articulated vehicle, works trucks and all trailers are excluded from the above classes and the following vehicles are also exempted from the test:

1.   Heavy locomotives.
2.   Light locomotives.
3.   Goods vehicles over 3.5 tonnes gross weight.
4.   Articulated vehicles other than articulated buses.
5.   Vehicles exempt from duty under section 7(1) of the Vehicles (Excise) Act 1971.
6.   Works trucks.
7.   Pedestrian-controlled vehicles.
8.   Vehicles kept and used by invalids.
9.   Vehicles temporarily in Great Britain.
10.   Vehicles proceeding to a port for export.
11.   Vehicles provided for use by the police force.
12.   Imported vehicles owned or in the service of HM navy, army or airforce.
13.   Vehicles which have Northern Ireland test certificates.
14.   Electrically propelled goods vehicles not exceeding 1525kg unladen weight.
15.   Certain hackney carriages.

Many of these vehicles are subject to the heavy goods vehicle testing and plating scheme (see Chapter 16 for details).

There is also an exemption which applies to vehicles which come within the MoT test scheme while they are being driven to a place by previous arrangement for a test or bringing it away, if it fails, to a place to have work done on it. This means that such vehicles can be driven on the road without a valid test certificate being in force, but only in the circumstances mentioned and no other.

The Test

Testing is carried out in accordance with *The MoT Tester's Manual* – copies available from HMSO. (*Note: This publication should not be confused with the* Heavy Goods Vehicle Inspection Manual *which relates solely to the goods vehicle annual test and which is also available from HMSO.*)

When presenting a vehicle for test the following conditions must be observed:
1.   The vehicle registration document must be produced.
2.   The vehicle must be sufficiently clean so as not to make the test unreasonably difficult.

3. The vehicle must have sufficient petrol and oil to enable the test to be completed.
4. If the vehicle is presented for the test in a loaded condition (ie in the case of light goods vehicles) the load must be secured or else removed.

Refusal and Discontinuance of Test

Failure to observe any of the above conditions can lead to a refusal to test the vehicle, and the test fee will be refunded. Furthermore, if the tester finds a defect of a serious nature which, in his opinion, makes it essential to discontinue the inspection on the grounds of risk to his own safety, risk to the test equipment or to the vehicle itself, he may do so and issue form VT 30 showing the defects which caused the test to be discontinued.

Items Tested

The following items are tested and must meet the conditions specified in *The MoT Tester's Manual:*

Section I	Function of obligatory front and rear lamps
	Function of obligatory headlamps
	Function of stop lamps
	Obligatory rear reflectors
	Function of direction indicators
	Aim of headlamps
Section II	Steering wheel and column
	Steering mechanism
	Power steering
	Front wheel bearings
	Suspension
	All suspension types
	Suspension assemblies (springs, torsion bars, etc)
	Shock absorbers
Section III	Parking brake and operating lever
	Parking brake mechanism (under vehicle)
	Service brake operating pedal
	Service brake mechanism under vehicle
	Brake performance test
Section IV	Tyres
	Roadwheels
Section V	Seat belts
Section VI	Function of windscreen washers
	Function of windscreen wipers
	Exhaust system
	Function of audible warning device
	Condition of the vehicle structure.

Additional Test Items

Since 1 November 1991 the examination has covered additional items such as exhaust emissions (see below) – particularly in light goods vehicle tests; ABS (ie anti-lock braking system) warning lights; rear seat belts (ie where fitted); rear wheel bearings; more extensive corrosion checks.

Since 1 January 1993 additional items have been included in the MoT test in accordance with EU requirements – hence the further test fee increases from this date. These include checks on the following vehicle components and systems:

- mirrors – a motor car must have a mirror on the driver's side and either an exterior mirror on the passenger side or an interior mirror; they must be in good condition, capable of adjustment and be secure;

- windscreens – damage (ie a crack) to a windscreen, exceeding 10mm in zone A (an area 290mm wide centred on a vertical line passing through the centre of the steering wheel and within the area swept by the windscreen wipers), will result in test failure as will obscuring clear vision with stickers or other obstructions;

- fuel tanks and feed pipes – must be in good condition and secure (including the tank filler cap);

- bodywork and body security – must be sound;

- seat fixings – must be secure and backrests lockable;

- security of doors and other openings – all doors must be capable of being opened from the inside and the outside and they must latch securely;

- number (ie registration) plates – must be secure and in good condition and the numbers and letters must be correctly spaced (incorrect spacing of letters and numbers to make up names or other words will result in test failure);

- vehicle identification number – must be present and capable of being clearly read;

- number plate lamps (at rear) – must be fitted and working correctly;

- rear fog lamps – vehicles first used after 1 April 1980 must have one and it must work, and an interior warning light must be provided and work;

- hazard warning devices – must work with the ignition on or off and there must be an interior warning light which works.

Exhaust Emission Tests
A check is made on emissions from diesel-engined vehicles subject to the test. These tests were re-introduced in 1994 after being temporarily suspended.

Cars, light goods vehicles and particularly goods vehicles within the weight range 3001kg to 3500kg gross weight which fall within the Class VII MoT test are subjected to additional exhaust emission checks during their annual roadworthiness test, as well as checks for excessively smoking exhausts. This involves measurement of the levels of carbon monoxide (CO) and hydrocarbons (HCs) to ensure that engines are properly tuned. The current maximum level of carbon monoxide acceptable for post-1 August 1983 registered vehicles is 4.5 per cent (6 per cent for vehicles first registered prior to this date). The limit for hydrocarbons for post-1 August 1975 registered vehicles is 1200 parts per million (ppm).

Higher standards (ie lower permissible emissions) are to be imposed for light commercial vehicles from 1997 as follows:
- light goods vehicles in Class II (ie between 1250kg and 1700kg gross weight)
  carbon monoxide (CO) limit   5.17kg/km
  hydrocarbon/nitrous oxide   1.4g/km

particulates	0.19g/km

- light goods vehicles in Class III (ie 1700kg to 3500kg)
carbon monoxide (CO) limit	6.9kg/km
hydrocarbon/nitrous oxide	1.7g/km
particulates	0.25g/km

Emission standards tests as described above do not apply to motor cycles, three-wheelers or diesel-engined vehicles.

Regulations permit tests for exhaust emissions to be carried out on the roadside as well as at approved MoT test garages.

Test Failure

If the vehicle fails to reach the required standard of mechanical condition a 'notification of refusal of a test certificate' (Form VT 30) is issued. This indicates the grounds on which the vehicle failed the test (ie it names the faulty components or component area and the actual fault).

Re-Tests

If a vehicle fails the test, no further test fee is payable if it is left at the garage for the necessary repairs to be carried out. If the vehicle is taken away after the test failure but is returned to the original garage or another test garage within 14 days of the original test date for repairs and retest, only 50 per cent of the test fee is payable. Re-tests carried out more than 14 days after the original test are charged at the full rate (see above). The full rate is also charged if a vehicle is repaired other than at the original test garage or any other approved testing station and is returned for test even if it is within 14 days of the original test.

Issue of Test Certificate

On completion of the test, if the vehicle is found to be in satisfactory condition and in compliance with the law, a test certificate (form VT 20) is issued. Additionally, a copy of form VT 29 is issued to indicate to the vehicle owner the general state of the vehicle and to point out components which may need attention in the future to keep it in good, safe working order. Comments on form VT 29 are made against the following headings:
  I. Lighting equipment
  II. Steering and suspension
  III. Braking system
  IV. Tyres and wheels
  V. Seat belts
  VI. General items.

Appeals

A vehicle owner can appeal if he is not satisfied with the result of the test. He must complete form VT 17 and send it to the local Traffic Area Office within 14 days of taking the test.

Production of Test Certificates

It is necessary to produce a current, valid test certificate when taxing a vehicle to which the regulations apply, otherwise an excise licence will not be issued. A police officer can ask to see a vehicle test certificate. If the driver does not have it with him he can be asked to produce the certificate at a police station nominated by him within 7 days.

Vehicle Defect Rectification Scheme

Operators of light vehicles (up to 3500kg gross weight) found on the road with non-endorsable minor vehicle defects (ie lights, wipers, speedometer, silencer etc) in certain areas may be offered the VDRS procedure by the police whereby no prosecution will result if:
1. Immediate arrangements are made for the repair of the defect.
2. The repaired vehicle and VDRS notice are presented to an MoT garage for examination and certification that the defects have been rectified.
3. The certificate is sent to a local Central Ticket Office.

Failure to follow the procedure on receipt of a VDRS notice will result in prosecution.

18: Vehicle Maintenance

The requirements of the Road Traffic Act 1988 for the annual testing of goods vehicles and trailers, the requirements of the current construction and use regulations, and the parts of the Transport Act 1968 setting out the conditions relating to vehicle maintenance under which an 'O' licence will be granted create a situation whereby operators must ensure that their vehicles and trailers are always safe, are in a fit and roadworthy condition and that their maintenance, vehicle inspection and maintenance record systems meet the requirements laid down in the legislation.

In order for the operator to meet these requirements he must maintain his vehicles and trailers (no matter how old they are) to a sufficiently high standard to enable them to pass the stringent annual goods vehicle test which they should be able to do on the test day *and on every other day when they are on the road*. To be able to do this, it is not sufficient merely to take the vehicle off the road for a short period once a year just before test day and work frantically to get it up to scratch, and for the rest of the year allow it to run on the road in a condition which is something below the required standard.

The risk of the vehicle encountering a roadside check, or being on the operator's premises when Vehicle Inspectorate examiners or the police decide to make an inspection, as they are empowered to do, is too great to take when the penalties for failure to maintain vehicles are so high. Besides the risk of heavy fines, there is the possibility that the operator may lose his 'O' licence since a satisfactory state of maintenance is one of the factors taken into account by the Licensing Authority in considering applications and renewals for such licences.

Maintenance Advice

A DoT Vehicle Inspectorate (VI) team along with operator associations has produced a Code of Practice on vehicle maintenance for transport operators – *Guide to Maintaining Roadworthiness: Commercial Goods and Passenger Carrying Vehicles* – available from HMSO (ie Government Bookshops), Traffic Area Offices and Heavy Goods Vehicle Testing Stations. Additionally, there is the DoT's free guide *A Guide to Goods Vehicle Operators' Licensing* (reference GV 74), which contains some specific advice on vehicle maintenance arrangements for licence applicants. The following notes are included as an appendix to the *Guide*.

There are two separate vehicle checks and inspections which should be carried out:

- daily running checks
- vehicle safety inspections and routine maintenance at set intervals on items which affect vehicle safety, followed by repair of any faults.

Daily running checks are normally carried out by drivers before a vehicle starts its daily journey. They are checks on such basic items as engine oil, brakes, tyre pressures, warning instruments, lights, windscreen wipers and washers and trailer coupling.

Vehicle safety inspections and routine maintenance should be carried out at set intervals of either time and/or mileage whichever occurs first. *Note: the West Midland Licensing Authority insists on inspections being carried out on a time basis only.*

How often these inspections are done should be decided by the nature of the operator's business. A vehicle used on long-distance work will need inspecting at different intervals from one employed in heavy traffic on local work with frequent stops and starts. The items inspected should include wheels, tyres, brakes, steering, suspension, lighting, and so on. More detailed information can be found in the Department of Transport's publication the *Heavy Goods Vehicle Inspection Manual* (available from HMSO). Vehicle checks and inspections are extra to a routine maintenance schedule. It is vital to the vehicle's safety that both types of checks and inspections are done.

Staff doing inspection checks must be able to recognise faults they find, such as excessive wear of components. They should also be aware of the acceptable standard of performance and wear of parts. Trade associations offer regular inspections for their members' vehicles.

Records

Records must be kept of all safety inspections to show the history of each vehicle. These records must be kept for at least 15 months. If vehicles from several operating centres are inspected and repaired at a central depot, the records may be kept at that depot, although VI examiners are entitled to request inspection of records at the operating centres where vehicles are based.

If an outside garage does the inspections and repairs, you must still keep maintenance records. You are responsible for the condition of any vehicle or trailer on your licence.

Facilities

These will depend on the number, size and types of vehicles to be inspected. It must be possible to inspect the underside of a vehicle with sufficient light and space to examine individual parts closely. Ramps, hoists or pits will usually be necessary, but may not be needed if the vehicles have enough ground clearance for a proper underside inspection to be made on hard- standing. Creeper boards, jacks, axle stands and small tools should be available.

As well as providing facilities for checking the underside of vehicles, operators should whenever possible use equipment for measuring braking efficiency and setting headlights. If many vehicles have to be inspected, it may be worth while providing a roller brake tester. If this cannot be justified, a decelerometer (Tapley-meter) to measure braking efficiency might be worth while.

Drivers' Reports

Drivers must report vehicle faults to whoever is responsible for having them put right. The maintenance system should allow for these reports, which must be recorded in writing either by the driver himself or by the person responsible for maintaining the vehicle. Owner-drivers must note faults as they arise and keep these notes as part of their maintenance record.

Hired Vehicles and Trailers

In the case of hired, rented or borrowed vehicles or those belonging to other operators used in inter-working arrangements, it is the user who is responsible for their mechanical condition on the road. If disciplinary action is taken as a result of a mechanical fault, it is against the user's licence, not the company from whom the vehicle is hired, or the owner.

The Choice: To Repair or Contract Out

The first decision an operator has to make when planning his vehicle maintenance is to determine whether it is to be carried out in his own workshops, by his own staff, or whether it is to be contracted out to a repair garage. The size of the fleet and what existing facilities and premises he already has will usually determine the method to be used. It is unlikely to be an economic proposition for a very small operator with a fleet of less than five or six vehicles to establish his own workshop unless he actually does the work himself. On the other hand it is unlikely to be an economic proposition for a large operator to contract out the work. There are, however, many examples in industry where the opposites apply, and very successfully too; so it remains very much a matter for the operator to assess his own requirements, balance out the costs of the alternatives and make arrangements accordingly.

Choice of Repairer

When choosing a repairer to do the work preference should be given to a garage which is a main distributor for the make of vehicle operated or an agent for that make of vehicle. The reason for this is that such a firm, as part of its arrangement with the manufacturer it represents, will have had to send some of its mechanics to the factory for training in the repair and servicing of that particular make of vehicle. This ensures that skilled staff are working on the vehicle. Moreover, the distributors are also usually required by the manufacturer to hold considerable stocks of spare parts, with a predominance of fast moving items. By using such a garage the risk of a vehicle being kept off the road waiting for spare parts is therefore considerably reduced, and since vehicle downtime is a heavy cost burden these days this is an important consideration. If a garage of this type is not available locally and a second choice has to be made, this should be a firm which is experienced in heavy vehicle repair work and which has suitably trained staff and the necessary equipment. A garage which normally only handles motor car repair work should not be used.

Repair Arrangements

Any arrangements made with a repairer should be in writing. The 'O' licence application form requests that copies of any maintenance contract should be sent with the application but in any event the Licensing Authority (LA) will want to know what arrangements have been made with a repairer. Verbal arrangements or the practice of sending vehicles in for repair as necessary on an ad hoc basis are not acceptable to the LAs, some of whom have said that a verbal agreement for these purposes is no agreement at all.

Maintenance Agreements
The agreement should include provisions for the repairer to be responsible for supplying the operator with suitable records of the inspections and repair work carried out. The operator should ensure that he can escape reasonably quickly and easily from any agreement in the event of the standard of work deteriorating and thereby placing his 'O' licence at risk.

Care should be taken when discussing or making an agreement that the repairer is aware of the consequences of any negligent action on his part on the livelihood of the operator, remembering that it is the vehicle 'user' who remains responsible at all times for the safe and satisfactory mechanical condition of the vehicle, and it is he who is responsible for keeping and producing records of vehicle inspections and other maintenance work.

There is no standard or recommended form of maintenance agreement in universal use; it is left to the parties concerned to agree on terms. *NB: a suitable format for agreements is suggested by the DoT in its guide to 'O' licensing. A copy of this is reproduced here for reference.* The agreement should clearly set out in detail the work to be done, the intervals at which it is to be done, the responsibilities of the operator and the garage, the form which records should take and the action to be taken if defects are discovered or repairs over a certain value are found to be necessary.

Maintenance Contracts

Some of the prominent commercial vehicle repair specialists offer contract maintenance schemes. There are usually a number of options in such schemes which, for instance, give the operator a choice of having his vehicles inspected only at set intervals; inspected and serviced according to the manufacturer's recommendations; or inspected, serviced and all repair work carried out (excluding damage caused by the vehicle having been involved in

A model agreement between the operator and a garage or agent for safety inspections and/or repair of vehicles and trailers subject to operators' licensing

This Agreement is made the .... day of .... 19 .... between

   (a) .... whose address [registered office] is

      .... ("the operator") of the one part, and

   (b) .... whose address [registered office] is

      .... ("the contractor") of the other part.

1. The Contractor agrees that he [it] will, in relation to every vehicle mentioned in the Schedule below, on every occasion when that vehicle is submitted by the operator as mentioned in Article 2 below on or after the date of this Agreement —

   (a) inspect all the items specified in the maintenance record in the form for the time being approved by the Department of Transport which relate to the vehicle;

   (b) if the operator so consents, carry out such renewals and repairs as may be necessary to ensure that the vehicle and every part of it specified in that maintenance record is in good working order and complies with every statutory requirement applying to it; and

   (c) complete that maintenance record to show —

      (i) which items were in good working order and complied with the relevant statutory requirements when the vehicle was submitted and which remain in that condition;

      (ii) which (if any) items were not in good working order or failed to comply with those requirements when the vehicle was submitted but have been replaced or repaired so that those requirements are satisfied; and

      (iii) which (if any) items were not in good working order or failed to comply with those requirements when the vehicle was submitted and which has not been so replaced or repaired.

2. The operator agrees that he [it] will —

   (a) submit to the contractor each vehicle mentioned in the Schedule below in order that the contractor may, as regards that vehicle, comply with the provisions of Article 1 above —

      (i) within .... weeks of the date of this Agreement, and, thereafter;

      (ii) within .... weeks of the last submission or, in the case of a motor vehicle, when the mileage shown on the odometer has increased by .... miles, whichever is the sooner;

   (b) pay to the contractor such reasonable charges as the contractor may make pursuant to his [its] obligations under Article 1 above; and

   (c) retain, and make available for inspection by an officer mentioned in section 82(1) of the Transport Act 1968, every maintenance record mentioned in Article 1 above for a period of at least 15 months commencing with the date of its issue.

3. This Agreement shall be determinable by either party giving to the other .... months written notice of his [its] intention to determine it.

A model agreement between the operator and a garage or agent for safety inspections and/or repair of vehicles and trailers subject to operators licensing. With acknowledgement to the Department of Transport (reproduced from GV 74)

Schedule

(Motor Vehicles and trailers which [are authorised vehicles] [it is intended shall become authorised vehicles] under an operator's licence [held] [applied for] by the operator under Part V of the Transport Act 1968).

1.   Motor Vehicles (give registration numbers and brief descriptions).

2.   Trailers (give brief descriptions).

As witness etc.

(Signature(s), or seal operator)

(Signature(s), or seal, or contractor)

A model agreement between the operator and a garage or agent for safety inspections (cont.)

an accident) including the supply of materials except for such things as tyres and batteries.

The charges (both labour and materials) for the various options are usually incorporated in the contracts and these remain fixed for the period of the contract with a clause which enables the garage to make additional charges if the vehicle exceeds, by a large margin, the mileage estimated by the operator at the time of negotiating the contract. The additional charges are usually based on the excess mileage at an agreed figure per mile.

In offering a fixed charge agreement to include all normal repair work, the garage is dependent to quite an extent on the good faith of the operator in using his vehicles in a manner which is not likely to involve the garage in excessive costs above what may be normally estimated in advance. Again, a clause may be included in the contract stating that additional charges will be raised for the repair of damage or defects because of misuse by the owner.

If a full maintenance contract under one of these schemes is negotiated, the garage usually accepts the responsibility for ensuring that the vehicle is always in a fully roadworthy condition, able to pass through VI roadside checks and pass its annual test without difficulty. In the event of a vehicle getting a PG 9 prohibition for defects, the operator has some grounds for a claim against the

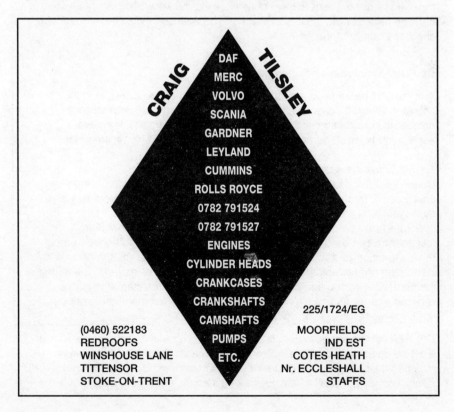

CRAIG TILSLEY

DAF
MERC
VOLVO
SCANIA
GARDNER
LEYLAND
CUMMINS
ROLLS ROYCE
0782 791524
0782 791527
ENGINES
CYLINDER HEADS
CRANKCASES
CRANKSHAFTS
CAMSHAFTS
PUMPS
ETC.

(0460) 522183
REDROOFS
WINSHOUSE LANE
TITTENSOR
STOKE-ON-TRENT

225/1724/EG

MOORFIELDS
IND EST
COTES HEATH
Nr. ECCLESHALL
STAFFS

garage but, unfortunately, he has no defence to present to the LA if he is called to explain why the vehicle was not maintained to the required standard because of the condition of 'O' licensing which makes the vehicle user totally responsible for the condition of his vehicles on the road (see below).

The garage offering a full fixed price contract maintenance scheme will prefer, wherever possible, to negotiate a contract for a vehicle from the day it is new because it is much easier to plan and cost the amount of work likely to be necessary for an estimated annual mileage and the spare parts required.

The effective working of a contract agreement needs the co-operation of both parties if it is to be successful. The operator may feel that in return for paying the garage a fixed price for full maintenance he is left with little responsibility. But he must make it his business to ensure that whatever other considerations may be pressing, the vehicle is sent to the garage when it is required if the contract is one in which the garage calls the vehicle in at specified intervals for the work to be done.

Although armed with what appears to be a fair agreement with a reputable repairer, the operator should not become complacent. It is strongly recommended that wherever possible he should arrange for a physical check of the work that the repairer claims to have carried out. While not suggesting that a repairer is likely to claim to have done work that he knows has not been done, it must be remembered what is at stake and an operator should therefore doubly check the quality of the work.

Contract Maintenance Schemes

A number of package contract maintenance schemes are available to the operator. Which he chooses, if indeed he chooses any, will depend on a number of considerations: the make of the vehicles he owns, the trade association to which he belongs, the availability of a suitable repairer, etc.

Freight Transport Association Maintenance Services
Under the FTA quality control maintenance inspection service a member operator contracts to have his vehicles inspected one or more times a year by Association inspectors (only skilled and highly experienced people are appointed) to see that they meet the requirements of the law. The inspections can be used by the operator either as a second check on his own inspection and maintenance system, or as the sole means of inspection of his vehicles to comply with the law in this respect, the actual maintenance work being carried out by his own staff. Alternatively, the scheme can be used as a means of checking the maintenance work carried out by a repair garage or agent.

The FTA vehicle check system is not purely a maintenance scheme as such. It is an inspection and maintenance advisory service available to Association members as a do-it-yourself scheme using the documents, checklists and advice provided. It can include contract inspections of vehicles by a qualified inspector.

After carrying out a vehicle check, the inspector completes a checklist indicating whether items are satisfactory within reasonable tolerances, or whether attention is needed. When the inspector returns to carry out the next check on the vehicle he will expect to see that his previous recommendations have been followed. If they have not, the operator is told that there is little point in paying for a service which he is not using to its fullest advantage.

Following the experience of many operators who have received PG9s on brand new vehicles, the FTA offers a service whereby it will conduct a full inspection of a brand new vehicle before it goes into service. One of the obvious advantages of this independent check is that it provides the operator with evidence to support claims to the vehicle supplier and manufacturer for new vehicles delivered in faulty condition.

Responsibility for Maintenance

The major point to remember when making any arrangements with a garage is that it is the operator (ie the vehicle 'user' – see pp 4–5 for definition), not the repairer, who remains responsible for the mechanical condition of the vehicle, even where defects are due to negligence by the repairer. In *A Guide to Operators' Licensing* available from local Traffic Area Offices the following warning is given:

> 'Operators who contract out their inspections and maintenance work are still the legal 'users' of their vehicles and as such will be held fully responsible by the LA for the arrangements they make and the condition of their vehicles. If either are unsatisfactory it is the operator's licence which will be placed in jeopardy.'

The same responsibility applies in the case of vehicles hired without drivers and trailers even if the hire company, as part of the agreement, carries out the inspection and repair work.

Responsibility for Records

Operators are also responsible for ensuring that proper records of maintenance work are kept. Even if the repairer makes the records and holds them on file it is up to the operator to ensure, first, that they are properly kept with all the necessary information recorded and, second, that they are retained on file, available for inspection, for a minimum period of 15 months. In completing the 'O' licence application 'declaration of intent' (see p 333) the vehicle operator promises to 'keep records'. This can be interpreted as meaning that the operator keeps the record rather than the repair garage keeping it on his behalf (see also p 333 and the rest of Chapter 19).

Negligence by Repairers

Unfortunately, however satisfactory the arrangements made with a garage may be in other ways, there is very little that the operator can do contractually to protect himself completely from negligence on the part of the repairer or the repairer's employees. It is unlikely that the repairer would agree to be party to a contract in which he has fully to indemnify the operator against failure of his employees to carry out work to a required standard, even though he may agree that morally he should be held responsible (moral responsibility, incidentally, has no standing in law). Moreover, most small operators are not likely to be in a position to have sufficiently persuasive powers to get the repairer to agree to such terms.

The operator has no protection against poor workmanship other than relying on the reputation of the garage. However, if he is a member of one of the trade associations (the Freight Transport Association for own-account operators and the Road Haulage Association for haulage contractors) he could try to enlist their help in pressing a claim.

In-House Repairs

The advantages of the operator having his own workshop and being able to do his own safety inspections and repairs are many, provided that sufficient vehicles are operated to justify the overheads involved. Principal among them is that by employing the staff he has direct control over the work carried out, the standard of the work and the record keeping which is so important.

If the operator provides his own maintenance facilities they must be of a suitable standard, although once again no specific details are given in the regulations. The main requirement is for a covered area with hard-standing and facilities including adequate lighting for conveniently inspecting the underside of vehicles. Ideally this means that either a pit in the ground or an hydraulic lift should be provided. The former is the most commonly used and it is very much the cheaper of the two alternatives. Suitable lighting, either fixed in the pit shining upwards to the underside of the vehicles or by means of portable inspection lamps, is necessary.

The remainder of the tools and machines with which a workshop should be equipped are left entirely to the operator's choice, but such items as a beam-setter and a portable Tapley brake-efficiency recorder are useful to check that vehicles comply with the test requirements. Servicing equipment such as jacks, high-pressure greasing equipment, high-pressure washing or steam cleaning equipment make maintenance work much less of a chore. Hand tools are, of course, essential and in general the better the equipment available (including the availability of the special tools often needed to carry out work on today's sophisticated vehicles) the more likelihood there is of the work meeting the required standard.

Vehicle Inspections

Vehicle inspections are an essential part of the maintenance programme, and legislation covers this aspect. Section 74 of the Road Traffic Act 1988 requires operators of goods vehicles to have them regularly inspected by a 'suitably qualified person' to ensure that they comply with construction and use regulations (Chapter 13) and, of course, it is in the operator's best interests to have vehicles regularly inspected to ensure that they are kept in a fully safe and roadworthy mechanical condition.

Frequency of Inspection

Legislation does not specify the intervals at which vehicles should be inspected, what form the inspection should take or what is meant by a 'suitably qualified person'. In the case of the first-mentioned, it is very much a question of the type of operation on which the vehicle is used. A tipping vehicle, for example, which spends much of its time on rough sites, with perhaps a fairly high mileage on the road as well, certainly should be inspected at least weekly. The same applies to a vehicle which, although it remains on normal roads, does perhaps 800 to 1200 miles a week. On the other hand, a monthly inspection may be quite sufficient for a local delivery vehicle doing low weekly mileages on good roads and spending a great deal of its time standing while deliveries are being made.

The operator's own experience of his type of operation should indicate the intervals between which wear and tear takes place and defects become apparent. It has even been suggested that vehicles standing out of use in depots should be checked at least monthly. Although there is some suggestion that inspection intervals can be based on either a time or miles/kilometres alternative it is becoming clear that some LAs at least will not accept anything other than a time based frequency from 'O' licence applicants. The VI Code of Practice on vehicle maintenance referred to on p 317 recommends that the maximum time interval between safety inspections should be six weeks with no mileage alternatives. It also recommends operators to have flow charts covering at least 12-monthly periods to indicate when vehicle safety inspections are due.

Items for Inspection

A full list of the items to be inspected at regular intervals is not laid down in regulations but it is obviously necessary that the inspection at the very least covers all the items set out in the *Heavy Goods Vehicle Inspection Manual* (see p 318), with particular emphasis on those items (eg brakes, steering, wheels, tyres, suspension systems and lights) that have special relevance to the safe operation of the vehicle. Tyres on vehicles used on site work or on local delivery work should receive particularly careful examination for damage.

The Inspector

There are no specified qualifications for a vehicle inspector. Clearly, the most obvious one is wide experience in the repair of heavy commercial vehicles. A person with such experience would know where and how to look for wear and for defects and would recognise the symptoms of hidden faults, such as uneven tyre wear indicating that the steering is out of alignment or that king-pins or wheel bearings are worn.

It is possible, however, to train a person specifically as a vehicle inspector, and this is being done in the industry. The emphasis in training in such cases must be on following a predetermined list, such as the *Heavy Goods Vehicle Inspection Manual*, examining every individual item carefully and methodically, testing the wear in components and measuring the tolerances of moving parts accurately.

It is desirable, although not a legal requirement, that the person carrying out the vehicle inspection should not be expected to carry out repairs, however small, at the time of making the inspection. To have to do this would cause a lack of concentration and could lead to other items being missed if repairs took up too much of the inspector's time. In the case of an owner-driver carrying out his own inspections and repairs this can be a difficult situation and in such instances it is useful to have the work verified and an audit-type check carried out by an outside agency, such as the FTA, to ensure that the vehicle is kept up to a high standard.

Authority to Stop Use of Vehicles

The inspector, besides being suitably qualified and experienced to spot defects, should have the authority to prevent a vehicle being taken on the road in the event of a serious or potentially dangerous defect being found.

Vehicle Servicing

Regular servicing as opposed to specific inspections and repairs is an important part of vehicle maintenance and as such it should be carried out with unfailing regularity at predetermined intervals of time or mileage. The importance of servicing cannot be too highly emphasised for two reasons. First, because the vehicle must meet the requirements of the law and second, because the operator will benefit by always having his vehicles ready for work and able to carry out a job without breakdowns and delays. It also increases the life of the working parts and consequently of the whole vehicle as well as reducing down-time costs and disrupted delivery schedules.

The intervals at which servicing should be carried out are left to the owner's discretion depending on the work on which the vehicle is employed, in much the same way as the intervals for inspection are decided. To give the owner some guidance, however, vehicle manufacturers usually provide a service schedule which the owner can use in his own workshop or which his agent will use when vehicles are sent in for servicing.

A useful guide to service intervals is 5000/6000 mile services carried out at least monthly with more extensive services at 15,000/20,000 mile/3-monthly intervals and 30,000/40,000 mile/6-monthly intervals. Progressively, service intervals are being extended with the use of longer life components and particularly improved filters and lubricating oils which can go for very much longer periods these days without detriment to their lubricating properties.

Cleaning of Vehicles

Particular reference is made in the instructions to operators submitting vehicles for DoT heavy goods vehicle tests, that those which are not sufficiently clean will be refused the test. Again, it is difficult to specify a standard of cleanliness for the underside of a vehicle, but the main point is that all the components listed for the examination must be easily visible so that inspectors can see without difficulty if wear or damage exists. To achieve and keep a suitable standard of cleanliness it is desirable for vehicles to be washed with either a steam cleaner or a high-pressure water washer at regular intervals and certainly immediately before they are taken to the test station.

If the maintenance of the vehicles is contracted out to a garage on a maintenance scheme, the operator should arrange either for the garage to clean the vehicle before inspection or for one of the many specialist vehicle cleaners to do it.

Enforcement of Maintenance Standards

There has been increasing concern in recent times about the standards of vehicle maintenance. As a result of this concern, the VI has stepped up the levels of checking on vehicles by enforcement staff particularly at night and at weekends. The purpose of these additional checks is to catch vehicles operating outside normal working hours – many legitimately but some possibly deliberately running the gauntlet.

19: Maintenance Records

There is a legal requirement under the Road Traffic Act 1988 for goods vehicle operators to keep records of maintenance work carried out on their vehicles. When completing form GV 79, 'O' licence application, operators have to make the statutory 'declaration of intent' in which they promise to fulfil undertakings made at that time throughout the duration of the 'O' licence. A number of items in the declaration of intent relate to maintenance records. From this can be determined what records are needed by law.

In the declaration of intent the operator promises to ensure that the following records will be kept:
1. Safety inspections.
2. Routine maintenance.
3. Repairs to vehicles.

A promise is also made that drivers will report 'safety faults' in their vehicles and the Licensing Authorities insist that these reports should be in writing and therefore they become part of the vehicle record-keeping system. The operator promises to keep all these records for a minimum period of 15 months and to make them available on request by Vehicle Inspectorate (VI) examiners or the Licensing Authority (LA).

Additionally, operators are frequently asked by vehicle examiners and the LAs to keep a wall chart showing vehicles in the fleet and when they are due for inspection, service and annual test. The VI Code of Practice on vehicle maintenance (see p 317) recommends annual flow-charts showing vehicle inspection-due dates.

Driver Reports of Vehicle Faults

It is a specific requirement (and part of the declaration of intent on an 'O' licence application form) that arrangements must be made for drivers to have a proper means of reporting 'safety faults' (ie defects) in the vehicle they are driving as soon as possible. As already mentioned, the LAs expect these defect reports to be made in writing, not verbally. Ideally, the report should be made either on an individual form which is completed and handed in by the driver, or in a defect book reserved for recording defects found on a vehicle which is kept in a convenient place where all drivers have easy access to it, and where whoever is responsible for ensuring that repair work is carried out can also easily reach it. To reiterate, it has been made abundantly clear that verbal reporting of defects is not in itself a system acceptable to the LAs.

Whichever method is used it is important that the repair of the defects is recorded on the form or in the book by a note of the work done and the signature of the person who has done it (see also under repair records). It is also important for the operator to ensure that whatever system of defect reporting is used, he makes regular checks on drivers and repair staff to see that the procedure is being followed correctly. This requirement has been pointed out by the LAs on a number of occasions. Using separate pads of defect sheets (see Figure 19.1) is the best alternative and where these can be made out in duplicate they provide the driver with his own copy of the report for future reference.

Systems of defect reporting which rely on a centrally located defect book or on verbal reports by drivers are open to the risk of drivers forgetting to report defects when they return to base. If a driver's attention is distracted by his manager who wants to talk to him, for example, just as he is about to report a defect, the defect may not be reported and another driver may take the vehicle out next day with a defect which could result in a prohibition notice being issued in a roadside check. A driver may also forget to report a defect when he returns late from a journey and is in a rush to get home or if he cannot find the defect book.

It is the operator's responsibility to make sure that the system used is infallible in all these circumstances firstly because it is an offence to fail to cause the defect to be reported and secondly because the vehicle could be found on the road subsequently with a safety fault not reported and not repaired.

Inspection Reports

The Road Traffic Act 1988 requires that records of regular safety inspections to vehicles must be made. For this purpose the vehicle inspector (see p 327) should have a sheet on which are listed all the items to be inspected (preferably in accordance with the contents of the *Heavy Goods Vehicle Inspection Manual* – see pp 285–8).

The inspection sheet should identify the necessary items for inspection with a cross-check reference number to the *Inspection Manual* to enable, if necessary, full details of the method of inspection of that item – and the reason for its rejection as not being within acceptable limits – to be determined. The sheet should have provision for the inspector to mark against each item whether it is 'serviceable' or 'needs attention' and space to comment on defects for immediate rectification and other items for attention at a future date or on which a watch should be kept if attention is not required immediately. The form should contain space for the inspector to sign his name and add the date.

Defect Repair Sheets

Besides ensuring that proper records are kept of vehicle safety fault reports made by drivers and of regular safety inspections, the operator must also keep a record showing that any defects reported or found on inspection are rectified in order to keep the vehicle in a fit, serviceable and safe condition. Records of such repairs may be added to the driver defect report or the inspection report to provide combined records or a separate repair or job sheet may be used.

DRIVER'S DAILY VEHICLE DEFECT REPORT

Date:	Driver's name:
Vehicle No:	Trailer Fleet/Serial No:

Note: Drivers are responsible for the safe condition of their vehicle and load. You are required by law to report defects to your employer.

Tick items on this check list that are in order; put cross against defective items.

DAILY CHECK		Tick or cross	
Fuel:		Lights:	
Oil:		Reflectors:	
Water:		Indicators:	
Battery:		Wipers:	
Tyres:		Washers:	
Brakes:		Horn:	
Steering:		Mirrors:	
Security of body:		Markers:	
Security of load:		Sheets/ropes/chains:	
Artics/Lorry & trailer combinations			
Brake hoses:		Electrical connections:	
Coupling secure:		Trailer No. plate:	

REPORT DEFECTS HERE

WRITE NONE HERE IF NO DEFECTS

Driver's signature:

Action Taken By:

Signature:

Figure 19.1 *A typical example of a driver's vehicle defect report available in pad form*

The important points about repair records are, first, that they should show comprehensive details of the actual repair work carried out, identifying components which were repaired or replaced and new parts added, and, second, that there should be a matching repair sheet for every defect reported or found on inspection so that the LA's vehicle examiners, when they visit to examine records, can see the report of the defect and then subsequently a report of the repair work carried out to rectify it. Reports of defects which do not have a corresponding repair record can arouse suspicion in the examiner's mind that perhaps the necessary repair has not been carried out and that the defect still exists. This is a good reason for him then to consider examining that vehicle, or perhaps the whole fleet.

Service Records

In addition to the records of defects and vehicle inspections which have to be kept, a record should also be kept of all other work carried out on the vehicle, whether it is repair or replacement of working parts or normal servicing (oil changes and greasing, etc).

Retention of Records

It is a legal requirement that records of maintenance must be retained by operators. The original inspection report, or a photocopy of it, with the inspector's comments, the date and the mileage at which the inspection was carried out, must be retained and kept available for inspection if required by the VI examiners for 15 months from the date of the inspection. Work sheets showing the repair work carried out following the inspection and all other work done on the vehicle, including repairs following defect reports by drivers, as well as defect reports themselves, must be kept for 15 months, available for inspection by VI examiners or the LA, if requested.

Repair Records from Garages

When vehicle safety inspections, servicing and repair work are carried out for the operator by repair garages, the operator should obtain from the garage comprehensive documentation to enable him to meet the legal requirements detailed above. In many cases the VI examiners are quite happy if the garage retains the records of inspection and repair, so long as they can be made available for examination when required. The operator must be certain, in these instances, that the garage is keeping proper records (for a period of 15 months) which satisfy the legal requirements and that they are being kept available for inspection, not bundled away out of easy reach in a store with thousands of others.

In the event of failure of the garage to keep records as required, the operator's licence would be at risk but there would be no penalty imposed on the garage. On the GV 79 declaration of intent (see p 330) the operator promised that he would '. . . make proper arrangements so that records are kept for (15 months) of all safety inspections, routine maintenance and repairs to vehicles and make them available on request'.

It is important to note that in this respect, invoices, or copies of invoices from garages, for repair work are not in themselves sufficient to satisfy the record-keeping requirement. It is the actual inspection sheet and repair sheets, or photocopies of them, which are needed because of the greater and more precise detail which they contain. Similarly, maintenance records in computer print-out form are unlikely to satisfy the requirement of enforcement staff to examine actual records – they will still want to see the original inspection sheets. This is an important point to consider with the increasing application of computers to transport operations and vehicle maintenance functions.

Location of Records

Where companies hold operators' licences in a number of separate Traffic Areas the maintenance records for the vehicles under each licence should be kept in the area covered by the individual licence (preferably at the vehicle operating centre). With the sanction of the local LA, records may be kept centrally at a head office or central vehicle workshop, although the vehicle examiners may ask for them to be produced for inspection at the operating centre of the vehicles in the Traffic Area – probably giving 3 days' to 1 week's notice to enable the records to be obtained from the central files.

Wall Planning Charts

While it is not strictly a legal requirement, many vehicle examiners (and the current West Midland LA) like to see operators using wall planning charts (see note above about use of flow charts in accordance with new VI Code of Practice on vehicle maintenance) to provide a visual reminder of important dates such as:
1. Vehicle/trailer due for inspection
2. Vehicle/trailer due for service
3. Vehicle/trailer due for DoT annual test
4. Excise duty due.

Such charts usually provide facilities for a whole year's recording of these items for the fleet (either shown by vehicle registration number or by fleet number).

Vehicle History Files

For efficiency in record keeping, a system of vehicle history files – one for each vehicle and trailer in the fleet – is most useful. This provides the facility for keeping all relevant records relating to individual vehicles and trailers together and in one place. Individual files can have all the important details of the vehicle/trailer on the front cover for easy reference, as follows:
1. Registration number/fleet number
2. Make/type
3. Date of original registration
4. Price new plus extras/options
5. Annual test date

6. Taxation (ie VED) date
7. Base/location
8. Model designation
9. Wheelbase (in/mm)
10. Body type
11. Special equipment
12. Chassis number
13. Engine number
14. Gearbox type/number
15. Rear axle type
16. Electrical system – 12V/24V
17. Plated weights – gross/axle
18. Supplier's name and address.

20: Safety – Vehicle, Loads and at Work

Transport operators, along with all other sectors of business and industry, are under constant pressure to become ever more conscious of the need for safety in their operations, in the provision of facilities for their employees and in the way their employees work and conduct themselves on work premises. Predominantly, this is influenced by the demands of the Health and Safety at Work etc Act 1974 and the stringent requirements which it imposes on employers and employees alike. But in transport, the requirements of the Road Vehicles (Construction and Use) Regulations 1986 (as amended), regarding the safety of vehicles and loads, place additional legal burdens on operators and drivers alike. The problems of safe loading and avoidance of vehicle overloading are not new but increased enforcement activity has accelerated concern in these areas. There is concern also on the wider front of safety in load handling and in the use of loading aids (fork-lift trucks for example) and with regard to vehicle manoeuvring in depots and works premises.

C&U Requirements

The Road Vehicles (Construction and Use) Regulations 1986 require that all vehicles and trailers, and all their parts and accessories, and the weight, distribution, packing and adjustment of their loads, shall be such that no danger is caused or likely to be caused to any person in or on the vehicle or trailer or on the road. Additionally, no motor vehicle or trailer must be used for any purpose for which it is so unsuited as to cause or be likely to cause danger or nuisance to any person in or on the vehicle or trailer or on the road.

Under the regulations, provisions relating particularly to bulk and loose loads make it an offence if a load causes a nuisance as well as a danger to other road users and such loads must be secured, if necessary by physical restraint, to stop them falling or being blown from a vehicle.

These regulations include two notable terms relating to load safety: one is the use of the term 'nuisance' in addition to 'danger'; so that to commit an offence the operator does not have to go so far as causing danger, merely causing nuisance is sufficient to land him in trouble. The other term is 'physical restraint' which clearly implies the need for sheeting and roping any load, such as sand or grain, hay and straw and even builder's skips carrying rubble, which may be blown from the vehicle.

The Safety of Loads on Vehicles

A Code of Practice – The Safety of Loads on Vehicles published by the DoT (available from HMSO – *NB: a new version, which was in draft form when this edition of the* Handbook *was being prepared, is expected in 1994–95*) sets out general requirements in regard to the legal aspects of safe loading, information on the forces involved in restraining loads, the strength requirements of restraining systems and load securing equipment, and then details specialised requirements for containers, pallets, engineering plant, general freight, timber, metal and loose bulk loads. It provides a list of dos and don'ts for drivers and others concerned with the loading of vehicles as shown below. The new version will include advice on anchor points and restraint methods.

For particular note is the basic principle on which the Code is based which is that:

'the combined strength of the load restraint system must be sufficient to withstand a force not less than the total weight of the load forward and half the weight of the load backwards and sideways'.

SAFE LOADING
Your own life and the lives of others may depend upon the security of your load

DOS

1. Do make sure your vehicle's load space and the condition of its load platform are suitable for the type and size of your load.

2. Do make use of load anchorage points.

3. Do make sure you have enough lashings and that they are in good condition and strong enough to secure your load.

4. Do tighten up the lashings or other restraining devices.

5. Do make sure that the front of the load is abutted against the headboard, or other fixed restraint.

6. Do use wedges, scotches etc., so that your load cannot move.

7. Do make sure that loose bulk loads cannot fall or be blown off your vehicle.

DON'TS

1. Don't overload your vehicle or its individual axles.

2. Don't load your vehicle too high.

3. Don't use rope hooks to restrain heavy loads.

4. Don't forget that the size, nature and position of your load will affect the handling of your vehicle.

5. Don't forget to check your load:
a. Before moving off;
b. After you have travelled a few miles;
c. If you remove or add items to your load during your journey.

6. Don't take risks.

Taken from DoT Code of Practice on the Safety of Loads on Vehicles

Safety Report

The Health and Safety Executive has produced a report entitled *Transport Kills* (HMSO) based on a study of fatal accidents in industry in which it indicates that motor vehicles are one of the biggest causes of industrial deaths. These motor vehicle related deaths occur during vehicle loading, unloading, maintenance and, of course, movement and are mainly caused by poor management, failure to provide safe working systems and inadequate training.

Included in the report is a checklist which is intended to help transport operators and others to reduce unnecessary risks and dangers. Companies should use it to help examine their current practices and to institute new and safer procedures. The checklist (reproduced below with acknowledgement to the HSE) is only a general guide, and it is emphasised that safety requirements vary with different types of operation. *However, it is important to note that firms could face prosecution if they do not meet the minimum safety standards outlined in the list.*

This checklist is intended as a general guide only. It will not necessarily be comprehensive for every operation and all points will not be relevant for all work.

Organisation, Systems and Training

- Have all health and safety aspects of the transport operation been assessed?
- Has an organisation (and arrangements) for securing such safety been detailed in the safety policy?
- Has a person been appointed to be responsible for the transport safety?
- Have safe systems of work been set up?
- What monitoring is carried out to ensure that the systems are followed?
- Have all drivers been adequately trained and tested?
- Is there a satisfactory formal licensing or authorisation system for drivers?
- Have all personnel been trained, informed and instructed about safe working practices where transport is involved?
- Is there sufficient supervision?

External Roadways and Manoeuvring Areas

- Are they of adequate dimensions?
- Are they of good construction?
- Are they well maintained?
- Are they well drained?
- Are they scarified when smooth?
- Are they gritted, sanded, etc, when slippery?
- Are they kept free of debris and obstructions?
- Are they well illuminated?

- Are there sufficient and suitable road markings?
- Are there sufficient and suitable warning signs?
- Are there speed limits?
- Is there a one-way system (as far as possible)?
- Is there provision for vehicles to reverse where necessary?
- Are there pedestrian walkways and crossings?
- Are there barriers by exit doors leading on to roadways?
- Is there a separate vehicle parking area?
- Is there any storage positioned close to vehicle ways?
- Is the yard suitable for internal works transport, eg smooth surface, hard ground, no slopes?

Internal Transport

- Are internal roadways demarcated and separated where possible from pedestrian routes with crossings and priority signs?
- Are there separate internal doors for trucks and pedestrians? Have these vision panels?
- Are blind corners catered for by mirrors, etc?
- Are trucks kept apart from personnel where possible?
- Do the trucks use a satisfactory warning system?

Vehicles

- Is there a maintenance programme for vehicles and mobile plant?
- Is there a fault reporting system?
- Are there regular checks to ensure that the vehicles are up to an acceptable standard?
- Are keys kept secure when vehicles and mobile plant are not in use?
- Are vehicles and mobile plant adequate and suitable for the work in hand?
- Is suitable access provided to elevated working places or vehicles?
- Are tractors and lift trucks equipped with protection to prevent the driver being hit by falling objects and from being thrown from his cab in the event of overturning?
- Are there any unfenced mechanical parts on vehicles, eg power take-offs?
- Are there fittings for earthing vehicles with highly flammable cargoes?
- Are loads correctly labelled (especially hazardous substances)?
- Are the vehicles suitable for use in all the areas they enter? Do they need to be to Division I or II standards, etc?
- If passengers ride on vehicles, do they have a safe riding position?

Loading and Unloading

- Do loading positions obstruct other traffic? Do pedestrian ways need diverting?

- Are there special hazards, eg flammable liquid discharge? Do pedestrians need to be kept clear?
- Is there a yard manager to supervise the traffic operation, to control vehicular movement and to act as a banksman during reversing?
- Has he received satisfactory training? Does he use recognised signals, and has he cover during his absences?
- Are there loading docks? Will the layout prevent trucks falling off or colliding with objects or each other?
- Are there any mechanical hazards caused by dock levellers, etc?
- Are methods of loading and unloading assessed? Are loads stable and secured?
- Are safe arrangements made for sheeting?
- Is there a pallet inspection scheme?

Motor Vehicle Repair

- Are appropriate arrangements made for tyre repair and inflation?
- Are arrangements made for draining and repair of fuel tanks?
- Is access available to elevated working positions?
- Are arrangements made to ensure brakes are applied and wheels checked?
- Is portable electrical equipment low voltage and properly earthed?
- Are moving vehicles in the workshop carefully controlled?
- Are vehicles supported on both jacks and axle stands where appropriate?
- Are engines only run with the brakes on and in neutral gear?
- Are raised bodies always propped?

Health and Safety at Work

There was a notable strengthening of the law on health and safety at work matters in 1993 with the introduction of a number of new pieces of UK legislation as part of the implementation of the EU's 'Framework Directive', namely EC Directive 391/1989. The overall objective of this Directive is to impose on employers a duty to encourage improvements in the health and safety of people at work. Individual Directives within the so-called framework cover:

- The workplace
- Personal protective equipment
- The use of display screen equipment
- Manual handling of loads.

The new UK regulations which came into force on 1 January 1993 specifically to implement the EU requirements were as follows:

1. Management of Health and Safety at Work Regulations 1992
2. The Workplace (Health, Safety and Welfare) Regulations 1992
3. Manual Handling Operations Regulations 1992
4. Health and Safety (Display Screen Equipment) Regulations 1992

5.   Provision and Use of Work Equipment Regulations 1992
6.   Personal Protective Equipment at Work Regulations 1992.

The first three items in the above list are described in outline in this chapter because of their more direct relevance to transport operations (see following pages and p 354). But this is not to say that the other legislation listed is not important – it must still be observed in road transport offices and workplaces.

The Management of Health and Safety

The new 'management' regulations (item 1 in the list above) are both wide ranging and general in nature, overlapping with many existing regulations. They have to be viewed as a 'catch-all' regulation, sitting astride other more specific health and safety provisions (eg the COSHH regulations – see pp 347–8). The Health and Safety Executive (HSE), in its Approved Code of Practice, advises that where legal requirements in these management regulations overlap with other provisions compliance with duties imposed by the specific regulations will be sufficient to comply with the corresponding duty in the management regulations. However, the HSE says, where duties in the Management of Health and Safety at Work Regulations 1992 go beyond those in the more specific regulations, additional measures will be needed to comply fully with the management regulations.

Specifically, these regulations cover:
1.   Requirements for employers (and self-employed persons) to make assessments of the risks to the health and safety of
     –   employees while they are at work
     –   other persons not in their employ
     arising out of or in connection with their conduct or the conduct of their undertaking.
2.   Employers with five or more employees must record the significant findings of their risk assessment (ie make an effective statement of the hazards and risks which lead management to take relevant actions to protect health and safety).
3.   Employers and self-employed persons must make and give effect to such arrangements, as are appropriate to the nature of their activities and the size of their undertaking, for the effective planning, organisation, control, monitoring and review of preventive and protective measures.
4.   Health surveillance must be provided for employees as appropriate to the risks to their heath and safety as identified by the assessment.
5.   Employers must appoint one or more (competent) persons to assist them in undertaking the measures necessary to comply with the requirements and prohibitions of this legislation.

The Workplace Regulations

These new regulations (see item 2 in the list of new legislation above) add further to existing legislation on workplaces. They apply to new workplaces as of 1 January 1993 and to any modifications or conversions to existing workplaces started after this date. Existing workplaces (ie which are unaltered) must comply from 1 January 1996.

Specifically the regulations impose requirements on the:

–   maintenance of workplaces
–   ventilation of enclosed workplaces and temperatures of indoor workplaces and the provision of thermometers
–   lighting (including emergency lighting)
–   cleanliness of the workplace, and of furniture, furnishings and fittings (also the ability to clean floors, walls and ceilings) and the accumulation of waste materials
–   room dimensions and unoccupied space
–   suitability of workstations (including those outside) and the provision of suitable seating
–   condition of floors, and the arrangement of routes for pedestrians or vehicles
–   protection from falling objects, and from persons falling from a height, or falling into dangerous substances
–   material of and protection of windows and other transparent or translucent walls, doors or gates, and to them being apparent
–   way in which windows, skylights or ventilators are opened and their position when left open and the ability to clean these items
–   construction of doors and gates (including the fitting of necessary safety devices), and escalators and moving walkways
–   provision of suitable sanitary conveniences, washing facilities and drinking water (including cups and drinking vessels)
–   provision of suitable accommodation for clothing and for changing clothes, for rest and for eating meals

Manual Handling

New UK legislation on the manual handling of loads (the Manual Handling Operations Regulations 1992) came into effect on 1 January 1993 following the introduction of EC Directive 269/1990 which requires national governments to make the necessary domestic legislation to reduce back and other injuries suffered through the manual handling of loads. Employers are required to take steps to reduce the risk of injury by using more mechanical aids, by providing more training in load handling and, specifically, by providing employees with precise information about load weights, centres of gravity and the heaviest side of eccentrically loaded packages.

The regulations themselves require the employer to avoid the need for employees to undertake manual handling operations at work which involve a risk of their being injured, or, where it is not possible to avoid the need for employees to undertake such operations, the employer must assess the operations to reduce the risk of injury to employees to the lowest level reasonably possible. Employees at work must make proper use of systems of work provided by the employer.

The HSE booklet *Manual Handling* (available from HMSO, price £5) provides excellent guidance on the regulations and their application. Additionally, these notes will provide useful information to employers on many aspects of manual handling.

The Health and Safety at Work etc Act 1974

Since the introduction of the Health and Safety at Work etc Act 1974 employers have had to take positive steps to draw up policy statements regarding health and safety at work, appoint safety representatives and establish safety committees in addition to ensuring that work places meet all the necessary safety requirements of the law. These responsibilities apply equally to employers in transport, and it should be remembered that here the requirements of the law apply to the transport operator's premises (ie offices, workshops, warehouses and yard) and to his vehicles which constitute the work place of drivers.

The Act replaced certain parts of the Factories Act and the Offices, Shops and Railway Premises Act, and added other provisions. There are four parts to the Act:

- Part I relates to health, safety and welfare at work
- Part II relates to the Employment Medical Advisory Service
- Part III amends the law regarding building regulations
- Part IV covers a range of general and miscellaneous provisions.

The main effects of the Act are:

1. To maintain and improve standards of health and safety for people at work.
2. To protect people other than those at work against risks to their health or safety arising from the work activities of others.
3. To control the storage and use of explosives, highly flammable or dangerous substances, and to prevent their unlawful acquisition, possession and use.
4. To control the emission into the atmosphere of noxious or offensive fumes or substances from work premises.
5. To set up the Health and Safety Commission and the Health and Safety Executive.

Duties of Employers

The Act prescribes the general duties of all employers towards their employees by obliging them to ensure their health, safety and welfare while at work. This duty requires that all plant (including vehicles) and methods of work provided are reasonably safe and without risks to health. A similar injunction relates to the use, handling, storage and transport of any articles or substances used in connection with the employer's work.

Provision of Necessary Information
In order that employees are fully conversant with all health and safety matters, it is the duty of the employer to provide all necessary information and instruction by means of proper training and adequate supervision.

Condition of Premises
Workplaces generally, if under the employer's control, must be maintained in such a condition that they are safe and without risks to health, have adequate means of entrance and exit (again this applies equally to vehicles as it does to 'premises') and must provide a working environment that has satisfactory facilities and arrangements for the welfare of everybody employed in the premises.

Statements of Safety Policy

It is necessary for an employer of five or more employees to draw up and bring to the notice of all his workforce *a written statement of company policy* regarding their health and safety at work with all current arrangements detailed for the implementation of such a policy. Stress is laid on the necessity of updating the 'statement' as the occasion arises and of communicating all alterations to the personnel employed.

Appointment of Safety Representatives

Involvement of all employees in health and safety activities is envisaged by the appointment (by a recognised trade union) or election of safety representatives from among the workforce. A safety representative should be a person who has been employed in the firm for at least *2 years* or who has had 2 years' similar employment 'so far as is reasonably practicable'. The broad duties of safety representatives are concerned with the inspection of workplaces, investigating possible hazards and examining the cause of accidents, investigating employees' complaints regarding health and safety matters and making representations to their employer on health and safety at work matters.

Safety Committees

At the written request of at least two safety representatives employers must establish a safety committee to review health and safety at work matters. The establishment of a safety committee creates a joint responsibility with the employer for concern with all health and safety measures at the workplace, together with any other duties arising from regulations or codes of practice.

Duty to Public

Employers and self-employed persons are required also to ensure that their activities do not create any hazard to members of the general public. In certain circumstances, information must be made publicly available regarding the existence of possible hazards to health and safety.

Summary of Duties of Employers to their Employees

1. It shall be the duty of every employer to ensure so far as is reasonably practicable the health, safety and welfare at work of all his employees. That duty includes in particular:
    (a) The provision and maintenance of plant and systems of work that are so far as is reasonably practicable safe and without risks to health.
    (b) Arrangements for ensuring so far as is reasonably practicable safety and absence of risks to health in connection with the use, handling, storage and transport of articles and substances.
    (c) The provision of such information, instruction, training and supervision as is necessary to ensure so far as is reasonably practicable the health and safety at work of his employees.
    (d) So far as is reasonably practicable the maintenance of any place of work that is under the employer's control in a condition that is safe and without risks to health, and the provision and maintenance of means of access to and egress from it that are safe and without such risks.
    (e) The provision and maintenance of a working environment for his employees that is so far as is reasonably practicable safe and without

risk to health and adequate as regards facilities and arrangements for their welfare at work.

2. Except in such cases as may be prescribed it shall be the duty of every employer to prepare and as often as may be appropriate revise a written statement of his general policy with respect to the health and safety at work of his employees and the organisation and arrangements for the time being in force for carrying out that policy, and to bring the statement and any revision of it to the notice of all employees.

3. It shall be the duty of any person who erects or installs any article for use at work in any premises where the article is to be used by persons at work to ensure, so far as is reasonably practicable, that nothing about the way in which it is erected or installed makes it unsafe or a risk to health when properly used.

4. No employer shall levy or permit to be levied on any employee of his any charge in respect of anything done or provided in pursuance of any specific requirement of the relevant statutory provisions.

Duties of Employees

The Act states in general terms the duty of an employee to take reasonable care for the safety of himself and others and to co-operate with others in order to ensure that there is a compliance with statutory duties relating to health and safety at work.

In this connection no person shall interfere with or misuse anything provided in the interests of health, safety or welfare either intentionally or recklessly.

Summary of Duties of Employees at Work

1. It shall be the duty of every employee while at work:
    (a) to take reasonable care for the health and safety of himself and of other persons who may be affected by his acts or omissions at work, and
    (b) as regards any duty or requirement imposed on his employer or any other person, to co-operate with him so far as is necessary to enable that duty or requirement to be performed or complied with.

2. No person shall intentionally or recklessly interfere with or misuse anything provided in the interests of health, safety or welfare.

General Duties of Employers and Self-Employed to Persons other than their Employees

1. It shall be the duty of every employer and of every self-employed person to conduct his undertaking in such a way as to ensure so far as is reasonably practicable that persons not in their employment are not thereby exposed to risks to their health and safety.

2. It shall be the duty of every employer and of every self-employed person to give to persons not in their employment who may be affected the prescribed information about such aspects of the way in which he conducts his undertaking as might affect their health and safety.

Improvement and Prohibition Notices

Improvement Notice
An improvement notice may be served on a person by a health and safety inspector in cases where he believes (ie is of the opinion) that the person is contravening or has contravened, and is likely to continue so doing or will do so again, any of the relevant statutory provisions. Such a notice must give details of the inspector's reason for his belief and requires the person concerned to remedy the contravention within a stated period.

Prohibition Notice
A prohibition notice with immediate effect may be served on a person under whose control activities to which the relevant statutory provisions apply are being carried on, or are about to be carried on, by an inspector if he believes that such activities involve or could involve *a risk of serious personal injury*. A prohibition notice must specify those matters giving rise to such a risk, the reason why the inspector believes the statutory provisions are, or are likely to be, contravened, if indeed he believes that such is the case. The notice must direct that the activities in question shall not be carried on unless those matters giving rise to the risk of serious personal injury, and any contravention of the regulations, are rectified.

Remedial Measures
Both improvement and prohibition notices may include directions as to necessary remedial measures and these may be framed by reference to an approved code of practice and may offer a choice of the actions to be taken. Reference must be made by the inspector to the Fire Authority before serving a notice requiring, or likely to lead to, measures affecting means of escape in case of fire.

Withdrawal of Notices and Appeals
A notice, other than a prohibition notice with immediate effect, may be withdrawn before the end of the period specified in the notice, or an appeal against it made. Alternatively, the period specified for remedial action may be extended at any time provided an appeal against the notice is not pending.

Electric Storage Batteries

The Health and Safety Executive has issued guidance on the safe charging and use of electric storage batteries in motor vehicle repair and maintenance. This is to help reduce the number of injuries which occur annually when batteries explode through mishandling and improper use, generally causing acid burns to face, eyes and hands as well as other injuries.

In a freely available leaflet the HSE firstly warns of the danger of charging batteries, particularly those described as maintenance free but which still give off flammable hydrogen gas which on contact with a naked flame will burn and cause the battery to explode. The advice it gives is as follows.

General Precautions
- Always wear goggles or a visor when working on batteries.
- Wherever possible, always use a properly designated and well-ventilated area for battery charging.
- Remove any metallic objects from hands, wrists and around the neck (eg rings, chains and watches) before working on a battery.

Disconnecting and Reconnecting Batteries
- Turn off the vehicle ignition switch and all other switches or otherwise isolate the battery from the electrical circuit.
- Always disconnect the earthed terminal first (often the negative terminal, but not always – CHECK) and reconnect it last using insulated tools.
- Do not rest tools or metallic objects on top of a battery.

Battery Charging
- Always observe the manufacturer's instructions for charging batteries.
- Charge in a well-ventilated area. Do not smoke or bring naked flames into the charging area.
- Make sure the battery is topped up to the correct level.
- Make sure the charger is switched off or disconnected from the power supply before connecting the charging leads, which should be connected positive to positive, negative to negative.
- Vent plugs may need to be adjusted before charging. Carefully follow the manufacturer's instructions.
- Do not exceed the recommended rate of charging.
- When charging is complete, switch off the charger before disconnecting the charging leads.

Jump Starting
Preparation
- Always ensure that both batteries have the same voltage rating.
- If starting by using a battery on another vehicle, check the earth polarity on both vehicles.
- Ensure that the vehicles are not touching.
- Turn off the ignition of both vehicles.
- Always use purpose-made, colour-coded jump leads with insulated handles – red for the positive cable and black for the negative cable.

Connection for vehicles with the *same* earth polarity
- First connect the non-earthed terminal of the good battery to the non-earthed terminal of the flat battery.
- Connect one end of the second lead to the earthed terminal of the good battery.
- Connect the other end of the second lead to a suitable, substantial, unpainted point on the chassis or engine of the other vehicle, away from the battery, carburettor, fuel lines or brake pipes.

Connection for vehicles with *different* earth polarity

The HSE warns that in view of the potential for confusion this should only be attempted by skilled and experienced personnel.

- First connect the earthed terminal of the good battery to the non-earthed terminal of the flat battery.
- Connect one end of the second lead to the non-earthed terminal of the good battery.
- Connect the other end of the second lead to a suitable, substantial, unpainted point on the chassis or engine of the other vehicle, away from the battery, carburettor, fuel lines or brake pipes.

Starting

- Ensure that the leads are well clear of moving parts.
- Start the engine of the 'good' vehicle and allow to run for about 1 minute.
- Start the engine of the 'dead' vehicle and allow to run for about 1 minute.

Disconnection

- Stop the engine of the 'good' vehicle.
- Disconnect the leads in the reverse order to which they were connected.
- Take great care in handling jump leads: do not allow the exposed metal parts to touch each other or the vehicle body.

Control of Substances Hazardous to Health (COSHH)

1988 regulations under this heading (commonly referred to as the COSHH regulations) are designed to further protect the health and safety of people at work and place additional responsibilities on employers to assess the risks to employees' health of working with hazardous substances. Employers must undertake an assessment of their work environment to determine the potential hazards and to take steps to minimise any such hazards. They must inform employees of any risks which exist and train them in safety procedures such as the handling of hazardous materials and the monitoring of the work environment. Failure to comply with the regulations will result in prosecution under these regulations and under the Health and Safety at Work Act.

Notification of Accidents

The Reporting of Injuries, Diseases and Dangerous Occurrences Regulations 1985 (commonly referred to as RIDDOR) require employers to notify fatal accidents, major accidents and those causing more than 3 days' incapacity for work, work related diseases, gas incidents and any dangerous occurrence whether or not anybody is injured.

The report of any notifiable accident or occurrence must be made to the appropriate authority (see below) by a 'responsible person'. Usually this is the employer himself in the case of his employees, but if the accident or occurrence involves a member of the general public, the responsible person is the person who controls the premises.

In cases where an accident occurs to an employee when he is away from base (as may be the case with lorry drivers and sales representatives, for example), although it remains the responsibility of his employer to report the accident or occurrence, the Health and Safety Executive suggests that it would be helpful if the owner or occupier of the premises where the event occurred were to advise the person's employer as soon as possible (although there is no legal obligation for him to do so). Where a dangerous occurrence involves the carriage of dangerous goods by road the person responsible is the 'O' licence holder.

Reporting Authorities

Reports of fatal and injury accidents and notifiable dangerous occurrences must be made to the 'enforcing authority' (ie the authority responsible for enforcing the Health and Safety at Work Act). Principally, the authority is the Health and Safety Executive but reports should be made to its individual Inspectorates as listed below, depending on the type of premises activity:

Premises by main activity	To whom to report
1. Factories and factory offices	HM Factory Inspector
2. Mines and quarries	HM Mines and Quarries Inspector
3. Farms (and associated activities), horticultural premises, and forestries	HM Agricultural Inspector
4. Civil engineering and construction sites	HM Factory Inspector
5. Statutory and non-statutory railways	Inspecting Officer of Railways
6. Shops, offices, separate catering services, launderettes	District Council (or equivalent)
7. Hospitals, research and development water supply, postal services and telecommunications, entertainment and recreational services, local government services, educational services and road conveyance	HM Factory Inspector services

Reporting an Accident

Report should be made to the local office of the enforcing authority as soon as possible, preferably by telephone:

1. any accident causing death or major injury to an employee;
2. any accident which occurs on premises which are under a person's control causing death or major injury to a self-employed person or to a member of the public;
3. any notifiable dangerous occurrence – incidents of these types affecting the work or equipment either of a firm or self-employed persons working on premises which are under a member of the firm's control should be reported even if nobody is injured.

'Major injury' is defined as follows:

   (a) fracture of the skull, spine or pelvis;

   (b) fracture of any bone:

       (i) in the arm other than a bone in the wrist or hand;

      (ii) in the leg other than a bone in the ankle or foot;

   (c) amputation of a hand or foot;

   (d) the loss of sight of an eye; or

   (e) any other injury which results in the person injured being admitted into hospital as an inpatient for more than 24 hours, unless that person is detained only for observation.

If the person responsible does not know whom to report to, he should tell the nearest office of the Health and Safety Executive who will pass on the report.

Details of the accident/dangerous occurrence should be entered in the record book (or other record system). Form F2509 may be used for this purpose. Within 7 days a written report must be sent, on form F2508 (available from Health and Safety Executive offices), to the enforcing authority. This form should be used to report on all accidents causing fatal or major injury and notifiable dangerous occurrences. In the case of a reportable disease form F2580A is used.

Other injuries to employees are also notifiable if they result in more than 3 days' absence from normal work, but all that is needed is to:

   (a) enter details of the accident in the accident record book (or other record system) – form F2509 may be used for this purpose.

   (b) complete form B176 (relating to a claim for Industrial Injury Benefit) when requested to do so by the local DSS office.

Dangerous Occurrences

The list of dangerous occurrences in the regulations is selective, the aim being to obtain information about incidents with a high potential for injury but with a low frequency of occurrence.

It is important to note that dangerous occurrences must be reported even though no injury was actually caused to any person.

Examples of incidents which constitute dangerous occurrences are as follows:

1.   Failure, collapse or overturning of lifts, hoists, cranes, excavators, tail-lifts, etc.
2.   Explosion of boiler or boiler tube.
3.   Electrical short circuits followed by fire or explosion.
4.   Explosion or fire which results in stoppage of work for more than 24 hours.
5.   Release of flammable liquid or gas (ie over one tonne).
6.   Collapse of scaffolding.
7.   Collapse or partial collapse of any building.
8.   Failure of a freight container while being lifted.
9.   A road tanker, to which the Hazchem regulations apply, either overturning or suffering serious tank damage while carrying a hazardous substance.

Note 1: This list is abbreviated. Where appropriate the regulations themselves should be consulted.

Note 2: A useful leaflet on the subject is available free from local offices of the Health and Safety Executive.

First Aid

Regulations require employers to train members of their staff in first aid techniques and to provide first aid equipment and facilities under The Health and Safety (First Aid) Regulations 1981. An approved Code of Practice established by the Health and Safety Commission provides both employers and the self-employed with practical guidance on how they may meet the requirements of the regulations. Further, HMSO has published a booklet in their health and safety series entitled *First Aid at Work* (HS(R) 11; price, £2.50).

The regulations state that 'an employer shall provide, or ensure that there are provided, such equipment and facilities as are adequate and appropriate in the circumstances for enabling first aid to be rendered to his employees if they are injured or become ill at work'.

Trained First-Aiders

The employer must provide suitable persons to administer first aid and these persons must have had specific (see Code of Practice) training or hold appropriate qualifications – occupational first-aiders are no longer considered to be 'suitable persons' for this purpose.

First Aid Boxes

First aid boxes must be provided. These should be properly identified as first aid containers, preferably with a white cross on a green background, and contain sufficient quantities of first aid material *and nothing else*. In particular the boxes should contain only material which a first-aider has been trained to use. The old style of contents list according to the numbers of persons covered has been scrapped in favour of a single specification list as follows:

    1 x Guidance card
    20 x Individually-wrapped sterile, adhesive dressings of assorted sizes
    2 x Sterile eye pads, with attachment
    6 x Individually-wrapped triangular bandages
    6 x Safety pins
    6 x Sterile, unmedicated wound dressings (approx 13cm x 9cm)
    2 x Sterile, unmedicated wound dressings (approx 28cm x 17.5cm).

Soap and water and disposable drying materials, or suitable equivalents, should also be available. Where tap water is not available, sterile water or sterile 9 per cent saline, in sealed, disposable containers (refillable containers are banned) each holding at least 300ml, should be kept easily accessible, and near to the first aid box, for eye irrigation.

The contents of first aid boxes should be replenished as soon as possible after use and items which deteriorate will need to be replaced from time to time. Items should not be used after the indicated expiry date on the packet. For this reason boxes and kits should be examined frequently to make sure they are fully equipped.

It is unwise to store non-listed medical supplies (ie proprietary items such as pain killer tablets etc) or indeed any other items in first aid boxes. This may contravene the COSHH or other regulations and could result in prosecution – as in one case when a firm was prosecuted because an HSE inspector discovered its staff had placed nail varnish and acetate varnish remover in the box 'for safety'!

Travelling First Aid Kits

An employer does not need to make first aid provisions for employees working away from his establishment. However, where the work involves travelling for long distances in remote areas, from which access to NHS accident and emergency facilities may be difficult, or where employees are using potentially dangerous tools or machinery, small travelling first aid kits should be provided.

The contents of such kits may need to vary according to the circumstances in which they are to be used. The regulations suggest that in general the following items should be sufficient:

    1 x Guidance card
    6 x Sterile adhesive dressings
    1 x Large sterile unmedicated dressing
    2 x Triangular bandages
    2 x Safety pins
    A supply of individually-wrapped, moist cleaning wipes.

First Aid Rooms

When the siting of a new first aid room is under consideration it should be borne in mind that there should, where possible, be toilets nearby and ready access to transport. Any corridors and lifts which lead to the first aid room may need to allow access for a stretcher, wheelchair, carrying chair or wheeled carriage. Consideration should also be given to the possibility of providing some form of emergency lighting. Such rooms should be cleaned every day and clearly identified.

The following facilities and equipment should be provided in first aid rooms:

1. Sink with running hot and cold water always available.
2. Drinking water when not available on tap.
3. Paper towels.
4. Smooth topped working surfaces.
5. An adequate supply of sterile dressings and other materials for wound treatment. These should be at least equivalent in range and standard to those listed above, and kept in the quantities recommended.
6. Clinical thermometer.
7. A couch with pillow and blankets (frequently cleaned).
8. A suitable store for first aid materials.
9. Soap and nail brush.
10. Clean garments for use by first-aiders and occupation first-aiders.
11. Suitable refuse container.

Safety Signs

A safety sign is defined as one which combines geometrical shape, colour, and a pictorial symbol to provide specific health or safety information or an instruction whether or not any text is included on the sign. Regulations concerning the specification of safety signs in work premises has applied to all such signs since 1 January 1986. Specifications for various types of safety sign are given in BS 5378 (Part I). They are briefly described as follows:

Prohibition Sign: round in shape with a white background and a circular band and cross-bar in red. The symbol must be black and placed in the centre of the sign without obliterating the cross-bar. Typical examples of such signs are those prohibiting smoking, pedestrians or the use of water for drinking purposes.

Warning Sign: triangular in shape with a yellow background and black triangular band. The symbol or words must be black and placed in the centre of the sign. Typical examples of such signs are those warning of the danger of fire, explosion, toxic substances, corrosive substances, radiation, overhead loads, industrial trucks, electric shocks, proximity of laser beams, etc.

Mandatory Sign: round in shape with a blue background with the symbol or words placed centrally and in white. Typical examples of such signs are those advising that certain pieces of protective clothing must be worn (eg goggles, hard hats, breathing masks, gloves, etc).

Safety Sign: square in shape with a green background and white symbol or words centrally placed. Typical examples of such signs are those for fire exits, first aid posts and rescue points.

Vehicle Reversing

Among the statistics of industrial accidents and road accidents, those resulting from vehicles reversing feature significantly. According to the HSE, nearly one-quarter of all deaths involving vehicles at work occur while the vehicle is reversing. In a new booklet entitled *Reversing Vehicles* (available from the HSE) the HSE gives the following advice to those concerned:

1. Identify all of the risks and decide how to remove them.
2. Remove the need for reversing.
3. Exclude people from the area in which vehicles are permitted to reverse.
4. Minimise the distance vehicles have to reverse.
5. Make sure all staff are adequately trained.
6. Use a properly trained banksman or guide (the booklet illustrates the signals such a person should use to ensure safety when vehicles are reversing).
7. Decide how the driver is to make and keep contact with the banksman.
8. Make sure all visiting drivers are briefed.
9. Make sure all vehicle manoeuvres are properly supervised.

Additionally:
10. Increase the area the driver can see.

11. Fit a reversing alarm (see below).
12. Use other safety devices (ie trip, sensing and scanning devices, and barriers to prevent vehicles over-running steep edges).

It is in consequence of dangers of reversing that legislation was changed to permit the voluntary fitment of reversing bleepers on certain goods and passenger vehicles (see p 261 for full details).

With the introduction of these voluntary provisions there is the risk that any operator deciding not to fit such equipment on a voluntary basis who then has one of his vehicles involved in a reversing accident could face proceedings under the Health and Safety at Work Act for not taking sufficient care in safeguarding the health of others. A number of successful prosecutions on this account have been reported and in some cases very heavy fines were imposed.

Safe Tipping

Another area where concern has been expressed over safety measures involves the use of tipping vehicles (and vehicles with lorry-mounted cranes and such like), whereby elevated bodies (or crane jibs) come into contact with overhead power cables or are sufficiently close for arcing to occur in wet conditions. The specific danger lies in touching the vehicle body or tipping controls while in contact with the power cable. Drivers are safe when a cable is touched, provided they remain in the cab where the vehicle tyres prevent completion of an electrical circuit. A spokesperson for the electricity supply industry advises drivers to remain in their cabs and drive clear. If this is not possible they should jump from the cab and NOT touch any part of the vehicle, remaining well clear until an electricity engineer has been contacted and reports that it is safe to return to the vehicle.

Safe Parking

Many accidents occur with parked vehicles and trailers. In particular, within transport depots there should be proper procedures for parking, especially in regard to parking areas and level standing for detached semi-trailers. A common problem is failing to ensure that semi-trailers are dropped on to hard and level ground, allowing the landing gear on one side to sink and the trailer to topple over to the side. Nose-diving is another common accident where nose-heavy semi-trailers are not supported with trestles when left detached. Ground sinkage can also be a cause of difficulty when recoupling tractive-units.

Reported cases of unsafe parking practices resulting in driver deaths (in *Commercial Motor*) concern the braking of parked semi-trailers. When recoupling tractive units, drivers insert the red air-hose (which releases the emergency brakes on the semi-trailer) without checking that both the semi-trailer ratchet brake and the tractive-unit handbrake are fully applied. In such circumstances the vehicle combination can move while the driver is out of the cab. Other potential sources of accidents are when articulated combinations are coupled near to walls and loading bays where there is always a danger of an unseen person walking behind the trailer. Again, if brakes are not properly set the semi-trailer is likely to shunt backwards as the tractive unit is driven under the coupling plate.

Fork-Lift Truck Safety

There has been much concern in recent years about the high level of industrial accidents which are caused or which result from fork-lift truck misuse. In consequence, a system of fork-lift truck driver licensing is to be established in order to ensure a safe standard of operation.

The scheme is voluntary but a Code of Practice will be established which employers will be bound to follow if they wish to avoid conflict with the Factory Inspectorate who have powers to order an employer to have drivers trained, to order that the use of fork-lift trucks be stopped immediately if they believe that danger is being caused, and to take an employer to court if an untrained driver causes an accident with a fork-lift truck.

Freight Container Safety Regulations

Owners and lessees and others in control of freight containers must ensure that they comply with the International Convention for Safe Containers – Geneva 1972. The Freight Containers (Safety Convention) Regulations 1984 apply to containers designed to facilitate the transport of goods by one or more modes of transport without intermediate reloading, designed to be secured or readily handled or both, having corner fittings for these purposes and which have top corner fittings and a bottom area of at least 7 square metres or, if they do not have top corner fittings, a bottom area of at least 14 square metres.

Containers must have a valid approval issued by the Health and Safety Executive or a body appointed by the HSE (or under the authority of a foreign government which has acceded to the Convention) for the purpose confirming that they meet specified standards of design and construction and should be fitted with a safety approval plate to this effect. If they are marked with their gross weight such marking must be consistent with the maximum operating gross weight shown on the safety approval plate. Containers must be maintained in an efficient state, in efficient working order and in good repair. Details of the arrangements for the approval of containers in Great Britain are set out in a document *Arrangements in GB for the Approval of Containers* available from the HSE.

The safety approval plate (issued by the HSE) as described in the regulations must be permanently fitted to the container where it is clearly visible and not capable of being easily damaged and it must show the following information:

CSC SAFETY APPROVAL
1.   Date Manufactured
2.   Identification Number
3.   Maximum Gross Weight   . . . kg   . . . lb
4.   Allowable Stacking Weight for 1.8g   . . . kg   . . . lb
5.   Racking Test Load Value.

Operation of Lorry Loaders

The use of hydraulically operated lorry loaders or lorry mounted cranes as they are more commonly called (in fact the name Hiab is becoming a generic term for this equipment) 'has reduced the risk of accident from the arduous and potentially injurious manhandling of loads and they reserve the strength of the driver for safe conduct of the vehicle', according to the Association of Lorry Loader Manufacturers and Importers of Great Britain (ALLMI). But, the Association says, despite the inherent safety of a properly designed and installed lorry loader, accidents still occur through lack of knowledge and understanding. For this reason ALLMI has published an excellent booklet called *Code of Practice for the Safe Application and Operation of Lorry Loaders.* Copies are available from the Association, at 38 Shaw Crescent, Sanderstead, South Croydon, Surrey CR2 9JA: Tel 081-657 6285.

See also Chapter 7 dealing with driver training for lorry-loader operatives.

Safety in Dock Premises

Under the Docks Regulations 1988 made under the Health and Safety at Work etc Act 1974, when goods vehicle drivers work in or visit docks premises including roll-on/roll-off ferry ports they must be provided with high visibility clothing to be worn when they leave the vehicle cab. The clothing may take the form of fluorescent jackets, waistcoats, belts or sashes and must be worn at all times when out of the cab on such premises including when on the vehicle decks of the ferry. Protective headgear (hard hats) must be supplied and worn in such areas where there is likely to be danger of falling objects from above (eg where cranes are working).

Drivers must leave the vehicle cab when parked on a straddle-carrier grid or where containers are being lifted on to or off the vehicle.

The HSC has published an approved Code of Practice – *Safety in Docks* – which is available from HMSO.

21: Loads – General, Animals, Food etc

In addition to the regulations referred to in Chapter 13 regarding the way in which vehicles are constructed and used, the operator will find many more regulations imposed on him which depend on the types of load he carries. Some of these provisions are included in the Road Vehicles (Construction and Use) Regulations 1986, and its many amendments, but others are to be found elsewhere. This chapter deals with normal loads and also covers some of the special points applicable to carrying food, animals, sand and ballast, solid fuel and containers.

Distribution of Loads

When loading a vehicle, care must be taken to ensure that the load is evenly distributed to ensure stability of the vehicle and to conform to the vehicle's individual axle weights as well as the overall gross weight. It is important on multi-delivery work to make sure that when part of the load has been removed in the course of a delivery, none of the axles has become overloaded because of the transfer of weight. This can happen even though the gross vehicle weight is still within permissible maximum limits and in such cases it is necessary for the driver to attempt to correct the situation by shifting the load, or part of it.

All loads should be securely and safely fixed, roped and sheeted, and chained or lashed if necessary. It is an offence to have an insecure load or a load which causes danger to other road users. Furthermore, it is a legal requirement of the C&U regulations that loads must not cause or be likely to cause a danger or nuisance to other road users and that they should be physically restrained (ie roped and sheeted) if necessary to avoid parts of the load falling or being blown from the vehicle. *NB: Tipping vehicles (and others) that carry loose loads which could emit dust into the atmosphere must be sheeted before travelling on the road under the separate provisions of the Environmental Protection Act.* It is an offence also for the securing ropes or other devices and sheets to flap and cause nuisance or danger to other road users. Heavy penalties, with maximum fines of up to £5000, can be imposed on conviction for offences relating to these matters.

Axle Load Calculations

Imposed axle loads can be calculated to determine whether a vehicle is operating legally in particular circumstances by using the following formula:

1. Determine the vehicle wheelbase.
2. Determine the weight of the load (ie payload).

3. Calculate the front loadbase (ie centre line of front axle to centre of gravity of load).
4. Apply the formula as follows.

Example of axle weight calculation:

$$\text{weight on rear axle} = \frac{\text{front loadbase}}{\text{wheelbase}} \times \text{payload}$$

weight on front axle = payload — weight on rear axle

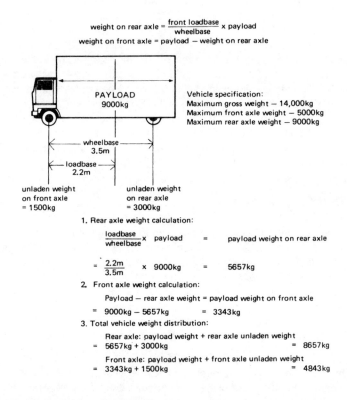

PAYLOAD
9000kg

Vehicle specification:
Maximum gross weight — 14,000kg
Maximum front axle weight — 5000kg
Maximum rear axle weight — 9000kg

wheelbase
3.5m

loadbase
2.2m

unladen weight
on front axle
= 1500kg

unladen weight
on rear axle
= 3000kg

1. Rear axle weight calculation:

$$\frac{\text{loadbase}}{\text{wheelbase}} \times \text{payload} \quad = \quad \text{payload weight on rear axle}$$

$$= \frac{2.2\text{m}}{3.5\text{m}} \times 9000\text{kg} \quad = \quad 5657\text{kg}$$

2. Front axle weight calculation:

Payload — rear axle weight = payload weight on front axle

= 9000kg — 5657kg = 3343kg

3. Total vehicle weight distribution:

Rear axle: payload weight + rear axle unladen weight
= 5657kg + 3000kg = 8657kg

Front axle: payload weight + front axle unladen weight
= 3343kg + 1500kg = 4843kg

Length and Width of Loads

Loads on normal goods vehicles (ie which come within the C&U regulations) in excess of the actual vehicle dimensions of length and width may be carried provided that certain special conditions are met, as indicated here. Details of the requirements relating to abnormal and projecting loads are given in Chapter 22.

Length

When moving a vehicle which complies with C&U regulations and its load is more than 18.65 metres long, the police must be notified 2 clear days in advance and a statutory attendant must also be carried. If a long load is carried on an articulated vehicle which is specially designed to carry long loads but in all other respects complies with the C&U regulations the 18.65 metre dimension is measured excluding the length of the tractive unit. Notice must also be given to the police and an attendant carried if the combination of a number of vehicles carrying one load is more than 25.9 metres long.

An overall limit of 27.4 metres is set for the length of a trailer and its load (the length of the towing vehicle is excluded from this dimension) above which movement can only be allowed by special order from the Secretary of State for Transport.

Further details of the requirements for markers, police notification and attendants on long loads are given in Chapter 22.

Width

The overall width of a normal load (ie not an indivisible load – see p 366 for definition) carried on a vehicle complying with the C&U regulations must not be more than 2.9 metres. The load itself must not project more than 305 millimetres on either side of the vehicle. There is an exception to this requirement when loose agricultural produce is carried.

If an indivisible load is carried on a normal vehicle which complies in all respects with the C&U regulations and the load is more than 2.9 metres wide, 2 days' notice must be given to the police of every district through which it is to pass. If such a load exceeds 3.5 metres width then the police must be notified as stated and an attendant must be carried.

If the load on a vehicle is more than 5 metres wide, 2 days' notice must be given to the police, a statutory attendant must be carried and a Special Order must be obtained from the Secretary of State for Transport on form VR1.

Loads in London

Before moving loads in central London which are more than 2.6 metres wide or 10.98 metres long or one and a quarter times the length of the vehicle, between 10.00 am and 7.00 pm on weekdays, police permission must be obtained.

Animals

New British legal requirements under The Welfare of Animals During Transit Order 1992 (which enforce in Britain the requirements of the EU directive, the Protection of Animals During Transport – EC Directive 628/1991) came into effect on 1 January 1993. This order requires livestock hauliers to register with the Ministry of Agriculture, Fisheries and Food (MAFF) and comply with strict rules to ensure that animals do not suffer distress or discomfort during transport by road and that the risk of the spread of disease is minimised. In the event of an outbreak of foot and mouth disease regulations place severe restrictions on the movement of cloven-hoofed animals. Pigs are not allowed to be moved away from sale premises without a local authority inspector licensing the movement. Separate regulations cover the carrying of horses and calves.

In general, both the British and the EU rules require that vehicles used for carrying animals should have non-slip floors with footholds, and the interior should not have any projections such as bolt-heads. Adequate ventilation is a necessity, proper loading ramps must be used and a roof must be provided to protect animals from the weather – including for the carriage of sheep on the top deck of multi-deck transporters. An exemption to this applies where the journey does not exceed 50 kilometres or where farmers are seasonally moving their own stock (eg between grazing or stock-holding locations). Vehicles must not be overcrowded and animals must not be mixed. Even horses of different breeds must be separated. There are minimum heights laid down for the decks of two-tier vehicles used for carrying sheep and pigs.

Those in charge of animals being transported must ensure that they have been provided with water and appropriate food (suitable for the species) before the journey, and water, food and rest at suitable intervals during the journey – in the case of sheep, pigs, cattle and horses this means at intervals of no longer than 15 hours. Other provisions apply when domestic dogs and cats, poultry and rabbits are transported.

Vehicles must be cleaned and disinfected after animals have been carried and before the next lot are loaded unless more than one load of the same class of animals is carried on the same day between the same points. It is an offence under the Movement and Sale of Pigs Order 1975 to use any vehicle whether owned by a farmer or hired which has been carrying pigs, unless the vehicle is cleansed and disinfected immediately after the pigs have been unloaded and before it is used again for carrying pigs or any other animal or thing. The disinfectant to be used for this purpose must be one approved officially for swine vesicular disease.

Different categories of pigs must not be carried together in the same vehicle. It is important to ensure that all pigs being picked up from one or more farms have the same kind of licence, or are slaughter pigs clearly marked with a red cross being moved to a slaughter market or slaughterhouse.

During the movement of animals, each consignment must be accompanied by a certificate signed by the person having control of the transport undertaking and giving the following details:
−   the origin and ownership of the animals;
−   their place of departure and destination;
−   the date and time of departure.

The operator must draw up plans for any journey exceeding 24 hours showing the arrangements for resting, feeding and watering the animals, and any alternative arrangements should the journey be changed or disrupted. This plan must be carried on the journey.

Copies of both the certificate and any journey plan required as stated above must be kept for a period of at least 6 months following the end of the journey and must be produced at the request of an inspector.

Food

Special regulations (the Food Hygiene [Markets, Stalls and Delivery Vehicles] Regulations 1966) apply to vehicles used for the carriage of food, excluding milk and drugs. All mobile shops and food delivery vehicles and the equipment carried by such vehicles must be constructed and maintained so that the food carried can be kept clean and fresh.

The driver of a food vehicle should wear clean overalls and if meat or bacon sides are carried, which the driver has to carry over his shoulder, he should wear a hat to prevent the meat touching his hair. He must not smoke while loading or unloading or serving the food (but he may do so in the cab of the vehicle if this is separate from the part of the vehicle in which the food is carried), and any cuts or abrasions on his hands must be covered with waterproof dressings.

If a driver or any other person concerned in the loading and unloading of food develops any infectious disease, his employer must notify the local authority health department immediately.

Food vehicles must have the name and address of the person carrying on the business shown on the nearside and the address at which the vehicle is garaged if this is a different address. If, however, a vehicle based in England or Wales has a fleet number clearly shown and is garaged at night on company premises, then the garage address is not required.

A wash hand basin and a supply of clean water must be provided on vehicles which carry uncovered food (except bread) unless the driver can wash his hands at both ends of his journey before he has to handle the food. When meat is carried, soap, clean towels and a nail brush must be provided on the vehicle. In Scotland all food-carrying vehicles must be provided with these items.

Mobile shops and food delivery vehicles must not be garaged with food still inside unless it can be kept clean.

An authorised officer of a council may enter and detain (but not stop while it is in motion) any food-carrying vehicle except those owned by British Rail or vehicles operated by haulage contractors.

The carriage of meat is covered by the Food Hygiene (General) Regulations 1970, and these require high standards of cleanliness, both for personnel involved in transporting and handling meat, and for vehicles used to carry meat.

Perishable Food

A whole range of legislation covers the carriage of food and particularly perishable food both in the UK and in Europe. When perishable foodstuffs are carried on international journeys to and through Austria, Belgium, Bulgaria, the CIS, the Czech Republic and Slovakia, Denmark, Finland, France, Germany, Italy, Luxembourg, Morocco, the Netherlands, Norway, Poland, Spain, Sweden and Yugoslavia the conditions of the Agreement on the International Carriage of Perishable Foodstuffs (known as the ATP agreement) must be observed.

The ATP agreement also applies in Britain (viz, the International Carriage of Perishable Foodstuffs Act 1976) and under its provisions vehicles used to carry perishable food must be constructed and tested (at 6-yearly intervals) to certain specified standards and must display an ATP approval plate to this effect. The principal requirement is that vehicles carrying certain specified perishable foodstuffs must comply with the body and temperature control equipment test standards and must be certified to this effect. The ATP agreement applies broadly to quick frozen, deep frozen, frozen and non-frozen foodstuffs but not fresh vegetables and soft fruit.

Information relating to food transport may be obtained from the Department of Health which has published a number of Codes of Practice, and from the Ministry of Agriculture, Fisheries and Food and the Royal Society of Health.

Chilled Food Controls (ie Temperature Controlled Food)

Following a series of food poisoning (salmonella and listeria) outbreaks in 1989 changes in legislation resulted in the introduction of the Food Hygiene (Amendment) Regulations 1990 which control the maximum temperature at which chilled food can be transported. The regulations apply to goods vehicles exceeding 7.5 tonnes gross weight carrying 'relevant food' products (which are listed in the schedules to the regulations). Such vehicles must have equipment capable of maintaining the temperature of the food at or below the specified temperature (from 1 April 1993, −5°C for certain foods as listed in the schedules and −8°C for other 'relevant' foods) or at or above 63°C as appropriate.

Vehicles up to 7.5 tonnes used for local deliveries from 1 April 1992 must be capable of keeping relevant food at or below the specified temperature or at or above 63°C as appropriate. However, in a case where the regulations specify that food is to be kept at 5°C, provided it is not kept in the vehicle for more than 12 hours, it may be at a higher temperature not exceeding 8°C.

There are exemptions to these requirements where food is to be sold within 2 hours (ie if prepared at 63°C or over) or 4 hours (ie if prepared below 63°C). Other variations (ie not exceeding 2°C for a maximum of 2 hours) may be permitted in certain specified circumstances such as variations in processing; while equipment is defrosted; during temporary breakdown of equipment; while moving food from one place or vehicle to another or any other unavoidable reason.

In due course EU legislation (ie commonly referred to as 'The Quick Frozen Food Directive') will require refrigerated transport operators to fit temperature recorders to vehicles carrying frozen foodstuffs (but not chilled foods). In the meantime the UK Quick Frozen Food (QFF) Regulations are effective in setting general conditions for the quality and use of equipment for storing and transporting food labelled 'Quick-Frozen' (but not ice-cream). These regulations require certain temperatures to be maintained within a percentage range of –18°C, or colder, as follows:

- during transport (other than local distribution) a tolerance for brief periods of 3°C (but not warmer than –15°C);
- during local distribution – no restriction at present but, after 9 January 1997, 6°C (but not warmer than –12°C).

Waste Food

Vehicles used to collect unprocessed waste food, intended for feeding to livestock and poultry, must be drip-proof, covered and enclosed with material capable of being cleansed and disinfected.

Vehicles must be thoroughly cleansed and disinfected on the completion of unloading. No livestock or poultry or foodstuff or anything intended for use for any livestock or poultry may be carried in any vehicle which is carrying unprocessed waste food intended for feeding to livestock or poultry. Furthermore, processed and unprocessed waste food may not be carried in the same vehicle at the same time.

Sand and Ballast Loads

The movement of sand and ballast comes to the attention of the Department of Trade and Industry and Trading Standards Inspectors principally because under the Weights and Measures Act 1985 these materials must be sold in weighed quantities (normally by volume in metric measures – in multiples of 0.2 cubic metres) and carried in calibrated vehicles which display a stamp placed on the body by the Trading Standards department of the local authority.

When sand and ballast (including shingle, ashes, clinker, etc) loads are carried, the driver must have a signed note (ie conveyance note) from the supplier indicating the following facts:
1. The name and address of the sellers.
2. The name of the buyer and the address for delivery of the load.
3. A description of the type of ballast.
4. The quantity by net weight or by volume.

5. Details of the vehicle.
6. The date, time and place of loading the vehicle.

The document containing these details must be handed over to the buyer before unloading, or if he is not there it must be left at the delivery premises. Where a delivery is to be made to two or more buyers each must be given a separate document containing the details shown above. Similar requirements apply to the carriage of ready-mixed cement.

Solid Fuel Loads

When solid fuel is carried a document giving similar details to those mentioned above for sand and ballast carrying must be held by the driver of the vehicle. When solid fuel is carried for sale in open sacks a notice on the vehicle in letters at least 60mm high must contain the following words: 'All open sacks on this vehicle contain 25kg or 50kg'.

Container Carrying

Container carrying has come to the fore in recent years, and while there are no specific regulations on this subject apart from the general safety provisions of the C&U regulations and the vehicle weight limitations, the authorities are concerned about the dangers arising from inadequate securing of containers. Containers, ideally, should be carried only on vehicles fitted with proper twistlocks or, failing this, should be secured by chains of sufficient strength with tensioners for adjustment and taking account of the recommended strength of restraint systems given in the DoT Code of Practice *Safety of Loads on Vehicles* (see pp 336–7).

The use of ropes for securing containers should be avoided because the corner castings through which they are passed are rough and will cut through the rope. Containers where possible should be loaded against the headboard and directly on the platform of the vehicle, not on timber packing which is liable to move.

Under the Freight Containers (Safety Convention) Regulations 1984 owners and lessees and others in control of freight containers used or supplied must ensure that they comply with the conditions of use as stated in the International Convention for Safe Containers 1972 (see also Chapter 20).

Fly Tipping

The Control of Pollution (Amendment) Act 1989, implemented in 1991, tightens up existing measures to prevent the illegal tipping (fly tipping) of waste, demolition rubble and such materials in unauthorised places. Among the specific measures legislated for under regulations called The Controlled Waste (Registration of Carriers and Seizure of Vehicles) Regulations 1991 are requirements for registration of waste transporters and tipper operators, the licensing of authorised operators, restriction of tipping to authorised sites and further powers to impound (and possibly sell) vehicles belonging to operators who dump loads illegally (such powers also exist under the Criminal Justice Act 1988).

NB: See Chapter 23 for more information on the carriage and disposal of waste materials.

22: Loads – Abnormal and Projecting

The construction and use (C&U) regulations, as previously described in Chapter 13, relating to the lengths, widths and weights of vehicles do not apply to heavy vehicles specially designed, constructed and used solely for the carriage of abnormal indivisible loads (commonly known as 'Special Types' vehicles). Certain conditions apply to these Special Types vehicles when abnormal indivisible loads are carried under the provisions of the Motor Vehicles (Authorisation of Special Types) General Order 1979 and 1981, and the Amendment Order 1987 which raised the threshold above which this legislation applies to 38 tonnes from 1 October 1989, divide such vehicles into three weight categories and set speed limits for such operations. This chapter details the legal requirements that apply when abnormal loads are carried and for the carriage of projecting loads.

Abnormal Indivisible Loads

For the purpose of regulations, abnormal indivisible loads (sometimes abbreviated to AILs) are loads which cannot, without undue expense or risk of damage, be divided into two or more loads for the purpose of carriage on the road and which cannot be carried on a vehicle operating within the limitations of the C&U regulations as described in Chapter 13.

Number of Abnormal Loads

While normally the carriage of only one abnormal load is permitted, two such abnormal loads may be carried on one vehicle within Category 1 or Category 2 (see below) provided the loads are from the same place and are destined for the same delivery address.

Engineering Plant
In the case of engineering plant, such plant and parts dismantled from it may be carried on the same vehicle (ie to constitute more than one or two loads) provided that the carriage of the parts does not cause the overall dimensions of the vehicle and the main load to be exceeded and that the parts are loaded and discharged at the same place as the main load.

Special Types Vehicles

Dimensions

Width
Special Types vehicles, locomotives and trailers and their loads are normally

permitted to be up to 2.9 metres wide but, if necessary to ensure the safe carriage of large loads, they may be up to a maximum of 6.1 metres wide.

Length

The overall length of a Special Types vehicle and its load must not exceed 27.4 metres which applies normally but where the abnormal load is carried on a combination of vehicles and trailers or on a long articulated vehicle the dimension of 27.4 metres is measured excluding the drawing vehicle.

Weight

The permissible maximum weight of a Special Types vehicle must not exceed 150,000kg. There is a limit on the maximum weight which may be imposed on the road by any one wheel of the vehicle, of 8250kg and the maximum weight imposed by any one axle must not exceed 16,500kg. These limits may be exceeded only if authorisation by Special Order is obtained from the Secretary of State for Transport (see p 371).

Vehicle Categories

The Special Types Order specified three separate weight categories for abnormal load vehicles as follows:

Category 1 – up to 46 tonnes gcw
Category 2 – up to 80 tonnes gcw
Category 3 – up to 150 tonnes gcw

Category 1 Vehicles

Vehicles within this category will normally fall within the C&U regulations in regard to permissible maximum weight, axle spacings and axle weights but where it is a five-axle articulated vehicle the weight may exceed 38 tonnes up to a maximum of 46,000kg (ie 46 tonnes) provided the following minimum relevant axle spacings are observed:

Relevant axle spacing	Maximum Weight
At least 6.5 metres	40,000kg
At least 7.0 metres	42,000kg
At least 7.5 metres	44,000kg
At least 8.0 metres	46,000kg

Category 2 Vehicles

Vehicles within this category may operate up to a maximum weight of 80,000kg (ie 80 tonnes) but they must have a minimum of five axles with a maximum weight of 50,000kg on any group of axles and they must meet the minimum axle spacing requirements specified below.

Individual wheel and axle weight weight limits are as follows:

Distance between adjacent axles	Maximum axle weight	Maximum wheel weight
At least 1.1 metres	12,000kg	6000kg
At least 1.35 metres	12,500kg	6250kg

Minimum axle spacings and applicable maximum weights are as follows:

Distance between foremost and rearmost axles	Maximum weight
5.07 metres	38,000kg
5.33 metres	40,000kg
6.00 metres	45,000kg
6.67 metres	50,000kg
7.33 metres	55,000kg
8.00 metres	60,000kg
8.67 metres	65,000kg
9.33 metres	70,000kg
10.00 metres	75,000kg
10.67 metres	80,000kg

Category 3 Vehicles

A minimum of six axles is needed on Category 3 vehicles operating up to a permissible maximum weight of 150,000kg (ie 150 tonnes) with a limit of 100,000kg on any group of axles or 90,000kg on any group of axles where the distance between adjacent axles is less than 1.35 metres. Vehicles in this category must meet the minimum axle spacing requirements specified below.

Individual wheel and axle weight limits are as follows:

Distance between adjacent axles	Maximum axle weight	Maximum wheel weight
At least 1.1 metres	15,000kg	7500kg
At least 1.35 metres	16,500kg	8250kg

Minimum axle spacings and applicable maximum weights are as follows:

Distance between foremost and rearmost axles	Maximum weight
5.77 metres	80,000kg
6.23 metres	85,000kg
6.68 metres	90,000kg
7.14 metres	95,000kg
7.59 metres	100,000kg
8.05 metres	105,000kg
8.50 metres	110,000kg
8.95 metres	115,000kg
9.41 metres	120,000kg
9.86 metres	125,000kg
10.32 metres	130,000kg
10.77 metres	135,000kg
11.23 metres	140,000kg
11.68 metres	145,000kg
12.14 metres	150,000kg

VED for Special Types Vehicles

A separate VED taxation class applies to Special Types vehicles – see Chapter 8 for details.

Braking Standards

Vehicles operating within Category 1 must meet the C&U regulation braking standards requirements and those operating within Categories 2 and 3 from 1 October 1989 must meet the EU braking standards (ie EC 230/71, 132/74, 524/75, 489/79).

Identification Sign

Vehicles operating under the Special Types Order must display an identification sign at the front. This sign on a plate at least 250mm x 400mm will have white letters on a black background as follows:

STGO	letters 105mm high
CAT...	letters and figures 70mm high

NB: A figure 1, 2 or 3 must follow the word 'CAT' as appropriate, depending on the category of vehicle.

Special Types Plates

Vehicles falling within Category 2 and 3 which have been manufactured since 1 October 1988 must display Special Types plates (in a conspicuous and easily accessible position) showing the maximum operational weights recommended by the manufacturer when travelling on a road at varying speeds as follows: 12, 20, 25, 30, 35, 40mph. The weights to be shown are the permissible maximum gross and train weights and the maximum weights for each individual axle. Plates on trailers (including semi-trailers) must show the permissible maximum weight for the trailer and the maximum-weights for each individual axle. The plates must be marked with the words 'Special Types Use'.

Speed Limits

Maximum permitted speeds* are specified for Special Types vehicles as follows:

Vehicle category	Motorways	Dual-carriageways	Other Roads
1. Up to 46 tonnes	60mph	50mph	40mph
2. Up to 80 tonnes	40mph	35mph	30mph
3. Up to 150 tonnes	30mph	25mph	20mph

Speeds for Wide Loads
Vehicles carrying loads over 4.3 metres but not over 6.1 metres wide are restricted to 30mph on motorways, 25mph on dual-carriageways and 20mph on other roads.

NB: It is important to note that tyre equipment on vehicles and trailers must be compatible with both the gross weight of the vehicle and the authorised maximum speed of operation.

Attendants

An attendant must be carried on Special Types vehicles:
1. when the vehicle or its load is more than 3.5 metres wide;
2. if the overall length of the vehicle is more than 18.3 metres (not including the length of the tractive unit in the case of articulated vehicles);
3. if the length of a vehicle and trailer exceeds 25.9 metres;
4. if the load projects more than 1.83 metres beyond the front of the vehicle;
5. if the load projects more than 3.05 metres beyond the rear of the vehicle.

If three or more vehicles carrying abnormal loads or other loads of dimensions that require statutory attendants to be carried travel in convoy, attendants need only be carried on the first and last vehicles in the convoy.

Police Notification

The police of every district through which a Special Types combination is to be moved must be given 2 clear days' notice (excluding Saturdays, Sundays and bank holidays in the case of notification under the STGO and excluding Sundays and bank holidays for notification under the C&U regulations) if:
1. the vehicle and its load is more than 2.9 metres wide;
2. the vehicle and its load (or trailer and load) is more than 18.3 metres long;
3. a combination of vehicles and trailers carrying the load is more than 25.9 metres long;
4. if the load projects more than 3.05 metres to the front or rear of the vehicle;
5. the gross weight of the vehicle and the load is more than 80,000kg.

The notice given must include details of the vehicle and the weight and dimensions of the load, the dates and times of the movement through the police district and the proposed route to be followed through that district.

In the case where notice has been given to the police of the movement of an abnormal load, they have the power to delay the vehicle during its journey if it is holding up other traffic or in the interests of road safety.

NB: At the time of preparing this edition of the Handbook *there is talk of privatising the escort duties for abnormal loads movements. Normally a police function, this task could be given over to private escort firms to relieve the police for more important work and to free the taxpayer from the burden of subsidising what is, after all, a commercial operation.*

Notification of Highway and Bridge Authorities

If a Special Types vehicle and its load weighs more than 80,000kg, or the weight imposed on the road by the wheels of such a vehicle exceeds the maximum limit laid down in the C&U regulations (see pp 217–19), 5 clear days' notice must be given to the Highway and Bridge Authorities for the areas through which the vehicle is to pass. Two days' notice is required when only the

C&U axle weight limit is exceeded. The operator of such a vehicle is also required to indemnify the Authorities against damage to any road or bridge over which it passes. These requirements also apply to vehicles which exceed the C&U gross weight limits and those which exceed their plated axle weight limits, plus mobile cranes and engineering plant which exceed the limits specified.

Stopping on Bridges
The driver of a vehicle carrying an abnormal load must ensure that no other such vehicle and load are on a bridge before he drives on to the bridge and once on the bridge he must not stop unless forced to do so.

Where a vehicle weighing more than 38,000kg gross has to stop on a bridge for any reason it must be moved off the bridge as soon as possible. If it has broken down, the advice of the bridge authority (usually the Highways Department of the local authority) must be sought before the vehicle is jacked up on the bridge. In the event of damage being caused to a road or bridge by the movement of a heavy or large load over it the highway authority can take steps to recover from the vehicle operator the costs of repairing the damage.

Special Orders

Written approval has to be obtained from the Secretary of State for Transport in cases where a Special Types vehicle and its load exceeds 5 metres in width. Application is made on form VR1 (ie the movement order) and a copy of this completed form must be carried on the vehicle. The route specified, and the date and timings for the journey notified in the application, must be adhered to, otherwise further approval will have to be sought.

Dump Trucks

When dump trucks (ie vehicles designed for moving excavated material) are used on the road the maximum permissible gross weight is limited to 50,800kg and the maximum axle weight allowed for such vehicles is 22,860kg. They must not exceed a speed of 12mph on normal roads and an attendant must be carried if the width exceeds 3.5 metres. If a dump truck is more than 4.3 metres wide, permission in the form of a Special Order from the Secretary of State for Transport is required before it is moved. Where three or more such vehicles over 3.5 metres wide travel in convoy only the first and last vehicles need carry attendants.

Other Plant

The Special Types General Order also makes special provision for other items of plant such as grass cutting machines and hedge trimmers, track-laying vehicles, pedestrian-controlled road maintenance vehicles, vehicles used for experimental trials, straddle carriers, land tractors used for harvesting, mechanically propelled hay and straw balers, vehicles fitted with moveable platforms and engineering plant. Any person proposing to move such items on public roads is advised to check that the appropriate legal requirements are met.

High Loads

While there are no general legal height restrictions on vehicles or loads (except those specific instances mentioned on pp 216–17), clearly vehicles must be loaded so that they can pass under bridges on the routes to be used.

In particular it should be noted that motorway bridges are built to give a clearance of 16ft to 16ft 6ins and overhead power cables crossing roads are set at a minimum height of 19ft (5.8 metres) where the voltage carried does not exceed 33,000 volts and at 6 metres where this voltage is exceeded (see also note on p 353 about tipping vehicle dangers in regard to contact with overhead power cables). When planning to move loads above 19ft it is a legal requirement to make contact with the National Grid Company or the appropriate regional electricity distribution company (ie previously the regional electricity boards) beforehand. It is also recommended that contact should be made with British Telecom regarding the presence of its overhead lines on the routes to be used. Some police forces now report that they will not action notification of abnormal load movements until both National Power and British Telecom have been notified.

Projecting Loads

Projecting loads may be carried on normal C&U regulation vehicles as well as on Special Types vehicles as described above. A projecting load is one that projects beyond the foremost or rearmost points or the side of the vehicle and, depending on the length or width of the projection, certain conditions apply when such loads are carried on normal vehicles including requirements for lighting and marking.

Side Projections

Loads more than 4.3 metres wide cannot be moved under the C&U regulations although they can be carried on vehicles which comply with those regulations. In such cases the provisions of the Special Types General Order apply (see above).

The normal width limit for loads is 2.9 metres overall or 305mm on either side of the vehicle except in the case of loose agricultural produce and indivisible loads. Where an indivisible wide load extends 305mm or more on one or both sides of the vehicle or exceeds 2.9 metres, the police must be given 2 clear days' notice, and if it exceeds 3.5 metres, police notification, end marker boards and an attendant are required.

Forward Projections

A load projecting more than 2.0 metres beyond the foremost part of the vehicle (on C&U vehicles) must be indicated by an approved side and end marker board (Figure 22.1) and an attendant must be carried. On Special Types vehicles only, where the load projects more than 1.83 metres to the front, an end marker must be displayed and an attendant carried.

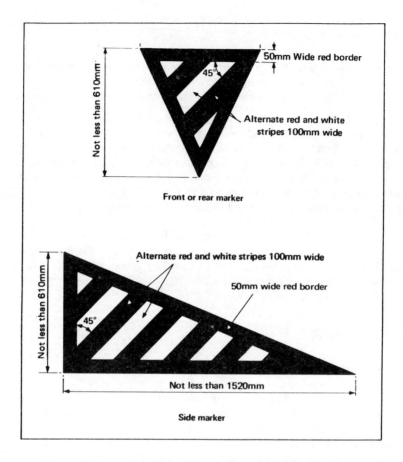

Figure 22.1 *The approved type marker boards which must be displayed when projecting loads are carried*

If a load projects more than 3.05 metres beyond the front (on both C&U and Special Types vehicles), the police must be given 2 days' notice of its movement, both side and end approved marker boards must be displayed and an attendant must be carried.

If a load projects more than 4.5 metres beyond the front of the vehicle the provisions mentioned above must be observed and additional side marker boards must be carried within 2.5 metres of the first set of side markers.

Rearward Projections

Where a load projects more than 1.0 metre beyond the rear of a C&U vehicle (1.07 metres on Special Types vehicles) it must clearly be marked (the form which this must take is not specified, but a piece of rag tied to the end is usually sufficient). If the real projection is more than 2.0 metres on C&U vehicles (1.83 metres on Special Types vehicles) an end marker board must be displayed.

Where the rearward projection is more than 3.05 metres the police must be notified, an attendant carried and approved side and end marker boards displayed. An end marker is not required if the projecting load is fitted with a rear reflective marker.

If the rearward projection exceeds 5.0 metres additional side marker boards must be carried within 3.5 metres of the first set of side markers.

Marker Boards

Marker boards carried in accordance with the requirements described above must conform to the dimensions and colours shown in Figure 22.1 and they must be indirectly illuminated at night.

Lighting on Projecting and Long Loads

Rearward Projections

When carrying a load projecting more than 1.0 metre beyond the rear end of the vehicle an additional red rear position light must be carried within 1.0 metre of the end of the load or if the projecting load covers the rear lights and reflectors of the vehicle, additional lights and reflectors must be fixed to the load. This is usually best accomplished by having a complete lighting set and reflectors fitted to a board which can be fixed to the load.

Side Projections

When a load projects sideways more than 400mm beyond the front and rear position lights of a vehicle, side position lights must be carried within 400mm of the outer edges of the load and additional rear lights must also be carried within 400mm of the outer edges of the load. White front and red rear reflectors must also be carried within 400mm of the outer edges of the load.

Long Vehicles

A vehicle or combination of vehicles which, together with their load, are more than 18.3 metres long must, when on the road during the hours of darkness, carry side marker lights on each side positioned within 9.15 metres of the front of the vehicle or load and within 3.05 metres of the rear of the vehicle or load and other lights positioned between these at not more than 3.05 metre intervals. These requirements do not apply if approved illuminated marker boards are carried or if the combination is formed of a towing vehicle and a broken-down vehicle.

In the case of a combination of vehicles carrying a supported load (a load not resting on a vehicle except at each end) when the total length of the combination exceeds 12.2 metres but not 18.3 metres, side marker lights must be carried when on the road during the hours of darkness, positioned not more than 1.53 metres behind the rear of the drawing vehicle and if the load extends more than 9.15 metres beyond the drawing vehicle an additional side marker light must be carried not more than 1.53 metres behind the centre line of the load.

23: Loads – Dangerous, Explosive and Waste

Five sets of regulations currently cover most of the legal requirements relating to the packaging, labelling and carriage by road of dangerous and explosive substances whether in bulk loads or small consignments. The relevant legal requirements are to be found in the following regulations (in abbreviated form):
1. The packaging and labelling regulations
2. The bulk tanker regulations
3. The carriage of dangerous substances in packages regulations
4. The explosives regulations
5. The driver training regulations.

Progressively, UK regulations in this field are moving towards the framework of the ADR agreement (see p 137) which relates to the international carriage of dangerous goods by road and which is less strict than current British legal requirements.

Under proposals put forward by the United Nations Economic Commission for Europe, transport operators who carry dangerous substances could be held liable to pay compensation for any damage or pollution resulting from spillages or vehicle accidents. Specifically, any carrier who, in transporting dangerous substances causes loss of life, personal injury, loss or damage to property or damage to the environment by contamination will be liable to meet compensation claims from victims who, because the legal principle of *strict* liability will apply, will not have to prove that the carrier is liable.

United Nations Classes and Packing Groups for Dangerous Goods

It is useful for the purposes of understanding the regulations to identify first the nine Classes into which the United Nations has divided dangerous substances and then the gradings (ie Packing Groups) by which the risks are identified.

The nine Classes are as follows:

Class 1   Explosives
Class 2   Compressed gases
Class 3   Flammable liquids
Class 4   4.1 Flammable solids
           4.2 Spontaneously combustible substances
           4.3 Dangerous when wet substances
Class 5   5.1 Oxidising agents
           5.2 Organic peroxides
Class 6   6.1 Toxics
           6.2 Infectious substances

Class 7   Radioactive

Class 8   Corrosives

Class 9   Miscellaneous substances.

The UN Packing Groups are as follows:

Group I   materials which are highly dangerous (ie which will kill rapidly)

Group II   medium danger substances (ie which can kill but take longer to do so)

Group III of lesser danger (which will not kill but could cause serious injury).

Packaging and Labelling

The Classification, Packaging and Labelling of Dangerous Substances Regulations 1984 (commonly referred to as the CPL regulations) which came into effect on 1 January 1986 were replaced in September 1993 by The Chemicals (Hazard Information and Packaging) Regulations 1993 (known as the CHIP regulations). These in turn have now been further replaced by The Carriage of Dangerous Goods by Road and Rail (Classification, Packaging and Labelling) Regulations 1994 (referred to as the CDG-CPL regulations) which came into effect on 1 April 1994.

These regulations serve a number of purposes:
1. They prepare the way for the introduction of the international ADR agreement on the carriage of dangerous goods by road as the basis for national legislation.
2. They make consequential amendments for the above purpose to existing UK legislation, namely the carriage in packages, carriage in road tankers and driver training regulations as described in detail in this chapter.
3. They impose requirements relating to the classification, packaging (in accordance with United Nations [UN] specifications) and labelling of dangerous goods for carriage by both road and rail modes of transport.

The regulations prohibit the supply or consignment of dangerous goods unless the classification for the goods and the specified particulars have been ascertained in accordance with legal requirements and the specified particulars are shown on the package. Suppliers and consignors may use only receptacles which are designed, constructed and maintained to prevent leakage when being handled and moved; to use receptacles of suitable material which will not be adversely affected by the chemical properties of the substances put into them; and to use only receptacles with replaceable closing devices where these devices are capable of being repeatedly reclosed without leakage of the contents.

Packages and receptacles containing substances must be properly labelled (on the container or receptacle and on any outer packaging in which they are enclosed) to show the following information:

1. Name and address of manufacturer, importer, wholesaler or supplier
2. The official designation of the substance
3. Indications of the nature of risk associated with the substance
4. The risk and safety phrases associated with the substance
5. Additional information relating to pesticides.

The regulations specify details concerning the size and nature of the labels to be used, that labels must be securely fixed and capable of being clearly seen against the background of that package, even if the package is not in its usual position (ie if it is turned on its side or upside down).

Dangerous Loads in Bulk Tankers/Tank Containers

The Dangerous Substances (Conveyance by Road in Road Tankers and Tank Containers) Regulations 1992 (replacing regulations under the same title dated 1981) are concerned with the carriage by road of dangerous substances in bulk either in conventional road tanker vehicles or in tank containers which can be lifted on and off road vehicles and conveyed by other transport modes (eg rail and ship).

The regulations are obviously important in their own right but there are four very important pieces of supporting literature. The first is the *Approved Substance Identification Numbers, Emergency Action Codes and Classifications for Dangerous Substances Conveyed in Road Tankers and Tank Containers* published by the Health and Safety Executive and commonly referred to as the 'Approved List'. The second document is the *Operational Code of Practice* which explains to transport operators what the regulations are really all about and, more particularly, it indicates to them in very clear language precisely what they must and must not do to comply with the law.

The third document is the *Code of Practice on Road Tanker Testing* and the fourth is the *Guide to the Regulations*. Additionally, there is a Health and Safety Executive explanatory booklet called *Transport of Dangerous Substances in Tank Containers* and another Code of Practice dealing with the construction of road tankers. All of these documents can be obtained from Her Majesty's Stationery Office.

Definitions

An understanding of the definitions of the main terms used is important to appreciate how the law applies.

Conveyance by Road

The term 'conveyance' which is used in the title of the regulation is precisely defined and it means rather more than would be generally expected in that it applies 'from the commencement of loading . . . until the (tank) has been cleaned or purged . . . whether or not the vehicle is on a road at the material time'. During the whole of this time the full weight of the legislation applies so that, for example, if a trained driver delivered a load of chemicals but another driver, untrained in the handling of that substance, drives the vehicle before the empty tank has been cleaned, then clearly an offence has been committed.

Tanker Operator

Operators of road tankers and other vehicles are defined in the regulations along with operators of tank containers. The 'operator' of a road tanker is the person who holds or who is required to hold an Operator's Licence under the provisions of section 60 of the Transport Act 1968. Where there is no requirement to hold an 'O' licence the operator is the keeper of the vehicle. The 'operator' of a tank container is defined as the tank owner or his agent or, in the absence of either of these, the operator of the vehicle on which the container is conveyed. In the case of leased or hired tanks the operator is the person to whom the tank is leased or hired. The significance of the definition of the term 'operator' is that it is he who is fully responsible for ensuring that the regulations are met.

Approved List

The List, in its up-to-date form (see below), shows for each substance the exact chemical name, the substance identification number, the emergency action code and the classification (eg flammable liquid, corrosive substance, toxic gas). The relevance of the identification numbers, emergency action codes and classification will be seen by reference to the text on the Hazchem marking scheme – see pp 383–4.

Dangerous Substance

'Dangerous substance' means a substance specified in the Approved List plus any substance which 'by reason of its characteristic properties creates a risk . . . which is comparable with the risk created by substances specified in the approved list'.

Road Tanker/Tank Container

For the purposes of the regulations, a road tanker is a goods vehicle with a tank structurally attached to or which is an integral part of the frame of the vehicle. A tank container is defined as being one (whether or not divided into separate compartments) which has a capacity greater than 3 cubic metres.

Restriction on Substances to be Carried

Strict limits are placed on the carriage of certain substances in that the regulations prohibit the carriage in a tanker or tank container of any organic peroxide unless the substance is named in the approved list or, if it is not named in the list, if its characteristic properties create no greater risk than other substances of similar characteristic properties which are specified in the list. The regulations also stipulate that if a substance is greater in concentration than that specified as the maximum concentration in the list it must not be carried.

Where a substance is not included in the Approved List it must not be carried. The omission may be due to the fact that the substance is too dangerous to travel in a road tanker or tank container or it may be because the substance is new and has not yet been classified. Advice in these circumstances can be obtained from the Health and Safety Executive.

Exemptions

The regulations do not apply to vehicles carrying non-dangerous bulk powder, granule or liquid goods (eg certain food products such as sugar and glucose, cement and sand, gas oil, derv and so on). They also do not apply to vehicles which are carrying dangerous goods in the following circumstances:

1. When engaged in international transport under the International Convention for the Conveyance of Goods by Rail (CIM).
2. When engaged in international transport under the European Agreement concerning the International Carriage of Dangerous Goods by Road (ADR).
3. When under the control of the armed forces of the Crown or of visiting forces.
4. When exempted from excise duty (VED) under the Vehicles Excise Act 1971, section 7(1).
5. When a road construction vehicle (other than a road tanker) is used for conveying liquid tar (ie substance identification number 1999 in the Approved List).

The regulations also do not apply when the substance being carried is a radioactive substance within the meaning of the Ionising Radiations (Unsealed Radioactive Substances) Regulations 1968 (SI 780/ 1958).

Condition of Tanker Vehicle/Tank Container

A road tanker vehicle or tank container must not be used for the conveyance of dangerous goods unless it is properly designed; is of adequate strength and good construction from sound and suitable material which would not be adversely affected by contact with the substance being carried. It must be designed, constructed and maintained so that the contents cannot escape and be suitable for the purpose for which it is being used. It must have adequate and correct ancillary fittings (eg flanges, fittings and hoses).

In accordance with the *Road Tanker Testing Code of Practice* the tank should be periodically tested and a tank certificate should have been issued indicating the substances that may be carried. Furthermore, following accident damage, repair or modification, the effects on the tank should be assessed and it should be re-tested and certified.

Information in Writing

Operators must not convey dangerous substances by road in a road tanker or tank container, unless they have first obtained from the consignor information which enables them to comply with the regulations and to be aware of the risks created by the substance to the health and safety of any person. It is the responsibility of the person supplying such information to ensure that it is accurate and sufficiently detailed.

Transport operators in turn must give to their vehicle drivers information *in writing* to enable them to know the identity of the substance, the nature of the dangers which may arise with that substance and the emergency action to be taken in the event of spillage, accident or other dangerous incident. The regulations place a duty on drivers to ensure that the written information they are given is kept in the cab of the vehicle and is readily available at all times while the substance in question is being carried.

While the legal requirement to supply information in writing can be satisfied by typing the details on a piece of the company's letter-headed paper, normally the most convenient method of providing drivers with written information on the products carried and which conforms with legal requirements is by means of commercially produced, re-usable, laminated 'Tremcards'. These transport emergency cards cover either a single product or a group of products and detail the nature of the hazard and describe what protective devices should be used or worn to deal with any spillage of the substance. The emergency action to be taken is detailed and first aid advice is also given on the card along with a reference telephone number. Standard Tremcards are prepared by the Conseil Europeen des Federations de l'Industrie Chimique (ie European Council of Chemical Manufacturers Federations, commonly known and referred to as CEFIC).

It is an important requirement that written information about substances which are not being carried at the time are removed from the vehicle, destroyed or placed in a 'securely closed container clearly marked to show that the information does not relate to a substance then being conveyed'. This is to avoid any confusion of the emergency services should they have to attend an incident involving spillage, leakage or a vehicle accident as to what substance is and is not currently being carried on the vehicle.

Safety Measures

The regulations specify precautions for preventing fire and ensuring that tanks are not overfilled. In particular, the operator should notify his drivers the maximum quantity of the dangerous substance to be loaded. There must be no smoking in the vicinity of the vehicle especially during loading and discharging and smoking materials and other sources of ignition such as torches and radios should also be kept well away at such times. The vehicle should be equipped with suitable serviceable fire extinguishers and with master switches and earthing leads which must be used at all times. The driver should be aware of all the potential dangers and should take all possible precautions to ensure the safety of the load, the vehicle, premises and, most importantly, all persons in the vicinity of the vehicle especially when on the public highway.

Parking of Dangerous Goods Vehicles

Dangerous goods vehicles must be parked in a safe place or they must be supervised at all times by the driver or another competent (ie trained) person over 18 years of age. This applies to tankers displaying hazard warning panels on which the emergency action code ends with the letter E. However, it does not apply if the driver can show that the tank is empty or that, if the identification code number 1270 is displayed, no petrol is being conveyed and the tank is empty; or if the number 1268 is displayed that no toluene or petroleum distillate (flash point less than 21°C) is being conveyed and the tank is empty.

Labelling of Road Tankers/Tank Containers

The regulations impose requirements for notices to be displayed on road tankers and tank containers which are being used to carry prescribed dangerous substances. References in the regulations (and this text) to hazard warning notices refer in fact to what are commonly known as Hazchem labels or markers.

The relevant substances to which the regulations refer and which are the subject of tanker labelling are specified in the Health and Safety Commission document *Approved Substance Identification Numbers, Emergency Codes and Classifications for Dangerous Substances conveyed in Road Tankers and Tank Containers* referred to earlier together with a substance identification number and appropriate emergency action code.

Emergency Action Codes
The emergency action codes are as follows:

1. By numbers 1 to 4 indicating the equipment suitable for fire fighting and for dispersing spillages (ie 1 = water jets, 2 = water fog, 3 = foam, 4 = dry agent)
2. By letters indicating the appropriate precautions to take as follows:

Letter	Danger of violent reaction	Protective clothing and breathing apparatus	Measures to be taken
P	Yes	Full protective clothing	Dilute
R	No	Full protective clothing	Dilute
S	Yes	Breathing apparatus	Dilute
S*	Yes	Breathing apparatus for fire	Dilute
T	No	Breathing apparatus	Dilute
T*	No	Breathing apparatus for fire	Dilute
W	Yes	Full protective clothing	Contain
X	No	Full protective clothing	Contain
Y	Yes	Breathing apparatus	Contain
Y*	Yes	Breathing apparatus for fire	Contain
Z	No	Breathing apparatus	Contain
Z*	No	Breathing apparatus for fire	Contain

*These symbols are shown as orange (or can be white) letters on a black background.

Where a letter 'E' is shown at the end of an emergency action code this means that consideration should be given to evacuating people from the neighbourhood of an incident.

The Hazchem Label
The Hazchem label must be orange with black borders, figures and letters except for the space where the individual hazard warning sign is placed which is white (see Figure 23.1). Hazard warning panels and Hazchem labels must be displayed at all times in accordance with the regulations, they must be kept clean and free from obstruction (except that a rear mounted panel may be

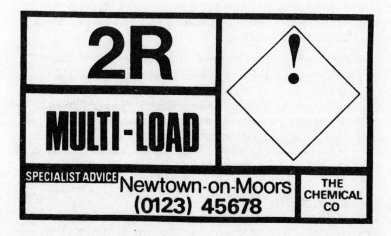

Figure 23.1 *Illustrations showing hazard warning labels for: top: full/single loads; bottom: multi-loads*

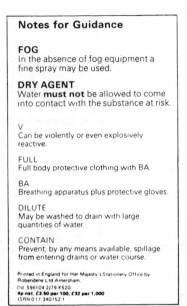

Figure 23.2 *Hazchem pocket reference card*

fitted behind a ladder 'of light construction which does not prevent the panel from being easily read') and when the tank is emptied of the substance indicated the label must be covered or removed. For cross reference a pocket card (Figure 23.2) is provided for firemen and other emergency service personnel which indicates the nature of the hazard and advises whether protective clothing and/or breathing apparatus should be used (see above for details of emergency action codes).

Hazchem labels are divided into a number of individual panels as follows:

1. Top left – a combined code which identifies the equipment to be used for fire fighting and dispersal of spillage (see codes above).
2. Second down on left – the substance identification number from the Approved List (also the name of the substance can be added).
3. On right – appropriate hazard warning symbol on white background.
4. Across bottom – telephone number for specialist advice and, if desired, the manufacturer's/supplier's name.

Single/Multi-Load Labelling

All relevant vehicles must be labelled in accordance with the regulations, depending on whether a single or multi-load is carried, as follows:

For Single loads: An orange/black Hazchem warning label containing the following details:

1. Emergency action code for the substance.
2. Substance identification number.
3. Appropriate hazard warning sign.
4. Telephone number where specialist advice can be obtained plus, if required:
   - the name of the substance;
   - the name of the manufacturer.

For Multi-loads: An orange/black Hazchem warning label containing the following details:
1. Appropriate multi-load emergency action code.
2. The word 'multi-load'.
3. The appropriate hazard warning sign.
4. Telephone number where specialist advice can be obtained.
5. If required, the name of the manufacturer.

Additionally, for multi-loads, on each tank or each compartment there should be a label showing:
1. The appropriate substance identification number (and the name if desired).
2. The hazard warning sign applicable to the contents of that tank or compartment where this is different from the hazards of any other substance on the vehicle.

Hazard Warning Signs/Symbols

Individual hazard warning signs or symbols (sometimes also called diamonds) are coloured as follows:

Description of signs	Symbol	Lettering	Background
Explosives	Black bomb blast	Black	Orange
Compressed gases	Black gas cylinder	Black	Green
Flammable liquids	Black flame	Black	Red
Flammable solids	Black flame	Black	White/vertical red stripes
Spontaneously combustible	Black flame	Black and white	White top, red bottom
Dangerous when wet	Black flame	Black	Blue
Organic agents	Black 'O' and flame	Black	Yellow
Oxidising peroxides	Black 'O' and flame	Black	Yellow
Toxics	Black skull/crossbones	Black	White
Infectious substances	Black symbol	Black	White
Radioactives	Black symbol	Black	Yellow top, white bottom
Corrosives	Black symbol	White	White top, black bottom
Miscellaneous	–	–	Black/white vertical stripes at top/white bottom

NB: It is useful to note that the Hazchem labels and some of the individual hazard warning signs are shown, in colour, in the new edition of the Highway Code.

Exemptions to Marking

There are certain exemptions that apply regarding the marking of vehicles. These are as follows:

1. If a vehicle is carrying dangerous substances transferred from another tanker or tank container which has broken down or has been damaged in an accident provided the vehicle is escorted by the police or fire brigade or alternatively if it displays the hazard warning signs shown to the regulations.
2. If the road tanker or tank container is travelling to a port for carriage by sea or from a port having been carried by sea and it is marked in accordance with the provisions of the International Maritime Dangerous Goods Code (IMDG Code) of the International Maritime Organisation (IMO).
3. If the road tanker or tank container is in the service of armed forces of the Crown or visiting forces engaged in manoeuvres or training subject to proper authority.

There is an exemption to the requirement for multi-load marking where such loads comprise only substances specifically listed in the approved list as being suitable for treatment as a single bulk load, in which case the bulk load labelling requirements will apply.

When a tanker or tank container has been emptied and cleaned, the hazard warning panels must be completely removed or completely covered except for the telephone number which may be left exposed. When two or more substances have been carried and one or more have been discharged and the respective tank compartments cleaned, the hazard warning labels referring to those substances must be removed or completely covered and the panels changed to those appropriate to the remaining load.

Driver Training

The Road Traffic (Training of Drivers of Vehicles Carrying Dangerous Goods) Regulations 1992, which implement in the UK the requirements of EC Directive

684/1989, require vocational training certificates for drivers of dangerous goods tankers and tank-container carrying vehicles over 3.5 tonnes gross weight (with capacities over 3000 litres) and explosives-carrying vehicles from 1 July 1992, and for drivers of other dangerous goods carrying vehicles from 1 January 1995. Such certificates will be issued after drivers have completed a suitable (government-approved) training course and passed an examination. These certificates will be renewable every 5 years following refresher training, but those drivers who have had at least 5 years' dangerous load driving experience prior to these dates will be exempt from the training requirement.

Full details of these dangerous goods driver training requirements are given in Chapter 7 of the *Handbook*.

Exemption Certificates

The regulations provide scope for Exemption Certificates to be issued by the Health and Safety Executive in special cases where there is a need because the regulations cannot be practically applied provided there is no risk to health or safety. A number of such certificates have been issued.

Enforcement

This legislation is enforced by two agencies, namely officers of the Health and Safety Executive and by the police in the course of their normal enforcement procedures against vehicles on roads.

Defence

Provision is made for any person to claim in his defence in any proceedings against him under the regulations that 'he took all reasonable precautions and exercised all due diligence to avoid . . .' committing the offence.

Risk Prevention Officers

Draft regulations being circulated by the Health and Safety Commission suggest the need for transport operators carrying more than 50 tonnes of dangerous goods a year to appoint a risk prevention officer in accordance with EU requirements. (*NB: This requirement was still not confirmed at the time this 1995 edition of the* Handbook *was in preparation.*)

The appointee will need to have passed an approved examination following specialised training and hold a relevant certificate, and his/her name will have to be notified to the authorities. Such person will have to be consulted by their employer prior to the purchase or hiring of vehicles and they will have the responsibility for checking the compatibility of vehicles and loads, the information provided in regard to dangerous loads, and the documentation prepared for and accompanying loads.

Conveyance of Dangerous Substances by Road in Packages

Regulations broadly similar to those described previously for the bulk carriage of dangerous substances by road, control the carriage of dangerous goods by road in packages, drums and other individual containers (referred to in the regulations as 'receptacles'). The Road Traffic (Carriage of Dangerous Substances in Packages etc) Regulations 1992, (commonly referred to as the 'PGR'), which replaced regulations of the same title enacted in 1985, apply:

1.   When self-reactive organic peroxides, flammable solids and other substances of similar risk (ie toxic gases, flammable gases, organic peroxides, asbestos or asbestos waste) are carried in any quantity and when dangerous substances are carried in bulk other than in tankers and tank containers (eg tippers).
2.   When United Nations Packing Group I substances are carried in receptacles with a capacity of 5 litres or more.
3.   When UN medium risk substances (ie packing groups II and III) are carried in receptacles with a capacity of 200 litres (ie 44 gallons) or more. They also apply when notionally empty receptacles (having previously carried specified substances) are carried.

NB: It should be noted that where substances which fall into these categories are carried in smaller receptacles, even if they comprise a whole vehicle load, the requirements of the regulations do not apply.

Code of Practice

In addition to the regulations mentioned above, there is an Approved List (see below), an *Operational Code of Practice* and a guide to the regulations. All these documents are available from Her Majesty's Stationery Office.

The Operator

For the purposes of these regulations, as with the tanker regulations, the term 'operator' is defined as the person who holds the Operator's Licence where such a licence is needed but otherwise the keeper of the vehicle.

Substances Covered

The substances covered by the regulations are those included in the 'Approved List' referred to in the text on Packaging and Labelling of dangerous substances (see p 377) – namely, *Information Approved for the Classification, Packaging and Labelling of Dangerous Substances* (not to be confused with the Approved List referred to in the text on the tanker regulations on p 379) – which are toxic or flammable gases, organic peroxides, substances listed in UN Packing Groups I and II, asbestos in any form, other hazardous wastes plus any other substances which are not specified in the list but which have similar chemical properties and present similar hazards to those in the list, and mixtures of both types of substance (ie listed and non-listed similar substances).

The regulations also cover LPG carried in cylinders with a capacity (ie water capacity) of 5 litres or more except where such a cylinder is carried to provide fuel for a piece of equipment.

Limitations on Carriage

There are strict limitations on the carriage of certain dangerous substances. Particularly, operators must not carry substances in greater concentrations than those specified in the Approved List or in contravention of any conditions specified in the list about inerting or stabilising substances.

When carrying organic peroxides and flammable solids (specified as having a 'self-accelerating decomposition temperature' of 50°C or below) which can only be safely conveyed under controlled temperature conditions the driver must have adequate means to ensure that the temperature is maintained at a safe level. The regulations apply to these particular substances irrespective of the quantity being carried. Further advice on the carriage of these substances is to be found in the HSC Code of Practice.

Construction of Vehicles

Vehicles used for carrying dangerous goods in packages must be properly designed, constructed and maintained and be suitable for the purpose for which they are used. Specialised vehicles are *not* required and no special provisions relating to the construction of vehicles apart from that just mentioned are included in the regulations.

NB: It has been suggested by both the police and a magistrates' court that the use of flat-bed lorries for carrying packaged hazardous loads is not satisfactory. This follows a serious accident on the M1 motorway in October 1990 when such a vehicle carrying a load of drums of acid was involved in an accident. However, this suggestion is not to be taken as a legal ruling, only advice in the interests of added safety.

Information

Vehicle operators have responsibility under the regulations for ensuring that they receive from consignors of dangerous substances adequate information about the substances to be carried to enable them to comply with the regulations in giving the driver instruction, in marking vehicles correctly, and in being aware of the risks created by the substances to the health or safety of any person. They should not commence the carriage of any substance until such time as they have received the necessary information described here – it is illegal to do so.

Drivers must be given appropriate information *in writing* on the substances in their charge (ie identification of the substances, the dangers inherent in carrying and handling the substances and the action to be taken in the event of emergency) by their employer and must at all times carry in the vehicle cab these written details of the substances on board. They have a duty to ensure that written notices for substances no longer on the vehicle are removed, destroyed or locked away securely. The provision of Tremcards (although not obligatory – the information in writing can be in any form) satisfies this requirement (see p 381).* Alternatively, the requirement for written information can be met by the labels on the packages (and the accompanying written statements which are required in any case where single receptacles of 250 litres or more are carried) so long as they comply with legal requirements. Information carried on vehicles must be given to the police and traffic examiners on request.

NB: The Health and Safety Executive has said that Multi-Load Tremcards can be illegal and should not be used unless the entire load is carried for the whole of the journey. Such cards can conflict with the legal requirement to remove or lock away information in writing relating to dangerous goods no longer on the vehicle.

Driver Training

Drivers of vehicles carrying dangerous substances in packages (ie those included in the Approved List and in appropriate quantities) must have received proper training (and refresher training where necessary) in the hazards of handling and conveying dangerous substances, in loading and unloading procedures, in the checks they should make before commencing a journey, in the use of safety equipment, in the emergency procedures to be followed in the event of accident or dangerous occurrence, about vehicle marking and generally about the requirements of the regulations. They should also be given instruction regarding their duties under the regulations and about the particular load they are to carry. Failure to comply with this requirement is an offence and the HSC has warned that operators can face fines of up to £2000 on conviction.

From 1 January 1995 drivers of vehicles carrying dangerous goods in packages will have to meet the same minimum training requirements currently imposed on tanker and tank container drivers – see p 386 and Chapter 7.

Supervision of Vehicles

Similar requirements apply for the supervision of vehicles loaded with dangerous substances to those already mentioned for tanker vehicles. In particular, if the products carried amount to 3 tonnes or more, the vehicle must be parked in a safe place or must be supervised by the driver at all times or by some other competent (ie trained) person over the age of 18 years.

Labelling of Vehicles

Vehicles used for carrying 500kg or more of dangerous goods in packages must be correctly marked with reflectorised orange-colour plates at both the front and rear. These plates are rectangular and plain orange in colour surrounded by a broad black border. They must be kept clean and free from obstruction and may remain in place when an amount less than 500kg of dangerous substance is being carried. *However, they must be removed or covered when no such substances are being carried.* Additionally, hazard warning symbols (see p 383) *may* be displayed – they are not legally required.

Loading, Stowage and Unloading

Everybody involved in the loading, stowage and unloading of dangerous goods on vehicles must ensure that the requirements relating to health and safety are observed at all times and that goods are secured in the vehicle safely with minimum risk of damage to the packages. In general, the requirements for safe loading contained in the C&U regulations should be carefully observed (see pp 336–7).

Safety Precautions

The driver has a duty to observe precautions against fire or explosion, to ensure that loading, stowage and unloading is carried out safely and generally to be aware of the need for care particularly when on the highway and when near to members of the public.

Carriage of Explosives

Regulations which mainly came into force on 3 July 1989 control the movement of explosives by road. The Road Traffic (Carriage of Explosives) Regulations 1989 were made under the Health and Safety at Work etc Act 1974 and replace parts of the requirements of the Explosives Act 1875.

For the purposes of the regulations the term 'explosives' means explosive articles or substances which have been classified under the Classification and Labelling of Explosives Regulations 1983 (SI 1140/1983) as being in Class 1 or those which are unclassified. An 'explosive article' is an article which contains one or more explosive substances and an 'explosive substance' is a solid or liquid substance or a mixture of solid or liquid substances, or both,

> which is capable by chemical reaction in itself of producing gas at such a temperature and pressure and at such speed as could cause damage to surroundings or which is designed to produce an effect by heat, light, sound, gas or smoke or a combination of these as a result of non-detonative self-sustaining exothermic chemical reactions.

The term 'Compatibility Group' also has a meaning assigned to it by the 1983 regulations referred to above. The term 'carriage' means from the commencement of loading explosives into a vehicle or trailer until they have all been unloaded, whether or not the vehicle is on a road at the time. However, the term carriage is not applied where explosives are loaded on an unattached trailer or semi-trailer; carriage begins and ends when the trailer is attached to and later is detached from the towing vehicle or when the explosives have been unloaded whichever is the sooner.

The new regulations prohibit the carriage of any explosive in Compatibility Group K in a vehicle and any unclassified explosive except where it is being carried in connection with an application for their classification and in accordance with conditions approved in writing by the Health and Safety Executive (or, in the case of military explosives, by the Secretary of State for Defence).

Explosives must not be carried in any vehicle being used to carry passengers for hire or reward except that a passenger in such a vehicle may carry explosives under the following conditions:

- the substance carried is an explosive listed in Schedule 1 to the regulations (eg certain cartridges, fireworks distress-type and other signals, certain flares, fuses, igniters, primers and other pyrotechnic articles), gunpowder, smokeless powder or any mixture of them
- the total quantity of such explosive carried does not exceed 2kg
- the explosives are kept by that person and are kept properly packed
- all reasonable precautions are taken by the person to prevent accidents arising from the explosives.

The person carrying the explosives on the vehicle remains totally responsible for them and no responsibility for them is legally attached to the driver or the vehicle operator.

Suitability of Vehicles and Containers

It is the vehicle operator's duty under the regulations to ensure that any vehicle or any freight container used for the carriage of explosives is 'suitable' to ensure the safety and security of the explosives carried bearing in mind their type and quantity. The operator is also responsible for ensuring that the specified maximum quantities of any particular class of explosive carried on a vehicle or in a freight container are not exceeded and that no greater quantity of explosive is carried than that for which the vehicle or container is 'suitable'.

The limits on quantities of explosives permitted to be carried are shown in the following table:

Type of explosives		Maximum quantity
Division	*Compatibility Group*	
1.1	A	500kg
1.1	B, F, G or I	5 tonnes
1.1	C, D, E or J	16 tonnes
1.2	Any	16 tonnes
1.3	Any	16 tonnes
Unclassified explosives carried solely in connection with an application for their classification		500kg

The operator must ensure that explosives in different Compatibility Groups are not carried unless permitted as shown in Schedule 3 to the regulations.

Marking of Vehicles

Vehicles used for the carriage of explosives must be marked at the front and rear with a 400mm x 300mm rectangular reflectorised orange plate with black border (max 15mm wide). Additionally a square placard set at an angle of 45° must be displayed on each side of the vehicle container or trailer containing the explosives (see Figure 23.3). The placard must conform to minimum dimensions as shown in the illustration below and have an orange-coloured background with a black border and with a 'bomb blast' pictograph and any figures or letters denoting classification and Compatibility Group shown in black.

Certain exemptions apply to the display of markings as described above where the quantities of explosives of particular categories carried are below limits set out in Schedule 4 to the regulations.

Markings on vehicles and containers must be clearly visible, be kept clean and free from obstruction and must be completely covered or completely removed when all explosives have been removed from the vehicle or container. Both the vehicle driver and operator are responsible for ensuring these marking provisions are complied with.

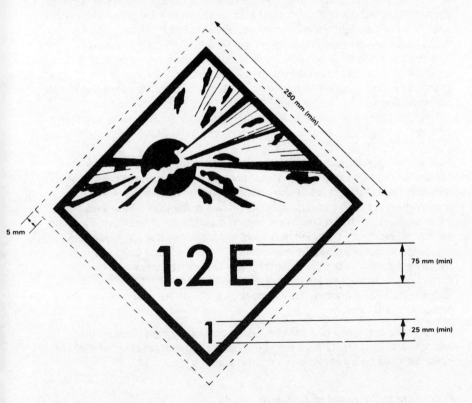

Figure 23.3 *Placard to be displayed on either side of vehicle or freight container carrying explosives*

Duty to Obtain Information

Operators must obtain information in writing from consignors of explosives which enable them to comply with the regulations. The consignor's duty is to ensure the information given is both accurate and sufficient to allow the operator to comply.

Information to be given to Drivers

The operator must give the driver or the vehicle attendant the following information in writing:
1. The division and Compatibility Group for classified explosives
2. The net mass (in tonnes or kg) of each type of explosive carried (or the gross mass if the nett mass figure is not available)
3. Whether the explosives carried are explosive articles or explosive substances (in the case of Group C, D or G explosives)
4. The name and address of the consignor, the operator of the vehicle and the consignee
5. Such other information as necessary to enable the driver to know the dangers which may arise and the emergency action to be taken.

This information must be carried on the vehicle at all times during the carriage from the start of the journey and must be shown on request to a police officer or goods vehicle traffic examiner. It must also be shown to a fire brigade officer or inspector if required. Information must not be carried on a vehicle when the explosives it refers to are no longer on that vehicle. It must be removed, destroyed or placed in a securely closed container marked to show that that the contents do not relate to explosives then being carried. If the necessary information is not available to the driver or vehicle attendant the explosives must not be carried.

Safe and Secure Carriage

The vehicle operator and the driver and any other person involved in the carriage of explosives must take all reasonable steps to prevent accidents and minimise the harmful effects of any accident. They must also prevent unauthorised access to, or removal of, all or part of the load. The operator and driver must ensure that a competent person is constantly in attendance with the vehicle whenever the driver is not present except during stops in a safe and secure place (as defined in the regulations – namely within a factory or magazine licensed under the Explosives Act 1875 or at a place with an exemption certificate granted under the Explosives Act 1875 (Exemptions) Regulations 1979 and when the vehicle is on a site where adequate security precautions are taken.

The operator and driver of a vehicle used to carry more than 5 tonnes of explosives in Division 1.1 must follow a route agreed with the chief officers of police for each area through which it is to pass.

Procedure in the Event of Accident

The driver or vehicle attendant must contact the police, fire brigade and vehicle operator as quickly as possible in the event of the following circumstances:

1. Spillage of explosives such as to constitute a safety risk
2. Damage to the explosives or their packaging such as to constitute a safety risk
3. If the vehicle overturns
4. If a fire or explosion takes place on the vehicle.

When such circumstances arise the driver, vehicle attendant and the operator must take all proper precautions to ensure the security of the explosives and the safety of persons likely to be affected and the vehicle operator must immediately notify the Health and Safety Executive.

Duration of Carriage and Delivery

Both the vehicle operator and the driver are responsible for ensuring that the carriage of explosives is completed within a reasonable period of time having regard to the distance involved, that explosives are unloaded as soon as reasonably practicable on arrival, that the explosives are delivered to the consignee or his agent or to another person who accepts them in custody for onward despatch provided they are delivered to a safe and secure place or a designated parking area in an airport, a railway transhipment depot or siding, a harbour or harbour area. If they cannot be delivered as required they must be returned to the consignor or his agent. If loaded in a trailer the trailer must not be detached except in a safe place or in an emergency.

Training of Drivers and Attendants

Vehicle operators must ensure that drivers and vehicle attendants have received adequate training and instruction to enable them to understand the nature of the dangers which may arise from the carriage of the explosives, the action to be taken in an emergency and their duties under these regulations and under the Health and Safety at Work etc Act 1974.

Vocational training certificates are required by drivers of explosives vehicles in accordance with the requirements of the Road Traffic (Training of Drivers of Vehicles Carrying Dangerous Goods) Regulations 1992 and EU legislation – see pp 386–7 and Chapter 7 for full details.

Minimum Ages

The regulations specify a minimum age of 18 years for those engaged in the carriage of explosives as a driver or vehicle attendant, for being made responsible for the security of explosives or for travelling in a vehicle carrying explosives unless in the presence of and supervised by a competent person over 18 years of age.

Radioactive Substances

Complex legislation controls the carriage of radioactive substances. In particular, the Radioactive Substances (Carriage by Road) (Great Britain) Regulations 1974 apply to the carriage of such materials, and the Ionising Radiations

Regulations 1985 cover controls on radiation doses received by persons working with such materials. Further, *A Code of Practice for the Carriage of Radioactive Materials by Road* is available (from HMSO) which sets out the law and gives advice on all aspects of transporting these materials.

Petroleum

The carriage of petroleum spirit in bulk tankers comes within the tanker regulations previously described. Petroleum spirit vehicles, when loaded, must not be left unattended except in a place approved by a local authority. The person in attendance must be either the driver or a competent person who is not less than 18 years old. A similar person must be in attendance during discharge to ensure there is no spillage, to ensure that all connections have been properly made and in particular to confirm that there is sufficient capacity in the storage tank to accept the delivery. At the time of petroleum spirit delivery a Certificate of Delivery must be completed. One copy is retained by the driver and returned to his employer, the other is retained by the owner of the storage tank.

Petroleum spirit carrying tanker vehicles must be constructed to conform with the regulation specification which includes such items as the provision of a fire screen between the vehicle power unit and the tank, location of the exhaust, the wiring specification and a master power cut-off switch.

Schedule 4 to the Dangerous Substances (Conveyance by Road in Road Tankers and Tank Containers) Regulations 1981 specifies details regarding the transfer of petroleum spirit into storage tanks at filling stations or other premises licensed for keeping 100,000 litres or less in storage.

CIA Voluntary Code

The Chemical Industries Association has a voluntary code of practice for road transport operators who carry dangerous substances. This code is the *Road Transport of Hazardous Chemicals – A Manual of Principal Safety Requirements.*

The manual is concerned with setting guidelines for good operating practice and covers the following items:
1. Responsibilities of the manufacturer, the transporter and the receiver.
2. Factors to be considered before chemicals are consigned.
3. Driver selection and training.
4. Operational standards, including vehicle construction and maintenance, loading and discharge procedures, vehicle routeing, safety equipment, the issue of instructions in writing and the marking of vehicles.
5. Dealing with emergencies.

A suggested general safety checklist for the road transport of chemicals is included in the manual as follows:
1. Is the vehicle roadworthy and suitable for the load to be carried?
2. Is the load secure and correctly stowed on the vehicle?
3. Are the containers of hazardous chemicals suitably labelled?

4. Does the vehicle display the appropriate statutory and other information relating to the chemical(s) being carried?

5. Has the driver been informed of the name, nature and hazardous properties of the chemical(s) being carried and the procedure to be followed in the case of emergency such as the spillage or escape of chemicals?

6. Have instructions in writing been placed in the cab of the vehicle or on the packages by way of a suitable label?

7. Is the vehicle properly equipped with first aid requirements, fire extinguishers and protective clothing and equipment?

8. Has consideration been given to routeing the vehicle to avoid traffic hazards and areas of high population density and areas especially vulnerable to contamination?

9. Has the driver been informed of any particular procedures which apply to the loading and discharge of the product involved?

10. Where haulage contractors are employed, has the contractor been checked for competency and informed of:
    (i) The nature and hazards of the load to be carried?
    (ii) Whether the conveyance of the chemical in question is controlled by legislation?
    (iii) The action to be taken in the event of emergency?

11. Has the driver been given specific instructions as to the procedure to be followed in bad weather, eg fog, ice, heavy rain etc?

Controlled and Hazardous Waste

Many transport operators (and skip-hire firms) are concerned with the disposal of 'waste' and, as such, they are affected by strictly enforced legislation effective from 1 April 1992. For these purposes waste may be considered in two forms:

– controlled waste which comprises 'household, industrial and commercial waste or any such waste' (ie including waste paper, scrap metal and recyclable scrap);
– hazardous waste which is material defined as 'special waste' in the controlled waste legislation or material which falls within the classification of dangerous substances for the purposes of the road tanker or packaged dangerous goods legislation described earlier in this chapter.

A range of legislation applies in this area of activity including the Disposal of Poisonous Waste Act 1972 (which makes it an offence to dispose of poisonous waste in an irresponsible way), the Control of Pollution (Special Waste) Regulations 1980, the Criminal Justice Act 1988 (which provides powers for the authorities to impound vehicles engaged in illegal fly tipping) and, more recently, the Control of Pollution (Amendment) Act 1989, the Controlled Waste (Registration of Carriers and Seizure of Vehicles) Regulations 1991 and the Controlled Waste Regulations 1992. Offences against the regulations can result in fines of £2000 on conviction.

Controlled Waste

Controlled waste should not be disposed of, to, or transported away by, a person or firm who is not legally authorised for this purpose.

Registration of Operators

The law (see above) requires operators who transport controlled waste within Great Britain to register with the appropriate waste regulation authority (ie in the area in which they have their business). It is an offence to fail to register or to carry controlled waste when not registered. Besides fines following prosecution and conviction for such offences, legislation provides powers for the seizure and disposal of vehicles used for such illegal purposes. Exemptions are provided for charities, voluntary organisations, domestic householders disposing of their own waste, waste collection authorities and producers of controlled waste. Builders and demolition companies are not exempt and must register and otherwise comply with the legislation.

Registration costs £95 initially and, currently, £65 at each 3-yearly renewal period. Registration may be refused to operators (ie companies or individuals) convicted of relevant offences as set out in the regulations.

Duty of Care

Under the Environmental Protection Act 1990 firms and individuals who produce, import, store, treat, process, recycle, dispose of or transport controlled waste (see above for definition) have a statutory 'duty of care'. This places responsibility for the completion of paperwork (ie 'Waste Transfer Notes') and to take all reasonable steps to stop waste escaping and ensuring its safety and security.

Waste Transfer Notes

Waste trasfer notes comprise written descriptions of waste handed over to other persons to transport and/or dispose of, and a transfer note signed by both parties (allowable as a single document) containing the following details.
1.   What the waste is and the quantity
2.   The type of container in which it is carried
3.   The time and date of transfer
4.   The place where the transfer took place
5.   The names and addresses of both parties (ie consignor and recipient)
6.   Detail as to which category each falls into (eg producer and registered waste carrier)
7.   A certificate number if either or both parties hold waste licences and the name of the authority from whom it/they was/were issued
8.   Reasons for any exemption from registration or waste licensing.

Copies of documents (ie descriptions of waste and/or transfer notes) given and received must be retained for at least 2 years. Both or either party may be required to produce these and prove in court where particular consignments of controlled waste originated.

Seizure of Vehicles

The seizure of vehicles aspects of the legislation are a new method of penalty in the UK – and, incidentally, are also being sought in regard to 'O' licence offenders. The law gives powers to waste regulation authorities to seize the vehicles of offenders, remove and separately store or dispose of loads as necessary, and dispose of (or destroy) vehicles following set procedures to publicise details of the seizure in local newspapers. Attempts will be made to seek out legitimate owners who may reclaim their vehicles on satisfactory proof of entitlement and identification.

Hazardous/Poisonous Waste Disposal

Broadly the law on hazardous waste requires that waste which is poisonous, noxious or polluting is not deposited on land where its presence is liable to give rise to an environmental hazard. And it is necessary for anyone removing or depositing poisonous material to notify both the local authority and the river authority before doing so.

An environmental hazard is defined as waste that is deposited in a manner or in such quantity that it would subject persons or animals to material risk of death, injury or impairment of health or threaten the pollution or contamination of any water supply. A booklet *Guidelines on the Responsible Disposal of Wastes* is available from the Confederation of British Industry. The European Commission has also produced a booklet on this subject for local authorities in Community countries.

There are additional controls over the carriage and disposal of particularly hazardous waste. The Control of Pollution (Special Waste) Regulations 1980 were introduced in order to comply with EU directives:

The main requirements of these regulations are outlined below:

1. Certain types of waste are to be regarded as special waste and subject to the additional controls. These are wastes that are regarded as dangerous to life as set out in the regulations.
2. Waste producers have to give not less than 3 days' and not more than 1 month's prior notice to Waste Disposal Authorities of their intention to dispose of a consignment of special waste.
3. A set of consignment notes must be completed when special wastes are transported. This means that a consignment of special waste can be transported from the producer to the disposal site only if each person has signed for it and taken on responsibility for it. This is to ensure that Waste Disposal Authorities know who is carrying the waste and they have to be informed within 24 hours of when it reaches the disposal site. Waste producers should take particular note of the requirement that all notices must be made on the statutory forms. Each form contains a unique reference number to assist the Authority in making sure that waste is safely disposed of.
4. A record of the location of the point of disposal on site of all special wastes must be kept in perpetuity. This is to ensure that proper arrangements can be made to bring the site back into use after the waste disposal operation has ceased.
5. Proper registers of consignments must be kept by the producers, carriers and disposers.
6. There will be a 'Season Ticket' arrangement for regular consignments of special wastes of similar composition disposed of at the same site. The

Waste Disposal Authorities will decide which producers and disposers in their areas qualify.

7. The Secretary of State will have emergency powers to direct receipt of special wastes at a particular site. This is likely to be rarely used.

8. Radioactive waste which also has the characteristics of special waste will be subject to the new controls.

Failure to comply with any of the requirements of the regulations is an offence.

Advice on particular problems can be obtained from Waste Regulation Authorities (usually the District or County Council).

Waste Site Licensing

The site licensing provisions of the Control of Pollution Act 1974 require all commercial, industrial or domestic wastes to be disposed of at a site licensed for that purpose and nowhere else. The licences are issued by the Waste Disposal Authorities (these are the County Councils in England and the District Councils in Wales and Scotland). These licences, which are a matter of public record, set conditions for the operation of each site including the types of waste which can be disposed of at that site, the manner of disposal, site supervision, boundary fences, notice boards etc. Waste Disposal Authorities must inspect sites regularly and make sure the operators are following the conditions of the licence. Failure to comply with licence conditions is a criminal offence. In addition to prosecution, the Authority can amend or even revoke the licence. See also section dealing with unauthorised tipping of waste (ie fly tipping) on pp 364–5.

Glass's Guide to Valuations

Glass's Guide is the market leading valuation service, selling well over twice as many copies as any other available, creating a true industry standard.

Glass's Guide to Car and Commercial Vehicle Values are available in a range of formats, including books, DOS, windows and raw data to suit your business needs.

"The great thing about Glass's is that it gives a single value for a well defined, clean car. It worries me that other guides, giving three values, could encourage managers to let vehicles go for less than their true worth"

David Turner, Fleet Manager, London Borough of Lambeth

To find out more about the whole range of services available from Glass's, simply call **01932 823 823** and ask for a copy of Glass's 1995 catalogue.

Glass's Information Services Limited, St Martin's Court, 37 Queens Road, Weybridge, Surrey KT13 9TU

24: Fleet Car and Light Vehicle Operations

Transport managers and other staff responsible for the operation of large goods vehicles within companies frequently have additional responsibilities for the operation of company-owned motor cars used by management, sales and service personnel, and light commercial vehicles which are outside the general scope of much of the legislation explained in this book. It is equally important that such vehicles should be operated strictly within the law. Most of the offences and penalties described in earlier chapters apply when operating such vehicles and the consequences of operating illegally can be serious. As already mentioned, for example, Licensing Authorities will take account of failure to operate light vehicles safely and within the law when deciding whether an applicant is a fit person or is of sufficient good repute to hold an 'O' licence for larger vehicles.

For the purposes of this chapter, light goods vehicles are vehicles with a gross plated weight not exceeding 3.5 tonnes (ie below the 'O' licensing, lgv plating and testing and the EU drivers' hours and tachograph requirement thresholds).

Much of the legislation applicable to the use of private motor cars and light goods vehicles has already been dealt with under such headings as excise duty, insurance and traffic regulations. The same system as described previously in Chapter 8 applies to the registration and excise licensing of such vehicles and the legal requirements for insurance cover have been outlined in Chapter 9.

Most of the traffic restrictions, particularly with regard to parking and waiting, road signs, motorway lights, breath tests and zebra crossings which apply equally to the private car and light vehicle driver and to the large goods vehicle driver, have been dealt with in Chapter 10. Nevertheless, some of these items of legislation are worth emphasising here for the benefit of the light vehicle fleet manager along with other matters applicable to such operations within firms.

Excise Duty

Private-type motor cars owned by firms and used for business purposes can be licensed at the private/light goods (PLG) rate of duty – currently £130 per year. Similarly, estate cars can be taxed at the £130 duty rate. If a private car is adapted for carrying goods, or an estate car is used for carrying goods, tools or samples, the appropriate goods rate of duty must be paid if the permissible maximum weight exceeds 3500kg.

In determining whether an estate car or dual-purpose vehicle (see p 144 for definition) is subject to the goods vehicle rate of duty, if its permissible maximum weight exceeds 3500kg consideration has to be given to the nature of the goods or burden carried. No specific guidelines are laid down but local Vehicle Registration Offices take the view that if any goods or samples, service equipment or spare parts are carried in connection with a business from which profit may result, then the vehicle should be taxed as a goods vehicle. Samples which can be accommodated in a normal briefcase would not generally constitute goods for this purpose.

The tax (ie VED) disc must be displayed on the windscreen of the vehicle on the near side where it can be easily seen. See Chapter 8 for details of vehicle excise duty requirements.

Insurance

Private-type motor cars and light goods vehicles, like all other vehicles, are required by law to be covered for third-party insurance risks as a minimum, but clearly in the case of fleet cars additional cover would be taken out, and in most instances comprehensive cover is advisable. Most insurance policies contain a variety of option clauses, some of which are included in the basic premium and others are available at extra cost.

Correct Cover

Of importance to the fleet manager is the need to ensure that company cars are fully insured for business use by employees of the company. The difference in insurance classification for business cars used for commercial travelling and for those which are not used for this purpose is an important point for the fleet manager to consider. Most policies differentiate between such use, and in a car fleet where vehicles are used by both salesmen and other staff and management it is essential to ensure that either the policy covers the whole of the car fleet for commercial travelling or, if it does not do so, the sales force should be restricted to driving only those cars which have this cover. If a salesman, or indeed any other staff member, uses a car for a purpose which can in any way be described as soliciting an order, then in insurance terms this use is classed as commercial travelling and the cover on the car must include this clause.

Dual-purpose vehicles registered and licensed as goods vehicles must be insured with cover permitting goods to be carried for business purposes.

Cover for Special Cars

There can be problems in a fleet with certain individual cars on insurance cover which may not be driven by certain employees. For example, if the managing director has a high powered sports-type car it is likely that the insurers would impose severe restrictions on the driving of that car, for whatever purpose (chauffeuring, ferrying back and forth for service or repairs, for example), by relatively inexperienced drivers or by, say, young drivers under 25 years of age. Alternatively, such cars may be restricted to named drivers only.

Private Use of Company Cars

A further point which needs consideration is cover for the employee's use of a company car for private purposes. It is general practice for firms to allow their employees this concession both for commuting to work and for family motoring at the weekends and for holidays. The cover taken out for fleet cars should specifically include this provision if such use is permitted, otherwise the vehicle owner (the firm) may be guilty of permitting the use of an uninsured car. In these circumstances, if an accident and claim were to result, the consequences for the employee could be very serious in terms of meeting damage claims and in prosecution for using an uninsured car. The same provisions should be applied if the firm permits the employee's wife (or husband) or even his children (if they are qualified drivers) to drive the car at any time.

Employees' Use of Own Cars

If employees are ever required, or likely to be required, to use their own private car for business purposes, even for only the occasional errand, the fleet manager should ensure that the employee has adequate insurance cover on his own vehicle for such purpose. This usually means that the employee's own policy must include provision for his car to be used in connection with the business of his employer.

The fleet manager, confronted with this type of situation, should ask to see evidence of the cover (ie a valid Certificate of Insurance or a temporary cover note showing the conditions for use covered by the policy) and not just rely on the word of the employee. Similarly, he would be wise to inform all other persons in the firm, management and staff alike, that employees should not use, or be requested to use, their own cars on company business without first having the insurance cover verified.

To avoid the dangers which could arise from an employee using his own private car for business purposes when it was not covered for such use, when the policy was not in force at all because the premium had not been paid, or because the policy was invalid as a result of the employee giving incorrect information at the time of completing the policy application, the company can take out a motor contingency policy which will indemnify it against any claims arising from an accident involving an employee undertaking company business in his own uninsured car. This type of policy is very cheap to obtain and well worth the cost when considered against the risks. For example, if the employee had an accident causing serious injury to one or more third parties and his own insurance proved to be void, the third parties would look to the employer on whose business the employee was engaged at the time of the accident to meet their claims.

Dual Car Use

When an employee who is provided with a company car also owns a car which his wife uses, an anomalous situation on insurance cover can arise if his own car is insured in his, and not his wife's name. This problem occurs because the cover provided on his own car automatically provides him with third-party cover while driving any other car not belonging to him. Consequently, when driving the firm's car for pleasure purposes he has the double cover provided by both the firm's policy and his own policy. If he should then be involved in an accident resulting in a third-party claim, his own insurers could be held partially liable for the damages arising out of the
claim.

In order to overcome this particular problem the major insurance companies have made an undertaking indicating that it is not their intention to take advantage of the cover provided by the driver's own personal insurance policy in such circumstances. However, not all insurance companies are party to this undertaking.

Payment by Passengers

Payment by passengers towards the cost of petrol consumed on a journey no longer infringes the 'hiring' exclusion clause on insurance cover. However, payment towards other motoring expenses such as parking fees or depreciation could still fall within the 'prohibition of hiring' clause in most insurance policies. In cases where this situation arises a check should be made with insurers to ascertain the current position.

Company Cars and Income Tax

Directors and higher paid employees are assessed for schedule E income tax on the 'personal benefit' which they derive from the provision of a car by their employer for their business and private use.

Higher paid employment for the purposes of company car tax assessment is employment where total remuneration including all expense payments and benefits in kind or cash amounts to £8500.

If the business use of the car is 'insubstantial' (the annual business mileage must be over 2500 to avoid a car being treated as substantially for private use) compared with private use, the benefit of having the car will be assessed for tax purposes at one-and-a-half times the scale figure shown in the table. The same one-and-a-half times increase in assumed benefit will apply to a second car provided to a person by a company.

If business use of the car is 'substantial' tax is based on the benefit derived according to a scale which covers both the capital value and the running costs of the car. If the business use of the car exceeds 18,000 miles per year the assumed taxable benefit will be reduced by half.

Company Car Taxable Benefits for 1994–95

	Age of car	
	Under 4 years annual benefit	4 years or more annual benefit
Cars having a market value not exceeding £19,250 when first registered		
Cars having a cylinder capacity:		
1,400cc or under	£2310	£1580
1,401cc to 2,000cc	£2990	£2030
2,001cc or more	£4800	£3220
Cars having a market value exceeding £19,250 when first registered:		
Original market value		
£19,251 to £29,000	£6210	£4180
More than £29,000	£10,040	£6660

Company Car Fuel Benefits for 1994–95

Cylinder capacity of car in cubic centimetres	*Original market value*	*Annual benefit*	
		Diesel	Non-diesel
1400cc or under	or less than £6000	£580	£640
1401cc to 2000cc	or £6000 to £8499	£580	£810
2001cc or more	or £8500 or more	£750	£1200

Taxable Benefits from 6 April 1994

From 6 April 1994 the taxable benefit for the use of a company car is based on its retail price and the annual mileage as follows.

Tax Benefit for 1994/95

Annual business mileage	Percentage of retail price
Up to 2500	35%
2501 to 17,999	23.33%
18,000 and over	11.66%

NB: A reduction of one-third of the relevant figure above applies if the car is 4 or more years old at the end of the tax year.

Private Use of Vans

The Inland Revenue is reported to be checking and clamping down on van drivers earning more than £8500 annually, who are permitted to use their vehicles for private purposes (including their use as transport between home and their place of work). Such private use should be declared on the relevant tax form (ie P11D) and the employee can be liable for tax based on 20 per cent of the vehicle's value. Safeguards against such tax liabilities include where the vehicle is taken home *en route* to a next-morning delivery, or where it is necessary for overnight security reasons, because the employee's own tools and personal work equipment is kept on board.

Since 6 April 1993, employees who have private use of a company van will be taxed by reference to a scale charge – for 1993/94, £500, but only £350 if the van is more than 4 years old – which will include the provision of private fuel. Vehicles over 3.5 tonnes will be exempt from this arrangement, as will any pooled van which is not normally kept at or near an employee's home.

National Insurance on Private Fuel Usage

Where employees are provided with free fuel for private use there is a liability to pay National Insurance contributions on the value of the benefit to the employee. Where such value has not been taken into account by the employer there is a risk of claims by the DSS for retrospective payment of contributions going back 1 year.

Construction and Use Regulations

Apart from the sections of the C&U regulations which apply to private cars concerning their construction, lighting, noise, silencing and all the provisions which require the motor car to be maintained in accordance with the regulations, they are not involved in the special requirements relating to goods vehicles, unless the car has in any way been adapted to carry goods, for example, by removing seats or fitting racks on which goods can be carried. In this case the goods vehicle rate of excise duty applies and the vehicles become subject to certain aspects of the C&U regulations regarding goods vehicles (see Chapter 13).

It should be remembered that the overriding requirement of the C&U regulations for all vehicles to be maintained at all times in such a condition that they shall not cause danger to people carried on the vehicle and other road users applies equally to private cars and to goods vehicles irrespective of their size or weight.

New cars manufactured for sale in Britain have to meet the braking system requirements of EC Directive 71/320 unless the system is of the dual-line type. This directive specifies maximum stopping distances equivalent to a braking efficiency of 27 per cent.

Also as a result of EU directives, reversing lamps and four-way hazard warning flashers are compulsory on all cars made in Britain. Regulations require rear fog lights to be fitted to new vehicles manufactured since 1 October 1979 and first used since 1 April 1980 (see pp 270–1 for more details).

Vehicles for Unleaded Petrol

Petrol-engined vehicles first used on or after 1 April 1991 must be designed and constructed to run on unleaded petrol – it is illegal to re-convert such vehicles to run on leaded petrol only. Vehicles which fall into this category must not be capable of being fitted with a petrol pump nozzle greater than 23.6mm in diameter (unless certain other conditions are met and the fuel filler is clearly marked with the word 'UNLEADED' or the symbol for unleaded fuel use).

Drivers' Hours and Records

Light Goods Vehicles

Light goods vehicles coming within the scope of this chapter are exempt from the EU regulations (ie because they do not exceed 3.5 tonnes gross weight) and in consequence of this only the relevant British provisions will apply.

The applicable limits to goods vehicles not exceeding 3.5 tonnes gvw are as follows:

Maximum daily driving time	10 hours
Maximum daily duty time	11 hours

There are no specified break or daily or weekly rest period requirements, no limits on continuous duty or weekly limits on duty or driving.

Drivers of light goods vehicles of not more than 3.5 tonnes gross plated weight and dual-purpose vehicles (see p 144 for definition) of any weight used for certain specialised duties are required to observe only a daily maximum driving time of 10 hours. This applies to light goods vehicles used:
1. By doctors, dentists, nurses, midwives or vets.
2. For any service of inspection, cleaning, maintenance, repair, installation or fitting.
3. By a commercial traveller and carrying only goods used for soliciting orders.
4. By an employee of the AA, the RAC or the RSAC.
5. For business of cinematography or of radio or television broadcasting.

Record Keeping
Light goods vehicle drivers (ie vehicles under 3.5 tonnes gross weight) and drivers of dual-purpose vehicles are exempt from the requirements to keep records of their driving, duty and rest periods.

Private Cars

Employees driving company-owned private-type motor cars are exempt from the requirements of the goods vehicle drivers' hours and record requirements, but only so long as they do not drive vehicles to which these requirements do apply. If at any time an employee who normally drives only motor cars also as part of his work needs to drive a goods vehicle to which the drivers' hours regulations do apply then any time spent driving a private car (his own or the company's) for his firm's business must be counted within his total daily working time (ie duty time). If the goods vehicle exceeds 3.5 tonnes gross weight the driving of the car should be shown on his record sheet (ie tacho- graph chart) for the day on which he drives a goods vehicle as 'other work'.

Full details of the goods vehicle drivers' hours and records are given in Chapters 3 and 4.

Tachographs

Light goods vehicles of not more than 3.5 tonnes gross plated weight are exempt from the EU tachograph regulations requiring tachograph fitment and use, but if such a vehicle is coupled to a goods carrying trailer so that the total of the combined gross weights exceeds 3.5 tonnes then the vehicle will come within the scope of the EU tachograph regulations, unless it is otherwise exempt due to special use. This means that a DoT calibrated tachograph must be fitted and must be used by the driver when the trailer is drawn (see Chapter 5 for full details of the tachograph regulations) and the EU drivers' hours rules must be followed (see Chapter 3 for details).

Speed Limits

Private cars and dual-purpose vehicles not drawing trailers are restricted to maximum permitted speeds on certain roads in accordance with the restriction sign-posted on the section of road and to overall maximum speeds of 60mph on single-carriageway roads and 70mph on dual-carriageways and motorways.

Speed limits for cars and light goods vehicles are as follows:

		Motorways	Dual-Carriageways	Other Roads
1.	Cars and car-derived vans	70	70	60
2.	Cars and car-derived vans towing trailer	60	60	50
3.	Rigid goods vehicles not exceeding 7.5 tonnes	70	60	50
4.	Articulated vehicles and rigid goods vehicles not exceeding 7.5 tonnes drawing trailer	60	60*	50

In Northern Ireland the limit for this category of vehicle is 50mph.

NB: In all cases of speed restrictions mentioned above, if specific lower limits are in force on any section of road, then it is the lower limit which must be observed.

Seat Belts

Compulsory fitment of seat belts applies in the case of the following light vehicles (see also p 195):

1. Goods vehicles not exceeding 1525kg unladen (first registered since April 1967).
2. Goods vehicles not exceeding 3500kg maximum gross weight (first registered since 1 April 1980).
3. Dual-purpose vehicles first registered since 1 January 1965.
4. Private cars first registered since 1 January 1965.

Vehicles to which the regulations apply first used since 1 April 1973 must be fitted with belts which can be secured and released with one hand only and must also be fitted with a device to enable the belts to be stowed in a position where they do not touch the floor. The belts must be maintained in a fit and serviceable condition and kept free from permanent or temporary obstruction which would prevent their being used by a person sitting in the seat for which the belt is provided. Failure to comply with these requirements is an offence and can lead to failure of the MoT test.

In cases where the regulations apply, as above, belts must be provided for the driver and one front seat passenger.

Wearing of Seat Belts

The wearing of front seat belts in vehicles fitted with them by law has been compulsory since 1 January 1983. Since 1 September 1989 children up to 14 years of age riding in the back seats of motor cars fitted with rear seat belts must wear those belts or restraints.

Since 1 July 1991 adult rear seat passengers travelling in cars or taxis fitted with rear seat belts must wear the belts provided, irrespective of the age of the vehicle.

Certain exemptions to the wearing of belts have been included in the regulations (see p 196 for list of exemptions).

Heavy fines may be imposed on conviction for failing to wear a seat belt as required by law. The responsibility for seat belt wearing rests with the person sitting in the seat for which the belt is provided except that responsibility for ensuring that children wear seat belts as required by law rests with the driver of the vehicle (ie not the parent or guardian who may be accompanying the child).

Fuel Consumption Tests

Concern about energy conservation led to the Energy Act 1976 and the Passenger Car Fuel Consumption Order 1977 being enacted. Because of interest among readers of the *Handbook* in this subject, details of the legal requirements and the scheme are outlined here.

Since 1 April 1978 new cars on display in showrooms and on forecourts must carry a label showing official fuel consumption figures for that model of car. Every dealer must have details of officially approved fuel consumption tests for all cars listed in this booklet available in his showroom for buyers to consult on request. In addition, where reference is made in promotional literature, such as advertisements, technical specifications and sales brochures, to the petrol consumption of a new car the test results must be included. In all such cases the results of all tests carried out must be quoted (urban cycle, 90km/h – and 120km/h where appropriate) in both miles per gallon and litres per 100km. These requirements do not apply in the Channel Islands or the Isle of Man.

Official Tests

The official tests are carried out in approved laboratories or on test tracks. They have been designed to be representative of real-life driving situations and the results achieved provide a guide to the models which are likely to be more economical in their fuel use.

The test results do not guarantee the fuel consumption of any particular car. Each new car has not itself been tested and there will inevitably be differences between cars of the same model. The driver's style, the loading of the car, road, weather and traffic conditions, the overall mileage of the car and its standard of maintenance will all affect its fuel consumption. For all these reasons the fuel consumption achieved on the road will not necessarily accord with the tests results.

The Standard Test
The tests follow an internationally agreed procedure and consist of two compulsory parts:
1. A cycle simulating urban driving
2. A constant speed test at 55mph (90km/h).

Models Included in the Tests
Almost all types of new passenger cars are covered by the tests. However, certain types are excluded as follows:
1. Cars manufactured before 1 January 1978
2. Second-hand cars
3. Cars adapted to carry more than eight passengers (excluding the driver)
4. Three-wheelers
5. Invalid carriages
6. Van-derived passenger cars
7. Cars built specially for export
8. Cars operating on four-wheel drive only
9. Cars whose engines run on diesel, liquefied petroleum gas or other such fuels.

These vehicles will not, therefore, be labelled in the showrooms. Heavy goods vehicles, vans and motorcycles are also excluded from these tests. Also, a small number of manufacturers and importers have been granted exemption from testing because of the low volume of production involved. Consequently, these particular makes and models are not likely to be of significant interest or concern to the fleet user.

Urban Test Cycle
The urban test cycle is carried out in a laboratory where equipment simulates the loads experienced under normal driving conditions and the standard patterns of urban driving. The car is driven from a fully warmed-up start and is taken through a cycle of acceleration, deceleration and idling with a maximum speed not exceeding 31mph (50km/h).

Constant Speed Test
The constant speed test at 56mph (90km/h) is intended to be representative of open road driving. It may be carried out in the laboratory or on a test track (under strictly controlled road and weather conditions).

Optional Constant Speed Test

This test is carried out at 75mph (120km/h) in a laboratory or on a test track. Although it is recognised that this test exceeds the UK maximum speed limit, it is included to illustrate to car drivers the worsening fuel consumption at higher speeds. It may also be useful to manufacturers exporting to some parts of Europe where speed limits are higher.

Only one production car is tested as a representative of each model. It must have been run in and have been driven for at least 1800 miles (3000km) before testing. In some cases several models, which do not differ significantly in certain technical characteristics thought to be important in determining fuel consumption, may be grouped together into a 'class'. Only one car in the class needs to be tested.

The Testing

The responsibility for testing lies with the manufacturers and importers themselves. They must either carry out the tests themselves or arrange for them to be carried out on their behalf. Department of Energy officials have the right to inspect the test laboratories and to witness tests in progress to ensure that they are being carried out correctly.

Manufacturers must submit their fuel consumption test results to the Department of Energy who record the results in an official fuel economy certificate.

DRIVER HIRE NATIONWIDE
Supplying workforce to transport and industry

As one of the country's major suppliers of commercial drivers and blue collar workers the Driver Hire network of over 40 franchised offices, covering the U.K. from Aberdeen to Exeter, has an in-depth understanding of the problems faced by distribution, warehouse, storage, packing and collection/delivery controllers and the regulations and procedures to which they must adhere.

We are able to offer a genuine 24 hour, 7 day, 52 week PERSONAL service at each of our locations backed up by both local people and knowledge and a proven BS 5750 (Part 2) operational system, which means that we can normally respond to your needs within an hour of your telephone order - often even faster.

This dedication to customer service, to which can be added the further benefits of Insurances, F.R.E.S. Membership, Union Approval and our unique charging structure can result in substantial savings of both time and money. Further cost savings arise as Tax, N.I. and general payroll administration are all dealt with by Driver Hire and, of course, you pay for operatives only when you need them - no sickness, holiday, or pension costs to consider.

Our selection and screening process includes stringent interviewing and reference checking as well as successful completion of a job-related competence test prior to acceptance onto our register. This degree of in-depth investigation at the outset ensures that only the most reliable and experienced personnel will be offered to fill your requirements. The Driver Hire aim is to provide a professional total quality service to all customers, tailored to individual needs.

The Driver Hire Network:	
Aberdeen:	01224 899449
Birmingham:	0121 6048001
Blackburn:	01254 682401
Bolton:	01204 381023
Bournemouth:	01202 895200
Bradford:	01274 370787
Bristol:	01454 320128
Chelmsford:	01277 356110
Cleveland:	01642 253123
Coventry:	01203 633465
Derby:	01332 799994
Edinburgh:	01506 855665
Enfield:	0181 443 7058
Exeter:	01392 490115
Glasgow:	0141 221 6000
Grimsby:	01472 821300
Guildford:	01483 750740
Heathrow:	0181 893 1450
Hereward:	01945 474138
Ipswich:	01473 233299
Leeds:	0113 2557522
Leicester:	0116 2516700
Manchester:	0161 877 9312
Milton Keynes:	01908 216700
Motherwell:	01698 275444
Newcastle:	0191 491 0503
Northampton:	01604 670199
Nottingham:	0115 9249259
Oldham:	0161 633 3368
Paisley:	0141 848 6070
Pontefract:	01977 600383
Preston:	01772 722721
Reading:	01734 770710
Romford:	01708 526789
Sheffield:	0114 2763838
Southampton:	01703 232995
Spalding:	01775 766068
Stockport:	0161 337 8862
Stoke:	01782 274312
Warrington:	01925 244469
Wigan:	01942 820515
Wolv'hampton:	01902 310124
Worcester:	01905 619597
York:	01904 430655

25: Rental, Hiring and Leasing of Vehicles

Rental, hiring and leasing of commercial vehicles is now seen as a major and very cost effective alternative to fleet ownership. Transport operating companies have shown increasing interest in the advantages of these means of vehicle acquisition. In many cases operators have reduced their owned fleets to the bare minimum required to service basic and predictable delivery requirements, topping up with short-term rental vehicles to meet peak trading demands, or using this source for the replacement of vehicles off the road for service, repair or annual test preparation. At the other extreme, firms have disposed completely of their owned fleets in favour of contract hire arrangements where vehicles are provided by a third-party contractor and maintenance is taken care of as part of the contractual arrangement. Between these extremes are the firms who use a combination of owned vehicles, long-term hired vehicles and short-term rental, to ensure the most economical and efficient overall transport operation whatever the seasonal fluctuations or other trading exigencies of their business.

These alternative means of adding vehicles to the fleet have both legal and operational implications; hence the reason for including this basic outline of the subject in the *Handbook*.

The Vehicle User

One of the most significant legal points in connection with the renting, hiring or leasing of vehicles concerns the status of the vehicle 'user'. If the person or company renting or hiring a vehicle provides the driver, then that person or company, as the employer of the driver and consequently as the user of the vehicle, carries the full weight of legal responsibility for both the safe mechanical condition and the safe operation of the vehicle when it is on the road.

It is the user's responsibility to ensure that the vehicle complies fully with the construction and use regulations in all respects but especially with regard to safety items such as brakes, lights, steering, horn, tyres, speedometer/ tachograph and vehicle markings. The fact that the rental or hiring company which owns the vehicle *should* ensure that all these items are in order (and indeed usually proclaims that it does ensure they are in order) makes no difference to where the blame lies and where the prosecution will be aimed if they are found not to be in order when the vehicle is being used on the road.

Furthermore, it is the user's responsibility to ensure that operator's licence provisions where applicable are fully complied with and that the drivers' hours and tachograph requirements are observed where these are applicable.

Rental

Rental of vehicles on a short-term basis of a few days or a few weeks, which is the usual arrangement, does not impose onerous contractual obligations on the hirer.

'O' Licensing Provisions
However, it does involve other legal obligations in respect of the vehicle itself and its use. For a start, much depends on the gross weight of the vehicle. If it is over 3.5 tonnes permissible maximum weight and has been rented for use in connection with a trade or business, then the person or firm renting it must hold an 'O' licence and there must be a margin on that licence to cover the renting of one or more additional vehicles.

There is no need to advise the Licensing Authority (LA) of details of the vehicle unless it is to be retained on hire for more than 28 days, after which time the LA must be notified so an 'O' licence windscreen disc can be issued for the vehicle. If the vehicle is rented for a shorter period and then returned to the rental company to be replaced by another vehicle, the LA does not have to be notified if the combined total of the two rental periods exceeds 28 days unless both are part of the same rental agreement.

If the over 3.5 tonnes vehicle is rented by a firm for use in another traffic area different from the one in which the 'O' licence is held, then an 'O' licence must be obtained in that other traffic area before a vehicle is permitted to operate from a base there.

Where the gross weight of the rented vehicle does not exceed 3.5 tonnes there are no legal obligations in respect of 'O' licensing unless it is used to tow a goods-carrying trailer with an unladen weight over 1020kg when the combined weights may take it over the 3.5 tonnes limit and into 'O' licensing.

Whether or not the rented vehicle comes within the scope of 'O' licensing, the person or firm renting it carries the user responsibility for its safe mechanical condition when it is on the road. Consequently, if vehicle faults result in prosecution the user will have to pay any fines imposed (not the rental company) and the user's 'O' licence will be put in jeopardy (even if the vehicle is not specified on his 'O' licence). Therefore careful selection of a reputable rental company with high maintenance standards is essential.

Charges and Payments

Charges for rented vehicles are usually on a time plus mileage basis in accordance with published scales so there is little scope for improvement in prices, although large users can sometimes negotiate discounts.

The particular advantage of rental is that payment for a vehicle is only made when the use of the vehicle is really required, and then the payment is out of revenue and not out of capital reserves. Further, because rental is normally for short periods only, it is easy to establish the total costs involved because there are no additional costs for the upkeep of the vehicle. Maintenance, tyre replacements, licences and most other costs apart from fuel and insurance are built into the rental price.

The disadvantage of rental is its high price if vehicles are taken for longer periods, because the rental companies try to recover their costs over a short time and the price covers the fact that only a certain number of hire days can be sold in a period.

Hire and Contract Hire

Hiring of vehicles (or more specifically contract hire), as opposed to rental, implies a longer term arrangement with a more rigid agreement as to the obligations of the parties involved. Hiring arrangements vary considerably since the vehicle provider and the customer draw up a contract to incorporate the services required. There are two principal forms of contract hire – vehicles supplied with drivers and vehicles supplied without drivers.

The important difference is that in the former case the contract hire company, as the employer of the driver, is the 'user' of the vehicles in law and therefore holds the 'O' licence and shoulders the legal responsibilities previously described, while the hirer merely operates the vehicles exclusively to suit his requirements. However, in the latter case the hirer is the 'user' and 'O' licence holder and, as with vehicles purchased and leased with his own employee drivers at the wheel, he carries the full legal responsibilities.

Between these two categories, a package is made up to suit individual company needs. A complete package normally includes full maintenance (ie safety inspections, service and repairs), fuel, licensing, insurance, parking areas, administration (ie checking of records, etc), replacement vehicles to cover downtime at no extra cost and free driver replacement to cover holidays and sickness.

Contract hire charges are normally made on a time and mileage basis (ie a fixed or standing charge and running charge). These charges are either increased annually to cover the hiring company's increased costs or linked to a published index. Sometimes such items as fuel surcharges are raised.

Advantages of Contract Hire
This method of vehicle acquisition offers a number of advantages. Principally, there is no investment of capital (generally not even an initial deposit to be found) and cash flow for transport services is predictable throughout the year, thus allowing easy budgeting. One regular monthly invoice covers all capital and operating costs. The hire charges are fully allowable against tax.

Overall, full contract hire with driver is advantageous to the operator, because it relieves him of the burdens of capital expenditure on an ancillary activity and of a welter of legal responsibilities and yet provides him with the right vehicles for his exclusive use to fulfil his delivery requirements as he wishes. He thus has the best of both worlds - all his transport needs met without the major burdens usually encountered by own fleet operators.

A further financial advantage can arise for a firm operating its own fleet but wishing to switch to contract hire to gain the benefits outlined. Contract hire companies will usually purchase a whole existing fleet and then contract-hire it back to the operator, thus still giving him resources to meet his transport needs and yet providing him with an immediate refund of the capital tied up in vehicles. This proposition can be used to advantage in relieving cash flow pressure.

Transfer of Undertakings Provisions
In recent years there has been considerable activity in firms switching from their own, in-house transport operations to contract hire, and from one contract hire firm to another. This trading of contracts has raised a number of problems, not least the contractural obligations of transferring employees between one employer and another. At one time it was thought that a loophole had been found in the law but now it is clear that the stringent provisions of the Transfer of Undertakings (Protection of Employment) Regulations 1981 apply in most such cases. These regulations are notoriously complex but failure to comply with their provisions could lead firms into substantial financial penalties and compensation claims.

Broadly, the regulations require that a firm acquiring a contract by which a transport operation is transferred, is obliged to take on the existing staff on payment terms and conditions which match those of the previous employer, or pay redundancy on terms equal to those which the employees would have secured under their contracts of employment with the employer they previously worked for.

Leasing

Leasing is a totally different concept from outright purchase or hire purchase in that the operator (ie the lessee) never actually owns the vehicle but he has the full use of it as though it was his own. It is also a different concept from rental and hiring arrangements in that it is purely a financial means of acquiring vehicles. In other words those putting up the money are not transport or vehicle operators, they are finance houses.

Several different forms of leasing are available (basically divided by the assumption of risk with the lessee taking the risk with a pure finance lease and the lessor retaining the risk with an operating lease) and legislation governing leasing arrangements is subject to change. Also, the way in which the accountancy profession treats leasing is subject to variation, so it is important to discuss any proposed leasing arrangement with a professional accountant before commitment to an agreement.

The general concept of leasing is that a finance house (ie the lessor) purchases a vehicle, for which the operator has specified his requirements and negotiated the price and any available discount from the supplier, and then it spreads the capital cost, interest charges, overhead costs and its profit margin over a period of time to determine the amount of the periodic repayments.

The three basic types of finance lease are as follows:
1. Full amortisation lease which runs for an agreed fixed period (the primary lease) followed by an optional secondary period (if the vehicle is still required) when the lease is extended for a nominal (ie 'peppercorn') rental.
2. Open-ended lease which allows the lessee to terminate the agreement on payment of a previously agreed settlement figure at any time after a fixed period (normally 1 year).
3. Balloon lease which has one large payment and a number of relatively low rental payments. The balloon can be at the beginning with a large payment to start the lease or at the end of the lease term with a large pre-calculated final payment reflecting a forecast residual value for the vehicle.

Usually this full range of choice only applies in the case of smaller vehicles. For heavy vehicles it is customary to apply 'full pay-out' types of lease, whereby the vehicle is fully paid for in the rentals with no residual value.

Because no capital outlay is involved in a leasing agreement, beyond the initial lease payment which is sometimes a number of monthly rentals lumped together, the lessee is able to obtain the vehicles he needs and yet still invest his own capital in more profitable business avenues.

Changes in accounting practice brought into effect (from the start of accounting years beginning after 30 June 1987) by the introduction of the accounting profession's Statement of Accounting Practice 21 (SSAP21) mean that no longer is it a case where leased vehicles do not appear as assets in the lessee's balance sheet. Consequently leased vehicles and plant appear on the balance sheet as assets matched by outstanding lease payments showing on the other side as liabilities.

Leasing is a complex financial area and, as already mentioned, it is important that proper professional advice is obtained before signing any agreement, otherwise the promised tax and other benefits may not materialise.

Where leasing is purely a financial arrangement, the advantages and disadvantages from an operational viewpoint are the same as for outright purchase. Because in principle the lessee operates the vehicle as though he owns it and he employs the driver, the full weight of legal responsibility, as already outlined, applies to him so he needs to have a full transport back-up of administration and operational staff, maintenance facilities and policies for selection of the correct vehicles and for replacement at the most economic intervals.

Review of Alternative Acquisition Methods

An excellent review of the merits of various alternative methods of vehicle acquisition is to be found in the National Freight Consortium's* publication *Financing the Acquisition of Commercial Vehicles* prepared independently by Andrew B. Jones CA, Partner in Charge, Tax Department, Ernst & Young, London.

NB: The NFC through its various operating companies is one of the UK's leading providers of vehicles on hire, contract hire, contract distribution and vehicle rental.

26: Vehicle Fuel Economy

Next to wage costs, fuel is the most expensive vehicle operating cost item. It is a high-cost commodity which is a major budget feature of all goods vehicle fleet operations. It is also subject to occasional and dramatic shortages as a result of political unrest in some of the major oil-producing countries, as we have seen with the Middle East. Scientists predict total extinction as world supplies of crude oil are consumed ever more rapidly by developed nations which have become increasingly dependent on transportation systems powered by oil-based fuels. Even the once much-heralded finds of oil in offshore waters of the British Isles and Eire are now known to have limited life expectancy.

While the search for and research into acceptable alternative fuels and power units goes on, it is important to take steps to minimise consumption of our present fuel supplies. As well as being a problem for nations, this is a problem which concerns all fleet operators, whatever their size. Besides any conscience they may have about energy conservation they will readily appreciate that fuel consumption must be reduced in the campaign to keep vehicle operating costs down.

Fuel and the Vehicle

Fuel consumption is substantially related to the type of vehicle, its power unit and drive line, its mechanical condition, the use to which it is put and how it is driven. In recent times manufacturers have offered fuel economy models within their ranges so the cost-conscious operator can choose between economy or outright performance.

The fuel-conscious operator who is in a position to buy new vehicles will undoubtedly choose fuel economy models where these are suited to his particular needs. However, for the most part, fleet operators have to stick with the vehicles they already have in their fleets and are faced with the need to consider how improved fuel economy can be achieved with existing vehicles.

Three principal areas exist for improvement in the vehicle itself. These are as follows:
1. Mechanical condition
2. Efficient use
3. Addition of fuel economy aids.

Mechanical Condition

A vehicle which is poorly maintained will inevitably consume more fuel.

Particular attention should be paid to efficient maintenance of the following items:

1. *Fuel system* (fuel tank, pipe lines, filters, pump and injectors). There should be no leaks and the vehicle should not emit black smoke. Both of these are causes or consequences of excessive consumption as well as matters which could result in test failure and prosecution. Fuel pumps and injectors should be properly serviced as recommended by the manufacturers.
2. *Wheels and brakes.* Wheels should turn freely and without any brake binding. Front wheels should be correctly aligned. Brake binding and misalignment cause unnecessary friction which is only overcome by the use of more fuel. Axles on bogies should be correctly aligned because tyres running at slip angles have a high rolling resistance and therefore are a source of increased fuel consumption.
3. *Driving controls.* Throttle cables, clutch and brake pedals should be correctly adjusted so the driver has efficient control over the vehicle. In particular, engine tickover should be accurately adjusted to save throttle 'blipping' to keep it running when the vehicle is stationary.

Efficient Use

Inefficient use of vehicles constitutes the greatest waste of fuel. The following activities should be avoided by careful route planning, scheduling and prior thought about the cost consequences:

1. Vehicles running long distances when only partially loaded.
2. Vehicles covering excessive distances to reach their destination.
3. Large vehicles being used for running errands or making small item deliveries which could be accomplished more efficiently, and certainly more economically, by other means.
4. Vehicles running empty.

It is frequently argued that traffic office staff have little control over these matters, since customer demands for orders and the need to give drivers freedom to choose routes are dictates which overrule efficient planning. Nevertheless, attempts should be made to persuade those concerned of the need for restraint in the quest for saving fuel and thereby reducing costs.

Fuel Economy Aids

The quest for fuel saving has led to a market for economy aids which can be added to existing vehicles. These aids fall into three general categories:

1. Streamlining devices such as cab-top air deflectors, under-bumper air dams, front corner deflectors for high trailers and box vans, in-fill pieces for lorry and trailer combinations and shaped cones for addition to the front of van bodies.
2. Road speed governors which restrict maximum speed – one of the greatest causes of excessive fuel consumption (speed limiters are now a mandatory requirement on certain heavy vehicles – see Chapter 13).
3. Engine fans and radiator shutters which are designed to ensure that diesel engines are always operating at the correct temperature to give the most efficient performance and fuel economy.

All these types of device can be justified economically to a varying degree, but it is important to note that fitting streamlining devices in isolation only reduces fuel consumption if vehicle speeds are kept down. If the driver is able to use the few extra miles per hour which these devices provide – which he will do unless otherwise restricted – then there will be little fuel saving and the cost of fitting would not wholly be justified.

Fuel and Tyres

The type and condition of tyres on a vehicle play a significant part in its fuel consumption. It is a proven fact that the lower rolling resistance inherent in radial ply tyres adds considerably to the fuel economy of the vehicle compared to the greater resistance of cross ply tyres.

Improvements in fuel consumption of 5 to 10 per cent can be expected from the use of radial ply tyres. Low profile tyres which offer a number of operational benefits over conventional radial tyres – such as reduced platform height and reduced overall height – also offer further possibilities for fuel saving.

The savings mentioned will only be achieved if the tyres are in good condition, are correctly inflated to the manufacturer's recommended pressures and are properly matched, especially when used in twin-wheel combinations. Neglect of tyre pressures is common in fleets and under-inflation is one of the major causes of tyre failure. It is also a major contributor to excessive fuel consumption.

Fuel and the Driver

Driving techniques, above all else, influence the overall fuel consumption of vehicles. A driver with a heavy right foot will negate all fuel-saving measures and devices and destroy any expectations of acceptable fuel consumption. Poor driving which has these consequences falls into two categories:

1. High-speed driving
2. Erratic, stop-go driving.

Fast driving consumes excessive fuel: this fact is beyond question but the extent of the extra consumption is difficult to assess accurately. Tests carried out some time ago (but still making a valid point) by the National Freight Company with a 32-ton articulated vehicle on motorway operation indicated that fuel consumption increased quite dramatically when the vehicle was travelling at over 40mph. In the tests, at 40mph the fuel consumption was 10.5mpg, at 50mph this fell by 2.6mpg to 7.9mpg and at 60mph a further reduction of 1.5mpg was experienced, making a 3.75mpg difference between 40mph and 60mph. This represents a 35.7 per cent increase in fuel consumption. The real significance of these figures will be fully appreciated when annual motorway mileage is calculated and multiplied by the increase in consumption, by the number of vehicles in the fleet and by the cost per gallon of diesel fuel.

In round figures the magnitude of the results of such a calculation could be as follows:*

Vehicles travel 40,000 miles per year on motorways;

3809.5 gallons @£2.27 per gallon = £8647.56 pa
6250 gallons @£2.27 per gallon = £14,187.50 pa

Extra consumption = 2,441 gallons @£2.27 per gallon = £5541.07 pa

If five vehicles are involved the extra cost of fuel amounts to £27,705 per year.

*NB: In the calculation above, an October 1994 UK diesel price of 50.10 pence per litre was converted to an equivalent gallon price.

The effects of erratic driving are more difficult to determine in quantitative terms, but it is sufficient to say that it results in abnormally high fuel consumption as well as causing excessive wear and tear on vehicle components. Impatience behind the wheel and an inability to assess in time what is happening on the road ahead leads the driver to see-saw between fierce acceleration to keep up with the traffic and violent braking to avoid running into the vehicle in front. Hence, the excessive use of fuel.

More economical driving is achieved by concentration on the road and traffic conditions ahead, anticipating well in advance how the traffic flow will move, and what is happening in front so that acceleration and braking can be more progressive and a smooth passage assured.

One other fuel-saving practice which the driver can adopt is to stop the engine while the vehicle is stationary rather than letting it tick over for unnecessarily long periods. If he feels this is necessary because his battery is in poor condition, then it is much cheaper to deal with the battery and charging problems than pay for the extra fuel to compensate.

Fuel and Fleet Management

Fleet operators can take a number of steps to reduce fuel consumption besides ensuring that the measures already mentioned are implemented.

Bulk Supplies/Buying

Control over the buying of bulk supplies and issues and over the buying of supplies outside from filling stations at the higher pump prices are important aspects for management attention. So too is accurate record keeping without which it is impossible to compare the fuel consumption of vehicles or to see which vehicles are consuming excessive amounts of fuel. Without records to identify these problems, there is no hope of remedying high fuel costs.

Recording Issues

Overfilling of vehicle tanks causing spillage is a common occurrence which wastes fuel and prevents accurate consumption records being obtained, as well as causing a mess and a health hazard. Accurate recording of issues against individual vehicles is another area which demands close attention. Modern electronic and computerised fuel-dispensing systems are available which ensure security of bulk supplies by preventing access except with a known key or card. This stops unauthorised drawing of fuel and it monitors issues to identified vehicles or key holders. The cost of such systems can be quickly recouped through savings in missing or unaccounted fuel and through better record keeping which enables high vehicle consumption to be quickly identified and investigated.

Buying Away from Base

Where bulk supplies are available at base, drivers should be discouraged from buying supplies from outside sources at higher prices except when absolutely necessary and even then they should buy only sufficient to get them home. Commonly, drivers fill tanks to the top which is enough to cover the trip home and to do further journeys as well on fuel which cost much more than that which they could have drawn once they got back to base. It is important to keep watch on the purchase of outside fuel supplies especially for cash because of the incentives which are offered to drivers to fill to the top. If outside fuel drawings are necessary then recognised bunkering card systems should be established.

Long-Range Tanks

In the case of vehicles which operate regularly on long distances – on international work, for example – the fitting of long-range tanks is an economic proposition because of the savings achieved by using bulk-purchased supplies. A word of caution though, the extra fuel carried should not be at the expense of payload unless this can be justified and on international work there is the cost of fuel levies or taxes to be borne in mind. These are chargeable when entering some countries and are based on the amount of fuel in the tank (see also p 152). British hauliers returning to the UK may also find they have to pay an excess fuel tax on supplies bought outside the UK.

Route Planning/Scheduling

Better planning of vehicle schedules and routes offers the prospect of quite considerable fuel savings, besides other savings which may also result. A reduction in the miles travelled to fulfil particular delivery schedules will inevitably result in fuel savings. If the schedules can be planned so that fewer vehicles are needed to carry out the operation then, besides the broader savings in vehicle costs, the fleet as a whole will use less fuel. Therefore, the elimination of unnecessary trips or trips where vehicles are only partly loaded is a major priority in the search for fuel cost savings.

Tachographs in Fuel Saving

The use of tachographs in vehicles and detailed analysis of tachograph charts are steps which have provided many fleet operators with fuel savings if nothing else. The value of chart recordings in this connection should not be overlooked.

Agency Cards

There are a number of agency and fuel cards available to fleet operators. Generally they can be categorised between those from the major oil companies (eg Shell, Esso, BP) and those from other commercial organisations (eg All-Star, Overdrive, Dial Card, Petrocheck, BRS Transcard).

Principally these systems offer the opportunity to buy fuel at pump prices, paying only when fuel is actually purchased. The differences in the various schemes is in the levels of service which they offer.

Shell Scheme

Shell claims to lead the field with its portfolio of five fuel cards, each designed to suit a particular business need. For example, the Shell Gold Card is intended mainly for the car and light vehicle operator with 20 plus vehicles, but for the heavy truck operator the convenient Shell agency card is more appropriate.

The Shell scheme offers a complete fleet management service enabling the authorised card holder to purchase all the legitimate items a vehicle user would need on the road (eg fuel, oil, tyres, batteries, windscreens, other parts, repair assistance). Prices charged are those which reflect maximum fleet discount no matter who or where the supplier. Thus the small fleet operator gets the same benefit as the large operator. Purchases are consolidated onto one VAT invoice and these invoices are submitted twice monthly. A 25-day credit period is given.

In addition to the invoice, Shell sends the operator three computer 'reports' as follows:

1. Vehicle transaction report and analysis
2. Cost-centre summary
3. Fleet report.

Use of the Shell Gold Card automatically confers AA membership and this can be voluntarily extended to include AA's Relay and Home Start services. The balance of any existing AA membership at the time of joining the Shell scheme is refunded by the AA.

The Shell European agency card, EuroShell, enables holders to obtain fuel and service at some 11,500 locations in 27 countries, of which very many have facilities to accommodate large goods vehicles, and pay tolls on certain European motorways (eg in France, Portugal and Austria) and at a number of tunnels.

BP Scheme

The BP Agency Card can be used at some 5000 filling stations in the UK, but only for fuel and lubricants. There is no charge for the card and fuel is paid for at the average BP UK pump price for the appropriate grade of fuel irrespective of where the fuel is purchased. The system provides the operator with an average of 27 days of credit – a useful saving device in these needy days of stringent economies.

An international version of the card enables holders to obtain fuel at over 6000 service stations in 14 European countries.

Automatic Bunkering

A number of automatic diesel fuel bunkering systems have been established recently to improve fuel supply services to heavy truck operators. Notable among these are Kuwait Petroleum (GB) Limited (selling under the Q8 brand name) which has introduced its International Diesel Service (IDS) in the UK in addition to its Europe-wide network. The Kuwait scheme features fully-automatic diesel sites, operating round the clock every day of the year, at which card-holding truck drivers can obtain fuel at an agreed price. The scheme is secure because the cards have a unique and secret four-digit PIN code which prevent unauthorised use. The latest-technology pumps enable drivers to refuel rapidly in their own currency and their own language. The operator receives a comprehensive invoice at regular intervals with fuel drawn charged at a competitive price with VAT recorded and a facility to reclaim VAT where appropriate.

A basically similar scheme is operated by Mobil Oil with its Mobil Diesel Club (MDC). The company has about 100 outlets on key truck routes in 14 countries including the UK. A security-coded card is used as described above to obtain fuel from automatic high-speed pumps.

Both Phillips Petroleum (RouteMate) and Gulf Oil (GB) Limited (G-stop) have agency systems providing bunkering networks with competitive fuel price structures. Other systems include independents such as Keyfuels and Fuelink (NFC Group).

Fuel Economy Checklist

Check
- Fuel systems free from leaks.
- Fuel pump and injectors serviced and correctly adjusted.
- Exhaust not emitting black smoke.
- Air cleaners not blocked.
- Engine operating at correct temperature.
- Wheels turning freely.
- Controls properly adjusted and lubricated.

Tyres
- Condition and inflation pressures.
- Possibility of changing to radials on all vehicles.

Drivers
- Speed limits not being exceeded.
- Driving methods smooth and gentle.
- Engines stopped when vehicle standing.

Management
- Control over supplies and issues.

- Avoidance of spillage, loss and unauthorised use.
- Purchases from outside suppliers kept to a minimum.
- Record systems accurate and up to date.
- Possibility of installing fuel issue and monitoring systems.
- Possibility of fitting fuel economy aids (deflectors, engine fans, etc).
- Possibility of using long-range fuel tanks on vehicles.
- Routeing and scheduling practices to reduce wasted journeys and unnecessary mileage.
- Fuel economy programme to ensure all possible steps being implemented efficiently and recorded accurately.

27: Mobile Communications and Information Technology in Transport

There has been a significant increase in interest in all types of mobile communications in recent years. No longer is the in-vehicle radio or radio telephone an executive toy to be fitted only in the chief executive's car. These days such equipment has proved and continues to prove itself to be a very cost-effective and efficient aid to a wide range of vehicle users from the company chairperson down through all levels of executive vehicle users, to the sales representative, the service engineer and the delivery driver. In fact, wherever it can be established that there is a need and justification for the vehicle driver to be in contact with his base, with colleagues, with customers and with others, here are the requirements for vehicle-based mobile communications.

Before looking at the systems of communication available it is useful to consider just what the benefits are to have vehicle drivers – whatever their status or purpose – in contact with others outside the vehicle. It is widely recognised that once a person gets into a vehicle and drives off he is totally cut off from his workplace and the people with whom he normally deals in the way of business. And until he arrives at a known destination where a message can be relayed to him, or unless he manages to find a roadside telephone which operates (it is commonly felt that few of them do so when needed), he remains out of contact and out of touch with what is going on in his business or the business of his employer.

Clearly, in these days of high costs, competitive marketplaces and a fast pace of business life, and taking account of technological developments, this is becoming an unacceptable penalty of having people travel by road, especially during working hours. The inability to contact a top executive could, in the extreme, result in missed opportunities in business deals. The inability to contact a sales representative or a service engineer could mean a lost order or at least an irate customer; the same applies with a delivery driver. In addition, the inability to contact a driver could mean wasted journeys because of cancellations or changed plans which arise after they have set out on their journey. It could mean a driver getting back to base and having to go back to a customer visited earlier, simply because he could not be contacted *en route* to be warned of late, or forgotten, orders or items.

It is a common experience in transport operations for considerable cost to be wasted through late, changed, redirected, cancelled orders and instructions. Try as he might, the transport or fleet manager can rarely avoid his share of these annoying frustrations and there is nothing that he can do usually because he cannot contact the driver while he is travelling.

Vehicle-based mobile communications at today's level of sophistication can change all this and the wasted costs of the past can be turned into savings and even into profit quite simply by being able to contact the driver, relay details of the changed plans and generally divert vehicles to meet the current needs of the business. The savings in wasted time and miles, the avoidance of heavy vehicles returning home empty because they can be directed to pick up return loads, and the response to last-minute customer demands are significant benefits which in themselves, or with other benefits, add up to offset the capital costs of buying and installing communications equipment and the on-going costs of rentals and call charges.

Choice of Communications

Mobile communications can be reviewed under four broad headings as follows:
1. CB radio
2. Radio paging
3. Private mobile radio (PMR)
4. Cellular telephone.

CB Radio

Citizens' Band (ie CB) radio was officially inaugurated in Britain in 1982 when the law made it permissible to operate such systems which had hitherto been illegal and had caused difficulty because their use interfered with the radio links of the emergency services and other official networks. Initially, CB was used by enthusiasts as a means of 'friendly' communication, to chat to other users and generally communicate non-business information. In time, however, it proved capable of serving more important needs such as the reporting of accidents and other emergencies, breakdowns, road blockages, diversions and so on. Similarly, it proved to have some use in providing communication between vehicle drivers and their base – within limited range – for the purposes of passing information about loads, schedules and changed instructions, for example.

Despite its obvious use for these purposes, CB has significant disadvantages: the frequencies are cluttered with undisciplined, long, sometimes foul and frequently frivolous conversation and chatter which can be heard by all users due to lack of privacy. This in itself is another disadvantage, along with the general interference experienced and the congestion on channels. Furthermore, the equipment itself has its limitations in terms of power and range, and in some areas there is less than satisfactory reception.

CB Licences
CB users (who must be over 14 years of age) must, by law, obtain an annual radio licence from a post office (using form CB 01) at a cost of £10 which covers the operation of a maximum of three sets. Further, licences costing £10 each per annum, are needed where more sets are to be operated but where licences are required for more than 15 sets postal application has to be made to the CB Licensing Unit, Chetwynd House, Chesterfield, Derbyshire S49 1PF (the same CB 01 application form is used). The application form in both cases

requires only the name of the person applying (who becomes the licensee), and their address with post code.

The licence permits the licensee to operate only the specified number of CB 'sending and receiving stations for wireless telegraphy using only apparatus which conforms in all respects to the Department of Trade and Industry (DTI) specifications MPT 1320 and/or MPT 1321'. Suitable equipment may be recognised by the presence of one or other of the DTI approval marks.

Power Limits

Power limits must not be increased above that stated in the specification mentioned above and there are limits on the antenna which may be used. With 27 MHz apparatus, the maximum length permitted is 1.65 metres and maximum diameter is 55 millimetres and with 934MHz apparatus – with provision for connection to an external antenna – a maximum of four elements is permitted none of which must exceed 17 centimetres in length. Power amplifiers and antenna are not permitted.

Code of Practice

A Code of Practice (copies available free from post offices) has been established by the Radio Regulatory Division of the DTI along with representatives of CB groups and a Parliamentary Working Party on CB Radio. The following is extracted from this Code.

How to Operate

1. LISTEN BEFORE YOU TRANSMIT. Listen with the Squelch control turned fully down (and Tone Squelch turned off if you have Selective Call facilities) for several seconds, to ensure you will not be transmitting on top of an existing conversation.
2. KEEP CONVERSATIONS SHORT when the channels are busy, so that everyone has a fair share.
3. KEEP EACH TRANSMISSION SHORT and listen often for a reply – or you may find that the station you were talking to has moved out of range or that reception has changed for other reasons.
4. ALWAYS LEAVE A SHORT PAUSE BEFORE REPLYING so that other stations may join the conversation.
5. CB SLANG ISN'T NECESSARY – plain language is just as effective.
6. BE PATIENT WITH NEWCOMERS AND HELP THEM.

Emergencies and Assistance

7. AT ALL TIMES AND ON ALL CHANNELS GIVE PRIORITY TO CALLS FOR HELP.
8. LEAVE CHANNEL 9 CLEAR FOR EMERGENCIES. If you have to use it (for instance to contact a volunteer monitor service) get clear of it as soon as you can.
9. IF THERE IS NO ANSWER ON CHANNEL 9, then call for help on either channel 14 or 19 where you are likely to get an answer.
10. IF YOU HEAR A CALL FOR HELP - WAIT. If no regular volunteer monitor answers, then offer help if you can.
11. THERE IS NO OFFICIAL ORGANISATION FOR MONITORING CB AND NO GUARANTEE THAT YOU WILL ALWAYS BE IN REACH OF A VOLUNTEER MONITOR.

CB IS NOT A SUBSTITUTE FOR THE 999 SERVICE ASHORE OR FOR VHF RADIO (CHANNEL 16) AFLOAT

Choice of Channel

12. RESPECT THE FOLLOWING CONVENTIONS:

Channel 9:	Only for emergencies and assistance.
Channel 14:	The calling channel. Once you have established a contact, move to another channel to hold your conversation.
Channel 19:	For conversations among travellers on main roads. (Remember, if you are travelling in the same direction as the station you are talking to, not to hog this channel for a long conversation.) Give priority to the use of this channel by long distance drivers to whom it can be an important part of their way of life.
Other:	You may find that particular groups in particular areas also have other preferred channels for particular purposes.

Interference

13. INTERFERENCE can be caused by any form of radio transmission. Avoid the risks. Put your antenna as far away as possible from others, and remember that you are not allowed to use power amplifiers. In the unlikely event that your CB causes interference, co-operate in seeking a cure using the suggestions from a good CB handbook. Moving the set or antenna a few feet may cure the problem.

Safety

14. NEVER ERECT OR USE AN ANTENNA UNDERNEATH OR NEAR AN OVERHEAD ELECTRIC LINE. Several CB enthusiasts have been killed because they did not appreciate the dangers involved. Keep antennas well clear of overhead lines at all times. If an antenna is already sited near a power line, do not attempt to remove it but get in touch with the local electricity company for advice.

15. WHEN MOUNTING ANTENNAS ON HIGH VEHICLES make sure that the top of the antenna is not so high that it is likely to foul overhead electrified lines at railway level crossings.

16. USE COMMON SENSE WHEN USING CB and do not transmit when it could be risky to do so. For example, don't transmit:

    (a) when fuel or any other explosive substance is in the open – eg at petrol filling stations, when petrol or gas tankers are loading or unloading, on oil rigs or at quarries

    (b) when holding a microphone may interfere with your ability to drive safely

    (c) with the antenna less than 6 inches from your face.

Radio Pagers

Radio pagers (commonly called bleepers) are a portable means of contact but only on a one-way basis from the sender to the receiver. Pagers fall into two main categories, tone pagers and voice pagers. With either type a person can be alerted to the fact that he or she is required and with more sophisticated equipment can be advised by varying tones to follow specific predetermined instructions, for example, to ring one telephone number or another. Voice pagers or 'talking bleepers' convey a spoken message which is usually repeated twice, thereby alerting the user to follow specific courses of action. Some pagers are capable of alerting the user by vibration (silent pagers) so that outsiders are not aware of the sound or so that the user is not interrupted in mid-conversation by an audible 'bleep'. Others have the ability to display messages on a liquid crystal screen from a text memory of up to some 800 characters.

While the cost of pagers is relatively low (usually only a few pence per day) and the unit itself is small and unobtrusive to carry around in the pocket (some are no bigger than credit cards) there are disadvantages to their use. The first is the limited range (generally not more than about 15 miles) and the second is the need for the receiver to find a working telephone in order to make the call for which he has been alerted.

Private Mobile Radio (PMR)

Portable radios are used for private two-way communication usually between a base station and a number of mobiles (ie radio-equipped vehicles) with the added facility in some cases of the mobile units being able to talk to each other. Generally, mobile radio operates over a limited range of some 10 to 20 miles, depending on location and the height of the base station aerial. Much depends on the type of terrain between the base station and the mobile unit. In open country far greater range may be obtained than in a city with built-up areas and many tall buildings.

Many systems have the disadvantage that anybody in the vehicle or within earshot can hear the message being relayed and that all mobile units hear a message intended for one only. However, some equipment has a facility for selective calling so that only one unit need be contacted at a time.

In the main, mobile radio is restricted to a closed system but there are facilities whereby this can be extended by linking into one of the national relay networks which have many base stations throughout the country. The Securicor 'Relayfone' system is a good example. Developments are in hand to allow, in the future, interconnection of private mobile radio systems into the public switched telephone network (PSTN).

Typical of PMR systems are those used by local taxi services where the driver has a hand-held microphone with an 'ON/OFF' switch. This PTT (press to talk) switch must be depressed to enable the driver to talk to the base station and then released while he listens to the returning message. The disadvantage of this is the road safety risk created by a driver trying to control his vehicle and operate the microphone switch (the *Highway Code* warns against such practices, and the police are alert to this habit and will take action against offending drivers).

Mobile radio operates on a number of alternative radio frequencies as follows.

Low band VHF (25–50MHz) provides the greatest range but it suffers from high noise levels and heavy channel loadings (used by police and emergency services).

High band VHF (150–174MHz) provides good coverage in built-up areas with less noise but there is still heavy usage of the channels.

UHF (450–512MHz) has a much shorter range than VHF but there is much less congestion on the channels and it provides good penetration in urban areas where there are many buildings and tall structures.

Within these radio frequencies equipment may be obtained for AM or FM operation.

Band 3

Increased demand for mobile communication has led to the allocation by the government of the VHF slot left vacant by the old 405-line black and white TV network for use in providing communications services. This is now called Band 3 PMR and is operated by the two official franchise holders GEC and Band Three Radio each with 200 channels.

This system will be of interest to those requiring only brief communication (ie not full conversation) between base station and vehicle (mobile unit), but with little need for communication outside and who wish to avoid the relatively high call costs of cell-phones and the rather higher capital costs of the cellular telephone units themselves.

Cellular Telephone

The advent of the cellular telephone system has revolutionised mobile communications. Today it is possible to have a telephone interconnected to the national and international telephone networks from a vehicle-based (ie mobile) installation or from a set carried neatly in an executive briefcase or even in a jacket pocket. Such systems provide the user with the facility to dial direct to almost any telephone number in the UK or to reach such numbers via the British Telecom operator, and to make international calls and calls to any other cellular telephone. Similarly, any telephone user can dial direct to a mobile cell-phone number.

The principle of the Total Access Communications System (TACS) cellular telephone system is that instead of a connection by wire as with the conventional telephone, the link is made by radio airwaves in the 900MHz radio band frequency using only 50MHz bandwidth. The UK is now divided into a number of individual cells like a honeycomb. Each cell is anything from 2 kilometres to 30 kilometres across with a transceiver which relays cellular calls to and from the normal public switched telephone network (PSTN), as well as from one cellular telephone to another.

A central computer (the brains of the system) monitors all traffic in the system and switches calls from one transceiver to another as the mobile cell-phone user travels from one cell into another. Calls go through both the cellular network and the British Telecom network, which is why call charges are higher than with the normal PSTN system, especially in the London area.

The government has licensed two operators to provide the network, namely British Telecom (Cellnet) and Racal (Vodafone) and currently some 90 per cent of the UK population is covered by both systems.

Equipment for operation within the cellular telephone system can be divided into three groups.
1.   Mobile units (eg installed in vehicles).
2.   Transportable units (for use in vehicles or can be carried in a briefcase for example).
3.   Portable units (small units which can be readily carried around and even fitted into a jacket pocket).

Voice Activation

A relatively new development in vehicle-based mobile units is voice activation to overcome the problems (and illegalities*) of answering the telephone and dialling numbers for outward calls while actually driving the vehicle. This equipment is programmed to recognise a voice signal (usually just a single word) spoken into the handset which sets off the dialling of a predetermined number (eg base, office, home). Other equipment is dashboard mounted so calls can be received and made 'hands off' to avoid the road safety risks and contravention of the advice given in the *Highway Code* against using 'phones' while on the move.

*NB: While it is not strictly illegal to answer or dial-out when driving, a policeman may, however, prosecute for not exercising full control of the vehicle (but the same could apply when changing a cassette tape or unwrapping a toffee!) – see p 434.

Equipment and Charges

A wide range of equipment is currently available from many suppliers in the market, operating on either or both Cellnet and Vodafone systems. Equipment can be purchased outright, leased or rented for a short period. The costs and charges incurred fall under the following headings.
1.   Purchase/lease of equipment.
2.   Installation charges (ie fitting transceiver and aerial to vehicle).
3.   Connection charges (once only).
4.   Monthly subscription charge.
5.   Call charges.

Call charges are invoiced to the user on a monthly or quarterly basis along with the subscription charge. These call charges are not governed by distance in the same way as normal BT call charges, thus a local call costs the same as a long distance call of the same duration. Outsiders making calls from a BT (ie PSTN) telephone to a cellular telephone number also incur higher costs at the 'M' rate (currently equivalent to a call from the UK to the Republic of Ireland).

Developments

Such is the rapid pace of development in this field that already a wide range of 'add on' facilities are, or soon will be, available to users. Links to data transmission equipment to allow communication between computers, the ability to send telexes and to interface with fax machines are all current possibilities which substantially extend the use of a mobile or transportable cellular telephone.

Choosing Cell-Phone Suppliers
Potential cell-phone users should contact a number of suppliers for details and demonstrations of the variety of equipment available before selecting alternatives and making final decisions. The RHA has negotiated a special deal with Racal Vodac for its members. In particular, attention should be given to the strength of signal obtained on each system (Cellnet or Vodafone) in certain areas, the efficiency of installation and the back-up services provided as well as the equipment itself. There is a great deal of choice and competition in this field so keen prices should be obtained.

Warning
It is useful to caution users and potential users of cellular telephones about three particular matters as follows.
1. Proper installation of equipment in vehicles is essential for both efficient operation and for safety reasons. Special care is needed in the case of installation in heavy vehicles with 24 volt electrical systems to avoid wiring faults and other electrical problems.
2. Insurance of equipment in vehicles is important; it is highly attractive to thieves. If stolen or lost in a vehicle accident or fire an insurer may decline to accept a claim if the installation of the equipment had not been notified beforehand. In general, motor insurance does not cover a cellular phone. It is relatively easy for the subscriber to prevent fraudulent use of a stolen set by notifying the air-time retailer. The retailer can disconnect the stolen unit remotely, thus preventing further use. Mobile cell phones are not easy to steal because the handset, transceiver and wiring loom have to be removed.
3. Users should be aware that foreign customs officials may impound mobile communications handsets when entering certain countries because use of the equipment is not compatible with overseas telecommunications networks and might interfere with emergency services.

Safety

The *Highway Code* now includes a special section on the use of microphones and car telephones. It advises against using a hand-held microphone or telephone handset while driving, except in an emergency. It says the driver should only speak into a fixed, neckslung or clipped-on microphone when it would not distract attention from the road. Drivers should not stop on the hard shoulder of a motorway to answer or make a call, no matter how urgent.

Satellite-Based Communications Systems

Technological developments in satellite-based communication systems provide facilities for long-range telephone, fax and paging links between base and vehicle, as well as positive vehicle tracking systems. Such systems are now widely used in North America and are of increasing interest to UK and European transport fleet operators. Using the same basic technology that puts instantaneous live pictures from sporting events and news reports on our television screens, the precise location of vehicles can be pinpointed and messages passed, but on a one-way basis only from base to vehicle, not vice versa, so drivers cannot abuse the system by calling friends and relatives world-wide.

It is not likely that satellite communications will replace other two-way mobile communications systems but with the potential proliferation of Euro-wide transport operations following the opening of the Single European Market in 1993, it will provide operators with a reliable and spontaneous means of contacting their drivers thousands of miles from base and at a price which, in terms relative to the cost of the driver constantly telephoning home to see if he is wanted, would be considered cheap.

Currently a number of organisations are running trial systems including British Telecom Mobile Communications (BTMC) via the Inmarsat satellite system, a consortium in which DAF Trucks has an interest (Roadacom), Locstar (a French company backed by Daimler-Benz and British Aerospace), and a joint US/French operation called Qualcomm-Alcatel in conjunction with Eutelsat (the European satellite consortium).

Information Technology in Transport

Reports prepared by Management Consultants KPMG Peat Marwick McLintock for the Institute of Logistics indicate that the use of information technology in distribution is growing rapidly with expenditure on this element amounting to some 2 per cent of turnover.

Information technology (IT) is quite simply the means by which information is collected accurately, fully analysed, transmitted to all relevant functions within and without an organisation, and disseminated by those charged with decision making. Many efficient transport and distribution operations these days are totally dependent on reliable IT systems (for example 'just-in-time' stocking principles hinge on rapid and efficient communications) for communication with customers, suppliers and contractors.

Within IT systems, one of the fastest-growing areas is that of Electronic Data Interchange (EDI) where transport and distributor firms receive orders, delivery documentation and invoices direct from their customers' computers into their own (compatible) systems via direct communications links, completely eliminating the delays, errors and other difficulties associated with the creating and movement of paper documents via postal and courier systems. EDI systems provide the benefits of rapid and accurate order passing and processing (at less direct cost) and with the potential for reducing stock levels.

KPMG noted in its reports to the Institute of Logistics the trend away from centralised computer systems for IT and a growth in the use of personal computers operated in linked systems or networked to a central mainframe computer.

28: Transport and the Environment

One of the most important and widely discussed issues in transport in the early 1990s has been (and will continue to be) the impact which the industry as a whole, and heavy lorries in particular, have on the environment. So-called 'green' issues feature in every aspect of transport operation from the siting of vehicle depots, to the routeing of heavy goods traffic and the disposal of certain loads, especially waste.

This *Handbook* charts in earlier chapters the legal requirements regarding choice of vehicle operating centres to satisfy 'O' licensing requirements, and the need for licence applicants to advertise their proposals in this regard and defend their premises against potential environmental representations from local residents. In the technical section, legal requirements concerning the emission of noise, smoke, fumes and exhaust pollutants are detailed, and in Chapter 23 the subject of waste carriage is covered with descriptions of the legal requirements regarding hazardous waste and the more recent legislation on 'controlled' waste – which is basically everyday waste emanating from commercial and industrial premises which is now subject to stringent control as to its carriage and disposal.

Currently, two of the UK's major transport groups, Excel Logistics (part of the NFC Group) and the Transport Development Group have developed clear environmental policies. Wincanton Distribution Services, too, has determined a policy 'to develop, maintain and operate [its] resources in an environmentally responsible manner'. These initiatives are concentrated mainly on fuel conservation at this stage which has a direct payback benefit by way of cost savings and an indirect benefit in the shape of an improved company 'image', as well as achieving actual reductions in the amount of carbon dioxide which their respective company vehicles discharge into the atmosphere. This in turn helps to reduce global warming which scientists have identified as being a problem of catastrophic proportions. However, there are many other ways in which these and other firms are contributing to the environmental effort, even down to the use of recycled paper for routine stationery needs, for example.

Another major step towards improving the environmental impact of transport and distribution operations has been taken by the Institute of Logistics with the publication of a three-part practical handbook* covering three distinct aspects of the subject – namely, Volume 1 which deals with transport, Volume 2 which covers non-transport and specialised logistics operations and Volume 3 which concerns environmental management.

NB: Copies of the handbook are available from the Institute at Douglas House, Queens Square, Corby, Northamptonshire NN17 1PL (price £195).

The Department of Transport foresaw the need for initiative on the environmental front by publishing its booklet *Transport and the Environment*, to coincide with the European Year of the Environment. This booklet was seen as a useful introduction for school pupils to the complex issues of environmental decision making and it provided a guide to environmentalists as to the DoT's activities in this field. It considers the impact of the major transport modes (including road freighting) and examines the effects of the Channel Tunnel and the problems of transport in towns, as well as outlining the work of the Transport and Road Research Laboratory (TRRL).

Yet another publication devoted to the subject is produced by IBM UK Limited from a text researched and written by Dr Peter Davis of the National Materials Handling Centre, Cranfield, Bedford MK43 0AL, under the title *Gearing up for the Environment – A Guide for Managers in Distribution*. More recently, the Road Haulage Association (RHA) published an environmental *Code of Conduct* following publication of its *Care of the Environment* leaflet published in 1989.

Such is the concern for this topic that some of the trade associations now have full-time officers concentrating on environmental issues and many firms have linked responsibility in this area with those of safety, health matters and quality.

Impact of Transport

Transport affects the environment in a variety of ways, of which some are more distinctly controllable by fleet operators than others and some produce more tangible benefits both to the vehicle operator and to the community than others, but it has to be remembered that, inevitably, any form of transport, serving any and every need imaginable, has an adverse impact on the environment. There is no such thing as a totally environmentally acceptable form or means of transport. What there can be, however, are means and systems of transport which are more 'friendly' towards the environment, and which can be controlled and managed in such a way that the environmental impact is minimised.

So far as transport operations are concerned the following list identifies some of the main areas of environmental impact which can be, and should be, challenged by management.

Vehicle depots
- siting
- noise, fumes, vibration and light emitted
- disposal of waste

Vehicle operations
- engine, exhaust, tyre, body and load noise
- smoke, fumes, gases, spray emitted
- fuel/oil consumption
- visual impact
- routes and schedules
- load utilisation

In addition to these more environmentally controllable aspects of transport, there are other aspects which have a powerful impact, but over which the transport manager has virtually no control, namely vehicle design and manufacture, road planning and building, legislative controls which are not necessarily environmentally oriented and customer demand which is influenced more by commercial pressure and financial consideration than by the vehicle operator's quest to, among other things, reduce fuel consumption, deliver during non-congested times, or combine loads to improve vehicle efficiency.

Possible Solutions

Transport managers and small fleet operators are undoubtedly limited in the steps they can individually take towards improving the environment, but this does not mean they should take no steps at all. Simple measures are available to them which will make a valuable contribution. The following list provides just a few examples.

In the Depot

- Examine the way that waste material is stored and disposed of
- Ensure that controlled waste (see Chapter 23) is correctly and safely stored on site and then handed over to licensed disposal contractors
- Avoid burning of waste which can cause pollution and lead to complaints
- Ensure that recyclable material is identified and saved for proper disposal – including waste paper and packing materials from office and stores
- Take steps to ensure that vehicle washing does not result in dirty (ie grease-laden) water draining on to neighbouring properties as well as into sewage systems
- Ensure that oil and fuel spillages do not pollute drains
- Consider the use of recycled products such as paper for administrative uses and packing
- Undertake regular depot clean-up campaigns (in particular, ensuring that the outside appearance of the depot is 'environmentally friendly' to local residents, business visitors and others).

On the Vehicle

- Ensure that legal requirements regarding noise, smoke, exhaust emissions and spray suppression are fully complied with
- Take steps to economise on fuel consumption*
- Provide driver training to ensure courtesy and consideration on the road and the use of defensive driving methods* (which saves on wear costs)
- Fit speed limiters (now a legal requirement on certain vehicles – see Chapter 13)
- Ensure that drivers obey rules about parking and causing obstruction with their vehicles, and are aware of the problem of visual intrusion, noise and vibration on domestic properties (contravention of these matters can jeopardise 'O' licences – see Chapter 1)

- Route vehicles and plan journeys to avoid congestion – ensure full utilisation of vehicles to avoid extra or unnecessary journeys (which add to congestion, air pollution, and the operator's own costs)
- Consider the visual impact of vehicles in terms of their general appearance and livery (change aggressive liveries to present a 'softer' image).

NB: These matters are discussed in more detail in Chapter 26.

In the Community
- Consider the sponsorship of local community efforts to improve the environment and encourage staff to undertake environmental protection projects.

29: Quality Management in Transport

Increasingly, road hauliers are facing demands from customers, and from principal contractors, to meet recognised standards of quality assurance in the form of certification to the British Standard BS 5750, or alternative national standards in the United Kingdom – or to EN 29000 (Europe) or ISO 9000 (International). This chapter outlines the basic requirements for meeting quality assurance standards to achieve recognised certification to these standards.

A great deal has been written about the subject of 'quality', and many myths have been spread, but it is quite simply the concept of doing things right first time, and right every time. This saves having to repeat operations at extra cost, and annoyance to the customer, whether in production of goods or the provision of a service – road haulage, for example. It means supplying customers with the service they need, not what the supplier thinks he can best provide. In haulage, it means, particularly, providing cost-effective deliveries – on time, to the right address with the load intact and undamaged and delivered by a courteous driver in a presentable vehicle. It does not mean a service which causes customer complaint.

The concept of 'quality' is not new. It has been applied to production for many years, especially by the Japanese, who have captured world markets for cars, motorcycles, electronics and cameras due to the inherent quality, reliability, durability and desirability of their products. Only more recently has the quality concept been applied to service industries, and particularly to road haulage.

Achieving quality assurance (QA) certification involves complex steps, changed ideas, new thinking and acceptance that 'old ways' must be replaced by new methods, despite extra paperwork, form filling, writing of manuals, checking and re-checking of standards, and visits from inspectors.

What are quality systems and quality management? A quality system is one where problems, queries, faults, and anything which could give rise to customer dissatisfaction or complaint, are identified and eliminated. Every aspect of operating procedures is critically examined to ensure that nothing unexpected (short of pure accident – and contingencies can even be established for these) can arise to jeopardise service to the customer. Quality management is the management of quality systems – a totally new way of doing business, hence the expression 'total quality management' (TQM).

Quality Assessment and Accreditation

Quality assessment and accreditation is the process by which a firm demonstrates to an accredited certification body that its services meet pre-established quality standards, followed by certification of this fact.

Accreditation

The British Standards Institution (BSI) standard for quality management systems is designated BS 5750. This incorporates the provisions of both the International and European standards and is fully accepted in the 12 Member States of the European Community and the seven European Free Trade Association (EFTA) members – Norway, Sweden, Finland, Iceland, Austria, Switzerland and Liechtenstein. There are a number of separate constituent parts of BS 5750 and ISO 9000 but those which specifically relate to road haulage are BS 5750 part 2 and ISO 9002.

Firms whose quality systems meet specified standards may register with an approved body and, on satisfactory completion of the formalities, receive accreditation – 'Registered Firm' status – when they may use the accreditation body's symbol of approval on its company literature (ie letterheads, brochures) and on vehicles. A road haulier accredited under BS 5750 in the UK is additionally accepted as meeting both European and International quality standards, an essential ingredient for trading in the single European market.

Standards

The essence of a quality haulage service is that every step in fulfilling customer orders is undertaken in accordance with a documented standard – a set of rules governing the best way to operate and against which day-to-day operations are compared. Thus, performing to standard means performing as set out in the rules. BS 5750 is a standard for quality systems which identifies the basic disciplines and specifies the procedures and criteria to be applied to ensure that services are of a quality that will always meet specified customer requirements.

Assessment by Accredited Certification Bodies

Assessment for accreditation involves a number of stages. Among these, one is the need to establish, document and maintain a quality system demonstrating a commitment to quality and to meeting customer service needs. Another is the selection of an appropriate accreditation body to which application for registration is made. This involves providing information about the business, its size and scope, the number of employees, how many locations are involved, the particular nature and manner of its operation and the services provided. Applicants must submit their documented quality system for examination and approval. The accreditation body checks this to ensure it covers all aspects of the quality system standard and then follows up with site visits to thoroughly review the operation in practice. These inspections – made by specialists in quality assessment – are a key element in quality assurance. Documented procedures are examined in detail and systems observed in operation to ensure that day to day procedures follow the documented quality system in every aspect and comply with the laid-down standards of the accreditation body.

Monitoring

A process of continuous monitoring for compliance with standards is maintained through inspectors from the accreditation body making regular visits – up to four times each year. Any drop in performance will require renewed efforts to bring procedures back to standard. Continued failure to meet the standard will result in withdrawal of registration.

Establishing a Quality System

To establish a quality system within a firm, it is necessary to identify key aspects of the business where quality principles are to be applied and record precisely how these should be carried out. This is achieved by translating individual tasks into descriptive text in manual form, and by making step-by-step checklists or by establishing Codes of Practice. The key aspects will include:

- Setting company policy and objectives for quality systems
- Determining the structure of the organisation and responsibilities
- Preparing instructions and Codes of Practice for quality work standards
- Monitoring subcontracted supplies and services for quality
- Establishing operational methods and controls to meet specified standards
- Inspection and monitoring of the transport service
- Controlling defective work
- Determining corrective actions before problems arise
- Measuring and recording quality performance
- Determining training requirements
- Establishing quality audit and review procedures.

Determining Customer Needs

Quality starts with the customer, the need to determine exactly what every customer wants. Never assume that one customer's service requirements are identical, or even remotely similar, to those of another. Each must be asked individually to define his precise expectations when making enquiries and placing orders. They will differ widely: a customer may even have differing service requirements in differing circumstances. These requirements, whether few or many, whether standard in all circumstances or varying widely for individual consignments, must be clearly understood by the haulier, leaving no doubt whatsoever as to what is needed and expected in all circumstances. This information should be carefully recorded (and regularly updated) to form the basis for the quality system. It will become the standard against which all future responses to each customer's demands will be measured to ensure satisfaction.

The next step is to consider internal workings of the haulage operation to determine how these relate to the provision of a quality service. This task should be split into individual components or identifiable activities and for each there should be a documented set of procedures to be followed. The following are some examples of such activities:

- Receipt of customer orders
- Planning daily work schedules
- Allocation of work to vehicles and drivers
- Preparation of collection/delivery note sets
- Instruction of drivers
- Preparation of vehicles
- Confirmations to customers
- Driver conduct on arrival
- Driver debriefing on completion of deliveries
- Recording/confirmation of work completed
- Invoicing.

Documenting the System

The individual tasks mentioned above – and many more if appropriate – should be written down in the form of checklists to ensure that no matter who takes to order, who allocates the vehicle or driver, who prepares the delivery notes or who carries out any of these functions, the established step-by-step procedure is followed and no key element is missed which could lead to customer dissatisfaction. Documenting the system fully effectively means preparation of a 'Quality Manual'.

Quality Procedures Manual

A quality procedures manual is a statement of the firm's commitment to quality, a constant reminder to management of its obligations to the firm's customers based on these documented procedures. It becomes the bible of operating practice for the staff which has to be followed from when a customer first rings with an enquiry to when to job papers are finally filed away. It should detail the:

- Organisational structure of the company
- Relationships between, and the responsibilities of, individual operating departments and functional managers
- How the company's quality system is to work in practice:
  - every day,
  - every time an order is received,
  - every time a customer rings up,
  - every time a load is scheduled, and
  - every time a driver/vehicle is allocated to a job.

No matter what process or function is carried out, the manual should define it, who does it, how he/she does it, when, where and in what sequence, what follow up action is carried out, by whom, to whom they report in the event of difficulties or potential problems, what records are kept and so on. It is the ultimate guide book to the firm: every single company operation is described and cross-referenced to every other related function.

Mistakes may still be made, but the procedures manual should take account of such possibilities by detailing the action to be taken when errors are discovered (or pointed out by customers), and by whom, as well as who in the firm such matters should be reported too, what reports should be written and to whom are they are to be submitted. These 'failure reports' should be acted on promptly, and details of the corrective action taken should also be recorded for future reference. By documenting failures, they should progressively be eliminated from the system.

Manuals within Manuals

A simple manual would have an opening section containing basic reference manual material as follows:

- Firm's name
- Description of the business in which it is engaged (eg haulage of aggregates and excavated materials; also contract haulier to . . .)
- Number of locations (eg one only head office and vehicle base at . . .)
- General description of resources (eg 'x' number of vehicles/plant)
- Names/functions of key departments (eg traffic/administration/workshop).
- Names/titles of key managers/personnel
- Job descriptions for key positions (eg operations manager/fleet engineer/company accountant/marketing manager)
- Date(s) and name(s) of person(s) compiling manual
- Names of persons issued with the quality manual and those responsible for keeping it up to date
- Statement of firm's quality policy
- Name(s) of person(s) with ultimate responsibility for compliance with quality policy.

From this point, the manual could divide into sections covering each functional department (eg traffic operations/accounts and administration/workshop). For each there could be a general statement of departmental responsibility; the names, positions and individual responsibilities of key personnel; and the lines of reporting and communication (ie upwards to the company's top management/board of directors, downwards to supervisors, shift leaders etc and sideways by liaison with other departmental heads).

To compile the procedures manual, the work of each department must be examined in detail to see what current practices exist, who does what job (routine ones, special ones, urgent ones, etc), what controls are imposed and what safety procedures, if any, are followed to avoid failures (eg loads missed, jobs not invoiced, vehicles missing out on services etc). Every step should be categorised within a functional heading (eg receipt of customer orders, recording orders on daily work sheets, allocating orders to vehicles, planning vehicle loads and so on) and the correct procedure for carrying out each step needs to be recorded. It is necessary to describe how details relating to the job should be entered, in what form or sequence, and the safety measures necessary to ensure that no essential information (eg a customer's special instructions) are missed.

The manual must be capable of being read – and the procedures understood and followed – by any member of the firm and by the inspector from the quality accreditation organisation which checks the procedures. If the wording is tortuous and the manual littered with technical terms and unexplained abbreviations its point will be lost. For this reason manual writing demands the use of clear and simple language. Jargon should be avoided and technical terms kept to a minimum. When they are essential they should be defined or explained. Essential legal obligations (such as goods vehicles driver's hours, tachographs, safe loading etc) must be explained or cross-referenced to a source of detailed legal information.

Manuals in a form other than full-size pages in ring binders (eg pocket-size handbooks) may be more appropriate for drivers. These should detail the rules to be followed on driver's hours, the correct use of tachographs, safe loading, dangerous load procedures, routine daily vehicle inspections, what to do in an emergency or the event of an accident and many other instructions. Workshop staff could have their own handbooks concentrating on safety procedures, what to do in the event of injury or accident, use of tools and equipment, procedures for drawing spare parts and consumable supplies and so on. Fitters who are required to road test vehicles should also be issued with a copy of the drivers handbook.

Other Documentation

Besides the main manual, other documentation systems are needed – a 'day book' in which to record customer orders, collection notes, delivery notes, receipt notes or combined consignment note sets, invoices, statements and such like. Many of the forms may relate to ongoing or periodic events such as annual examination of employee driving licences; annual medical examinations and eyesight tests; inspections of safety equipment and safety signs, workshop equipment, fire fighting equipment and first aid kits. Other forms would record company meetings with staff, drivers, workshop staff, and training sessions on new procedures or new legislation. Examples of all forms used should be included in the procedures manual to show which form to complete in any particular set of circumstances, where supplies of the form can be found, how it should be completed and what to do with it following completion.

Training to Achieve Quality Standards

All staff within a firm seeking quality accreditation should be properly trained to operate in accordance with specified quality procedures (ie to perform to standard). Management, staff and workers should be updated on the latest techniques and new developments. For example, heavy vehicle drivers may need refresher courses in driving skills to eradicate bad habits and training in efficient and economical driving techniques for new vehicles, or vehicle equipment such as new types of gearbox; they need reminders and updating on essential safety procedures and the use of safety equipment, on legal requirements (such as on driver's hours rules and tachograph use) and on the use of mechanical loading aids.

At all levels from management to driver/operative, training improves job skills. Additionally, it contributes to the individual's motivation and job interest and provides an incentive to do that job better and take more of an interest in the firm's overall objectives – namely, to provide a quality service. The following list show examples of training from which various grades of management and staff may benefit:

Top management –
> Management techniques
> Sales/marketing/public relations
> Control over people
> Motivational skills
> Employment legislation
> Financial controls
> Taxation matters

Middle management –
> Administrative controls
> Business systems
> Computer familiarisation
> Management techniques
> Motivational skills
> Health and safety law

Functional management –
> Transport legislation
> Safety systems
> Engineering skills
> Computer techniques
> Accounting practices

Goods vehicle drivers –
> Economic driving
> Safety procedures
> Legal requirements
> Dealing with customers
> Dangerous loads requirements

Loaders –
> Safety matters

Fork-lift truck drivers –
> Driver training
> Safety procedures
> Truck maintenance

Vehicle workshop staff –
> Safety procedures
> Vehicle manufacturer training
> Component training
> First aid

Administrative and secretarial staff –
> Office procedures

Computer/word processing systems
Telephone techniques
Use of office equipment (photocopiers, fax machines etc)
First aid

While the range of courses and training opportunities is endless, there is a need to identify activities which demand acquired skills – for example, where special competence is legally required such as for goods vehicle vocational driving licences and dangerous goods training. These require priority over other training because failure could result in prosecution, or loss of operating licences.

Training Records

All training should be carefully recorded whether a 30 minute in-company explanation of new legislation or the desired telephone answering techniques, or a longer external training course for management. This should show dates, who attended, who presented the training session, the duration, the location, the facilities/aids used, the objectives of the session, the results achieved and any other relevant data, including any follow-up action necessary. Similarly, where staff achieve other qualifying standards as a result of company- sponsored incentives (eg attendance at evening or day-release classes or by home-study learning), or on a voluntary basis, these should be recorded.

Monitoring of Quality Standards

Successful operation of quality management systems involves a continuous monitoring process to ensure quality procedures are maintained and do not slip back into old, inferior methods. A variety of monitoring methods may be employed. The establishment of Quality Circles is one, with groups or teams of volunteer employees meeting to discuss the application of quality, particularly in relation to their own departments or work sections. They identify quality 'problem' areas and put forward suggestions as to possible solutions.

In the small firms, where the cost structure does not warrant full-time monitoring, two methods in particular will ensure that standards are maintained. First, regular monitoring of customer perception of the quality of service they are receiving. Customers should be asked to point out any deficiencies and their answers noted, analysed and corrective action taken. Second, maintaining full and clear communication in the firm will ensure that everybody from top to bottom is on the same quality wavelength and working towards the same goals. In this way the whole firm will feel united in the quest for total quality management.

Further Information

Further information on the establishment of quality systems in road haulage may be obtained from the following.

Road Haulage Association (RHA)
Tel: 0932 841515

British Standards Institution (BSI)
Business Development Advice
Tel: 0908 220908

Department of Trade and Industry (DTI)
Tel: 071 215 5000

30: UK Road Network Developments

There was a time when the provision of roads in the UK was taken for granted. In more recent years this has become a much more contentious issue as it becomes more clear as to the extent to which demand exceeds both the capacity and capability of our roads to meet modern traffic requirements, and as successive governments continue in their failing to spend on road development and maintenance what the exchequer extracts from motorists and transport fleet operators by way of vehicle excise duty (VED). According to the British Road Federation (BRF), only about 24 per cent of the tax revenue paid by road users in 1990/91 (£19 billion) was spent on roads – and the situation is getting worse, annually.

At the end of 1990 Britain had a network of 358,034 kilometres of roads. Since 1980, the network has grown by 18,400km, but over 15,000km of the growth has been on unclassified roads, mainly in housing developments. Except for 77km of local motorways, all other motorways and trunk roads are the direct responsibility of the government, making up about 4.4 per cent of the road system. However, this small proportion carries 32 per cent of all traffic and 54 per cent of heavy goods traffic.

Total roads by type and percentage of the UK network are shown in the following table:

Type of road	Length (km)	% of network
Motorways	3070	0.8
Trunk roads (excl. motorways)	12,674	3.5
Principal roads (excl. motorways)	35,149	9.8
Classified (non-principal)	110,554	31.0
Unclassified	196,558	54.9
Total	358,034	100.0

Traffic flows, taken across all roads, grew by 42 per cent between 1980 and 1990, with growth on motorways exceeding 73 per cent (sections of the M25 and M6 regularly carry more than 130,000 vehicles per day against a normal design capacity of only 79,000 vehicles per day).

Despite this growth in traffic, only some 23km of motorway are currently planned or under construction, and this to cater for further projected traffic growth of up to 36 per cent more cars and 27 per cent more lorry traffic by the year 2000.

NB: Information and table above courtesy of British Road Federation.

The UK road building and repair programme has gone on piecemeal over the years and looks set to continue to do so. Currently, around 5000 miles of our roads are in urgent need of structural repair according to the BRF and this includes some 40 per cent of our motorways.

In February 1993 the Department of Transport announced that construction work would start on 41 new national road schemes during 1993/94, involving record expenditure of more than £2 billion of which £1369 million is allocated to road construction and £550 million to capital maintenance. The DoT estimates that the value to industry and the community of these schemes alone in accidents avoided and time saved to road users will be £2.5 billion.

However, despite this short-term expenditure, no clear long-term strategy has been developed to cater for increasing domestic demand or the anticipated post-1993 boom in passenger and freight traffic movement between the UK and the rest of Europe. With new motorways taking at least 15 years to build, Britain is clearly faced with major road infrastructure problems which will, in particular, inhibit the potential for UK firms to take advantage of the new trading promises of the single market.

Road Pricing – Paying for Motorway Use

Currently, the government is toying with the idea that funds for road development can be generated by means of private investment, although the Road Haulage Association (RHA) is adamant that private sector involvement in the road programme is to be deplored, particularly as long as the disparity exists between the monies collected from VED and fuel tax (totalling £147 billion*) and road expenditure (£6 billion*).

*NB: RHA figures.

In May 1993 the Secretary of State for Transport issued a consultation paper (on which comments were required by 17 September 1993) on the government's plans for charging for the use of motorways. It took the form of a Green Paper entitled *Paying for Better Motorways: Issues for Discussion*. It was emphasised that the government had taken no decisions and had no existing blue prints, but conventional tolling with plazas, toll booths and barriers on existing motorways was said to be impracticable. The options include electronic tolling and motorway permits.

The Green Paper points out that motorway charging has the potential to offer important benefits, but it also raises a number of fundamental issues which it lists as follows:

- In the 1980s traffic on UK motorways doubled. There is already peak-hour congestion on the busiest sections. Without further action congestion will increase substantially as the economy grows.
- Government expenditure on the national road network is already at record levels. But there is a need to keep firm control over public expenditure. Given the competing pressures it is not realistic (the government says!) to expect higher levels of funding from traditional sources.
- Motorway charging is a possible solution. Motorways are expensive to

build and maintain, and offer a premium level of service albeit half of road users rarely or never use them.

- Motorway charging would provide another source of finance for improving roads. This would improve the service to road users and enable the economy to benefit from a faster rate of expansion of transport capacity than would otherwise be possible.
- Motorway charging would ensure that we make more effective use of the existing network; and it would help secure the efficiency and value for money that a market approach brings.
- Direct charging would create a stream of revenue which could open up new ways of involving the private sector, in particular in widening and upgrading existing motorways.

Three pricing bands are proposed in the Green Paper:

- 1.5 pence per mile or £250 for an annual permit.
- 3.0 pence per mile or £500 for an annual permit.
- 4.5 pence per mile or £750 for an annual permit.

The road haulage industry fears that cost increases at these levels for motorway use could not be absorbed in the normal course of events and would have to be passed on by way of increased road haulage rates.

Subsequent to the Green Paper, the government has announced that motorway charging would be introduced subject to the development of appropriate technology (see below) and parliament's approval in due course of the necessary legislation.

UK Road Development News

A number of positive and encouraging road developments have been announced by the DoT in the early part of 1994 as follows:

- From 1 April 1994 a new Highways Agency (an executive agency of the Department of Transport) became responsible for managing and maintaining the existing road network and the new road construction and improvement programme. The Agency will be required to achieve specified standards in delivering an efficient, reliable, safe and environmentally acceptable trunk road network under the provisions of a new Road User's Charter launched by the DoT (copies of the Charter can be obtained from most motorway service areas, AA shops and offices, regional offices of the DoT and the Highways Agency, or by writing to the Highways Agency, Room 12/23, St Christopher House, Southwark Street, London SE1 0TE).
- The 1994 Transport Report (published in March 1994) detailed plans for a 3-year spend over the period 1994–95 to 1996–1997 of £5.4 billion on the national motorway and trunk road network and an additional £3 billion in local roads and transport.
- The DoT announced that construction work is to start on 22 new national road contracts worth over £1 billion in 1994–95 – overseen by the new Highways Agency. A number of these schemes will be carried out under the 'design and build' scheme whereby the contractor is responsible for both the design and construction, thus ensuring increased certainty over timing and cost and greater efficiency of design.

- A new and radical prioritisation of the road programme was announced by the the the Secretary of State for Transport in March 1994. This is intended to get essential schemes built faster, concentrate effort on urgently needed bypasses and motorway widening schemes and remove those schemes from the programme which are no longer considered to be environmentally acceptable or are not needed for the foreseeable future.
- In June 1994 the new Highways Agency announced details of a £500 million road maintenance programme. This expenditure is to be split, £350 million for structural maintenance of carriageways and £150 million for structural maintenance of bridges and other structures. 150 major maintenance schemes within the programme are to be started in the year commencing June 1994.
- In July 1994 the final section of the A14 link road between the M1 and the A1 was opened. This completes a high-quality, dual-carriageway link from the motorway network in the Midlands, across East Anglia, to the east coast ports. In consequence of this, parts of the old A45 from Felixstowe to the west of Cambridge have been renumbered as the A14 and, similarly, the old A604 from Cambridge to Huntingdon also becomes the new A14, which then extends westwards to link with the M1 at junction 19.

Cones Hotline

A new Cones Hotline telephone number has been established to enable callers, for the price of a local call (from wherever the call is made), to report hold-ups on England's trunk roads and motorways. It was established to allow anyone who feels parts of the trunk road or motorway appear to be coned off unnecessarily or unreasonably to complain. The line is open 24 hours a day to note and look into all enquiries. If requested, a written explanation will be provided on why cones are in place in particular places. If it turns out that there is no good reason for the cones to be there, the DoT undertakes to get them removed

The Cones Hotline number is: 0345 504030.

Motorway Tolling

A step towards motorway tolling was taken in 1994 when the DoT invited over 350 electronics manufacturers, software houses and consultants from around the world to work with the government on the development of an electronic tolling system for Britain's motorway system. The government is searching for the technology to do the job without the use of conventional toll booths which have been ruled out for existing motorways because of the land they would require and the congestion they would cause.

TERN Routes

As part of the European Union's Trans-European Road Network (TERN) scheme for identifying nationally and internationally important routes, including links to ports, to meet the road transport needs of the 21st century, the DoT has named the roads which will form the UK's contribution. This is seen as a step towards the creation of an effective network of roads across Europe, essential for making the Single Market a reality, and to increasing the competitiveness of British industry in European and international markets.

31: Services

This final chapter in the *Handbook* provides an opportunity to include information on various topics which may be of interest or use to readers. For example, for this 1995 issue the subjects of Legal Services, Breakdown and Vehicle Recovery, Return Loading, Franchising and Motorway Services are included. In future years further suitable and appropriate subjects may be included to expand this service.

Legal Services

As the growth of this *Handbook* and the expansion of its main subject material clearly demonstrate, the transport industry in general, and the operation of heavy goods vehicles in particular, is fraught with a burden of legislation verging on leviathan proportions. But while the volume of the legislation is one thing, understanding and complying with its strictures is quite another, and it is here that transport managers and fleet operators alike find problems. While this book, and others like it, attempt to explain the law in simple terms, none of them are much help when, having failed to understand the law on a particular point, or even failed to realise that it even applied, the operator faces prosecution and a court appearance.

Deciding how to plead, guilty or not guilty; knowing what defences can and should be put forward; knowing when an absolute offence as been committed when putting forward feeble excuses will add nothing to the defence; knowing what mitigating circumstances may be put forward in given circumstances; and knowing how to plead for leniency in the penalty to be imposed when convicted, these are all things that books are not good at explaining. This is when a good solicitor is necessary to weigh up the case, determine whether the police and Crown Prosecution Service (CPS) have got the charges right to suit the alleged offence (it is a fact that they do not always do so), and decide what plea the accused should make.

It is clear that many firms facing prosecution for transport or vehicle-related infringements prefer to plead guilty and pay the fine rather than incur the extra costs of legal representation, and the possible attendant (ie adverse) publicity by pleading not guilty and making a case of it. It is clear also that in so doing, some of those are convicted on the wrong grounds or for offences where they could have raised a legitimate defence (eg by being exempt from a particular requirement - which the enforcement authorities are not always good at recognising).

What is important to remember, for big firms as well as small fleet operators, is that convictions for most of the offences relevant to transport operations will bring, besides fines, the probable jeopardy of 'O' licences, and the driving licences of proprietors and partners.

For these reasons, in most cases, the retention of a solicitor is strongly advised where cases are to be dealt with by a court (in the case of fixed penalty offences, taking the simple line of accepting that an offence has been committed and paying the penalty can be the easiest and cheapest way out). Importantly though, such is the complexity of transport, vehicle-related and road traffic law that any solicitor chosen should be one who is recognised as having expertise and a reputation in these areas of law. Any fear that a known 'name' may cost extra (which is probably not the case) should be weighed against his or her ability to present an experienced case. The legal reports columns of the transport press regularly highlight cases where good advocates have charges against their clients quashed or cases dismissed because they really do know the ins and outs and technicalities of the law.

The following is a list of well-known transport lawyers who will provide suitable advice or defend cases as necessary.

Jonathon Lawton
(Aaron & Partners)

Grosvenor Court, Forecourt Street, Chester CH1 1HG
Tel: 0244 315366

John Backhouse

23 Wellington Street, St Johns, Blackburn, Lancs BB1 8DE
Tel: 0254 677311

Michael Carless
(Carless, Davies & Co)

140 Stourbridge Road, Halesowen, Worcs
Tel: 021-550 2181/4429

Stephen Kirkbright
(Ford & Warren)

Westgate Point, Westgate, Leeds
Tel: 0532 436601

Malcolm Partridge
(Daynes Hill & Perks)

Paston House, Princes Street, Norwich,
Norfolk NR3 1BD
Tel: 0603 660241

Breakdown and Recovery Services

There has been a proliferation of heavy vehicle breakdown and recovery services in recent years providing roadside services to help the driver with a broken-down truck, tyre problems or accident-damaged vehicle. Many such services have a freephone contact arrangement. These services are operated by independent firms (many are members of AVRO – see below) and by the vehicle manufacturers. Examples of the former are AA Truck Rescue (previously AA-BRS Rescue), RAC Commercial Assistance (previously RAC-Octagon) and National Breakdown* and of the latter include DAF Aid, Action Service Volvo, Fodensure and MAN Rescue.

The Association of Vehicle Recovery Operators (AVRO) is a major player in this field and as such has developed a code of practice in conjunction with the Retail Motor Industry Federation (RMI), while the Road Rescue Recovery Association (RRRA) has its own code. Quite separately, and in conjunction with the British Standards Institution, the RAC has its own code of practice.

The AVRO Directory lists the following members:

Northern region

Alpha Auto Services	Cramlington, Northumberland. Tel: 091-250 0009
Tebay Vehicle Repair Ltd	Tebay, Cumbria. Tel: 058 74 241
Lynch Motors Ltd	Parkside, South Shields, Tyne & Wear. Tel: 091-456 4665
Ron Perry Test & Tune Ltd	Hartlepool, Cleveland. 0740 644223

South Midlands region

Brooks & Stratton Ltd	Welwyn Garden City, Herts. Tel: 0707 330678
3B's Rescue	Banbury, Oxfordshire. Tel: 0295 750236

Eastern region

John Canham Ltd	Clacton-on-Sea, Essex. Tel: 0255 432888
JS Holmes Ltd	Wisbech, Cambridgeshire. Tel: 0945 81243
J & A Recovery	Brandon, Suffolk. Tel: 0842 810146

Greater London

Arcade Motors Ltd	Tottenham, London. Tel: 081-363 2323
Kenfield Motors Ltd	Hayes, Middlesex. Tel: 081-569 2323

Queens Motors Ltd	Penge, London. Tel: 081-778 6666
MV Recovery	Croydon, Surrey. Tel: 081-686 1883
J Winfield Motors	Hayes, Middlesex. Tel: 081-848 7421

Southern region

McAllisters Recovery	Aldershot, Hampshire. Tel: 0252 22289
Brighton Recovery	Brighton, East Sussex. Tel: 0273 430420
Dawes of Swanley	Swanley, Kent. Tel: 0322 62211
Langley Vale Recovery	Epsom, Surrey. Tel: 0372 277021
Thames Valley Motors	Newbury, Berkshire. Tel: 0635 48772

Western region

Avon Commercial Recovery	Severn Beach, Bristol. Tel: 04545 2331
Walls Garage Services	Severn Bridge, Bristol. Tel: 04545 3472
Bristol Omnibus Co	Bristol, Avon. Tel: 0272 558211
Lamb Hill Recovery	Clumpton, Devon. Tel: 0884 38572
P G Hayes	Minehead, Somerset. Tel: 0643 705363
J & P Motors Ltd	South Petherton, Somerset. Tel: 0460 40553

Wales

Walls Truck Services Ltd	Newport, Gwent. Tel: 0633 246622

Information on National Breakdown's Trucklink and Trailerlink services can be obtained from: National Breakdown, FREEPOST, Leeds, Yorkshire LS99 2NB or by telephoning 0532 393666 (Fax no: 0532 573111).

Return Load Services

Goods vehicles running needlessly empty are a costly inefficiency on the part of the operator and wasteful of valuable natural resources (eg fuel), as well as contributing to the environmental problems outlined in Chapter 28. While it is not always possible, or indeed desirable for operational reasons, to seek and carry return loads, nevertheless many more vehicles travel the roads with empty space behind than is necessary or could be justified by an efficiency and cost-conscious industry.

It is recognised that return loading presents many problems of strategy as well as the uncertainties about securing payment (and at an acceptable price) for such work. However, there are now more reliable return load systems which allow hauliers to secure loads from reputable sources where the haulage rate has not been creamed in the way established by many disreputable clearing houses in the past.

A number of computerised load-matching systems are now operating of which the French Lamy Teleroute is probably the best known. This operates throughout Europe on a 24-hour/365-days basis and only requires the haulier to have a Videotex terminal or personal computer with a telephone line to connect to Teleroute's international network.

Other hi-tech return load services are provided by Returnline (Tel: 0455 233998) and CargoFile (Tel: 0277 363756).

Franchising

Franchising is a business activity in which an existing (usually large and successful) operator contracts with newcomers or smaller firms in the business to provide a service under its well-established name and to its established standards (eg of quality and reliability etc).

The benefit to the franchisor (ie the existing operator) is that it expands his business and spreads his name and reputation without significant capital investment on his part. The disadvantage is that if the franchisee proves to be unsatisfactory in the way he conducts business the franchisor's good name (and that of his products or services) will suffer.

The benefit to the franchisee is that he can start a new business (or convert his existing business) with the backing of an establised name (ie a recognised corporate identity) and reputation, and with proven systems of operating, marketing and administration already in place. The disadvantage is that once committed to the franchise contract, the franchisee will have little freedom or flexibility to exercise his own entrepreneurial flair, management skills or administrative know-how. He will have to conform to the rigidly laid-down system of his franchisor, although this can be of significant advantage in imposing the rigid disciplines necessary to establish a sound and successful business.

Common examples of well-known franchising operations are those of McDonald's, to be found in the main streets of most of the world's major cities, and Colonel Sanders' Kentucky Fried Chicken. But this type of business is not confined to fast-food chains. It covers many other fields as well, including Athena retails shops (selling pictures, prints, cards etc), and Ryman the Stationer retail outlets. In transport and related industries, franchised express parcels operations (such as Amtrak Express Parcels Limited) and workshop tools and equipment supply (eg Snap-on Tools) are two of the most popular franchise operations.

The Amtrak Franchise

Amtrak provides potential franchisees with a helpful information pack explaining the benefits of franchising, mainly as outlined above but including such other advantages as the franchisee not having to handle billing, or cash transactions (and, therefore, not incurring bad debts), these being a central, computerised function and the responsibility of the franchisor. Franchisees in this case are paid promptly, on a monthly basis, for the work they have done in the form of commission for every single collection and delivery made (and irrespective of whether or not the customer has actually paid his account).

Franchisees are allocated an exclusive territory (of which there are about 120 in mainland UK) in which they are licensed to trade and their role is to collect and deliver express parcels in these territories which are linked into the company's central sorting hub in the midlands by a fleet of heavy trunk vehicles owned by the company, not the franchisee.

Amtrak provides its franchisees with back-up support by means of a team of area managers who will assist with sales and marketing strategies and help the franchisee to operate within the company's guidelines. And, before starting, the candidate attends a training course to familiarise himself with the nature of the parcels business, the operational procedures to be followed and sales and marketing techniques.

Legal Considerations

While franchising as a means of starting in business provides many advantages (and possibly less risk of loss of initial capital – although there are no guarantees on this count), there are legal considerations which must be taken into account. Mainly these fall into two categories: one is the employment relationship between the franchisor and franchisee and the other concerns goods vehicle operator licensing where vehicles over 3.5 tonnes maximum gross weight are involved.

Employment Contracts

Normally, one would expect a franchised operation to involve a contract for the provision of service by an independent (possibly self-employed) contractor. In this case the franchisee is not an employee of the franchisor and is therefore responsible for (among other things) making his own National Insurance contributions, paying income tax and registering for VAT (subject to turnover being above the statutory limit at which registration is mandatory). However, should the contract be one of service (as opposed to one for the provision of . . .) then this could be held to be an employment contract where the franchisor would be liable for such matters as National Insurance (ie deducting the employee's contribution and paying his own employer's contribution) and deducting income tax from the franchisee's payments.

These are not the only legal considerations under this heading, but they are the main ones – it still leaves open the question as to liability for redundancy payments and such like. For this reason, the inexperienced franchising candidate should seek advice either from a solicitor or an accountant before signing (irrevocably) any documents.

Operators' Licensing

This subject is covered extensively in Chapter 1 of this book, but it should be said here that should the vehicle which the franchisee plans to operate exceed 3.5 tonnes maximum gross weight, then he will need to obtain an 'O' licence before commencing operations. In fact, he should do this before even committing himself to the franchise contact because, irrespective of what the franchisor may say, there is no guarantee of getting an 'O' licence. A franchisor cannot buy one for the franchisor or allow him, as a self-employed contractor, to operate under his own 'O' licence – this would be illegal.

Further Information

Further information on franchising can be obtained from the following organisations:

British Franchise Association
Thames View, Newton Road, Henley-on-Thames, Oxon RG9 1HG
Tel: 0491 578049

National Federation of Self-Employed and Small Business
32 St Anne's Road West, Lytham St Annes, Lancs FY8 1NY
Tel: 0253 720911

The following high Street banks have departments dealing specifically with franchise operations:

Barclays Bank Plc
Franchise Unit, PO Box 120, Longwood Close, Westwood Business Park, Coventry CV4 8JN
Tel: 0203 532451

Lloyds Bank Plc
Franchise Department, Commercial Banking, PO Box 112, Canons Way, Bristol BS99 7LB
Tel: 0272 433136

National Westminster Bank Plc
Franchise Section, Commercial Banking Service, 4th Floor, National House, 14 Moorgate, London EC2R 6BS
Tel: 071-728 1684

Midland Enterprise
Midland bank Plc, PO Box 2, 41 Silver Street, Sheffield S1 3GG
Tel: 0742 529037

Advice on the use of franchising consultants is available from:

Franchise Consultants Association
James House, 37 Nottingham Road, London SW17 7EA
Tel: 081-767 1371

Franchise World is published bi-monthly by the Franchise Consultants Association (see above for address) and a directory (*Franchising World*) is also available from the same address.

Motorway Services in the UK

The Department of Transport is planning to sell off (ie privatise) all motorways service areas (MSAs) and deregulate the planning and acquisition process for new service area ventures – this was further confirmed in June 1994 with the announcement that the new Highways Agency of the DoT would be selling its interest in 47 MSA sites in England. Service areas are currently operated under long-term leases from the DoT with the franchisees paying a premium initial payment and a token rental. While the DoT will continue to guarantee and maintain minimum standards of operation and service, responsibility for identifying potential sites and obtaining planning consent will rest with the private sector dealing directly with the relevant local authorities.

The following are the major service area operators who welcome heavy vehicle drivers and provide suitable services.

BP Truckstops
– Alconbury (Cambridgeshire) – on A604 near A1(M)
– Carlisle (Cumbria) – on M6 at junction 44
– Penrith Industrial Estate (Cumbria) – off M6 at junction 43

–   Birtley (Newcastle-upon-Tyne) – on A194(M) at junction with A1
–   South Mimms (Hertfordshire) – at junction of M25 with A1(M), M25 junction 23
–   Rugby (Northants) – on M1 at junction 18 (near M45 and M6)
–   Wolverhampton (Staffs) – on M54 at junction 1

Open:	24 hours/365 days
Parking:	For approx 150 heavy vehicles (parking fee includes free shower and towel and restaurant/shop voucher)
Accommodation:	Rooms with showers at Alconbury, Birtley, Carlisle and Penrith
Services:	24-hour diesel supply, fax, photocopier and shop

For further details: Tel 0272 872333

Granada
–   Carlisle – on M6 at junction 41/42
–   Washington – on A1(M)
–   Burton-on-Trent – on M6 between junctions 35 and 36
–   Manchester – on M62 between junctions 18 and 19
–   Wakefield – on M1 between junctions 38 and 39
–   Ferrybridge – at junction of M62/A1
–   Blyth – at junction of A1(M)/A614
–   Trowell – on M1 between junctions 25 and 26
–   Birmingham – on M5 between junctions 3 and 4
–   Tamworth – at junction of A5/M42
–   Leicester – at junction M1/A50
–   Monmouth – on A40
–   Magor (Wales) – on M4 at junction 23
–   Toddington – on M1 between junctions 11 and 12
–   Chippenham – on M4 between junctions 17 and 18
–   Chieveley – at junction of M4/A34
–   Heston – on M4 between junctions 2 and 3
–   Thurrock – on M25 at junction 31
–   Saltash – on A38 bypass
–   Exeter – on M5 at junction 30
–   Warminster – at junction of A36/A350
–   Grantham – at junction of A1/A151
–   Edingurgh – on A1, Musselburgh bypass

Open:	24 hours/365 days
Parking:	Varies approx 6–120 heavy vehicles (parking fee includes free shower and restaurant/shop vouchers)
Accommodation:	Rooms with bathroom, colour TV, tea/coffee maker
Services:	24-hour diesel supply, breakdown recovery, fax, photocopier and shop

For further details: Tel 0525 873881

Roadchef
–   Clacket Lane (Westerham) – on M25
–   Taunton Deane (Somerset) – on M5

- Sedgemoor (Somerset) – on M5
- Rownhams (Hants) – on M27
- Pont Abraham (Dyfed) – on M4
- Sandbach (Cheshire) – on M6
- Killington Lake (Cumbria) – on M6
- Hamilton (Lanarks) – on M74
- Bothwell (Lanarks) – on M74
- Harthill (Lanarks) – on M8

Open:	24 hours/365 days
Parking:	For approx 30–100 heavy vehicles (parking fee includes free shower and restaurant voucher)
Accommodation:	Rooms with bath/shower, tea/coffee maker, hair dryer
Services:	24-hour diesel supply, shop and cash dispensers.

For further details: Tel 0452 303373

European Road Parks
- Beckton Roundabout (London Dockland) – on A13

Open:	24 hours/365 days
Parking:	For approx 120 heavy vehicles (with security surveillance etc) (showers available)
Accommodation:	Rooms with showers
Services:	24-hour diesel supply, fax, photocopier and shop

For further details: Tel 081-594 8730

Pavilion Services Limited (formerly Rank Organisation)
- Medway Pavillion (formerly Farthing Corner) – on M2 between junctions 4 and 5
- Severn View (formerly Aust) – on M4 at junction 21
- Cardiff West – on M4 at junction 33
- Swansea – on M4 at junction 47
- Hilton Park (Wolverhampton) – on M6 between junctions 10A and 11
- Knutsford – on M6 between junctions 18 and 19
- Forton (Lancaster) – on M6 between junctions 32 and 33
- Rivington (Bolton) – on M61 between junctions 6 and 8
- Newark – at junction of A1/A46/A17
- Bangor – at junction of A5/A55.

Open:	24 hours/365 days
Parking:	For approx 30–60 heavy vehicles (with security patrol) (parking fee includes free shower and food voucher)
Accommodation:	Rooms with showers, TV, hair dryer, trouser press (food voucher)
Services:	24-hour diesel supply, breakdown recovery, fax, photocopier, shop and cash dispensers

For further details contact: Roadside Services, Barker's House, Baker's Road, Uxbridge, Middx UB8 1RG

Truckers Rest

– Located at West Bromich, just off junction 1 of the M5.

Services: Serves a variety of foods, often with a special 'national' flavour (eg Italian, Irish, Scottish, Spanish etc) on a weekly rotation.

For further details telephone: 021-500 5040.

Appendices

Note

UK telephone dialling codes and the international dialling-out code are to
change during 1995 as follows:

1. In most cases an additional figure 1 is to be inserted after the first zero
    (eg London 071 becomes 0171).

2. In the following cases the code is to be replaced by a new code and an
    extra digit is to be added to the beginning of the telephone number:

    Leeds 0532 becomes 0113 – numbers start with an additional 2
    Sheffield 0742 becomes 0114 – numbers start with an additional 2
    Nottingham 0602 becomes 0115 – numbers start with an additional 9
    Leicester 0533 becomes 0116 – numbers start with an additional 2
    Bristol 0272 becomes 0117 – numbers start with an additional 9

3. The international dialling-out code will change from 010 to 00

Appendix I

The Traffic Area Network

Traffic Area Office	**Counties Covered**
North-eastern Hillcrest House, 386 Harehills Lane, Leeds LS9 6NF Tel 0532 833533 Fax 0532 489607	South Yorkshire, Tyne and Wear and West Yorkshire, the counties of Cleveland, Durham, Humberside, Northumberland and North Yorkshire.
North-western Portcullis House, Seymour Grove, Manchester M16 0NE Tel 061-886 4000 Fax 061-886 4019	Greater Manchester and Merseyside, the counties of Cheshire, Clwyd, Cumbria, Derbyshire, Gwynedd and Lancashire.
West Midland Cumberland House, 200 Broad Street, Birmingham B15 1TD Tel 0121-608 1000 Fax 0121-608 1001	West Midlands, the counties of Hereford and Worcester, Shropshire, Staffordshire and Warwickshire.
Eastern Terrington House, 13–15 Hills Road, Cambridge CB2 1NP Tel 0223 358922 Fax 0223 532110	Counties of Bedfordshire, Buckinghamshire, Cambridgeshire, Essex, Hertfordshire, Leicestershire, Lincolnshire, Norfolk, Northamptonshire, Nottinghamshire and Suffolk.
South Wales Caradog House, 1–6 St Andrews Place, Cardiff CF1 3PW Tel 0222 395426 Fax 0222 371675	Counties of Dyfed, Gwent, Mid-Glamorgan, Powys, South Glamorgan and West Glamorgan.
Western The Gaunts' House, Denmark Street, Bristol BS1 5DR Tel 0272 755000 Fax 0272 755055	Counties of Avon, Berkshire, Cornwall, Devon, Dorset, Gloucestershire, Hampshire, Isle of Wight, Oxfordshire, Somerset and Wiltshire.
South-eastern and Metropolitan Ivy House, 3 Ivy Terrace, Eastbourne BN21 4QT Tel 0323 721471 Fax 0323 721057	Greater London, the counties of Kent, Surrey, East Sussex and West Sussex.
Scottish 83 Princes Street, Edinburgh EH2 2ER Tel 031-529 8500 Fax 031-529 8501	Scotland.

Appendix II

Transport Trade Associations and Professional Bodies

Freight Transport Association Regional Offices

Head Office:
Hermes House,
157 St John's Road,
Tunbridge Wells,
Kent TN4 9UZ
Tel 0892 526171/Fax 0892 534989

Midlands:
Hermes House,
Hall Street,
Dudley,
West Midlands DY2 7BQ
Tel 0384 237321/Fax 0384 456220

Northern:
Springwood House,
Low Lane,
Horsforth,
Leeds LS18 5NU
Tel 0532 589861/Fax 0532 586501

Scottish (including Northern Ireland):
Hermes House,
Melville Terrace,
Stirling FK8 2ND
Tel 0786 71910/Fax 0786 50412
(Belfast: Tel 0232 241616)

London and South-eastern:
Hermes House,
157 St John's Road,
Tunbridge Wells,
Kent TN4 9UZ
Tel 0892 526171/Fax 0892 534989

Western:
Hermes House,
Queen's Avenue,
Clifton,
Bristol BS8 1SE
Tel 0272 731187/Fax 0272 238269

Enquiries regarding management training should be made to Training and Personnel Services, Hermes House, Tunbridge Wells (see *Head Office* above)

Road Haulage Association District Offices

Head Office:
Roadway House,
35 Monument Hill,
Weybridge,
Surrey KT13 8RN
Tel 0932 841515/Fax 0932 852516

Scotland:
Roadway House,
17 Royal Terrace,
Glasgow G3 7NY
Tel 041-332 9201

North-western:
124 Market Street,
Farnworth,
Bolton BL4 9EP
Tel 0204 71521

North-eastern:
Roadway House,
Beaumont Street West,
Darlington,
Co. Durham DL1 5SY
Tel 0325 281495

Eastern:
Roadway House,
Rightwell,
Bretton Centre,
Peterborough PE3 8DR
Tel 0733 261131

Western:
Roadway House,
Cribbs Causeway,
Bristol BS10 7TU
Tel 0272 503600

Midlands:
Roadway House,
50 Sedgley Road West,
Dudley,
West Midlands DY4 8AL
Tel 021-557 4911

Other Trade Associations

Association of District Councils
26 Chapter Street,
London SW1P 4ND
Tel 071-233 6940

*Association of Vehicle Recovery
Operators (AVRO)*
201 Great Portland Street,
London W1N 6AB
Tel 071-580 9122/Fax 071-580 6376

British Association of Removers (BAR)
3 Churchill Court,
58 Station Road,
North Harrow HA2 7SA
Tel 081-861 3331/Fax 08-861 3332

*British International Freight Association
(BIFA)
(incorporating the Institute of Freight
Forwarders)*
Redfern House,
Browells Lane,
Feltham,
Middx TW13 7EP
Tel 081-844 2266/Fax 081-890 5546

British Industrial Truck Association
Scammell House,
High Street,
Ascot,
Berks SL5 7JF
Tel 0344 23800/Fax 0344 291197

*British Vehicle Rental and Leasing
Association (BVRLA)*
13 St Johns Street,
Chichester,
West Sussex PO19 1UU
Tel 0243 786782/Fax 0243 533851

Bus and Coach Council (BCC)
Sardinia House,
52 Lincoln's Inn Fields,
London WC2A 3LZ
Tel 071-831 7546

Electric Vehicle Association (EVA)
Aberdeen House,
Headley Road,
Grayshott, Hindhead,
Surrey GU26 6LA
Tel 0428 735536

*UK Warehousing Association
(formerly the National Association of
Warehouse Keepers)*
Walter House,
418–422 Strand,
London WC2R 0PT
Tel 071-836 5522

*National Association of Waste Disposal
Contractors (NAWDC)*
Mountbarrow House,
6–20 Elizabeth Street,
London SW1W 9RB
Tel 071-824 8882/Fax 071-824 8753

*National Tyre Distributors Association
(NTDA)*
Broadway House,
The Broadway,
Wimbledon,
London SW19 1RL
Tel 081-540 3859

Owner Operators UK Limited
Bittenham Springs,
Ewen,
Nr Cirencester,
Gloucestershire GL7 6BY
Tel 0285 770833

*Retail Motor Industry Federation (RMI)
(previously the Motor Agents Association
(MAA))*
201 Great Portland Street,
London W1N 6AB
Tel 071-580 9122/Fax 071-580 6376

*Society of Motor Manufacturers and
Traders (SMMT)*
Forbes House,
Halkin Street,
London SW1X 7DS
Tel 071-235 7000/Fax 071-235 7112

Tachograph Analysis Association
Merseyside Innovation Centre,
131 Mount Pleasant,
Liverpool L3 5TF
Tel 051-708 0123/Fax 051-709 5645

Transfrigoroute (UK)
Queensway House,
2 Queensway,
Redhill,
Surrey RH1 1QS
Tel 0737 768611/Fax 0737 761685

Transport Association
9th Floor,
City Centre Tower,
7 Hill Street,
Birmingham B5 4UU
Tel 021-643 5494

Transport Users Group
Transport House,
Stretford Motorway Estate,
Stretford,
Manchester M32 0ZH
Tel 061-866 8599

*Vehicle Builders and Repairers
Association (VBRA)*
Belmont House,
Finkle Lane,
Gildersome,
Leeds LS27 7TW
Tel 0532 538333/Fax 0532 380496

Professional Bodies in Transport

Chartered Institute of Transport (CIT)
80 Portland Place,
London W1N 4DP
Tel 071-636 9952/Fax 071-637 0511

Institute of Grocery Distribution
Letchmore Heath,
Watford,
Herts WD2 8DQ
Tel 0932 857141

Institute of Logistics
Douglas House,
Queens Square,
Corby,
Northants NN17 1PL
Tel 0536 205500/Fax 0536 400979

Institute of the Motor Industry (IMI)
'Fanshaws',

Brickendon,
Hertford SG13 8PQ
Tel 099 286 521

*Institute of Road Transport Engineers
(IRTE)*
1 Cromwell Place,
Kensington,
London SW7 2JF
Tel 071-589 3744/Fax 071-225 0494

*Institute of Transport Administration
(IoTA)*
32 Palmerston Road,
Southampton SO1 1LL
Tel 0703 631380/Fax 0703 634165

Institution of Mechanical Engineers (IME)
1 Birdcage Walk,
London SW1H 9JJ
Tel 071-222 7899/071-222 4557

*Institute of the Moving Industry
(formerly the Institute of the Furniture
Warehousing and Removing Industry
(IRFWI))*
3 Churchill Court,
58 Station Road,
North Harrow HA2 7SA
Tel 081-861 3331/Fax 081-861 3332

Appendix III

Other Organisations Connected with Transport and Transport Journals

Automobile Association (AA)
Fanum House,
Basing View,
Basingstoke,
Hants RG21 2EA
Tel 0256 493028/Fax 0256 493389

British Road Federation (BRF)
Cowdray House,
6 Portugal Street,
London WC2A 2HG
Tel 071-242 1285/Fax 071-831 1898

British Standards Institution (BSI)
Linford Wood
Milton Keynes MK14 6LE
Tel 0908 220022/Fax 0908 320856

Department of Transport (DoT)
2 Marsham Street,
London SW1P 3EB
Tel 071-276 3000

International Road Freight Office
Westgate House,
Westgate Road,
Newcastle upon Tyne NE1 1TW
Tel 091-261 0031/Fax 091-222 0824

International Road Transport Union (IRU)
Centre International
3 rue de Varembe,
PB 44,
1211 Geneve 20,
Switzerland
Tel (from UK) 010+41 22 734 13 30
Fax 010+41 22 733 06 60

National Breakdown Recovery Club
PO Box 300,
Leeds LS 99 2LZ
Tel 0532 393666

*Road Transport Industry Training Board
(RTITB) now called Centrex*
Capitol House, Empire Way,
Wembley,
Middx HA9 0NG
Tel 081-902 8880/Fax 081-903 4113

Royal Automobile Club (RAC)
RAC House,
M1 Cross,
Brent Terrace,
London NW2 1LT
Tel 081-452 8000/Fax 081-208 0679

Royal Society of Arts (RSA)
(Examinations Dept),
Westwood Way,
Westwood Business Park,
Coventry CV4 8HS
Tel 0203 470033/Fax 0203 468080

*Royal Society for the Prevention of
Accidents (RoSPA)*
Cannon House,
Priory Queensway,
Birmingham B4 6BS
Tel 021-200 2461

Vehicle Inspectorate – Executive Agency
Goods Vehicle Centre,
Welcombe House,
91–92 The Strand,
Swansea,
Glamorgan SA1 2DH
Tel 0792 458888

Transport Journals

Commercial Motor
Quadrant House,
The Quadrant,
Sutton,
Surrey SM2 5AS
Tel 081-661 3500

Distribution Business
Quadrant House,
250 Kennington Lane,
London SE11 5RD
Tel 071-924 5885/Fax 071-978 5515

Export and Freight
Carn Industrial Estate
Portadown BT63 5RH

Northern Ireland
Tel 0762 334272

Logistics Focus (Journal of Institute of Logistics)
Douglas House,
Queens Square,
Corby,
Northants NN17 1PL
Tel 0536 205500

Freight Management International
230–234 Long Lane
London SE1 4QE
Tel 071-403 4353
Fax 071-403 0233

Freight (Journal of FTA)
Hermes House,
St John's Road,
Tunbridge Wells TN4 9UZ
Tel 0892 26171

Headlight
PO Box 96
Coulsdon,
Surrey CR5 2TE
Tel 081-660 2811

International Freighting Weekly (IFW)
Maclean Hunter House,
Chalk Lane,
Cockfosters Road,
Barnet,
Herts EN4 0BU
Tel 081-975 9759

Logistics Europe
Castle Chambers
85 High Street
Berkhamsted
Herts HP4 2BR
Tel 0442 878787 Fax 0442 870888

Materials Handling News
Quadrant House,
The Quadrant,
Sutton,
Surrey SM2 5AS
Tel 081-661 3500

Motor Transport
Quadrant House,
The Quadrant,

Sutton,
Surrey SM2 5AS
Tel 081-661 3500

Removals and Storage (Journal of British Association of Removers)
3 Churchill Court,
58 Station Road,
North Harrow HA2 7SA
Tel 081-861 3331

Road Law
Barry Rose Law Periodicals
Little London,
Chichester,
Sussex PO19 1PG
Tel 0243 783637

Roadway (Journal of RHA)
Roadway House,
104 New King's Road,
London SW6 4LN
Tel 071-736 1183

Transport (Journal of Chartered Institute of Transport)
80 Portland Place,
London W1N 4DP
Tel 071-636 9952

Transport Engineer (Journal of the Institute of Road Transport Engineers)
1 Cromwell Place
London SW7 2JF
Tel 071-589 3744/Fax 071-225 0494

Transport Management (Journal of Institute of Transport Administration)
32 Palmerston Road,
Southampton SO1 1LL
Tel 0703 631380

Truck
Village Publishing Ltd,
24A Brook Mews North,
London W2 3BW
Tel 071-224 9242/Fax 071-402 3994

Trucking International
Messenger House,
33–35 St Michael's Square,
Gloucester GL1 1HX
Tel 0452 307181/Fax 0452 307170

Appendix IV

Department of Transport Goods Vehicle Testing Stations

South-east

Station No/Name	Address
25. Bicester	Launton Road, Bicester, Oxon OX6 0JG Tel 0869 243416/242562
05. Canterbury	Hersden, Canterbury, Kent CT3 4HB Tel 0227 710010/710852
35. Chelmsford	Widford Industrial Estate, Chelmsford, Essex CM1 3AE Tel 0245 259341/259209
12. Cowes* (Isle of Wight)	Prospect Road, Cowes, Isle of Wight PO3 7AD Tel 0983 293171
34. Crimplesham (Kings Lynn)	Bexwell Airfield, Crimplesham, King's Lynn, Norfolk PE33 9DU Tel 0366 382481/382866
30. Alvaston (Derby)	Off Raynesway, Alvaston, Derby DE2 7AY Tel 0332 571961
03. Edmonton	Anthony Wharf, Lea Valley Trading Estate, Angel Road, Edmonton, London N18 3JR Tel 081-803 7733
06. Gillingham	Ambley Road, Gillingham, Kent ME8 0SJ Tel 0634 232541/232754
33. Spitalgate (Grantham)	Spitalgate Airfield, Blue Harbour, Grantham, Lincs NG31 7TX Tel 0476 62012/65799
09. Guildford	Moorfield Road, Slyfield Industrial Estate, Guildford, Surrey GU1 1SA Tel 0483 65151-3
07. Hastings	Ivy House Lane, Ore, Hastings, East Sussex TN35 4NN Tel 0424 430248
40. Ipswich	Holbrook Road/Landseer Road, Ipswich IP3 0DF Tel 0473 259061-2
08. Lancing	Churchill Industrial Estate, Lancing, West Sussex BN15 8TU Tel 0903 753305/754276
31. Leicester	40 Cannock Street, Barkby Thorpe Road, Leicester LE4 7HT Tel 0533 760144/767405
37. Leighton Buzzard	Stanbridge Road, Leighton Buzzard, Bedfordshire LU7 8QG Tel 0525 373074

04. Mitcham	Redhouse Road, Croydon, Surrey CR0 2AQ Tel 081-684 1499
11. Newbury	Hambridge Lane, Newbury, Berkshire RG14 5TZ Tel 0635 47649
29. Watnall (Nottingham)	Main Road, Watnall, Nottingham NG16 1JF Tel 0602 382591-2
38. Norwich	Jupiter Road, Hellesdon, Norwich NR6 6SS Tel 0603 408128
39. Peterborough	Saville Road, Westwood, Peterborough PE3 6TL Tel 0733 263399/263423
02. Purfleet	Tank Hill Road, Purfleet RM16 1SX Tel 0708 866651-2
36. Royston	South Close, Orchard Road, Royston, Herts SG8 5HA Tel 0763 242697/244822
10. Southampton (Botley)	Hillsons Road, Bottings Trading Estate, Botley, Southampton SO3 2DY Tel 0489 785522
32. Weedon (Northampton)	Cavalry Hill Industrial Park, Weedon, Northampton NN7 4PP Tel 0327 40697
01. Yeading	Willow Tree Lane, Yeading, Hayes, Middlesex UB4 9BS Tel 081-845 9826/9828

South Wales and West

75. Aberystwyth*	Llanrhystyd, Dyfed SY23 5BT Tel 0974 202447
62. Ammanford	Tirydail Lane, Ammanford, Dyfed SA18 3AR Tel 0269 592875
26. Garrett's Green (Birmingham)	Garrett's Green Industrial Estate, Birmingham B33 0SS Tel 021-783 6560/6561
16. Bristol	Ashton Vale Road, Ashton Gate, Bristol BS3 2JE Tel 0272 661419/661472
13. Calne	Porte Marsh Road, Calne, Wiltshire SN11 9EW Tel 0249 812351
20. Redruth/Camborne	Wilson Way, Redruth, Cornwall TR15 3RP Tel 0209 216851/216023
18. Exeter	Grace Road, Marsh Barton Trading Estate, Exeter EX2 8PU Tel 0392 78267
23. Gloucester	Ashville Road, Gloucester GL2 6ET Tel 0452 529749/520401
77. Haverfordwest	Withybush, Haverfordwest, Dyfed SA62 4BN Tel 0437 764402
21. Hereford	Faraday Road, Westfield Trading Estate, Hereford HR4 9NS Tel 0432 267956
22. Kidderminster	Worcester Road, Kidderminster, Worcestershire DY11 7RD Tel 0562 745857

76. Llandrindod Wells*	Gun Park, Waterloo Road, Llandrindod Wells, Powys LD1 6DH Tel 0597 822788
60. Llantrisant	School Road, Miskin, Pontyclun, Mid-Glamorgan CF7 8YR Tel 0443 224771
19. Plymouth	Agaton Fort, Budshead Road, Ernesettle, Plymouth, Devon PL5 2QY Tel 0752 362294
61. Pontypool	Polo Ground Industrial Estate, New Inn, Pontypool, Gwent NP4 0YN Tel 0495 756001-3
15. Poole	Darby's Lane, Nuffield Industrial Estate, Poole BH17 7SA Tel 0202 674685/672844
73. St Austell*	Par Moor Road, Par, Cornwall PL24 2SQ Tel 0726 812218
14. Salisbury	Brunel Road, Churchfields Industrial Estate, Salisbury, Wilts SP2 7PU Tel 0722 322898
24. Shrewsbury	Ennerdale Road, Harlescott, Shrewsbury, Shropshire SY1 3LF Tel 0743 352530/351037
72. South Molton	Station Road, South Molton, Devon EX36 3LL Tel 0769 572248/573339
28. Swynnerton (Stoke-upon-Trent)	Station Road, Cold Meece, Stone, Staffordshire ST15 0OP Tel 0785 760213/760226
17. Taunton	Taunton Trading Estate, Norton Fitzwarren, Taunton TA2 6RX Tel 0823 282525-6
27. Featherstone (Wolverhampton)	Cat and Kittens Lane, Featherstone, Wolverhampton, West Midlands WV10 7JD Tel 0902 397722

North

70. Barrow-in-Furness*	C/o Corporation Transport Depot, Hindpool Road, Barrow-in-Furness, Cumbria CA14 2PE Tel 0229 824679
42. Beverley	Grove Hill Road, Beverley, North Humberside HU17 0JG Tel 0482 881522/881629
54. Bredbury (Manchester)	Lingard Lane, Bredbury, Woodley, Stockport SK6 2QX Tel 061-430 5160/7941
53. Bromborough	Dock Road South, Bromborough, Wirral, Merseyside L62 4SH Tel 051-643 1013
74. Caernarvon	Cibyn Industrial Estate, Caernarvon, Gwynedd LL55 2BD Tel 0286 672567

57. Carlisle	Brunthill Road, Kingstown Industrial Estate, Carlisle CA3 0HA Tel 0228 28106
49. Darlington	Banks Road, McMullen Road, Darlington, Co. Durham DL1 1YE Tel 0325 460547
44. Doncaster	Welsdyke Road, Adwick-le-Street, Doncaster DN6 7DU Tel 0302 724404
45. Grimsby	South Humberside Industrial Estate 1, Pyewipe, Grimsby, South Humberside DN31 2TB Tel 0472 353703/683785
55. Heywood (Manchester)	Ex-RAF Site, Middleton Road, Heywood, Lancs OL10 2LT Tel 0706 369913/369917
50. Keighley	Station Road, Steeton, Keighley, West Yorkshire BD20 6RW Tel 0535 653433
56. Kirkham (Preston)	Freckleton Road, Kirkham, Preston PR4 2RA Tel 0772 684809/683785
48. Leeds	Patrick Green, Woodlesford, Leeds LS26 8HE Tel 0532 825060/821156
58. Simonswood (Liverpool)	Stopgate Lane, Simonswood, Kirby, Liverpool L33 4YA Tel 051-547 4445
71. Milnthorpe	Milnthorpe Railway Station, Cumbria LA7 7LR Tel 05395 563751
51. Newcastle upon Tyne	Sandy Lane, Gosforth, Newcastle upon Tyne NE3 5HB Tel 091-236 5011
41. Scarborough	Cayton Road, Scarborough, North Yorkshire YO11 3BY Tel 0723 582695-6
46. Sheffield	Orgreave Way, Handsworth, Sheffield SL5 9LT Tel 0742 692334/692778
43. Walton (York)	Wighill Lane, Walton, Wetherby, West Yorkshire LS23 7DU Tel 0937 844560
69. Workington*	Pittwood Road, Lillyhall Industrial Estate, Lillyhall, Workington CA14 4JP Tel 0900 64456
59. Wrexham	Llay Road, Llay, Wrexham, Clwyd LL12 0TL Tel 0978 852422

Scotland

67. Aberdeen	Cloverhill Road, Bridge of Don Industrial Estate, Aberdeen AB2 8SF Tel 0224 702357/703774
52. Berwick-on-Tweed	Tweedside Trading Estate, Berwick-on-Tweed TD15 2XF Tel 0289 306004
79. Dumfries*	Heath Hall Industrial Estate, Locharbriggs, Dumfries and Galloway DG1 3PH Tel 0387 61141

81. East Fortune*	Building No 16, East Fortune Airfield, Drem, Lothian EH39 5LF Tel 0620 88350
64. Livingston (Edinburgh)	Houston Industrial Estate, Grange Road, Livingston, Lothian EH54 5DD Tel 0506 30053
86. Fort William*	Highland Omnibus Co, Fort William, Inverness PH33 6PP Tel 0397 702687
63. Bishopbriggs (Glasgow)	Crosshill Road, Bishopbriggs, Glasgow G64 2SA Tel 041-772 6321-2
65. Inverness	Seafield Road, Longman Industrial Estate,Inverness IV1 1RG Tel 0463 235505
84. Keith*	TA Depot, Banff Road, Keith, Grampian, Banffshire AB5 3ET Tel 0542 882819
66. Kilmarnock	216 Western Road, Kilmarnock, Strathclyde KA3 1LP Tel 0563 24312/27771
82. Kirkcaldy*	Park Road, Gallatown, Kirkcaldy, Fife KY1 3EL Tel 0592 51233/51493
90. Kirkwall*	BT Workshop, Haston Airfield, Kirkwall, Orkney KW16 1RE Tel 0856 872074
87. Lairg*	County Council Roads Depot, Laundery Road, Lairg, Highland IV27 4QE Tel 0549 2143
89. Lerwick*	County Council Depot, Gremista, Lerwick, Shetland ZE1 0PX Tel 0595 4900
85. Lochgilphead*	Unit 12, Kilmore Industrial Estate, Lochgilphead, Argyll PA31 8PR Tel 0546 603206
83. Montrose*	Building 61, Montrose Airfield, Montrose, Tayside DD10 9BB Tel 0674 73760
78. Newton Stewart*	Industrial Estate, Wigtown Road, Newton Stewart, Dumfries and Galloway DG8 6JZ Tel 0671 2516
68. Perth	North Muirton Industrial Estate, Arran Road, Perth PH1 3DZ Tel 0738 32037-8
91. Portree*	County Council Roads Depot, Dunvegan Road, Portree, Isle of Skye IV51 9HD Tel 0478 2306
80. St Boswells*	Charlesfield, St Boswells, Melrose, Borders TD6 0HH Tel 0835 23701
92. Stornoway*	Airport Building 29, Stornoway, Isle of Lewis PA86 0BN Tel 0851 703007

88. Wick*

Site 27A, Airport Industrial Estate, Wick, Highland
KW1 4QS
Tel 0955 2605

NB: All stations except those marked * are open from 08.00 to 17.00 Monday to Thursday and 08.00 to 16.30 on Fridays. Stations marked * are open only on demand.

Goods Vehicle Testing and operation of the Goods Vehicle Testing Stations is carried out by the Vehicle Inspectorate, an Executive Agency of the DoT.

Appendix V

Driving Standards Agency Regional Offices and LGV Driving Test Centres

Regional Office	LGV Driving Test Centres
North-eastern Westgate House, Westgate Road, Newcastle upon Tyne NE1 1TW	Berwick*, Beverley, Darlington, Grimsby, Keighley, Leeds, Newcastle, Sheffield, Walton (York).
North-western Portcullis House, Seymour Grove, Manchester M16 0NE	Bredbury (Manchester), Caernarvon*, Carlisle, Heywood (Manchester), Kirkham (Preston), Simonswood (Liverpool), Wirral (Birkenhead), Wrexham.
West Midland Cumberland House, Broad Street, Birmingham B15 1TD	Garretts Green (Birmingham), Featherstone (Wolverhampton), Shrewsbury, Swynnerton (Stoke-on-Trent).
Eastern (Nottingham) Birkbeck House, 14/16 Trinity Square, Nottingham NG1 4BA	Alvaston (Derby), Leicester, Watnall (Nottingham), Weedon.
Eastern (Cambridge) Terrington House, 13/15 Hills Road, Cambridge CB2 1NP	Chelmsford, Ipswich, Leighton Buzzard, Norwich, Peterborough, Waterbeach (Cambridge).
South Wales Caradog House, 1–6 St Andrews Place, Cardiff CF1 3PW	Llantrisant, Neath, Pontypool, Templeton (Havefordwest).
Western The Gaunts' House, Denmark Street, Bristol BS1 5DR	Bristol, Camborne, Chiseldon (Swindon), Exeter, Gloucester, Poole, Plymouth, Taunton.
South-eastern Ivy House, 3 Ivy Terrace, Eastbourne BN21 4QT	Canterbury, Culham, Gillingham, Hastings, Isle of Wight*, Lancing, Reading, Southampton.
Metropolitan PO Box 2224, Charles House, 375 Kensington High Street, London W14 8QH	Croydon, Enfield, Guildford, Purfleet, Yeading.
Scottish 83 Princes Street, Edinburgh EH2 2ER	Aberdeen, Bishopbriggs (Glasgow), Drem*, Dumfries*, Elgin (Keith)*, Galashiels*, Inverness, Kirkwall*, Lerwick*, Livingston (Edinburgh), Machrihanish (Kintyre)*, Oban*, Perth, Port Ellen*, Stornoway*, Wick*.

* Tests are conducted only occasionally at these centres.

The Driving Standards Agency is an Executive Agency of the Department of Transport.

Appendix VI

Tachograph Manufacturers and Approved Centres

This appendix lists the three main tachograph manufacturers, two independent tachograph repairers and the tachograph centres approved by the Secretary of State for Transport in the UK (ie 'approved workshops' as referred to in EC legislation), based on lists issued and updated from time to time by the DoT. This list is based on the DoT official listing dated April 1993.

Tachograph Manufacturers

Lucas Kienzle Instruments Limited,
36 Gravelly Industrial Park,
Birmingham B24 8TA
Tel 021-328 5533

Time Instrument Manufacturers Limited,
5 Alston Drive,
Bradwell Abbey,
Milton Keynes MK13 9HA
Tel 0908 220020

TVI Europe Limited,
Kilspindie Road,
Dundee DD2 3QJ
Tel 0382 833033

Independent Tachograph Repairers

Tachodisc Limited GB D 01R
Loomed Road,
Chesterton,
Newcastle-under-Lyme ST5 7QQ

Instrument Repair Services GB D 01R
35 Redcliffe Road,
West Bridgford,
Nottingham NG2 5FF

NB: It should be noted that the American-owned tachograph manufacturing firm of Veeder Root, based in Dundee, Scotland, has now been taken over by its management and operates under the name TVI Europe Limited.

Approved Tachograph Centres

ENGLAND

Avon

BRISTOL GB H 131
Lucas Service UK Ltd,
Short Street,
St Phillips,
Bristol BS2 0SW

BRISTOL GB H 118
Weston-Super-Mare Motors,
Days Road,
Barton Hill,
Bristol BS5 0AJ
Tel 0272 551571

BRISTOL GB H 202
Westrucks Plc,
Britannia Road,
Patchway Trading Estate,
Bristol BS2 0UD
Tel 0272 772671

BRISTOL GB H 216
S A Trucks (Bristol) Ltd,
Third Way,
Avonmouth,
Bristol BS11 9YL
Tel 0272 821241

BRISTOL GB H 222
Lex Tillotson Bristol,
Days Road,
St Phillips,
Bristol BS2 0OP
Tel 0272 557755

BRISTOL GB H 303
ATAC,
Unit 3B, Severnside Trading Estate,
St Andrews Road,
Avonmouth,

Bristol BS11 9YQ
Tel 0272 828583

BRISTOL GB H 314
ATAC,
C/o BRS Western Ltd,
Spring Street,
Bedminster,
Bristol BS3 4BG
Tel 0272 770411

CHIPPING SODBURY GB H 110
Dando's (Motor Services) Ltd,
Hatters Lane,
Chipping Sodbury BS17 6AS
Tel 0454 310136

WESTON-SUPER-MARE GB H 313
Weston-Super-Mare Motors,
Bridge Road Works,
Weston-Super-Mare BS23 3NF
Tel 0934 628127

WESTON-SUPER-MARE GB H 323
Coombs Travel,
Searle Crescent,
Winterstoke Commercial Centre
Weston-Super-Mare BS23 3YA

Bedfordshire

BEDFORD GB F 105
Charles King (Motors) Ltd,
Hudson Road,
Elms Farm Industrial Estate,
Bedford MK41 0JQ
Tel 0234 40041

BEDFORD GB D 208
Arlington Motor Co Ltd,
The Embankment,
Barkers Lane,
Bedford MK41 9SD
Tel 0234 270000

BEDFORD GB F 317
Banks Truck Centre Ltd,
3 Brunel Road,
Barkers Lane Industrial Estate,
Bedford MK41 9TL
Tel 0234 211241

DUNSTABLE GB N 323
Renault Truck Chiltern,
Luton Road,
Dunstable LU5 4QE

DUNSTABLE GB N 222
Trimoco Luton Ltd,
Skimpot Road,
Dunstable LU5 4JX
Tel 0582 597575

LEIGHTON BUZZARD GB F 217
Chassis Developments Ltd,
Grovebury Road,
Leighton Buzzard LU7 8SL
Tel 0525 374151

Berkshire

BRACKNELL GB K 307
John Lewis & Co Ltd,
Central Vehicle Workshop,
Doncastle Road,
Southern Industrial Area,
Bracknell RE12 4YB
Tel 0344 424080

READING GB K 104
Penta Truck & Van Centre,
Station Road,
Theale RG7 4AG
Tel 0734 323383

READING GB K 200
Anchor Trucks,
20 Commercial Road,
Reading RG2 0RN
Tel 0734 312660

READING GB K 122
Renault Trucks Reading,
Bennett Road,
Reading RG2 0QX
Tel 0734 752355

READING GB K 215
Lucas Services UK Ltd,
16-20 Long Barn Lane,
Reading RG2 7SZ
Tel 0734 861202

READING GB K 502
Reading Leyland DAF Ltd,
Station Road,
Theale,
Reading RG7 4AG
Tel 0734 323383

Buckinghamshire

AYLESBURY GB E 202
Perrys Ltd,
Griffin Lane,
Aylesbury HP19 3BY
Tel 0296 26162

COLNBROOK GB N 334
Scantruck West Limited,
Skyway 14,
Calderway (Off Horton Road),
Colnbrook SL3 0BQ
Tel 0753 686467

HIGH WYCOMBE GB N 306
Biffa Ltd,
Kingsmill,
London Road,
High Wycombe
Tel 0494 21221

MILTON KEYNES GB E 218
City Truck Sales Ltd,
10, Northfield Drive,
Milton Keynes MK15 0DE
Tel 0908 665152

MILTON KEYNES GBN 113
Leyland DAF Distribution,
Chesney Wold,
Bleak Hall,
Milton Keynes MK6 1LH
Tel 0908 663991

MILTON KEYNES GB E 306
Perrys Ltd,
Clarke Road,
Mount Farm Estate,
Milton Keynes MK1 1NP
Tel 0908 74011

MILTON KEYNES GB N 342
Dawson Freight Commercials Ltd,
Delaware Drive,
Tongwell,
Milton Keynes MK15 8JH
Tel 0908 74011

Cambridgeshire

CAMBRIDGE GB F 100
Marshall (Cambridge) Ltd,
Airport Garage,
Newmarket Road,
Cambridge CB5 8SQ
Tel 0223 293131

CAMBRIDGE GB F 212
Lucas Services UK Ltd,
442 Newmarket Road,
Cambridge CB5 8JU
Tel 0223 315931

CAMBRIDGE GB F 312
Gilbert Rice Ltd,
Commercial Vehicle Division,
375-381 Milton Road,
Cambridge CB4 1SR
Tel 0223 425959

HUNTINGDON GB F 307
Murkett Bros Ltd,
Ring Road,
Huntingdon PE18 6HX
Tel 0480 52697

PETERBOROUGH GB F 111
Ford & Slater of Peterborough,
316 Padholme Road,
Peterborough PE1 1BA
Tel 0733 47100

PETERBOROUGH GB F 215
T C Harrison Group Ltd,
Truck Division,
Oxney Road,
Peterborough PE1 5YN
Tel 0733 558111

PETERBOROUGH GB F 213
Sellers & Batty (Peterborough) Ltd,
Fengate,
Peterborough
Tel 0733 60591

PETERBOROUGH GB F 315
BRS Midlands Ltd,
Fengate Commercials,
Nursery Lane,
Fengate,
Peterborough PE1 5BG
Tel 0733 54481

Cheshire

CREWE GB C 116
Crewe Tachograph Centre,
Chamberlains Transport Ltd,
Bradley Road,
Haslington,
Crewe CW1 1PU
Tel 0270 581224

ELLESMERE PORT GB C 225
Tachograph Chester Ltd,
Rossfield Road,
Ellesmere Port L65 3AW
Tel 051-355 2101

NORTHWICH GB C 302
North West Truck Engineering Ltd,
Griffiths Road,
Lostock Gralam,
Northwich CW9 7NU
Tel 0606 48671

SANDBACH GB C 227
Sandbach Truck Centre Ltd,
Station Road,
Elworth,
Sandbach CW11 9JG
Tel 0270 763291

STALYBRIDGE GB C 233
Tameside Tachograph Centre,
Bayfreight Ltd,
Premier Mill,
Tame Street,
Stalybridge SK15 1ST
Tel 061-338 8700

SUTTON WEAVER GB C 125
Leyfield Commercial Services Ltd,
Ashville Way,
Sutton Weaver,
Nr Runcorn WA7 3EZ
Tel 0928 790098

WARRINGTON GB C 236
P & O Ferrymasters Ltd,
Leacroft Road,
Risley,
Warrington WA3 6NN
Tel 0925 810000

WARRINGTON GB C 330
Ryland Vehicle Group North West Ltd,
Winwick Street Factory,
John Street,
Warrington WA2 7UD
Tel 0925 33271

MIDDLEWICH GB C 241
Beechs Garage (1983) Ltd,
Brooks Lane,
Middlewich CW10 0JH
Tel 060-683 2930

WIDNES GB C 321
Sutton & Son (St Helens) Ltd,
6 Tanhouse Lane,
Widnes WA8 0RZ
Tel 051-424 3078

Cleveland

BILLINGHAM GB B 317
North East Leyland DAF Ltd,
Cowpen Bewley Road,
Haverton Hill,
Billingham TS23 4EX
Tel 0642 370555

BILLINGHAM GB B 132
Renault Trucks Northeast,
Nuffield Road,
Cowpen Bewley Industrial Estate,
Billingham TS23 4DA

STOCKTON ON TEES GB A 100
Hargreaves Vehicle (North East) Ltd,
Bowesfield Lane,
Stockton on Tees TS18 3HF
Tel 0642 614121

STOCKTON ON TEES GB A 111
BRS Limited,
Malleable Way,
Pontract Lane,
Stockton on Tees TS18 2QZ
Tel 0642 606333

STOCKTON ON TEES GB A 202
Auto Electrics (Teesside) Ltd,
Thornaby House,
Thornaby,
Stockton on Tees TS17 6BW
Tel 0642 607901

STOCKTON ON TEES GB A 208
Electro Diesel North East Ltd,
Portrack Grange Road,
Portrack Industrial Estate,
Stockton on Tees
Tel 0642 605050

Cornwall

HELSTON GB H 206
Arlington Motor Co (Helston) Ltd,
St Johns,
Helston TR13 8EX
Tel 03257 2561

LAUNCESTON GB H 214
Pannell Commercials,
Pennygillam Industrial Estate,
Launceston PL15 7ED
Tel 0566 3896/4361

LAUNCESTON GB H 319
T H Cawsey Commercials Ltd,
Newport Industrial Estate,
Launceston PL15 8EX
Tel 0566 2805

ST AUSTELL GB H 106
People 2000 Ltd,
Slades Road,
St Austell PL25 4HP
Tel 0724 672333

ST AUSTELL GB H 212
John Hewitt Group Ltd,
Westhaul Park,
Par Moor Road,
St Austell PL25 3RA
Tel 0726 812382/4/5

Cumbria

COCKERMOUTH GB A 206
Thomas Armstrong (Transport
Services) Ltd,
Workington Road,
Flimby,
Maryport CA15 8RY
Tel 0900 682111

CARLISLE GB A 102
Duncan's Tachograph Centre,
Kingstown Industrial Estate,
Carlisle CA3 0EP
Tel 0228 515284

CARLISLE GB A 303
Carlisle Commercials Ltd,
Kingstown Industrial Estate,
Carlisle CA3 0AH
Tel 0228 29262

CARLISLE GB A 304
C G Trucks,
Wakefield Road,
Kingstown Industrial Estate,
Carlisle CA3 0HE
Tel 0228 24234

CARLISLE GB C 235
Solway Leyland DAF Ltd,
Kingstown Industrial Estate,
Carlisle CA3 0HD
Tel 0228 39394

DISTINGTON GB A 311
Myers & Bowman Ltd,
Prospect Works,
Distington CA14 5XH
Tel 0946 830247/8/9

KENDAL GB C 504
Lakeland Commercials Ltd,
Mintsfeet Road North,
Mintsfeet Trading Estate,
Kendal LA9 6LZ
Tel 0539 723956

Derbyshire

BURTON ON TRENT GB C 239
Jeffries Haulage Limited,
Swadlicote Road,
Woodville,
Burton on Trent DE11 8DD

DERBY GB E 501
T C Harrison Group Ltd,
Chequers Road,
Derby DE2 6EN
Tel 0332 31188

DERBY GB E 223
F B Atkins & Son Ltd,
Burton Road,
Findern DE6 6BG
Tel 0332 516151

DERBY GB E 308
Kays Mackworth Ltd,
Ashbourne Road,
Mackworth DE3 4 NB
Tel 0332 824371

DERBY GB E 226
Sherwood Leyland DAF Ltd,
Berristow Lane,
Blackwell DE55 5HP
Tel 0773 863311

Devon

BARNSTAPLE GB H 130
P M Clarke (Commercials) Ltd,
Severn Brethren Trading Estate,
Barnstaple
Tel 0271 45151

BARNSTAPLE GB H 229
North Devon (Redbus) Ltd,
Coney Avenue,
Barnstaple EX32 8QJ
Tel 0271 25452

EXETER GB H 125
Westrucks Plc,
Chelsea Centre,
Heron Road,
Sowton EX2 7LL
Tel 0392 444660

EXETER GB H 204
Lucas Service UK Ltd,
Grace Road,
Marsh Barton,
Exeter EX2 8QE
Tel 0392 70235

EXETER GB H 213
Frank Tucker (Commercials) Ltd,
Peamore Garage,
Alphington,
Exeter EX2 9SL
Tel 0392 832662

EXETER GB H 300
Evans Halshaw SW Ltd,
Grace Road,
Marsh Barton,
Exeter
Tel 0392 76561

NEWTON ABBOT GB H 305
Newton Abbot Motors Ltd,
Bradley Lane,
Newton Abbot TQ12 1JT
Tel 0626 65081

PLYMOUTH GB H 116
Vospers Trucks Ltd,
Hobart Street,
Plymouth PL1 3LW
Tel 0752 668040

PLYMOUTH GB H 225
Lucas Service (UK) Ltd,
Plymouth City Bus,
Milehouse,
Plymouth

TOTNES GB H 224
Wincanton Distribution Services Ltd,
Dart Hills,
Babbage Road,
Totnes TQ9 5JA
Tel 0803 867892

Dorset

BOURNEMOUTH GB K 114
Lucas Service UK Ltd,
Elliot Road,
West Howe Industrial Estate,
Bournemouth BH11 8LN
Tel 0202 570507

DORCHESTER GB H 114
Trustees of R W Toop,
T/A Bere Regis & District Motor
Services,
Grove Trading Estate,
7 Bridport Road,
Dorchester DT1 1RW
Tel 0305 62992

POOLE GB H 223
English Ford Truck,
1 Yarrow Road,
Tower Park,
Poole BH17 7NY
Tel 0202 671122

POOLE GB H 318
Adams Morey Limited,
Yeomans Way,
Yeomans Industrial Estate,
Bournemouth BH8 0BJ

WEYMOUTH GB H 500
Marsh Road Garages,
Marsh Road,
Weymouth DT4 8JD
Tel 03057 76116

Durham

CHESTER-LE-STREET GB A 106
Millbay Truck & Van Centre,
Drum Road,
Chester-Le-Street DH3 2AF
Tel 091-410 4261

CHESTER-LE-STREET GB A 309
Transfleet Services Ltd,
Penshaw Way,
Portabello Trading Estate,
Birtley,
Chester-Le-Street DH3 2SA
Tel 091-410 4437

DARLINGTON GB A 104
Skipper of Darlington Ltd,
Allington Way,
Yarm Road Industrial Estate,
Darlington
Tel 0325 59242

DARLINGTON GB A 300
North Riding Garages Ltd,
Middleton St George,
Darlington DL2 1HR
Tel 0325 332941

DARLINGTON GB A 301
Darlington Commercials,
Lingfield Way,
Yarm Road Industrial Estate,
Darlington DL1 4PY
Tel 0325 55161

SPENNYMOOR GB A 105
Exel Logistics Regional Services,
Green Lane Industrial Estate,
Spennymoor DL16 6BW
Tel 0388 815900

SHOTTON COLLIERY GB B 131
Geoffrey Division,
Victoria Fleet Services,
Victoria Garage,
Black Friar Street,
Shotton Colliery DH6 2NX
Tel 091-517 0577

Essex

BARKING GB N 506
Dagenham Motors,
51 River Road,
Barking IG11 0SW

BASILDON GB N 232
Arlington Commercial Vehicle Sales &
Repairs,
Cranes Close,
Basildon SS14 3JD
Tel 0268 20223

BENFLEET GB F 102
W Harold Perry Ltd,
Stadium Way,

Benfleet SS7 3NU
Tel 0268 775544

CHELMSFORD GB N 339
Lucas Service UK Ltd,
3 Montrose Road,
Dukes Park Industrial Estate,
Chelmsford CM2 6TE
Tel 0245 466166

CHELMSFORD GB F 205
Trimoco Plc,
11 Montrose Road,
Dukes Park Industrial Estate,
Chelmsford CM2 6TF
Tel 0245 466619

CHELMSFORD GB F 502
County Motor Works (Chelmsford) Ltd,
Eastern Approach,
Springfield,
Chelmsford CM2 6PT
Tel 0245 466333

COLCHESTER GB F 209
Colchester Fuel Injections Ltd,
Haven Road,
Colchester CO2 8HT
Tel 0206 862049

COLCHESTER GB F 500
Candor Motors Ltd,
114 Ipswich Road,
Colchester CO4 4AB
Tel 0206 571171

DAGENHAM GB N 229
Dairy Products Transport,
Selinas Lane,
Chadwell Heath,
Dagenham RH8 1QH
Tel 081-517 5444

GRAYS GB N 329
Harris Commercial Repairs Ltd,
601 London Road,
West Thurrock,
Grays RM16 1AN
Tel 0708 864426

HARLOW GB N 205
Arlington Motor Co Ltd,
Potter Street,
Harlow CM17 9NP
Tel 0279 422391

PURFLEET GB N 206
ScanTruck Ltd,
Arterial Road,
Purfleet RM16 1TR
Tel 0708 864915

Gloucestershire

GLOUCESTER GB H 217
Watts Truck Centre Ltd,
Mercia Road,

Gloucester GL1 2SQ
Tel 0452 525721

GLOUCESTER GB H 502
Target of Gloucester Ltd,
Bristol Road,
Hempsted,
Gloucester GL2 5YS
Tel 0452 21581

GLOUCESTER GB H 120
BRS Western Ltd,
St Oswald Road,
Gloucester GL1 2RP
Tel 0452 20387/500529

GLOUCESTER GB H 227
Lucas Service (UK) Ltd,
Sparrows Wharf, Bristol Road,
Gloucester GL12 6DH
Tel 0452 524951

GLOUCESTER GB H 320
ATAC (Gloucester),
Unit 55a,
Gloucester Trading Estate,
Hucclecote,
Gloucester GL3 4AA
Tel 0452 371047

GLOUCESTER GB H 321
Richard Read (Transport) Ltd,
Longhope,
Gloucester GL17 0QG
Tel 0452 830456/7

TEWKESBURY GB H 126
Mudie Bond Ltd,
New Town Trading Estate,
Tewkesbury GL20 8JD
Tel 0684 295090

Hampshire

ALDERSHOT GB K 220
P D E (Farnham) Ltd,
Pavilion Road,
Aldershot GU11 3NX
Tel 0252 316504

BASINGSTOKE GB K 113
Jacksons (Baskingstoke) Ltd,
Roentgen Road,
Daneshill East Industrial Estate,
Basingstoke RG24 0NT
Tel 0256 461656

BASINGSTOKE GB K 317
Heathrow Commercials Ltd,
Bell Road,
Basingstoke RG24 0FB
Tel 0703 663500

EASTLEIGH GB K 100
Hendy Lennox Ltd,
Bournemouth Road,
Chandlers Ford,

Eastleigh SO5 3ZG
Tel 0703 271271

FAREHAM GB K 119
Spartrucks Ltd,
Standard Way,
Fareham Industrial Estate,
Fareham PO16 8XL
Tel 0329 286224

FAREHAM GB K 316
Southway Commercials Ltd,
26 Brunel Way,
Segensworth East,
Fareham PO15 5SD
Tel 0329 579880

PORTSMOUTH GB K 112
Wadham Kenning Commercials,
Burrfields Road,
Copnor,
Portsmouth PO3 5NN
Tel 0705 664900

PORTSMOUTH GB K 221
Lucas Service UK Ltd,
Airport Services Road,
Portsmouth PO19 5PY
Tel 0705 650776

PORTSMOUTH GB K 301
Arlington Motor Co (Portsmouth) Ltd,
Norway Road,
Hilsea,
Portsmouth PO2 9QZ
Tel 0705 661321

SOUTHAMPTON GB K 201
Morey Leyland DAF,
The Causeway,
Redbridge,
Southampton SO9 4YS
Tel 0703 663000

SOUTHAMPTON GB K 223
Lucas Service UK Ltd,
Oakley Road,
Southampton SO9 7PR

SOUTHAMPTON GB K 304
Bristol Street Motors (Southampton) Ltd,
2nd Avenue,
Millbrook,
Southampton SO1 0LP
Tel 0703 701500

SOUTHAMPTON GB K 121
Hendy Truck Cosham Ltd,
Southampton Road,
Cosham,
Portsmouth PO6 4RW
Tel 0703 370944

SOUTHAMPTON GB K 319
Trimtruk,
286 Weyhill Road,
Andover SP10 3LS

Hereford & Worcester

HEREFORD GB D 330
Hereford Tacho Centre,
Perseverence Road,
Hereford HR4 9SP

BROOMHALL GB D 300
Carmichael Trucks Ltd,
Bath Road,
Broomhall,
Worcester WR3 3HR
Tel 0905 820109

EVESHAM GB D 217
Coulters Garage (Evesham) Ltd,
Four Pools Lane,
Evesham WR11 5DW
Tel 0386 442525

HEREFORD GB D 211
Lucas Service UK Ltd,
Mortimer Road,
Hereford HR4 9JG
Tel 0432 265571/267678

Hertfordshire

BOREHAMWOOD GB N 503
W H Perry Ltd,
Stirling Corner,
Stirling Way,
Borehamwood WD6 2AX
Tel 081-207 3100

HATFIELD GB N 217
S & B Commercials Ltd,
Travellers Close,
Welham Green,
Hatfield AL9 7LA
Tel 0707 261111

HITCHIN GB N 341
Hitchin Recovery Limited,
48 Burymead Road,
Hitchin SG5 1RT
Tel 0462 421686

KINGS LANGLEY GB N 331
E J Masters Ltd,
Railway Terrace,
Kings Langley WD4 8JA
Tel 0923 268921

ST ALBANS GB N 310
Godfrey Davis (St Albans) Ltd,
Bricknell Park Industrial Estate,
Ashley Road,
St Albans
Tel 0727 59155

SOUTH MIMMS GB N 340
Scan Truck Ltd,
Bignells Corner,
20 St Albans Road,
South Mimms EN6 3NG
Tel 0707 49955

WATFORD GB N 219
Vales Truck Centre Ltd,
Tolpits Lane,
Watford WD1 8QP
Tel 0923 776688

Humberside

BROUGH GB B 208
Humberside Motors Ltd,
Junction 38, M62,
Newport,
Brough HU15 2RD
Tel 04342 2297

GRIMSBY GB E 214
Hartford Motors (Grimsby) Ltd,
Corporation Road,
Grimsby DN31 1UH
Tel 0472 358941

GRIMSBY GB B 315
T H Brown Ltd,
Estate Road One,
South Humberside Industrial Estate,
Grimsby DN31 2TA
Tel 0472 46913

HULL GB B 110
Crossload Commercials Ltd,
Valletta Street,
Hedon Road,
Hull HU9 5NP
Tel 0482 781831

HULL GB B 213
Lex Tillotson (Hull) Ltd,
Hedon Road,
Hull HU9 5PJ
Tel 0482 795111

HULL GB B 216
Crystal of Hull Ltd,
Little Fair Road,
Hedon Road,
Hull HU9 5LP
Tel 0482 703337

HULL GB B 313
Torridan Commercial Vehicles Ltd,
Ann Watson Street,
Stoneferry,
Hull HU8 0BJ
Tel 0482 839677

KINGSTON UPON HULL GB B 504
Crystal Truck Centre,
Clough Road,
Kingston upon Hull HU6 7PU

KINGSTON UPON HULL GB B 221
Scanlink Ltd,
Central Orbital Trading Estate,
Waverley Street,
Kingston upon Hull HU6 7PU

SCUNTHORPE GB E 112
BRS Northern Ltd,

Grange Lane North,
Scunthorpe DN16 1BY
Tel 0724 270462

SCUNTHORPE GB E 302
Lex Tillotson Scunthorpe,
Midland Industrial Estate,
Kettering Road,
Scunthorpe DN16 1VW
Tel 0724 282444

SOUTH KILLINGHOLME GB E 219
H & L Garages Ltd,
Humber Road,
South Killingholme DN40 3DL
Tel 0469 571666

Isle of Wight

NEWPORT GB K 120
Riverside Motors Ltd,
Riverside Works,
Little London,
Newport PO30 5BT
Tel 0983 522552

Kent

HYTHE GB K 320
McTruck Engineering Ltd,
Unit D 2, Lympne Industrial Estate,
Hythe CT21 4LR
Tel 0303 262888

ST MARYS HOO GB K 118
Squires & Knight Ltd,
Fenn Corner,
St Marys Hoo,
Nr Rochester ME3 8RF
Tel 0634 271987

ASHFORD GB K 209
Crouch's Garage Ltd,
Station Road,
Ashford TN23 1PJ
Tel 0233 623451

ASHFORD GB K 306
Channel Commercials Plc,
Brunswick Road,
Cobbs Wood Estate,
Ashford TN23 1EH
Tel 0233 629271

BROADSTAIRS GB K 116
Thanet Commercials Plc,
Unit 12,
Hornet Close,
Pysons Road Industrial Estate,
Broadstairs CT10 2TD
Tel 0843 603480/602194

CANTERBURY GB K 102
Invicta Motors Ltd,
134 Sturry Road,
Canterbury CT1 1DR
Tel 0227 762783

CANTERBURY GB K 204
Lucas Service UK Ltd,
Maynard Road,
Wincheap Industrial Estate,
Canterbury CT1 3RH
Tel 0227 453510

CRAYFORD GB N 335
Acorn Truck Sales Ltd,
Acorn Industrial Park,
Crayford Road,
Crayford DA1 4AL
Tel 0322 56415

DARTFORD GB N 105
K T Trucks Ltd,
Dartford Industrial Estate,
Hawley Road,
Dartford DA1 1NQ
Tel 0322 229331

ERITH GB N 226
South Eastern Auto Electrical Services Ltd,
Unit 26,
Manford Industrial Estate,
Erith
Tel 0322 342277

MAIDSTONE GB K 222
South Eastern Auto Electrical Service Ltd,
Wharf Road,
Tovil,
Maidstone ME15 6RT
Tel 0622 690010

MAIDSTONE GB K 303
Volvo (GB) Ltd,
T/A Maidstone Commercials Ltd,
Beddow Way,
Forstal Road,
Aylesford ME20 7BT
Tel 0622 710811

ROCHESTER GB K 315
S E Trucks Ltd,
Arnold Close,
Sir Thomas Longley Road,
Medway City Estate,
Rochester ME2 4DP
Tel 0634 711144

SITTINGBOURNE GB K 212
Sparshatts of Kent Ltd,
Unit 10,
Eurolink Industrial Estate,
Murston,
Sittingbourne ME10 3RN
Tel 0795 479571

SWANSCOMBE GB N 343
Thameside Truck Centre Ltd,
Manor Way,
London Road,
Swanscombe DA10 0LL
Tel 0322 384747

TONBRIDGE GB N 101
Stormont Enginering Co Ltd,

Commercial Vehicle Division,
Hildenborough,
Tonbridge TN11 8NN
Tel 0732 833005

TUNBRIDGE WELLS GB K 224
Lucas Service UK Ltd,
North Farm Road,
High Brooms Industrial Estate,
Tunbridge Wells TN2 3EA
Tel 0892 510800

TUNBRIDGE WELLS GB K 107
Kent & Sussex Truck Centre,
Longfield Road,
North Farm Industrial Estate,
Tunbridge Wells TN2 3EY
Tel 0892 515333

Lancashire

ACCRINGTON GB C 220
Gilbraith Commercials Ltd,
Market Street,
Church,
Accrington BB5 0DN
Tel 0254 31431

ACCRINGTON GB C 124
Lynch Truck Services Ltd,
Unit 3, Plot 8,
Newhouse Road,
Huncoat Industrial Estate,
Accrington BB5 6NT
Tel 0254 301331

BLACKBURN GB C 305
Fox Commercial Vehicles Ltd,
Navigation Garage,
Forrest Street,
Blackburn
Tel 0254 675111

BURNLEY GB C 329
Burnley & Pendle Transport
Company Ltd,
Queensgate,
Colne Road,
Burnley BB10 1HH
Tel 0282 25244

CHORLEY GB C 114
Gilbraith Parts and Service Ltd,
Ackhurst Road,
Chorley PR7 3EH
Tel 02572 76421

CLITHEROE GB C 229
Steadplan Ltd,
Salthill Industrial Estate,
Clitheroe BB7 1QL
Tel 0200 27415

FORTON GB C 128
Cabus Garage,
A6 Lancaster Road,
Forton,

Lancashire
Tel 0524 791417

HAYDOCK GB C 325
Haydock Commercial Vehicles Ltd,
Yew Tree Trading Estate,
Kilbuck Lane,
Haydock WA11 9XW
Tel 0942 714103

LANCASTER GB C 317
Pye Motors Ltd,
Parliament Street,
Lancaster LA1 1DA
Tel 0524 63553

LEYLAND GB C 500
SVAP Leyland DAF Ltd,
Leyland Service Centre,
Croyston Road,
Gate No 1,
Starter Works,
Leyland PR5 1SN
Tel 0772 421400

LITTLE HOOLE GB C 109
BRS Northern Ltd,
Liverpool Road,
Little Hoole,
Nr Preston PR4 5JT
Tel 0772 617668

MORECAMBE C 133
Carlisle Commercials Ltd,
White Lund Industrial Estate,
Morecambe LA3 3BY
Tel 0228 29262

PRESTON GB C 303
Ribblesdale Auto-Electrics Ltd,
Marsh Lane,
Preston PR1 8YN
Tel 0772 555011

PRESTON GB C 127
Leyland Auto Electrical & Diesel Ltd,
Unit 220,
Walton Summit Industrial Estate,
Bamber Bridge,
Preston PR5 8AL
Tel 0772 38583

PRESTON GB C 328
Lancashire Leyland DAF,
Unit 224,
Walton Summit Centre,
Bamber Bridge,
Preston PR5 8AL
Tel 0772 38111

ROCHDALE GB C 201
Rochdale Motor Garage,
T/A Kirkby Trucks,
Durham Street,
Rochdale OL11 1LR
Tel 0706 355355

Leicestershire

BLABY GB E 120
BRS Midlands Ltd,
Leycroft Road,
Bursome Industrial Estate

Blaby LE8 3DU
Tel 0533 340100

HINCKLEY GB E 210
Paynes Garages Ltd,
Watling Street,
Hinckley LE10 3ED
Tel 0455 38911

LEICESTER GB E 206
Cossington Commercial Vehicles Ltd,
System Road,
Cossington,
Leicester LE7 8UZ
Tel 0533 607111

LEICESTER GB E 217
A B Butt Ltd,
Frog Island,
Leicester LE3 5AZ
Tel 0533 513344

LEICESTER GB E 300
Batchelor Bowles & Co Ltd,
Freemens Common,
Aylstone Road,
Leicester LE2 7SL
Tel 0533 557567

LEICESTER GB E 322
Prestige Trucks,
Bilton Way,
Lutterworth LE17 4JA
Tel 0455 556221

LOUGHBOROUGH GB E 106
Charnwood Trucks Ltd,
M1 Truck Centre,
Ashby Road,
Shepshed,
Loughborough LE12 5BR
Tel 0509 502121

Lincolnshire

BOSTON GB E 204
C F Parkinson Ltd,
Fydell Crescent Workshop,
Fydell Crescent,
George Street,
Boston PE21 8XQ
Tel 0205 63008

LINCOLN GB E 102
John Longden Ltd,
PO Box 19,
Crofton Road,
Lincoln LN3 4PJ
Tel 0522 538811

LINCOLN GB E 203
C F Parkinson Ltd,
Outer Circle Road,
Lincoln LN2 4HU
Tel 0522 530176

LINCOLN GB E 224
Ford & Slater of Lincoln,
Sleaford Road,
Bracebridge Heath,
Lincoln LN4 2NQ
Tel 0522 522231

SPALDING GB F 308
R C Edmondson (Spalding) Ltd,
St Johns Road,
Spalding PE11 1JA,
Tel 0775 67651

SPALDING GB F 308
Ford and Slater Ltd,
58 Station Road,
Donnington,
Spalding PE11 4UJ,
Tel 0775 820777

Greater London

BARKING GB N 322
Gifford Tachograph Services &
Commercial Ltd,
Assured House,
Cheques Lane,
Dagenham
Tel 081-593 1550

CHINGFORD GB N 334
Nordic Commercials Ltd,
47 Sewardstone Road,
Chingford E4 7PU
Tel 081-529 8686

CROYDON GB N 200
Dees of Croydon Ltd,
Dees Commercial Centre,
Carlton Road,
Croydon CR2 0BP
Tel 081-680 4466

CROYDON GB N 338
C Barber & Sons Ltd,
Barbers Tachograph Centre,
87 Beddington Lane,
Croydon CR0 4TD
Tel 081-689 4414

DAGENHAM GB N 111
Lancaster Trucks Ltd,
Lakeside Estate,
Heron Way,
West Thurrock RM16 1WJ
Tel 0708 861321

ENFIELD GB N 100
Hunter Vehicles Ltd,
Crown Works,
Southbury Road,
Enfield EN1 1UD
Tel 081-805 5175

ENFIELD GB N 228
Arlington Enfield Truck Centre,
Mollison Avenue,
Brimsdown,
Enfield EN3 7NE
Tel 081-804 1266

FELTHAM GB N 223
Heathrow Commercials Ltd,
Staines Road,
Bedfont,
Feltham TW14 8RP
Tel 0784 243571

LEYTON GB N 316
Linear Truck & Electronic Sensor,
Auriol Drive,
Oldfield Lane,
Greenford UB6 0AY
Tel 081-578 5688

HAYES GB N 213
Dagenham Motors (Hayes) Ltd,
Dawley Road,
Hayes UB3 1EH
Tel 081-561 8888

ISLEWORTH GB N 224
Currie Trucks Ltd,
207/209 Worton Road,
Isleworth TW7 6DS
Tel 081-568 4343

PECKHAM GB N 336
The Londoners Tacho Centre Ltd,
1A Brabourn Grove,
Peckham SE15 2BS
Tel 071-639 1212

PARK ROYAL GB N 327
Transfleet Services Ltd,
17 Western Road,
Western Trading Estate,
Park Royal NW10 7LT
Tel 081-961 5225

WANDSWORTH GB N 318
Hunts Trucks (Wandsworth) Ltd,
2 Armoury Way,
Wandsworth SW18 1SH
Tel 081-871 3021

BRIXTON GB K 225
A23 Tacho Centre Ltd,
146/156 Brixton Hill,
London SW2 1SD
Tel 081-671 7781

ALPERTON GB N 508
Alperton Ford,
374 Ealing Road,
Alperton HA0 1HG
Tel 081-997 3388

UXBRIDGE GB N 345
Martin Endersby Truck & Trailer Services,
91 Cowley Road,
Uxbridge UB8 2AG
Tel 0895 239834

WILLESDEN GB N 235
Bel Truck and Van Ltd,
24 Salter Street,
Hythe Road Estate,
Willesden NW10 6UN
Tel 081-969 1616

Greater Manchester

BOLTON GB C 118
Contract Hire (Commercial Vehicles) Ltd,
Raikes Lane Industrial Estate,
Manchester Road,
Bolton BL3 2NS
Tel 0204 32441

BOLTON GB C 134
Bridge Mills Service Centre Ltd,
Rochdale Road,
Edenfield BL0 0RE
Tel 0706 826344

BOLTON GB C 238
Manchester Truck & Bus Ltd,
Lyon Industrial Estate,
Moss Road,
Kearsley BL4 8TZ
Tel 0204 707227

DUKINFIELD GB C 332
BRS Northern Ltd,
Tameside Branch,
Crescent Road,
Dukinfield SK16 4HZ
Tel 061-308 2827

HYDE GB C 120
D Hulme Ltd,
Broadway Industrial Estate,
Dukinfield Road,
Hyde SK14 4QZ
Tel 061-366 9400

MANCHESTER GB C 130
ERF Manchester Ltd,
Trafford Park Road,
Trafford Park M17 1NJ
Tel 061-848 8331

MANCHESTER GB C 309
Manchester Garages (Trucks) Ltd,
Gorton Lane,
Gorton M18 8BT
Tel 061-223 7131

MANCHESTER GB C 123
Chatfield Martin Walter Ltd,
T/A CGL Truck Services,
Ashburton Road East,
Trafford Park M17 1GT
Tel 061-872 6855

MANCHESTER GB C 327
Quicks Trucks Ltd,
Moseley Road,
Trafford Park M17 1PG
Tel 061-872 7711

MANCHESTER GB C 204
Chatfields of Manchester,
40-46 Ashton Old Road,
Ardwick M12 6NA
Tel 061-273 7351

MANCHESTER GB C 333
Manchester Truck & Bus Ltd,
Bredbury Parkway,
Bredbury,
Stockport SK6 2SN
Tel 061-406 6620

MIDDLETON GB C 324
West Pennine Trucks Ltd,
Stakehill Industrial Estate,
Middleton M24 2RL
Tel 061-653 9700

MIDDLETON GB C 334
Manchester Truck Centre,
Duncan Street,
Salford M5 3SQ
Tel 061-873 8048

SALFORD GB C 234
Salford Leyland DAF Ltd,
West Egerton Street,
Salford M5 4DY
Tel 061-872 7241

STRETFORD GB C 223
Lucas Service UK Ltd,
Unit 18,
Severnside Trading Estate,
Textilose Road,
Trafford Park M17 1WB
Tel 061-872 5521

STRETFORD GB C 129
BRS Northern Engineering,
Barton Dock Industrial Estate,
Barton Dock Road,
Stretford M32 0XP
Tel 061-872 7551

WIGAN GB C 237
Shearings Bus & Coach,
Lockett Lane,
Bryn,
Wigan WN3 4AG
Tel 0942 27220

WORSLEY GB C 119
Roy Braidwood & Sons,
Worsley Trading Estate,
Lester Road,
Little Hulton,
Worsley M28 6PT
Tel 061-799 3801

Merseyside

LIVERPOOL GB C 105
Woodwards SVS,
Merton Street,
Bank Road,

St Helens WA9 1HU
Tel 0704 820266

LIVERPOOL GB C 501
J Blake & Co Ltd,
178 Lodge Lane,
Liverpool L8 0QW
Tel 051-727 4501

LIVERPOOL GB C 205
Lucas Service UK Ltd,
Vandries Road,
Liverpool L3 7BJ
Tel 051-236 7063

LIVERPOOL GB C 502
Peoples (Liverpool) Ltd,
Hawthorn Road,
Bootle,
Liverpool L20 9DA
Tel 051-922 8481

LIVERPOOL GB C 314
Thomas Hardie Commercials Ltd,
Newstet Road,
Knowsley Industrial Park (North),
Nr Liverpool L33 7TJ
Tel 051-546 5291

LIVERPOOL GB C 215
Perris & Kearon Ltd,
173-175 Crown Street,
Liverpool L7 3LZ
Tel 051-709 4262

LIVERPOOL GB C 331
North West Truck Engineering,
Unit 22,
Interchange Estate,
Wilson Road,
Huyton
Tel 051-480 0098

ST HELENS GB C 113
Roberts Motors Ltd,
St Helens Ford,
City Road,
St Helens WA10 6NZ
Tel 0744 26381

Middlesex

GREENFORD GB N 234
Normand Commercial Vehicles Ltd,
Auriol Drive,
Oldfield Lane,
Greenford UB6 0AY
Tel 081-575 5688

Norfolk

DISS GB F 219
Trumber Truck Care Ltd,
57 Victoria Road,
Diss IP22 3JD
Tel 0379 652161

FAKENHAM GB F 501
R C Edmondson (Fakenham) Ltd,
Oak Street,
Fakenham NR21 9DU
Tel 0328 851122

GREAT YARMOUTH GB F 306
L G Perfect (Engineering) Ltd,
Jubilee Works,
Hafreys Road,
Great Yarmouth NR31 0JL
Tel 0493 675131

KING'S LYNN GB F 112
GDM Transport Engineering Ltd,
Leyland Truck Centre,
Maple Road,
Kings Lynn PE34 3AH
Tel 0553 761112

KING'S LYNN GB F 310
BRS Southern Ltd,
Oldmeadow Road,
Hardwich Trading Estate,
King's Lynn PE30 4TZ
Tel 0553 692414

NORWICH GB F 206
Lucas Service UK Ltd,
Weston Road North,
Norwich NR3 3TL
Tel 0603 410301

WISBECH GB F 313
Duffield of East Anglia Ltd,
Boleness Road,
Wisbech PE13 2RE
Tel 0945 463355

NORWICH GB F 313
Busseys Ltd,
Whiffler Road,
Norwich NR3 2AW
Tel 0603 424022

NORWICH GB F 116
Carrow Commercials Ltd,
Unit 8 Industrial Estate,
Kerrison Road,
Norwich NR1 1JA
Tel 0603 621415

NORWICH GB F 220
Gales Commercial Vehicles Ltd,
Ayton Road,
Wymondham NR18 0QQ
Tel 0953 601222

THETFORD GB F 309
Almaco Motors Ltd,
Caxton Way,
Thetford IP24 3RY
Tel 0842 752457

North Humberside

HULL GB B 109
Thompson of Hull Ltd,

230-236 Anlaby Road,
Hull HU3 2RR
0482 23681

HUMBERSIDE GB B 218
H & L Garage Ltd,
Grange Road North,
Scunthorpe DN16 1BT
Tel 0724 856655

HUMBERSIDE GB B 321
Jayserve Ltd,
Mariner Street,
Grimsby DN14 5BW
Tel 0405 767977

Northamptonshire

DAVENTRY GB E 119
ITR Transerv Ltd,
Unit 3, Plant House,
Royal Oakway North,
Daventry NN11 5PQ
Tel 0327 77009

KETTERING GB E 109
John R Billows (Sales) Ltd,
Pytchley Road Industrial Estate,
Kettering NN15 6JJ
Tel 0536 516233

NORTHAMPTON GB E 211
Northampton Diesel & Electrical Services
Ltd,
Holloway Industrial Estate,
Weedon Road,
Northampton NN5 5DG
Tel 0604 55321

NORTHAMPTON GB E 318
Arlington Motor Co Ltd,
Bedford Road,
Northampton NN1 5NS
Tel 0604 250151

NORTHAMPTON GB E 208
Airflow Streamlines Plc,
Truck Centre,
Letts Road,
Northampton NN4 9HQ
Tel 0604 581121

NORTHAMPTON GB E 321
Peter Brown Leyland DAF Ltd,
Gayton Road,
Milton Malsor,
Northampton NN7 6AB
Tel 0604 858923

WELLINGBOROUGH GB E 310
E Ward (Wellingborough) Ltd,
Truck Division,
Northampton Road,
Wellingborough NN8 3PP
Tel 0933 440110

WELLINGBOROUGH GB E 122
Harborne Group Ltd,

Stewarts Road,
Finedon Industrial Estate,
Wellingborough NN8 4RJ
Tel 0933 279211

Northumberland

BERWICK UPON TWEED GB A 308
Cochranes Garage (Berwick) Ltd,
Tweedside Trading Estate,
Berwick Upon Tweed TD15 2XF
Tel 0289 305585

CHOPPINGTON GB B 217
Heathline Commercials Ltd,
Stakeford Lane,
Choppington NE62 5QJ
Tel 0670 824006

Nottinghamshire

BEESTON GB E 116
Barton Transport Plc,
61 High Road,
Chilwell,
Beeston NG9 4AD
Tel 0602 221441

HOVERINGHAM GB E 110
Bellmore Transport Services,
Hoveringham Tachograph Centre,
Hoveringham,
Lowdham NG14 7JY
Tel 0602 663197

HUCKNALL GB E 107
K & M (Hauliers),
T/A Nevilles Garage Ltd,
The Aerodrome,
Watnall Road,
Hucknall NG15 6EN
Tel 0602 630630

NEWARK GB E 220
C F Parkinson (Notts) Ltd,
Brunel Drive,
Northern Industrial Estate,
Newark NG24 2EG
Tel 0636 72631

NOTTINGHAM GB E 317
Hooley's Garage Ltd,
Abbey Street,
Lenton,
Nottingham NG7 2PP
Tel 0602 786145

NOTTINGHAM GB E 121
Sherwood Leyland DAF Ltd,
522 Derby Road,
Lenton,
Nottingham NG7 2GX
Tel 0602 787274

NOTTINGHAM GB E 225
R H Commercial Vehicles Ltd,
Lenton Lane,

Nottingham NG7 2NR
Tel 0602 866571

NOTTINGHAM GB F 319
Transfleet Services Ltd,
Boulevard Industrial Estate,
Beacon Road,
Beeston NG9 2JR
Tel 0602 228271

STAPLEFORD GB E 209
Trent Trucks Sandcliffe Motor Group,
Nottingham Road,
Stapleford NG9 8AU
Tel 0602 395000

WORKSOP GB E 123
Scania Coach Sales Ltd,
Claylands Avenue,
Worksop S81 7DJ
Tel 0909 500822

WORKSOP GB E 313
Seafield Technical Services,
Claylands Avenue,
Worksop S81 7
Tel 0909 475561

Oxfordshire

BANBURY GB E 118
Banbury Trucks Ltd,
Transport Centre,
Unit One, Tuston House,
Station Approach,
Banbury OX16 8MB
Tel 0295 267528

BANBURY GB K 312
Hartford Motors (Banbury) Ltd,
98 Warwick Road,
Banbury OX16 7AH
Tel 0293 267711

KIDLINGTON GB N 112
Oxford Leyland DAF,
Langford Lane,
Kidlington OX5 1RY
Tel 08675 71511

OXFORD GB E 215
Evenlode Truck Centre Ltd,
Eynsham Road,
Cassington,
Oxford OX5 1DD
Tel 0865 881581

OXFORD GB K 311
Hartford Motors (Oxford) Ltd,
Besselsleigh Road,
Nr Wootton,
Abingdon OX13 6TP
Tel 0865 736377

OXFORD GB K 226
Tappins Coach World,
Collett Southmead Industrial Park,

Didcot OX11 7ET
Tel 0235 511115

Shropshire

SHREWSBURY GB D 116
Furrows Commercial Vehicles Ltd,
Ennerdale Road,
Harlescott Industrial Estate,
Shrewsbury SY1 3NP
Tel 0743 357971

SHREWSBURY GB D 212
Lucas Service UK Ltd,
Lancaster Road,
Harlescott,
Shrewsbury SY1 3NJ
Tel 0743 355061

SHREWSBURY GB D 326
Hartshorne (Shrewsbury) Ltd,
Ainsdale Drive,
Harlescroft,
Shrewsbury SY1 3SL
Tel 0743 236341

TELFORD GB D 320
Furrows Commercial Vehicles Ltd,
Halesfield,
Prince Street,
Telford
Tel 0952 684433

Somerset

BRIDGEWATER GB H 129
Hickley Valtone Ltd,
Unit 1, East Quay,
Bridgewater TA6 4DB
Tel 0278 423570

FROME GB H 115
J R Harding & Sons (Frome) Ltd,
Whitworth Road,
Marston Trading Estate,
Frome BA11 4BY
Tel 0373 64970

SHEPTON MALLET GB H 215
Tachograph Services (Shepton Mallet) Ltd,
Crowne Trading Estate,
Shepton Mallet BA4 5QU
Tel 0749 343963

TAUNTON GB H 107
White of Taunton) Ltd,
Commercial Division,
Bathpool,
Taunton TA2 8BA
Tel 0823 74957

TAUNTON GB H 203
H N Hickley & Co Ltd,
Castle Street,
Tangier,
Taunton TA1 4AU
Tel 0823 276041

TAUNTON GB H 307
National Carriers Fleetcare,
Canal Road,
Taunton TA1 1PL
Tel 0823 331151

WINCANTON GB H 219
Wincanton Truck Centre,
Aldermeads,
Wincanton BA9 9EB
Tel 0963 33800

YEOVIL GB H 101
Douglas Seaton Ltd,
West Hendford,
Yeovil BA20 2AG
Tel 0935 27421

YEOVIL GB H 309
Abbey Hill Truck Centre Ltd,
Boundary Road,
Lufton Trading Estate,
Yeovil BA22 8SS
Tel 0935 32399

Staffordshire

BURTON UPON TRENT GB D 108
Marley Transport Ltd,
Lichfield Road,
Branston,
Burton upon Trent DE14 3HG
Tel 0283 712264

BURTON UPON TRENT GB D 317
BRS Midlands Ltd,
Derby Street,
Burton upon Trent DE14 2LN
Tel 0283 63702

CANNOCK GB D 124
Cannock Tachograph Centre Ltd,
Units 2 & 3,
Cannock Industrial Centre,
Walk Mill Lane,
Bridgetown
Tel 0543 574487

NEWCASTLE UNDER LYME GB D 314
Hartshorne (Potteries) Ltd,
Rosevale Road,
Parkhouse Industrial Estate,
Newcastle under Lyme ST5 7EF
Tel 0782 566400

STAFFORD GB D 200
Lloyds Garage Ltd,
Stone Road,
Stafford ST16 2RA
Tel 0785 51331

STOKE-ON-TRENT GB D 101
Beeches Garage (1983) Ltd,
Leek Road,
Hanley,
Stoke-on-Trent ST1 6AD
Tel 0782 213836

STOKE-ON-TRENT GB D 500
Chatfields of Stoke,
Commercial Vehicle Division,
Clough Street,
Hanley,
Stoke-on-Trent ST1 4AR
Tel 0782 208155

STOKE-ON-TRENT GB D 315
Kay's Ltd,
Leek New Road,
Cobridge,
Stoke-on-Trent ST6 2DE
Tel 0782 264121

STOKE-ON-TRENT GB D 120
Seddon Atkinson Stoke,
Grove Road,
Fenton ST4 4PL
Tel 0782 575222

STOKE-ON-TRENT GB D 219
BRS Midlands Ltd,
Repair Centre,
Vernon Road,
Stoke-on-Trent ST4 2QF
Tel 0782 48281

Suffolk

BECCLES GB F 214
Gales Garages Ltd,
Common Lane North,
Beccles NR34 9QD
Tel 0502 717023

BURY ST EDMUNDS GB F 115
Chassis-Cabs Ltd,
Northern Way,
Bury St Edmunds IP52 6NL
Tel 0284 68570

IPSWICH GB F 114
Marshall (Ipswich) Ltd,
Lodge Lane,
Great Blakenham,
Ipswich IP6 0LB
Tel 0437 240200

IPSWICH GB F 210
Lucas Service UK Ltd,
Hadleigh Road Industrial Estate,
Ipswich IP2 0HB
Tel 0473 215931

IPSWICH GB F 320
Duffields of East Anglia,
Foxtail Road,
Ransomes Park,
Ipswich IP3 9RT
Tel 0473 718223

LOWESTOFT GB F 101
Days Garage Ltd,
Whapload Road,
Lowestoft NR32 1UR
Tel 0502 565353

STOWMARKET GB F 316
Ro-Truck Ltd,
Violet Hill Road,
Stowmarket IP14 1NN
Tel 0449 613553

SUDBURY GB F 113
Solar Garage (Ipswich),
Co-operative Society Ltd,
Cornard Road,
Sudbury CO10 6XA
Tel 0953 72301

WOODBRIDGE GB F 104
A G Potter (Framlington) Ltd,
New Road,
Framlington,
Woodbridge IP13 9AI
Tel 0728 723215

Surrey

CROYDON GB K 314
Heathrow Commercials Ltd,
Beddington Farm Road,
Croydon CR0 4XB
Tel 081-665 5775

EPSOM GB N 500
Cummings & Foster Ltd,
T/A Benhill Motors,
London Road,
Ewell-by-pass,
Epsom KT17 2PT
Tel 081-394 2196

GUILDFORD GB N 307
Grays Truck Centre Ltd,
Slyfield Industrial Estate,
Woking Road,
Guildford GU1 1RY
Tel 0483 571012

GUILDFORD GB N 207
F G Barnes & Sons Ltd,
Slyfield Industrial Estate,
Woking Road,
Guildford GU1 1RT
Tel 0483 37731

HORLEY GB N 107
Horley Services Ltd,
Salfords Industrial Estate,
Horley,
Nr Redhill RH1 5ES
Tel 0297 771481

Sussex

CHICHESTER GB K 308
Francis Transport Ltd,
Portfield Quarry,
Shopwyke Road,
Chichester PO20 6AD
Tel 0243 780011

EASTBOURNE GB K 211
Eurotrucks Ltd,
Eastbourne Road,
Pevency,
Westham,
North Eastbourne BN24 5NH
Tel 0323 767626

HORSHAM GB K 205
Evans Halshaw (Sussex) Ltd,
78 Billingshurst Road,
Broadbridge Heath,
Horsham RH12 3LP
Tel 0403 56464

HORSHAM GB N 303
Hancocks of Horsham Ltd,
53-55 Bishopric,
Horsham RH12 1QJ
Tel 0403 54331

PORTSLADE GB K 227
Lucas Service UK Ltd,
Unit 8,
Victoria Road Trading Estate,
Portslade BN4 1XD
Tel 0273 439955

SHOREHAM-BY-SEA GB K 503
Evans Halshaw (Sussex) Ltd,
44 Dolphin Road,
Shoreham-by-Sea BN4 1DY
Tel 0273 454887

WORTHING GB K 117
I G Bacon Commercials Ltd,
Meadow Road Industrial Estate,
Worthing BN11 2RU
Tel 0903 204127

Tyne & Wear

BLAYDON GB B 318
North East Leyland DAF Ltd,
Chainbridge Road,
Blaydon NE21 5TR
Tel 091-414 3333

GATESHEAD GB A 109
BRS Northern Ltd,
Earlsway Eastern Avenue,
Team Valley Trading Estate,
Gateshead NE11 0UU
Tel 091-487 8844

GATESHEAD GB A 200
Lucas Service UK Ltd,
Arlsway Valley Trading Estate,
Gateshead NE11 0RQ
Tel 091-491 1722

GATESHEAD GB A 323
Albany Motors,
Saltmeadows Road,
Gateshead NE8 3AH

NEWCASTLE UPON TYNE GB A 103
Union Trucks Ltd,

Mylord Crescent,
Killingworth,
Newcastle upon Tyne NE12 0UW
Tel 091-268 3141

NEWCASTLE UPON TYNE GB A 211
Henly's (Newcastle) Ltd,
Melbourne Street,
Newcastle upon Tyne NE1 2ER
Tel 091-261 1471

NEWCASTLE UPON TYNE GB B 322
Bell Truck Sales Ltd,
Bellway Industrial Estate,
Whitley Road,
Longbenton NE12 9SW
Tel 091-227 0787

NORTH SHIELDS GB A 110
Renault Trucks North East,
Unit b6,
Third Avenue,
Tyne Tunnel Trading Estate,
North Shields NE29 7SQ
Tel 091-296 2848

SUNDERLAND GB A 307
Cowies of Sunderland,
North Hylton Road,
Southwick,
Sunderland SR5 3HQ
Tel 091-549 1111

Warwickshire

ALCESTER GB D 105
Smiths Coaches (Sherington) Ltd,
Tything Road,
Arden Forest Industrial Estate,
Alcester B49 6EX
Tel 0789 764401

BARFORD GB D 117
Oldhams of Barford,
Wellesbourne Road,
Barford CV35 8DS
Tel 0926 624333

RUGBY GB D 123
Noden Truck Centre,
3 Avon Industrial Estate,
Butlers Leap,
Rugby CV21 3UY
Tel 0788 579535

West Midlands

BIRMINGHAM GB D 122
Evans Halsham Leyland,
31 Shefford Road,
Aston B6 4PQ
Tel 021-359 8261

BIRMINGHAM GB D 110
Gerard Mann,
2 Lichfield Road,

Aston B6 5SV
Tel 021-622 3031

BIRMINGHAM GB D 216
Birmingham Trucks Ltd,
292 Wharfdale Road,
Tyseley B11 2DT
Tel 021-707 9700

BIRMINGHAM GB D 220
Lucas Service UK Ltd,
171 Lichfield Road,
Aston B6 5SN
Tel 021-327 1525

BIRMINGHAM GB D 501
Bristol Street Motors Ltd,
Beacon House,
Long Acre,
Nechells B7 5JJ
Tel 021-322 2222

BIRMINGHAM GB D 324
Transfleet Services Ltd,
Bannerly Street,
Garretts Green Industrial Estate,
Birmingham B33 0SL
Tel 021-784 4000

BIRMINGHAM GB D 318
Midlands BRS Ltd,
Bromford Mills Branch,
Erdington B24 8DP
Tel 021-322 2200

BRIERLEY HILL GB D 119
Dudley Tachograph Centre Ltd,
Thorns Road Trading Estate,
Quarry Bank,
Brierley Hill DY5 2JS
Tel 0384 70301

BROWNHILLS GB D 322
Brownhills Tachograph Centre,
Lindon Road,
Brownhills WS8 7BW
Tel 05433 372528

COVENTRY GB D 319
Dawson Freight Commercials Ltd,
Unit 1,
Eden Street,
Coventry CV6 5HE
Tel 0203 683221

COVENTRY GB D 221
Carwood Motor Units Ltd,
Herald Way,
Binley,
Coventry CV3 2RQ
Tel 0203 449533

HALESOWEN GB D 209
Lex Tillotson Birmingham,
Park Road,
Halesowen B63 2RL
Tel 0384 424500

WALSALL GB D 111
Maybrook Trucks Ltd,
Coppice Lane,
Aldridge WS9 9AA
Tel 0922 56356

WALSALL GB D 502
Tildesley Ford,
Northgate,
Aldridge WS9 8TH
Tel 0922 743031

WALSALL GB D 305
Hartshorne Motor Services Ltd,
Bentley Mill Close,
Walsall WS2 0BN
Tel 0922 720941

WALSALL GB D 306
S Jones (Garages) Ltd,
Truck Centre,
Westgate,
Aldridge WS9 8ET
Tel 0922 743783

WARLEY GB D 312
Transfleet Services Ltd,
Pearsall Drive,
Oldbury B69 2RA
Tel 021-544 5125

WEDNESBURY GB D 213
Wincanton Transport Ltd,
Heath Estate,
Whitworth Close,
Darlaston,
Wednesbury WS10 8LJ
Tel 021-526 3833

WEST BROMWICH GB D 118
Browns Electro-Diesel Co,
C/o Millward Motor Engineering,
Church Lane Industrial Estate,
West Bromwich B71 1AL
Tel 021-525 4355

WEST BROMWICH GB D 328
Kelly Trucks,
Kendrick Way,
West Bromwich B71 4JW
Tel 021-525 7000

WEST BROMWICH GB D 307
Guest Motors Ltd,
Old Meeting Street,
West Bromwich B70 9SQ
Tel 021-553 2737

WOLVERHAMPTON GB D 103
Don Everall Motor Sales Ltd,
Bilston Road,
Wolverhampton WV2 2QE
Tel 0902 51515

WOLVERHAMPTON GB D 316
A F Glaze Ltd,
Dixon Street,
Wolverhampton
Tel 0902 55434

Wiltshire

CALNE GB H 121
Syms Engineering (Calne) Ltd,
Port Marsh Industrial Estate,
Calne SN11 9PZ
Tel 0249 814333

CHIPPENHAM GB H 123
Chippenham Truck Services Ltd,
Unit 4-5, Bumpers Farm Industrial Estate,
Chippenham SN14 6NQ
Tel 0249 443523

SALISBURY GB H 128
The Tachograph Centre,
Unit 9, Harnham Trading Estate,
Salisbury SP2 8NW
Tel 0722 322004

SWINDON GB H 109
Swift Transport Ltd,
'D' Building,
Parsonage Road,
Stratton St Margaret SN3 4RL
Tel 0793 827050

SWINDON GB H 220
Gardiners Auto Electrical Services Ltd,
Hawksworth Trading Estate,
Swindon SN2 1EE
Tel 0793 29254

SWINDON GB H 312
BRS Western Ltd,
Redway Park,
Europa Industrial Estate,
Swindon SN3 4ND
Tel 0793 83151

WESTBURY GB H 312
Rygor Commercials Ltd,
The Broadway,
West Wilts Trading Estate,
Westbury BA13 4HU
Tel 0373 864334

North Yorkshire

BOROUGHBRIDGE GB B 117
Boroughbridge Motors Ltd,
Bar Lane,
Boroughbridge YO5 9NN
Tel 0423 322741

MALTON GB B 212
Slaters Transport Ltd,
Kirby Misperton,
Malton YO17 0UE
Tel 0653 86275

NORTHALLERTON GB A 112
G Abbots and Sons,
Aumans House,
Leeming,
Northallerton D17 9RZ
Tel 0677 422858

THIRSK GB B 128
Crossroads Commercials Ltd,
Stockton Road,
Thirsk YO7 1AY
Tel 0845 522057

YORK GB B 207
York Autoelectrics Ltd,
58 Layerthorpe,
York YO3 7YN
Tel 0904 654513

YORK GB B 307
Polar Motor Co (York) Ltd,
Fulford Way,
Fulford,
York YO1 1YF
Tel 0904 625371

YORK GB B 126
BRS Northern Ltd,
7 James Street,
York YO1 3DW
Tel 0904 410955

South Yorkshire

BARNSLEY GB B 112
BRS Northern Ltd,
Shawfield Road,
Carlton Industrial Estate,
Barnsley S71 3HS
Tel 0226 298321

BARNSLEY GB B 305
The Polar Truck Centre,
Wombwell Lane,
Stairfoot,
Barnsley S70 3NX
Tel 0226 732732

DONCASTER GB B 114
Trailer Supermarket (Bawtry) Ltd,
Doncaster Road,
Bawtry,
Doncaster DN10 6NX
Tel 0302 710711

DONCASTER GB B 310
E G Charlesworth Ltd,
Truck Service Department,
Decoy Bank North,
Carr Orange Industrial Estate,
Doncaster DN4 5JE
Tel 0302 327111

SHEFFIELD GB B 312
Oughtibridge Tachograph Centre,
Station Lane,
Oughtibridge,
Sheffield S30 3HS
Tel 0742 863881

SHEFFIELD GB B 316
Sherwood Leyland DAF (Sheffield) Ltd,
Highfield Lane,

Sheffield S13 9DB
Tel 0742 693230

SHEFFIELD GB B 201
T C Harrison Group Ltd,
Sheffield Truck Operations,
Shepcote Lane,
Sheffield S9 1TX
Tel 0742 440051

SHEFFIELD GB B 211
Plaxton Parts & Services,
Ryton Road,
South Anston,
Sheffield S31 7ES
Tel 0909 551155

SHEFFIELD GB B 219
Lucas Service UK Ltd,
300 Savile Street,
Sheffield S4 7UD
Tel 0742 752522

SHEFFIELD GB B 133
South Yorkshire Trucks Ltd,
10 Rutland Way,
Sheffield S3 8DG
Tel 0742 795522

SHEFFIELD GB B 135
Marriott Bros Ltd,
Greenlane Depot,
Ecclesfield,
Sheffield S30 3WY

SHEFFIELD GB B 220
Lucas Service UK Ltd,
Unit 9,
Tenpound Walk,
Tenpound Industrial Estate,
Doncaster DN4 5HX
Tel 0302 342794

SHEFFIELD GB B 502
Cameron Motor Services Centre,
Barrow Field Road,
Platts Common Trading Estate,
Hoyland,
Sheffield S30 9SF
Tel 0226 742654

West Yorkshire

BATLEY GB B 210
Lucas Service UK Ltd,
232 Bradford Road,
Batley WF17 6LF
Tel 0924 472415

BRADFORD GB B 125
Northside Truck Centre,
Legrams Lane,
Bradford BD7 2HR
Tel 0274 577311

BRADFORD GB B 300
C D Bramall (Bradford) Ltd,
Truck Workshops,

Dagenham Road,
Dudley Hill,
Bradford BD4 9LS
Tel 0274 682224

BRIGHOUSE GB B 122
Reliance Commercial Vehicles Ltd,
Wakefield Road,
Brighouse HD6 1QQ
Tel 0484 712611

CLECKHEATON GB B 115
Crossroads Commercials Ltd,
Bradford Road,
Gomersal,
Cleckheaton BD19 4BE
Tel 0274 851111

HUDDERSFIELD GB B 309
Brockholes Ltd,
Leeds Road,
Huddersfield HD2 7UW
Tel 0484 428111

KEIGHLEY GB B 320
David Peel Auto Electricians Ltd,
Valley Road,
Keighley BD21 4LZ
Tel 0535 662078

LEEDS GB B 102
Archbold Truck Sales & Service Ltd,
Albert Road,
Morley LS27 8TT
Tel 0532 522333

LEEDS GB B 302
Trimoco Motor Group Ltd,
Trimoco Trucks,
54 Dolly Lane,
Leeds LS9 7NE
Tel 0532 421222

LEEDS GB B 104
WASS Ltd,
123 Hunslett Road,
Leeds LS10 1LD
Tel 0532 439911

LEEDS GB B 202
Sewell of Leeds Ltd,
49 Marshall Street,
Leeds LS11 9SU
Tel 0532 435101

LEEDS GB B 127
Transfleet Services Ltd,
Parkside Lane,
Dewsbury Road,
Leeds LS11 5TD
Tel 0532 773377

LEEDS GB B 130
Knottingley Truck Ltd,
Common Lane,
Knottingley WF11 8BG

LEEDS GB B 134
Pelican Engineering Co (Sales) Ltd,

Wakefield Road,
Rothwell Haigh,
Leeds LS26 0RU

MIRFIELD GB B 324
Northern Commercial (Mirfield) Ltd,
Roe Head Garage,
Far Common Road,
Mirfield WF14 0DQ

PUDSEY GB B 100
Chatfields of Leeds,
Grangefield Industrial Estate,
Richardshaw Lane,
Pudsey LS28 6SO
Tel 0532 571701

ROTHERHAM GB B 129
Crossroads Commercials Ltd,
Canklow Meadows Industrial Estate,
West Bawtry Road,
Rotherham
Tel 0709 365566

Scotland

Borders Region
GALASHIELS GB M 207
Chalmers McQueen Ltd,
Albert Place,
Galashiels TD1 3DL
Tel 0896 2729

Central
FALKIRK GB M 501
Scottish Road Services Ltd,
Burnbank Road,
Bainsford,
Falkirk FK2 7PD
Tel 0324 22805

FALKIRK GB M 209
Millar's of Falkirk Ltd,
Millar's Truck Centre,
North Main Street,
Carronshore,
Falkirk FK2 8HZ
Tel 0324 556211

FALKIRK GB M 113
Transfleet Services Ltd,
Whitehouse Road,
Springkerse Industrial Estate,
Stirling FK7 7SP
Tel 0786 51761

Dumfries & Galloway
DUMFRIES GB M 108
Gateside Commercials (DFS) Ltd,
Brownrigg Loaning,
Dumfries DG1 3JT
Tel 0387 61146/7

DUMFRIES GB M 307
Scottish Road Services Ltd,
Glasgow Road,
Dumfries DG2 ONY
Tel 0387 53171

STRANRAER GB M 211
Western SMT Co Ltd,
Lewis Street,
Stranraer
Tel 0776 5174

Fife
KIRKCALDY GB L 109
Drummond Motor Co Ltd,
Ferrard Road,
Kirkcaldy KY2 5RY
Tel 0592 201555

KIRKCALDY GB L 306
Laidlaw (Fife) Ltd,
Forth Avenue,
Kirkcaldy KY2 5PS
Tel 0592 261199

Grampian
ABERDEEN GB L 204
The Harper Motor Co Ltd,
218 Auchmill Road,
Bucksburn,
Aberdeen AB2 9NB
Tel 0224 714741

ABERDEEN GB L 311
Aberdeen Trucks Ltd,
Greenwell Road,
Tullos,
Aberdeen
Tel 0224 873641

ABERDEEN GB L 213
McPhersons Transport,
Units 3 & 4,
Girdleness Road,
Aberdeen AB1 4DQ
Tel 0224 895050

ABERDEEN GB L 107
Grampian Regional Transport Ltd,
395 King Street,
Aberdeen AB19 1SP
Tel 0224 637047

ABERLOUR GB L 206
McPherson Transport (Aberlour) Ltd,
Fisherton Garage,
Aberlour AB2 9LB
Tel 0340 5401

ELGIN GB M 116
Scottish Road Services Ltd,
Grampian Road,
Elgin,
Morayshire
Tel 0343 542171

HUNTLY GB L 304
Woodside Garage,
Knock,
Huntly AB5 5LJ
Tel 046 686 245

Highland

INVERNESS GB L 201
Macrea & Dick Ltd,
36 Academy Street,
Inverness IV1 1SG
Tel 0463 712000

INVERNESS GB L 310
Highland Bus and Coach,
1 Seafield Road,
Longman Industrial Estate,
Inverness IV1 1TN
Tel 0463 231404

INVERNESS GB L 312
W M Dunnett & Co Ltd,
Mansons Lane,
Thurso KW14 8EW
Tel 0847 63101

TAIN GB L 103
G Bannerham (Tain) Ltd,
Shore Road,
Tain,
Rosshire N19 1EH
Tel 0862 2480

Lothian

BROXBURN GB M 216
Foden Scotland,
East Mains Industrial Estate,
Broxburn EH52 5AU
Tel 0506 854834

DUNBAR GB M 219
Lowland Scottish Omnibus,
Countless Road,
Dunbar EH42 1JH
Tel 0368 62343

EDINBURGH GB M 318
Fulton Auto Electric Ltd,
15 Bowling Green Street,
Leith,
Edinburgh EH6 5PR
Tel 031-554 1571

EDINBURGH GB M 118
James Bowen & Son Ltd,
Newbridge Industrial Estate,
Newbridge,
Edinburgh EH28 8PJ
Tel 031-333 5333

EDINBURGH GB M 317
Reliable Vehicles Ltd,
Newbridge Industrial Estate,
Newbridge,

Edinburgh
Tel 031-333 2362

EDINBURGH GB M 319
Wylies (Edinburgh) Ltd,
Edgefield Road Industrial Estate,
Loanhead,
Edinburgh EH20 9TB
Tel 031-448 2333

LOANHEAD GB M 314
Lothian Leyland DAF Ltd,
Straiton,
Loanhead
Edinburgh EH20 9HQ
Tel 031-440 4100

Strathclyde

AIRDRIE GB M 213
Central Motors (Calderbank) Ltd,
Carlisle Road,
Airdrie ML6 8RD
Tel 023 67 62881

BARRHILL GB M 316
W & J Barr & Sons (Scotland) Ltd,
Braehead,
Barrhill KA26 0QR
Tel 0465 82300

BELLSHILL GB M 312
Laidlaw Trucks (Strathclyde) Ltd,
Rigghead Industrial Estate,
Bellshill ML4 3LG
Tel 0698 747015

BLANTYRE GB M 120
Ailsa Trucks (Northern) Ltd,
Whistleberry Road,
Blantyre ML3 0ED
Tel 0698 823300

CAMPBELTOWN GB M 212
Martin Maintenance Campbeltown Ltd,
Ben Mhor Garage,
Saddell Street,
Cambeltown,
Argyll
Tel 0586 53155

CUMBERNAULD GB M 301
Scotia DAF Trucks Ltd,
8 South Wordpark Court,
Wordpark South,
Cumbernauld G67 3HE
Tel 0236 727771

CUMNOCK GB M 103
Kerr & Smith (Cumnock) Ltd,
Riverside Garage,
Ayr Road,
Cummock KA18 1BJ
Tel 0290 22440

GLASGOW GB M 115
Lex-Tillotson (Leyland DAF) Ltd,
131 Bogmore Road,

Glasgow G51 4TH
Tel 041-425 1530

GLASGOW GB M 122
Renault Trucks Commercials Ltd,
Penilee Road,
Hillington Industrial Estate,
Glasgow G52 4AW
Tel 041-882 3277

GLASGOW GB M 204
Lucas Service Ltd,
200-210 Garscube Road,
Glasgow G4 9RR
Tel 041-332 6591

GLASGOW GB M 206
Wylies Ltd,
149 Kilbirnie Street,
Glasgow G5 8JH
Tel 041-429 6262

GLASGOW GB M 210
Leyland Bus,
Scottish Service Centre,
2121 London Road,
Glasgow G32 8XJ
Tel 041-778 3491

GLASGOW GB M 215
Ailsa Trucks (Northern) Ltd,
101 Kelburn Street,
Barrhead,
Glasgow G78 2LB
Tel 041-881 5851

GLASGOW GB M 303
Transfleet Services Ltd,
79 Hardgate,
Govan,
Glasgow G51 4SX
Tel 041-445 3913

GLASGOW GB M 218
Reliable Vehicles Ltd,
Clyde Street,
Renfrew,
Glasgow PA4 8SL
Tel 041-886 5633

GLASGOW GB M 110
Callenders Engineering Co Ltd,
47 Kirklee Road,
Kelvinside,
Glasgow G12 OSR
Tel 041-334 8155

KILMARNOCK GB M 221
J McKinnock (Truck Services) Ltd,
West Hillhead,
Western Road,
Kilmarnock KA3 1PH
Tel 0563 44888

KILMARNOCK GB M 313
Johnston & Drynan,
1 Fullarton Street,
Kilmarnock KA1 2RB
Tel 0563 2300

KILMARNOCK GB M 504
Ayrshire Tachograph Centre,
Nursery Avenue,
Kilmarnock KA1 3JD
Tel 0563 23084

PAISLEY GB M 306
Clanford Motors Ltd,
37-41 Lonead,
Paisley PA1 1SY
Tel 041-887 0191

Tayside

DUNDEE GB L 101
Dundee Plant Co Ltd,
T/A Dundee Truck Centre,
411 Clepington Road,
Dundee DD3 8ED
Tel 0382 813644

DUNDEE GB L 108
Camperdown Motor Co Ltd,
Kingway West,
Dundee DD2 4TD
Tel 0282 623111

DUNDEE GB L 300
Tayscot Truck Ltd,
Smeaton Road,
Wester Gourdie,
Dundee
Tel 0382 623263

FORFAR GB L 214
A M Phillips Ltd,
Muiry Faulds Garage,
Angus,
By Forfar DD8 1XP
Tel 0370 82255

PERTH GB L 215
Lucas Service UK Ltd,
165 Brook Street,
Dundee DD1 5DE
Tel 0382 27122

PERTH GB L 307
G Mutch Mechanical Services,
Shore Road,
Perth
Tel 0738 30291

PERTH GB L 500
Frews Cars Ltd,
Riggs Road,
Perth PH2 0NT
Tel 0738 25121

Wales

Clwyd

CHESTER GB C 108
Quirks Trucks Ltd,
Premier Garage,

Bretton,
Chester CH4 0DS
Tel 0244 660681

CLWYD GB C 132
Deeside Truck Services,
Pinfold Lane,
Alltami,
Mold CH7 6NY
Tel 0244 547202

COLWYN BAY GB C 107
Gordon Ford,
The Trading Estate,
Mochdre,
Colwyn Bay LL28 3HA
Tel 0492 46756

DEESIDE GB C 326
Thomas Hardie Commercials (North Wales),
Chester Road East,
Deeside CH5 1QA
Tel 0244 822707

WREXHAM GB C 231
Arlington Wrexham Truck Centre,
Kays Wrexham Ltd,
Wrexham Road,
Rhostyllan,
Nr Wrexham LL14 4DP
Tel 0978 291915

WREXHAM GB C 312
Border Tachograph Services Ltd,
Pentybont Industrial Estate,
Penybont Works,
Chirk,
Wrexham LL14 JAW
Tel 0978 823434

Dyfed

CARMARTHEN GB G 303
Western BRS Ltd (West Wales Branch),
Heol Alltycrap,
Johnstown,
Carmarthen SA31 3NE
Tel 0267 230650

HAVERFORDWEST GB G 105
Merlin Motor Co Ltd,
Fishguard Road Industrial Estate,
Haverfordwest SA62 4BT
Tel 0437 762468

LLANFYRNACH GB G 206
Mansel Davies & Son Ltd,
Taffyale Garage,
Station Yard,
Llanfyrnach SA35 0BZ
Tel 0239 831631

LLANRHYSTUD GB G 212
Lewis's Coaches,
Bryneithin,
Llanrhystud
Tel 0974 202495

Mid-Glamorgan

BRIDGEND GB G 210
BRS Western Ltd,
Westerton Road,
Bridgend CF31 3YS
Tel 0656 653331

CAERPHILLY GB G 104
R J Brown Ltd,
The Truck Centre,
Pontygwindy Industrial Estate,
Caerphilly CF8 3HU
Tel 0222 852222

TREFOREST GB G 301
Griffin Mill Garages Ltd,
Upper Boat,
Treforest,
Nr Pontypridd
Tel 044 384 2216

South Glamorgan

CARDIFF GB G 109
Tanner Tachograph Centre Ltd,
C/o W T Davies (Transport) Ltd,
232 Penarth Road,
Cardiff CF1 7XJ
Tel 0222 225580

CARDIFF GB G 214
Lucas Service UK Ltd,
Unit 2c,
Glynstell Close,
off Hadfield Road,
Cardiff CF1 8TR
Tel 0222 28361

CARDIFF GB G 307
Cardiff Truck Centre,
Whittle Road,
Leckwith Industrial Estate,
Cardiff CF1 8AT
Tel 0222 378581

West Glamorgan

SWANSEA GB G 108
City Electro Diesel Services,
Site F 3, Nantyffin Road South,
Llansamlet,
Swansea
Tel 0792 792010

SWANSEA GB G 110
Griffin Mill Garages Ltd,
21 Viking Way,
Winch Wen,
Swansea SA1 7DA
Tel 0792 469595

THE TRANSPORT MANAGER'S AND OPERATOR'S HANDBOOK

SWANSEA GB G 209
Shorts Auto Electrical Services Ltd,
43-49 Station Road,
Landore,
Swansea SA1 2JE
Tel 0792 469595

SWANSEA GB G 300
Bevan Commercial Vehicles Ltd,
18-20 Morfa Road,
Hafod,
Swansea SA1 2EH
Tel 0792 650646

Gwent

NEWPORT GB G 211
BRS Western Ltd,
Watch House Parade,
Alexandra Dock,
Newport NP9 5YG
Tel 0633 840083

NEWPORT GB G 111
Commercial Motors (Newport) Ltd,
Frederick Street,
Newport NP9 2DR

NEWPORT GB G 306
Newport Ford,
T/A Newport Truck Centre,
Leeway Industrial Estate,
Newport NP1 0QU
Tel 0633 278020

Gwynedd

CAERNARFON GB C 222
J T Jones & Sons,
Dulyn Motors,
Penygroes,
Caernarfon LL54 6NB
Tel 0286 880218

HOLYHEAD GB C 335
R J R Commercials,
Berwyn Garage,
Porthdafarth Road,
Holyhead
Tel 0407 762575

Powys

BRECON GB G 207
Williams Motors (Cwmdu) Ltd,
The Walton,
Rich Way,
Brecon LD3 4EG
Tel 0874 2223

KNIGHTON GB G 308
The Knighton Truck Company Ltd,
Station Road,
Knighton LD7 1DT

NEWTOWN GB C 216
Grooms Industries,
Pool Road,
Newtown SY16 1DL
Tel 0686 626731

Northern Ireland

The following list of Tachograph Centres are those approved for Northern Ireland by the Department of the Environment (DoE), Belfast, which fulfils the role of the Department of Transport, London, in the Province.

Agnew Commercials Ltd,
47 Mallusk Road,
Newtownabbey,
Co. Antrim
Tel 0232 342411

Cahill Motor Engineering (NI) Ltd,
5-7 Sandra Road,
Whitehouse,
Newtownabbey,
Co. Antrim
Tel 0232 365652

Cars Ltd,
16-19 Merchants Quay,
Newry,
Co. Down
Tel 0693 60606

J E Coulter (Trucks) Ltd,
Commercial Way,
Mallusk,
Newtownabbey,
Co. Antrim
Tel 0232 837171

J & J Curley Ltd,
Commercial Vehicle Services,
Tamlaght,
Enniskillen,
Co. Fermanagh
Tel 0356 87276

Dencourt Motors Ltd,
1-3 Mallusk Road,
Mallusk,
Newtownabbey,
Co. Antrim
Tel 0232 342221

Dennison Commercials,
37 Hillhead Road,
Ballyclare,
Co. Antrim
Tel 09603 52827

Eakin Bros Ltd,
48 Main Street,
Claudy,
Co. Londonderry
Tel 0504 338641

ETS Vehicles Ltd,
Dromore Road,
Omagh,
Co. Tyrone
Tel 0662 243491

Kearns & Murtagh,
52 Cecil Street,
Newry,
Co. Down
Tel 0693 65720

James Kingsberry Ltd,
1 Ballymacarret Road,
Belfast
Tel 0232 452527

Pedlow Tachograph Centre,
72 Annesborough Road,
Lurgan,
Co. Armagh
Tel 0762 328151

Roadside Motors Ltd,
1 Dromore Road,
Lurgan,
Co. Armagh
Tel 0762 321038

Road Trucks Ltd,
Circular Road,

Larne,
Co. Antrim
Tel 0574 275524

RK Truck Centre Ltd,
Edgar Road,
Comber Road,
Carryduff,
Belfast
Tel 0232 813600

Seven Towers Motor Co Ltd,
166-168 Larne Road,
Ballymena,
Co Antrim
Tel 0266 42272

T B F Thompson (Gervagh) Ltd,
Killyvalley Road,
Colleraine,
Co. Londonderry
Tel 0256 58353

The Tachograph Centre (Ulsterbus) Ltd,
Railway Place,
Coleraine,
Co. Londonderry
Tel 0256 54946

Appendix VII

Department of Transport Weighbridges

This list shows the geographical location of Department of Transport dynamic axle weighbridges within Traffic Areas. Locations marked * provide a self-weighing facility.

North-Western

A5, Holyhead, Anglesey, North Wales.

Wallasey Tunnel (East), Liverpool, Merseyside.

Wallasey Tunnel (West), Liverpool, Merseyside.

M62, Junction 20, Thornham, Greater Manchester.

A59, Salmesbury, Preston, Lancashire.

A556, Rostherne, Cheshire.

A5117/M56, Dunkirk, Cheshire.

A74, Harker, Carlisle, Cumbria.

A494, Ewloe, Clwyd, North Wales.

North-Eastern

A1, Scotch Corner, Middleton Tyas, North Yorkshire.

A19, Wellfield, County Durham.

A1/A659, Boston Spa, Wetherby, West Yorkshire.

A63, South Cave, North Humberside.

*Nepshaw Lane, Gildersome, Morley, Leeds, Yorkshire.

A1, Robin Hoods Well, Skellow, Doncaster, South Yorkshire.

Blackley New Road, Ainley Top, Huddersfield, West Yorkshire.

A61, Barnsley Road, Tankersley, Barnsley, South Yorkshire.

*King George V Docks, Kingston-upon-Hull, Humberside.

*A15, Approach Road, Humber Bridge, Hessle, South Humberside.

A46/52, Saxondale, The Old Nottingham Road, Bingham, Nottinghamshire.

West Midlands

A449, Link Road, Warndon, Nottinghamshire.

A5, Wall Island, Lichfield, Staffordshire.

M6, Doxey, Nr Stafford, Staffordshire.

A45/M45, Thurleston Island, Dunchurch, Rugby, Warwickshire.

M5, Junction 3, Quinton, Nr. Birmingham, West Midlands.

*A40, Three Crosses, Ross-on-Wye, Gloucestershire.

Eastern

A43, Towcester, Northamptonshire.

A17/A15, Holdingham, Sleaford, Lincolnshire.

A428/M1, Junction 18, Crick, Northamptonshire.

* Stone Grove Road, The Dock, Felixstowe, Suffolk.

Old A45 Road, Risby, Suffolk.

A1, Southbound, Connington Fen, Sawtry, Cambridgeshire.

A1, Northbound, Lower Caldecote, Sandy, Bedfordshire.

A414/M11, Junction 17, Harlow, Essex.

M1, Junction 14, Newport Pagnell, Buckinghamshire.

Southlands Road, Denham, Buckinghamshire.

South Wales

A40, Penblewin, Narberth, Dyfed.

M4, Coldra, Newport, Gwent.

Western

*Millbay Docks, Plymouth, Devon.

A303, Wylye, Wiltshire.

Off A361, Sampford Peverell, Tiverton, Devon.

*Continental Freight Ferry Terminal, New Harbour Road, Poole, Dorset.

A38, Anchor Inn Lay-by Kennford, Devon.

A35, Lay-by, Puddletown, Dorchester, Dorset.

*Portsmouth Docks, Portsmouth, Hampshire.

A46, Tormarton, Avon.

A34/A415, Marsham Road, Abingdon, Oxfordshire.

South-Eastern and Metropolitan

A30, Staines Bypass, Surrey.

M25, Junction 9, Leatherhead interchange, Surrey.

A13, Barking, Essex.

A3, Burpham, Nr Guildford, Surrey.

A1, Holloway Road, London (North).

*Dover, East Dock, Dover, Kent.

*Dover, West Dock, Dover, Kent.

A23, Handcross, West Sussex.

Stockers Hill, Boughton, Kent.

A27, Beddingham, East Sussex.

A26/A22, Millpond, Maresfield, East Sussex.

*A27, Withy Patch, Lancing, West Sussex.

Sheerness Docks, Sheerness, Kent.

Scottish

A75, Castle Kennedy.

*A90, Cramond.

*A92, Findon, Nr Aberdeen.

A74, Beattock Summit.

M9, Craigforth, Stirling.

*Tayside Truckstop, Smeaton Road, Dundee.

These weighbridges are operated by the Vehicle Inspectorate, an Executive Agency of the Department of Transport

Appendix VIII

Police Forces in Great Britain
(For Notification of Abnormal Loads Movements)

In most cases, Police Forces require fax notifications of abnormal load movements to be clearly identified as such in the heading or prefix to the message. Communications should be addressed to the Chief Constable, unless a specific department is indicated.

England

Constabulary/Force	Address/Tel/Fax Nos for Notification
Avon and Somerset	Abnormal Loads Department, Almondsbury Motorway Station, Gloucester Road, Almondsbury, Bristol BS12 4AG. Tel 0454 201559 Fax 0454 201344
Bedfordshire Police	Road Traffic Department, Halsey Road, Kempston, Bedford MK43 9AX. Tel 0234 842491 Fax 0234 842407
Cambridgeshire	Hinchingbrooke Park, Huntingdon PE18 8NP. Tel 0480 456111 Fax 0480 414448
Cheshire	Castle Esplanade, Chester CH1 2PP. Tel 0244 612215 Fax 0244 315539
City of London Police	Traffic Operations, 37 Wood Street, London EC2V 7HN. Tel 071-601 2190 Fax 071-601 2760 *(For City of London movements only – not Metropolitan area)*
Cleveland	PO Box 70, Police Headquarters, Ladgate Lane, Middlesborough, Cleveland TS8 9EH. Tel 0642 301109 Fax 0642 301115
Cumbria	Carleton Hall, Penrith, Cumbria CA10 2AU. Tel 0768 64411 Fax 0768 217216
Derbyshire	Operations Dept, Butterley Hall, Ripley, Derby DE5 3RS. Tel 0773 570100 Fax 0773 572225
Devon and Cornwall	Abnormal Loads Office, Middlemoor, Exeter EX2 7HQ. Tel 0392 452268 Fax 0392 452426
Dorset Police	Winfrith, Dorchester, Dorset DT2 8DZ. Tel 0305 223731 Fax 0305 223987
Durham	Uniform Operations Department, Aykley Heads, Durham DH1 5TT. Tel 091-386 4929 Fax 091-386 3913
Essex Police	PO Box 2, Springfield, Chelmsford CM2 6DA. Tel 0245 491491 Fax 0245 452424
Gloucestershire	Control Room, Holland House, Lansdown, Cheltenham GL51 6QH. Tel 0242 521321 Fax 0242 263987

Greater Manchester Police — Traffic Management Section, PO Box 47, Chester House, Boyer Street, Manchester M16 0SD. Tel 061-856 1678-9 Fax 061-856 1676

Hampshire — West Hill, Winchester, Hants SO22 5DB. Tel 0962 868133 Fax 0962 879683 For Isle of Wight movements: Tel 0983 528000

Hertfordshire — Stanborough Road, Welwyn Garden City, Herts AL8 6XF. Tel 0707 331177 Fax 0707 324805

Humberside Police — Queen's Gardens, Kingston upon Hull, Humberside HU1 3DJ. Tel 0482 26111 Fax 0482 226877

Kent County — Traffic Department, London Road, British Legion Village, Maidstone, Kent ME20 7SL. Tel 0622 882325 Fax 0622 608156

Lancashire — PO Box 77, Hutton, Preston PR4 5SB. Tel 0772 614444 Fax 0772 615009

Leicestershire — Traffic Management Department, St Johns, Narborough, Leicester LE9 5BX. Tel 0533 482442 Fax 0533 482327

Lincolnshire Police — PO Box 999, Lincoln LN5 7PH. Tel 0522 558125 Fax 0522 558098

Merseyside Police — Traffic Headquarters, Smithdown Lane, Liverpool L7 3PR. Tel 051-777 5512 Fax 051-777 5599

Metropolitan Police — New Scotland Yard, Broadway, London SW1H 0BG. Tel 071-230 2752/3031 Fax 071-230 3745 *(Covers Greater London but not City of London)*

Norfolk — Martineau Lane, Norwich NR1 2DJ. Tel 0603 768769 Fax 0603 761722

Northamptonshire Police — Abnormal Loads Office, Mereway, Northampton NN4 0JQ. Tel 0604 703422 Fax 0604 703417

Northumbria Police — Area Operations Room, Forum Way, Cramlington, Northumberland NE23 6SB. Tel 0661 872555 Fax 0661 863788

North Yorkshire Police — Force Communications Centre, Racecourse Lane, Northallerton DL7 8RB. Tel 0609 783131 Fax 0609 789213

Nottinghamshire Police — Force Control Room, Sherwood Lodge, Arnold, Nottingham NG5 8PP. Tel 0602 672144 Fax 0602 672145

South Yorkshire Police — Snig Hill, Sheffield S3 8LY. Tel 0742 523252 Fax 0742 523250

Staffordshire Police — Central Traffic Group, Western Road, Stafford ST18 0YY. Tel 0785 232674 Fax 0785 232673

Suffolk — Martlesham Heath, Ipswich IP5 7QS. Tel 0473 611611 Fax 0473 610577

Surrey — Mount Browne, Sandy Lane, Guildford, Surrey GU3 1HG. Tel 0483 571212 Fax 0483 300279 *NB: Certain parts of the county of Surrey fall within the jurisdiction of the Metropolitan Police.*

Sussex Police — Malling House, Lewes, Sussex BN7 2DZ. Tel 0273 475432 Fax 0273 480310

Thames Valley Police — Headquarters, Kidlington, Oxford OX5 2NX. Tel 0865 846422-3 Fax 0865 846726

Warwickshire	PO Box 4, Leek Wootton, Warwick CV35 7QB. Tel 0926 415260 Fax 0926 415388
West Mercia	Abnormal Loads Officer, Hindlip Hall, Hindlip, Worcester WR3 8SP. Tel 0905 723000 Fax 0905 756161
West Midlands Police	Central Motorway Patrol Group, Thornbridge Avenue, Birmingham B42 2AG. Tel 021-322 6018 Fax 021-322 6039
West Yorkshire Police	Operations Division, PO Box 9, Wakefield WF1 3QP. Tel 0924 293919 Fax 0924 293909
Wiltshire	London Road, Devizes SN10 2DN. Tel 0380 722341 Fax 0380 734135

Wales

Constabulary/Force	Address/Tel/Fax Nos for Notification
Dyfed-Powys Police	PO Box 99, Llangunnor, Carmarthen, Dyfed SA31 2PF. Tel 0267 236444 Fax 0267 234262
Gwent	Headquarters, Croesyceiliog, Cwmbran, Gwent NP44 2XJ Tel 0633 838111 Fax 0633 867717
North Wales Police	Operations Control, Glan-y-Don, Colwyn Bay LL29 8AW. Tel 0492 517171 Fax 0492 512720
South Wales	Communications Division, Bridgend, Mid-Glamorgan CF31 3SU. Tel 0656 648378 Fax 0656 648411

Scotland

Constabulary/Force	Address/Tel/Fax Nos for Notification
Central Scotland Police	Randolphfield, Stirling FK8 2HD. Tel 0786 456000 Fax 0786 462681
Dumfries and Galloway	Cornwall Mount, Dumfries DG1 1PZ. Tel 0387 52112 (extn 4551) Fax 0387 50763
Fife	Wemyss Road, Dysart, Kirkcaldy, Fife KY1 2YA. Tel 0592 52611 Fax 0592 51650
Grampian Police	Traffic Department, Nelson Street, Aberdeen AB2 3EQ. Tel 0224 624316/639111 Fax 0224 625735
Lothian and Borders Police	Fettes Avenue, Edinburgh EH4 1RB. Tel 031-311 3418 Fax 031-311 3038
Northern	Old Perth Road, Inverness IV2 3SY. Tel 0463 239191 Fax 0463 230800 *(For Highland Region, Orkney, Shetland and Western Isles)*
Strathclyde Police	Traffic Operations, Meiklewood Road, Glasgow G51 4EU. Tel 041-883 0451 Fax 041-810 4677
Tayside Police	PO Box 59, West Bell Street, Dundee DD1 9JU. Tel 0382 23200 Fax 0382 200449

Appendix IX

Local and Other Authorities
(For Notification of Abnormal Loads Movements)

Local Authorities (responsible for highways and bridges)

England

County Council	Contact Address/Tel/Fax
Avon	Highways and Engineering Department, Avon House North, St James Barton, Bristol BS99 7SG. Tel 0272 290777 Fax 0272 211023
Bedfordshire	Director Transport Engineering, County Hall, Bedford MK42 9AP. Tel 0234 363222 (extn 2697) Fax 0234 228770
Berkshire	Director of Highways and Planning, Shire Hall, Shinfield Park, Reading RG2 9AP. Tel 0734 234776 Fax 0734 310268
Buckinghamshire	County Engineer, Mount Pleasant, Wendover Road, Stoke Mandeville HP22 5TB. Tel 0296 387000 Fax 0296 387067
Cambridgeshire	Director of Transportation, Shire Hall, Castle Hill, Cambridge CB3 0AP. Tel 0223 317707 Fax 0223 317735
Cheshire	Highways Services, Backford Hall, Nr Chester CH1 6EA. Tel 0244 603683 Fax 0244 603810
Cleveland	County Surveyor and Engineer, PO Box 77, Gurney House, Gurney Street, Middlesbrough TS1 1JL. Tel 0642 262697 Fax 0642 220298
Cornwall	County Surveyor, Western Group Centre, Radnor Road, Scorrier TR16 5EH. Tel 0209 820611 Fax 0209 821808
Cumbria	Director of Higways and Transportation, Citadel Chambers, Citadel Row, Carlisle CA3 8SG. Tel 0228 234356 (extn 3223) Fax 0228 514974
Derbyshire	Planning and Highways Officer, Traffic Section, County Offices, Matlock DE4 3AG. Tel 0629 580000 (extn 7591) Fax 0629 580119
Devon	Information Room, Lucombe House, County Hall, Exeter EX2 4QN. Tel 0392 383329 Fax 0392 382321
Dorset	County Surveyor, County Hall, Dorchester DT1 1XJ. Tel 0305 251000 Fax 0305 225186
Durham	Director of the Environment, County Hall, Durham DH1 5UQ. Tel 091-386 4411 Fax 091-383 4096

East Sussex	Highways and Transportation Department, Sackville House, Brooks Close, Lewes BN7 1UE. Tel 0273 482272 Fax 0273 479536
Essex	Highways Department, County Hall, Chelmsford CM1 1QH. Tel 0245 492211 Fax 0245 345010
Gloucestershire	County Surveyor, Shire Hall, Gloucester GL1 2TH. Tel 0452 425555 Fax 0452 506065
Hampshire	County Surveyor, The Castle, Winchester SO23 8UD. Tel 0962 846710 Fax 0962 854045
Hereford and Worcester	County Engineer and Planning Officer, County Hall, Spetchley Road, Worcester WR5 2NP. Tel 0905 766149 Fax 0905 763000
Hertfordshire	Director of Transportation, 'Goldings', North Road, Hertford SG14 2PY. Tel 0992 556080 Fax 0992 556102
Humberside	Director of Technical Services, County Hall, Beverley HU17 9XA. Tel 0482 872110 Fax 0482 872170
Isle of Wight	County Surveyor, County Hall, Newport, IOW PO30 1UD. Tel 0983 823761 Fax 0983 521817
Kent	Highways and Transportation Department, Sandling Block, Springfield, Maidstone ME14 2LQ. Tel 0622 696955 Fax 0622 691028
Lancashire	County Surveyor, PO Box 9, Guild House, Cross Street, Preston PR1 8RD. Tel 0772 264477 Fax 0772 562537
Leicestershire	Director of Planning and Transportation, County Hall, Glenfield, Leicester LE3 8RJ. Tel 0533 657172 Fax 0533 314186
Lincolnshire	Engineering Consultancy Services, Witham Park, Waterside South, Lincoln LN5 7JN. Tel 0522 552910 Fax 0522 552925
Norfolk	County Surveyor, County Hall, Martineau Lane, Norwich NR1 2DH. Tel 0603 223287 Fax 0603 627258
Northants	Director of Planning and Transportation, Northampton House, Northampton NN1 2HZ. Tel 0604 236722 Fax 0604 236660 For loads through Northampton Borough contact: Technical Services Officer, Cliftonville House, Bedford Row, Northampton NN4 0NR. Tel 0604 34734
Northumberland	Technical Services Directorate, County Hall, Morpeth NE61 2EF. Tel 0670 533000 (extn 4274) Fax 0670 510098
North Yorkshire	Highways and Transportation Department, County Hall, Northallerton DL7 8AH. Tel 0609 780780 Fax 0609 779838
Nottinghamshire	Director of Transport and Planning, Trent Bridge House, Fox Road, West Bridgford, Nottingham NG2 6BJ. Tel 0602 774490 Fax 0602 772406
Oxfordshire	County Engineer, Speedwell House, Speedwell Street, Oxford OX1 1NE. Tel 0865 815741 Fax 0865 815085

Shropshire	County Surveyor, The Shirehall, Abbey Foregate, Shrewsbury SY2 6ND. Tel 0743 253144 Fax 0743 253134
Somerset	Director for the Environment, County Hall, Taunton TA1 4DY. Tel 0823 255670 Fax 0823 332773
Staffordshire	County Surveyor, Highway House, Riverway, Stafford ST16 3TJ. Tel 0785 266522 Fax 0785 211279 *(Also covers M6 between jncts 1-6 and 9-16 and M54)*
Suffolk	County Surveyor, St. Edmund House, Rope Walk, Ipswich IP4 1LZ. Tel 0473 265666 Fax 0473 230078
Surrey	County Engineer, Highway House, 21 Chessington Road, West Ewell, Epsom KT17 1TT. Tel 081-541 7132 Fax 081-393 3384
Warwickshire	Director of Planning and Transportation, Barrack Street, Warwick CV34 4SX. Tel 0926 412402 Fax 0926 412903 *(Also covers M6 jnct 1-5, M69 jnct 1-M6, M42 jnct 3-11, M40 jnct 11-17/3A)*
West Sussex	County Surveyor, Highways Management Division, Tower Street, Chichester PO19 1RH. Tel 0243 777574 Fax 0243 777257
Wiltshire	Department of Highways and Transportation, County Hall, Blythesea Road, Trowbridge BA14 8JD. Tel 0225 753641 Fax 0225 765196

Greater London (ie London Boroughs) and Metropolitan Borough Councils

Barking (LB)	Controller of Technical Services, Municipal Offices, 127 Ripple Road, Barking, Essex IG11 7PB. Tel 081-592 4500 Fax 081-594 8077
Barnet (LB)	Controller of Engineering Services, Barnet House, 125 High Road, Whetstone, London N20 0EJ. Tel 081-446 8511 Fax 081-446 6494
Bexley (LB)	Directorate of Engineering and Surveying, Sidcup Place, Sidcup, Kent DA14 6BT. Tel 081-303 7777 Fax 081-309 5803
Brent (LB)	Construction Division, Mahatma Gandhi House, Wembley Hill Road, Wembley, Middlesex HA8 8AD. Tel 081-904 1244 (extn 60057) Fax 081-900 5738
Bromley (LB)	Chief Engineer, Civic Centre, Stockwell Close, Bromley, Kent BR1 3UH. Tel 081-464 3333 Fax 081-313 0095
Camden (LB)	Chief Engineer, 4th Floor, Town Hall Extension, Argyle Street, London NC1H 8EQ. Tel 071-278 4444 Fax 071-413 6952
City of London Corporation	City Engineer, PO Box 270, Guildhall, London EC2P 2EJ. Tel 071-606 3030 Fax 071-260 1559
Croydon (LB)	Public Works Manager, Taberner House, Park Lane, Croydon, Surrey CR9 3RN. Tel 081-760 5566 Fax 081-760 5664
Ealing (LB)	Borough Engineer, 22-24 Uxbridge Road, Ealing, London W5 2BP. Tel 081-579 2424 Fax 081-567 1980

Enfield (LB)	Borough Engineer, PO Box 52, Civic Centre, Silver Street, Enfield, Middlesex EN1 3XD. Tel 081-366 6565 Fax 081-982 7405
Greenwich (LB)	Borough Engineer, Peggy Middleton House, 50 Woolwich New Road, London SE18 6HQ. Tel 081-854 8888 Fax 081-316 6095
Hackney (BC)	Head of Engineering, Joseph Priestley House, 49 Morning Lane, Hackney, London E8 1DS. Tel 081-986 3123 Fax 081-985 1638
Hammersmith and Fulham (LB)	Department of Environment, Town Hall Extension, King Street, London W6 9JU. Tel 081-748 3020 Fax 081-741 5052
Haringey (BC)	Chief Structural Engineer, Hornsey Town Hall, The Broadway, London N8 9JJ. Tel 081-975 9700 Fax 081-862 1793
Harrow (LB)	Director of Engineering, PO Box 38, Civic Centre, Harrow, Middlesex HA1 2UZ. Tel 081-863 5611 (extn 2424) Fax 081-424 1006
Havering (BC)	Borough Engineer, Spilsby Road, Harold Hill, Romford, Essex RM3 8UU. Tel 0708 772222 Fax 0708 773766
Hillingdon (LB)	Hillingdon Engineering Consultancy, Glovers Grove, Ruislip HA4 7YB. Tel 0895 250563/250444/677592 Fax 0895 250676
Hounslow (LB)	Director of Planning and Transport, Civic Centre, Lampton Road, Hounslow, Middlesex TW3 4DN. Tel 081-862 5417 Fax 081-862 5801
Islington (LB)	Chief Highways Engineer, 222 Upper Street, London N1 2UH. Tel 071-477 2299 Fax 071-477 2783
Kensington and Chelsea (RB)	Director of Highways and Traffic, Council Offices, 37 Pembroke Road, London W8 6PW. Tel 071-373 6099 Fax 071-370 3752
Kingston upon Thames (RB)	Director of Engineering, Guildhall 2, Kingston upon Thames, Surrey KT1 1EU. Tel 081-547 5915 Fax 081-547 5926
Lambeth (LB)	Director of Environmental Services, Courtenay House, 9-15 New Park Road, London SW2 4DU. Tel 071-926 7389 Fax 071-926 7082
Lewisham (BC)	Lewisham Engineering, PO Box 927, London SE14 6AP. Tel 081-695 6000 Fax 081-469 2715
Merton (LB)	Director of Environmental Services, Merton Civic Centre, London Road, Morden, Surrey SM4 5DX. Tel 081-545 3193 Fax 081-543 6085
Newham (LB)	Higways Department, 25 Nelson Street, London E6 4EH. Tel 081-472 1430 Fax 081-472 1815
Redbridge (LB)	Borough Engineer, Lynton House, 255/259 High Road, Ilford, Essex IG1 1NY. Tel 081-478 3020 Fax 081-478 9051
Richmond upon Thames (LB)	Chief Officer Planning, Civic Centre, 44 York Street, Twickenham, Middlesex TW1 3BZ. Tel 081-891 7342 Fax 081-891 7702
Southwark (BC)	Highway Engineer, Municipal Offices, Walworth Road, London SE17 1RY. Tel 071-525 5000 Fax 071-525 2213

Sutton (LB)	Borough Engineer, 24 Denmark Road, Carshalton, Surrey SM5 2JG. Tel 081-770 5000 Fax 081-770 6112
Tower Hamlets (LB)	Strategic Engineering, No 3 Millharbour, Isle of Dogs, London E14 9NJ. Tel 081-538 3011 Fax 081-538 4599
Waltham Forest (LB)	Director of Engineering, Municipal Offices, The Ridgeway, Chingford, Essex E4 6PS. Tel 081-527 5544 Fax 081-524 8960
Wandsworth (LB)	Director of Technical Services, Town Hall, Wandsworth High Street, London SW18 2PU. Tel 081-871 6668 Fax 081-871 7561
Westminster City Council	Director of Planning and Transportation, Westminster City Hall, 64 Victoria Street, London SW1E 6QP. Tel 071-798 2622 Fax 071-798 2658

NB: LB = London Borough; BC = Borough Council; RB = Royal Borough

Greater Manchester

Manchester City Council	City Engineer and Surveyor, PO Box 488, Town Hall, Manchester M60 2JT. Tel 061-247 3265 Fax 061-247 3239 *For movements on district roads in Greater Manchester.*
Motorways	Parkman Consulting Engineers, Cunard Building, Liverpool L3 1ES. Tel 051-236 6066 Fax 051-236 1276 *For movements on all motorways in Greater Manchester area* except *M6.*

Merseyside

Knowsley (MBC)	Director of Planning and Development, Municipal Buildings, PO Box 26, Archway Road, Huyton l36 9FB. Tel 051-443 2231 Fax 051-480 0267
Liverpool City Council	City Engineer, Steers House, Canning Place, Liverpool L1 8JA. Tel 051-225 4322 Fax 051-709 3029
St. Helens (MBC)	Director of Environment and Design Services, Wesley House, Corporation Street, St Helens WA10 1HF. Tel 0744 24061 Fax 0744 24055
Sefton (MBC)	Borough Engineer and Surveyor, 5th Floor, Balliol House, Stanley Precinct, Bootle L20 3NJ. Tel 051-934 4233 Fax 051-934 4215
Wirral (MBC)	Borough Engineer, Town Hall, Bebington, Wirral L63 7PT. Tel 051-645 2080 Fax 051-644 6731

NB: MBC = Metropolitan Borough Council.

Tyne & Wear

Gateshead (MBC)	Director Engineering Services, Civic Centre, Regent Street, Gateshead NE8 1HH. Tel 091-477 1011 Fax 091-478 8422
Newcastle upon Tyne (CC)	Department of Engineering and Environment, Civic Centre, Barras Bridge, Newcastle upon Tyne NE1 8PD. Tel 091-232 8520 Fax 091-261 7867

North Tyneside (MBC)   Head of Transport & Engineering, Graham House, Whitley Road, Benton, Newcastle upon Tyne NE12 9TQ.
Tel 091-201 0033 Fax 091-215 0641

South Tyneside (MBC)   Chief Engineer, Town Hall, Westoe Road, South Shields NE33 2RL.
Tel 091-427 1717 Fax 091-427 7171

Sunderland (MBC)   City Engineer, Civic Centre, Sunderland SR2 7DN.
Tel 091-563 2300 Fax 091-510 9104

NB: MBC = Metropolitan Borough Council; CC = City Council.

West Midlands

Birmingham (CC)   City Engineer, 1 Lancaster Circus, Queensway, Birmingham B4 7DQ.
Tel 021-300 7455 Fax 021-333 4705

Coventry (CC)   City Engineer, Traffic Management Section, Broadgate House, Coventry CV1 1NH.
Tel 0203 832107 Fax 0203 832150

Dudley (MBC)   Borough Engineer, The Council House, Mary Stevens Park, Stourbridge DY8 2AA.
Tel 0384 456000 Fax 0384 452455

Sandwell (MBC)   Director of Technical and Development Services, PO Box 42, Wigmore Buildings, Pennyhill Lane, West Bromwich B71 1EN.
Tel 021-569 4141 Fax 021-569 4072

Solihull (MBC)   Director of Technical Services, PO Box 19, Council House, Solihull B91 3QT.
Tel 021-704 6478 Fax 021-704 6404

Walsall (MBC)   Director of Engineering and Town Planning, Civic Centre, Darwall Street, Walsall WS1 1DG.
Tel 0922 652558 Fax 0922 23234

Wolverhampton (MBC)   Director of Technical Services, Bankfield House, 45 Waterloo Road, Wolverhampton WV1 1RP.
Tel 0902 27811 Fax 0902 20021

For movements on all motorways in West Midlands area notify the following:

Sir Owen Williams and Partners, Motorway Maintenance Compound, Bescot, Walsall WS1 4NG.
Tel 0922 21334 Fax 0922 32032

NB: MBC = Metropolitan Borough Council; CC = City Council.

Yorkshire

Barnsley (MBC)   Director of Public Services, Central Offices, Kendray Street, Barnsley S70 2TN.
Tel 0226 772123 Fax 0226 772099

Doncaster (MBC)   Head of Engineering Design, Barnsley Road, Scawsby, Doncaster DN5 7UD.
Tel 0302 782961 Fax 0302 390051

Rotherham (MBC)   Director of Engineering, Bailey House, Rawmarsh Road, Rotherham S60 1TD.
Tel 0709 822972 Fax 0709 379419

Sheffield (MDC)   The Director DBS, 2-10 Carbrook Hall Road, Sheffield S9 2DB.
Tel 0742 735844 Fax 0742 736128

For loads travelling on, over or under motorways in the Yorkshire area it is necessary to notify one of the following agency departments:

Sir Owen Williams and Partners, Aston Motorway
Maintenance Compound, Hardwick Lane, Aston,
Sheffield S31 0BE.
Tel 0742 872338 Fax 0742 873403
*(for M1 jnct 30 to 38, A1(M) Tinsley Viaduct, M18 to
jnct 5, Morehall Bridge, Wadworth Viaduct).*
Frank Graham Consulting Engineers Ltd, Dewsbury
Road, Tingley, West Yorkshire WS3 1SW.
Tel 0532 526188 Fax 0532 380289
*(for M1 jncts 38 to 47, M62 jncts 22 to 34, M621,
M606, A1).*
West Yorkshire Highways, Engineering Department,
Dudley House, Albion Street, Leeds LS2 8JX.
Tel 0532 476174 Fax 0532 476357
*(for movements on all roads and urban motorways
within Yorkshire District Council areas).*

NB: MBC = Metropolitan Borough Council.

Wales

Clwyd (CC)	Director of Highways and Transportation, Shire Hall, Mold CH7 6NF. Tel 0352 752121 (extn 2655) Fax 0352 756444
Dyfed (CC)	County Engineer and Surveyor, Llanstephan Road, Carmarthen SA31 3LZ. Tel 0267 233333 (extn 4323) Fax 0267 238378
Gwent (CC)	County Engineer, County Hall, Cwmbran NP44 2XN. Tel 0633 838838 (extn 2682) Fax 0633 832986
Gwynedd (CC)	Highways and Transportation Department, County Offices, Caernarfon LL55 1SH. Tel 0286 679426/679306 Fax 0286 5961
Mid-Glamorgan (CC)	Bridges Section, Highways and Transportation Department, Council Buildings, Greyfriars Road, Cardiff CF1 3LJ. Tel 0222 820820 Fax 0222 820777
Powys (CC)	County Surveyor, County Hall, Llandrindod Wells LD1 5LG. Tel 0597 826652 Fax 0597 826260
South Glamorgan (CC)	County Engineer, County Hall, Atlantic Wharf, Cardiff CF1 5UW. Tel 0222 873456 Fax 0222 873161
West Glamorgan (CC)	Director of Environment and Highways, County Hall, Swansea SA1 3SN. Tel 0792 471090 Fax 0792 652712

NB: CC = County Council

Scotland (including Orkney, Shetland and the Western Isles)

Borders (RC)	Director of Roads and Transportation, Regional Headquarters, Newtown St. Boswells TD6 0SA. Tel 0835 23301 Fax 0835 23998
Central (RC)	Director of Roads and Transportation, Viewforth, Stirling FK8 2ET. Tel 0786 442730 Fax 0786 443078
Dumfries and Galloway (RC)	Director of Roads and Transportation, Council Offices, English Street, Dumfries DG1 2DD. Tel 0387 61234 Fax 0387 60111

Fife (RC)	Director of Engineering, Fife House, North Street, Glenrothes KY7 5LT. Tel 0592 754411 Fax 0592 610385
Grampian (RC)	Director of Roads, Woodhill House, Westburn Road, Aberdeen AB9 2LU Tel 0224 682222 Fax 0224 662005
Highland (RC)	Director of Roads and Transport, Regional Buildings, Glenurquhart Road, Inverness IV3 5NX. Tel 0463 702696 Fax 0463 702606
Lothian (RC)	Director of Highways, 19 Market Street, Edinburgh EH1 1BL. Tel 031-469 3719 Fax 031-469 3737
Strathclyde (RC) (Argyll and Bute)	Chief Engineer, Department of Roads, Lochgilphead PA31 8RD. Tel 0546 602233 Fax 0546 3749
(Ayr)	Chief Engineer, Regional Offices, Wellington Square, Ayr KA7 1DR. Tel 0292 612327 Fax 0292 612143
(Dumbarton)	Chief Engineer, Regional Office, Dumbarton G82 3PU. Tel 0389 27636 Fax 0389 27070
(Glasgow)	Chief Engineer, Department of Roads, 20 India Street, Glasgow G2 4PF. Tel 041-227 2429 Fax 041-227 3944
(Hamilton)	Chief Engineer, Almada Street, Hamilton ML3 0AL. Tel 0698 454726 Fax 0698 454757
(Paisley)	Chief Engineer, Regional Offices, Cotton Street, Paisley PA1 1ST. Tel 041-842 5749 Fax 041-842 5050
Tayside (RC)	Director of Roads and Transport, Tayside House, 28 Crichton Street, Dundee DD1 3RE. Tel 0382 303922 Fax 0382 303015
Orkney Islands (IC)	Director of Engineering and Technical Services, Council Offices, Kirkwall KW15 1NY. Tel 0856 873535 Fax 0856 876094
Shetland Islands (IC)	Director of Roads and Transport, Grantfield, Lerwick ZE1 0NT. Tel 0595 2024 Fax 0595 4544
Western Isles (IC)	Director of Engineering Services, Council Offices, Sandwick Road, Stonoway, Isle of Lewis PA87 2BW. Tel 0851 703773 Fax 0851 705349

NB: RC = Regional Council; IC = Island Council

Tunnel and Bridge Authorities

Blackwall and Rotherhithe Tunnels	Tunnels Manager, Greenwich Borough Council, Blackwall Tunnel Office, Naval Row, London E14 9QE. Tel 071-987 4208/3601 Fax 071-987 6571
Dartford Tunnel	Dartford River Crossing Ltd, Tunnel Operating Division, South Orbital Way, Dartford, Kent DA1 5PR. Tel 0322 221603 Fax 0322 294224
Mersey Tunnels	General Manager, The Mersey Tunnels, Georges Dock Building, Georges Dock Way, Pierhead, Liverpool, Merseyside L3 1DD. Tel 051-236 8602 (extn 200) Fax 051-255 0610

Tyne Tunnel	Tyne and Wear PTA, Howdon, Wallsend, Tyne and Wear NE28 0PD. Tel 091-262 4451 Fax 091-263 1031

* * *

Erskine Bridge	The Bridge Manager, Erskine Bridge, Bishopton, Strathclyde PA7 5QD Tel 041-812 7022 Fax 041-812 6625
Forth Road Bridge	The General Manager, Forth Road Bridge Joint Board, South Queensferry, West Lothian EH30 9SF. Tel 031-319 1699 Fax 031-319 1903
Humber Bridge	The Bridgemaster, Humber Bridge Board, PO Box 6, Ferriby Road, Hessle, Humberside HU13 0JG. Tel 0482 647161 Fax 0482 640838
Tay Bridge	The Bridge Manager, Tay Road Bridge Joint Board, Marine Parade, Dundee, Tayside DD1 3JB. Tel 0382 21881-2

Electricity Authorities/Companies

(For notification of high loads – ie above 5.8 metres)

Area	Board/Company Address
Eastern	Engineering Director, Eastern Electricity, Wherstead Park, PO Box 40, Wherstead, Ipswich, Suffolk IP9 2AQ. Tel 0473 688688 Fax 0473 601036
East Midlands	Power Systems Manager, East Midlands Electricity, PO Box 4, North PDO, 398 Coppice Road, Arnold, Nottingham NG5 7HX. Tel 0602 269711 Fax 0602 671637
London	Transport Manager, London Electricity Board, Templar House, 81/87 High Holborn, London WC1V 6NU. Tel 071-242 9050 Fax 071-242 2815
Merseyside and North Wales	Chief Engineer, Merseyside and North Wales Electricity Board Plc, Head Office, Sealand Road, Chester CH1 4LR. Tel 0244 652234 Fax 0244 652353
Midlands	Chief Engineer, Midlands Electricity, PO Box 8, Mucklow Hill, Halesowen, West Midlands B62 8BP. Tel 021-423 2345 Fax 021-422 7977
North-western	Chief Engineer, Norweb Plc, Talbot Road, Manchester M16 0HQ. Tel 061-875 7113 Fax 061-875 7456
Northern	Director of Engineering, Northern Electric Plc, Carliol House, Market Street, Newcastle upon Tyne NE99 1SE. Tel 091-221 2000 Fax 091-235 2211
Scotland	Distribution Operations Manager, Scottish Hydro-Electric Plc, South Inch Business Centre, Shore Road, Perth PH2 8BW. Tel 0738 455050 Fax 0738 455055
South-eastern	Systems Operations Manager, Seaboard Plc, Grand Avenue, Hove, East Sussex BN3 2LS. Tel 0273 724522 Fax 0273 226373
South of Scotland	General Manager (Operations), Scottish Power Plc, 154 Montrose Crescent, Hamilton ML3 6LL. Tel 0698 281777 Fax 0698 268901

South Wales	Chief Engineer, South Wales Electricity Plc, St Mellons, Cardiff CF3 9XW. Tel 0222 792111 Fax 0222 777759
South-western	Operations Manager, Engineering, South-western Electricity Plc, Aztec West, Almondsbury, Bristol BS12 4SE. Tel 0454 201101 Fax 0454 616670
Southern	Operations Administration, Southern Electricity, Westacott Way, Littlewick Green, Maidenhead, Berkshire SL6 3QB. Tel 0628 822166 Fax 0628 584402
Yorkshire	Systems Operations Manager, Yorkshire Electricity, 161 Gelderd Road, Leeds LS1 1QZ. Tel 0532 415000 Fax 0532 415057

Rail and Water Authorities

(For notification of matters concerning rail crossings and bridges – eg abnormal load movements and damage to bridges caused by goods vehicles)

British Waterways Board	Technical Services Department, Wellington Park House, Thirsk Row, Leeds LS1 4DD. Tel 0532 450711 Fax 0532 451880
British Rail	*Eastern Region* – Room C2, Hudson House, York YO1 1HP. Tel 0904 522351 Fax 0904 522802
	Western Region – Western House, 1 Holbrook Way, Swindon, Wilts SN1 1BY. Tel 0793 515898 Fax 0793 515723
	Network SouthEast (South) – Wellesley Grove, Croydon, Surrey CR9 1DY. Tel 081-666 6789 Fax 081-666 6519
	Network SouthEast (North) – Grosvenor House, Room 305, 112 Prince of Wales Road, Norwich NR1 1NZ. Tel 0603 285567 Fax 0603 762791
	Regional Railways – 1st Floor, West Wing, Stanier House, 10 Holiday Street, Birmingham B1 1TG. Tel 021-654 4251 Fax 021-654 4528
	ScotRail – ScotRail House, 58 Port Dundas Road, Glasgow G4 0HG. Tel 041-335 3102 Fax 041-335 2862
London Underground	3rd Floor, East Wing, Ashfield House, 7 Beaumont Avenue, London W14 9UY. Tel 071-724 5600 (extn 28012) Fax 071-918 5485

Appendix X

Training Facilities for Dangerous Goods Drivers

Under provisions contained in the Road Traffic (Training of Drivers of Vehicles Carrying Dangerous Goods) Regulations 1992 (as amended 1993) which came into effect on 1 July 1992 – and in accordance with the requirements of EC Directive 684/89 – drivers of dangerous goods-carrying vehicles must receive training. The following establishments, approved by the City and Guilds of London Institute on behalf of the Department of Transport, provide training in the handling of various classes of dangerous goods for tanker vehicle drivers and for those drivers whose vehicles carry goods in packages – see Chapters 7 and 23 for full details of the relevant requirements.

NB: The list indicates the courses held (ie T = tanker, P = packaged goods) and the classes of dangerous goods covered.

Avon

Avonmouth

Chemfreight Training Ltd, The Avon Lode, Third Way, Avonmouth, Bristol BS11 9YP
Tel 0272 827841
(Courses T + P covering all classes).

Nailsea

National ADR Training Consultancy 6 Goss Barton Nailsea, Nr Bristol BS19 2XD
Tel 0275 857920
(Courses T + P , covering classes 2, 3, 4, 5, 6, 7 and 8).

Portbury

Training Force, Regional Freight Centre, Portbury, Bristol BS20 9XX
Tel 0275 374401
(Courses P only, covering classes 2, 3, 4, 5, 6, 8 and 9).

Bedfordshire

Cranfield

Dale Auto Training and Transport, 192 High Street, Cranfield MK43 0EN
Tel 0234 750131
(Courses T + P covering classes 2, 3, 4, 5, 6, 8 and 9).

Berkshire

Slough

Fullers Transport Training Services 475 Malton Avenue Trading Estate Slough SL1 4QU
Tel 0753 530829
(Courses P only covering classes 2, 3, 4, 5, 6, 8 and 9).

Sunningdale

Nationwide Transport Training, 2/3 Rise Road Sunningdale SL5 0BH
Tel 0344 875481
(Courses T + P covering classes 2, 3, 4, 5, 6, 8 and 9).

Cambridgeshire

Cambridge

East Anglian Driver Training, Granta Terrace, Stapleford, Cambridge CB2 5DJ
Tel 0223 845342
(Courses T + P covering classes 2, 3, 4, 5, 6, 8 and 9).

Peterborough

Mid Anglia HGV Training, 138 Peterborough Road, Whittlesey, Peterborough PE7 1PD
Tel 0733 202156
(Courses T + P covering classes 2, 3, 4, 5, 6, 8 and 9).

Cheshire

Crewe

Air Products, IGD Training Department, Weston Road, Crewe
Tel 0270 583131
(Courses T + P covering classes 2, 3, 4, 5, 6, 8 and 9).

Ellesmere Port

Calor Gas Ltd, Docklands Road, Ellesmere Port, South Wirral L65 4EG
Tel 051-355 3700
(Courses T + P covering class 2 only).

Shell UK Ltd, Operations Training Centre (Stanlow), PO Box 12, Ellesmere Port, South Wirral L65 4BD
Tel 051-350 0123
(Courses T + P covering classes 3 and 3 only).

Nantwich

Trainability UK, Headmasters House, 108 Welsh Row, Nantwich CW5 0PE
Tel 0270 610650
(Courses P only covering classes 2, 3, 4, 5, 6, 8 and 9).

Runcorn

Chemfreight Training Ltd, Cormorant Drive, Picow Farm Road, Runcorn WA7 4UD
Tel 0928 580505
(Courses T + P covering all classes).

Linkman Tankers, Picow Farm Road, Runcorn WA7 4UW
Tel 0928 580588
(Courses T + P covering classes 2, 3, 4, 5, 6, 8 and 9).

Warrington

North Cheshire Training Association, Unit 11/4, Palatine Industrial Estate, Causeway Avenue, Off Wilderspool Causeway, Warrington WA4 6QQ
Tel 0925 58342
(Courses P only covering classes 2, 3, 4, 5, 6, 8 and 9).

Cleveland

Middlesborough

Centre for Industrial and Commercial Training, Hamilton House, Cargo Fleet Lane, Middlesbrough TS3 8DJ.
Tel 0642 222821
(Courses T + P covering classes 2, 3, 4, 5, 6, 8 and 9).

Teeside Training Enterprise, Middlesborough Road East, South Bank, Middlesborough TS6 6TZ
Tel 0642 433295
(Courses T + P covering classes 2, 3, 4, 5, 6, 8 and 9).

Cumbria

Maryport

DJP Training (Cumbria), 25D Solway Estate, Maryport CA15 8NF.
Tel 0900 816460
(Courses T + P covering all classes).

West Cumbria Training Ltd, 23B Solway Estate, Maryport CA15 8NF
(Courses T + P covering classes 2, 3, 4, 5, 6, 8 and 9).

Penrith

Cumbria Fire Service Training Centre, Bridge Lane, Penrith CA11 8HY
(Courses T only covering class 3 only).

Devon

Exeter

Trans-Plant Training Ltd, Unit 2, Exeter Livestock

Centre, Matford Park Road, Exeter EX2 8FD
Tel 0392 426242
(Courses T + P covering classes 2, 3, 4, 5, 6, 8 and 9).

Dorset

Poole

Wessex Transport Training Ltd, 59 Old Wareham Road, Parkestone, Poole BH17 7NN
Tel 0202 717881
(Courses T + P covering classes 1, 2, 3, 4, 5, 6 and 8).

East Sussex

Hastings

EP Training Services Ltd, 2 Alexandra Parade, Park Avenue, Hastings TH34 2PQ
Tel 0424 432200
(Courses T + P covering classes 1, 2, 3, 4, 5, 6, 8 and 9).

Essex

Colchester

TDT Management Services Ltd, London Road, Kelvedon, Colchester CO5 9AU
Tel 0376 570863
(Courses T + P covering classes 2, 3, 4, 5, 6, 8 and 9).

Vocational and Safety Training Ltd, 4 Churchill Avenue, Colchester CO5 9HN
(Courses T + P covering classes 2, 3, 4, 5, 6, 8 and 9).

Grays

ADR Modular Training, Bushy Bit House, Back Lane, off Ship Lane, Aveley, Grays RM16 1NU
Tel 0708 861339
(Courses T + P covering classes 1, 2, 3, 4, 5, 6, 8 and 9).

Harlow

P&O Transport Training, River Way, Harlow CM20 2HB
Tel 0279 418341
(Courses T + P covering classes 2, 3, 4, 5, 6, 8 and 9).

Rochford

Rochford Training, 57 South Street, Rochford SS4 1BL
Tel 0702 542414
(Courses P only covering classes 2, 3, 4, 5, 6, 8 and 9).

Romford

Transport 4-2000, 29/33 Victoria Road, Romford RM1 2JT
Tel 0708 737003
(Courses T + P covering classes 1, 2, 3, 4, 5, 6, 8 and 9).

West Thurrock

Truckworld Thurrock, Oliver Road, West Thurrock RM16 1ED
Tel 0928 580505
(Courses T + P covering all classes).

Wickford

AMRAF Trining Plc, 4 Station Court, Station Approach, Wickford SS11 7AT
Tel 0984 856310
(Courses T + P covering classes 1, 2, 3, 4, 5, 6, 8 and 9).

Hampshire

Aldershot

MOD – Tri-Service Resettlement Organisation, Resettlement Centre, Gallwey Road, Aldershot GU11 2DG
(Courses T + P covering classes 2, 3, 4, 5, 6, 8 and 9).

Bordon

Transport 4-2000, 12 Spruce Avenue, Whitehill, Bordon GU35 8TA
Tel 0420 489854
(Courses T + P covering classes 2, 3, 4, 5, 6, 8 and 9).

Southampton

Freight Training Services Ltd, Nutsey Lane, Totton, Southampton SO4 3NB
Tel 0703 666661
(Courses T + P covering all classes).

Hertfordshire

St Albans

P&O Transport Training Ltd, VER House, Park Industrial Estate, Frogmore, St Albans AL2 2DR
Tel 0727 873878
(Courses T + P covering classes 2, 3, 4, 5, 6, 8 and 9).

Humberside

Hull

Training for Industry (Humberside) Ltd, Willow House, Clay Street, Hull HU8 8HA
Tel 0482 223520
(Courses T + P covering classes 2, 3, 4, 5, 6, 8 and 9).

Grimsby

Management and Industrial Training Services Ltd, 260 Macaulay Street, Grimsby DN31 2EY
Tel 0472 440512
(Courses P only covering classes 2, 3, 4, 5, 6, 8 and 9).

Immingham

Shell UK Downstream, SGL Ltd, West Riverside, Immingham DN40 1AA
Tel 0469 578076
(Courses T + P covering classes 2, 3, 4, 5, 6, 8 and 9).

Kent

Faversham

TDT Management Services Ltd, Kent & Sussex Training Services, London Road, Dunkirk, Faversham ME13 9LG
Tel 0227 750447
(Courses T + P covering classes 1, 2, 3, 4, 5, 6, 8 and 9).

Gillingham

Sandal Business Services, Wigmore Road, Gillingham ME88 0RT
Tel 0634 372380
(Courses T + P covering classes 1, 2, 3, 4, 5, 6, 8 and 9).

Maidstone

Kent Fire Service, Loose Road, Maidstone ME15 9QB
Tel 0892 526171
(Courses T + P covering all classes).

Sittingbourne

Wright Training Services, 66 The Quay, Conyer, Sittingbourne
Tel 0795 521054
(Courses T + P covering classes 1, 2, 3, 4, 5, 6, 8 and 9).

Tunbridge Wells

Freight Transport Association, Hermes House, St John's Road, Tunbridge Wells TN4 9UZ
Tel 0892 526171
(Courses T + P covering all classes).

Leicestershire

Earl Shilton

Driver Training and Management Services Ltd, West Street, Earl Shilton LE9 7EJ
Tel 0455 848578
(Courses T + P covering classes 2, 3, 4, 5, 6, 8 and 9).

Hinckley

Garage and Transport Training (S. Leics) Ltd, Jacknell Road, Dodwells Bridge Industrial Estate, Hinckley
Tel 0455 251516
(Courses T + P covering classes 2, 3, 4, 5, 6, 8 and 9).

Leicester

J. Coates (HGV Services) Ltd, 46/50 Great Central Street, Leicester LE1 4NF
Tel 0533 512962
(Courses T + P covering classes 1, 2, 3, 4, 5, 6, 8 and 9).

Stoney Stanton

Calor Gas Ltd, Occupation Road, Stoney Stanton LE9 6JJ
Tel 0455 272273
(Courses T + P covering class 2 only).

London

Petroleum Training Federation, Suite 1, Morley House,
314-322 Regent Street, London W1R 5AB
Tel 071-255 2335
(Courses T + P covering classes 2, 3 and 9 only).
Mobile training provided.

Greater Manchester

Bolton

Red Rose Training, Europa Way, Bolton M26 9HE
Tel 0204 862999
(Courses T + P covering classes 2, 3, 4, 5, 6, 8 and 9).

Manchester

Air Products (UK) Ltd, Manchester Road, Carrington,
Manchester M31 4AG
Tel 061-755 4381
(Courses T + P covering classes 2, 3, 4, 5, 6, 8 and 9).

P&O Transport Training, Barton Dock Road,
Manchester M32 0YJ
Tel 061-865 0886
(Courses T + P covering classes 2, 3, 4, 5, 6, 8 and 9).

Trafford Park

Hargreaves Training Ltd, Units 5-6 Guiness Road, Har
Trading Estate, Trafford Park, Manchester M17 1SR
Tel 061-872 7916
(Courses T + P covering classes 2, 3, 4, 5, 6, 8 and 9).

Merseyside

Liverpool

ACL HGV Training Ltd, 87 Greenodd Avenue, West
Derby, Liverpool L12 0HE
Tel 051-256 8088
(Courses T + P covering classes 2, 3, 4, 5, 6, 8 and 9).

Trafalgar Training Centre, Bridge House, South
Trafalgar Dock, Liverpool L3 0AG
Tel 051-236 4592
(Courses T + P covering classes 1, 2, 3, 4, 5, 6, 8 and 9).

St Helens

Sutton & Son (St Helens) Ltd, Sutton Heath, St Helens
WA9 5BW
Tel 0744 811611
(Courses T + P covering classes 2, 3, 4, 5, 6, 8 and 9).

Norfolk

Norwich

Norfolk Training Services, Harford Centre, Hall Road,
Norwich NR4 6DG
Tel 0603 259900
(Courses T + P covering classes 2, 3, 4, 5, 6, 8 and 9).

Nottinghamshire

Nottingham

Trent Transport Training, The Watson Centre, Artic
Way, Kimberley, Nottingham NG16 2HS
Tel 0602 384982
(Courses P only covering classes 2, 3, 4, 5, 6, 8 and 9).

Retford

PM Training Centre, Haygarth House, Batworth,
Retford DN22 8ES
(Courses T + P covering classes 2, 3, 4, 5, 6, 8 and 9).

Worksop

North Notts Training Group, Claylands Avenue,
Dukeries Industrial Estate, Worksop S81 7DJ
Tel 0909 475745
(Courses P only covering classes 2, 3, 4, 5, 6, 8 and 9).

Oxfordshire

Abingdon

Thames Valley Training, 4 Foster Road, Abingdon
OX14 1YN

Tel 0235 530085
(Courses P only covering classes 2, 3, 4, 5, 6, 8 and 9).

Didcot

Air Products (UK) Ltd, Hawkesworth Road, Smithmead Industrial Estate, Didcot OX11 7PG
Tel 0235 510771Wallingford

Ridgeway International Ltd, 69 High Street, Wallingford OX10 0BX
Tel 0491 39780
(Courses P only covering classes 1 and 7).

Shropshire

Telford

Freight Transport Association, MOTEC, High Ercall, Telford TF6 6RB
Tel 0892 526171
(Courses T + P covering classes 2, 3, 4, 5, 6, 8 and 9).

West Midlands Training Group Ltd, 38 Paddock Mount Offices, Dawley, Telford TF4 3PR
Tel 0384 401199
(Courses T + P covering classes 2, 3, 4, 5, 6, 8 and 9).

Somerset

Bridgewater

VTT Ltd, Suprema Estate, Edington, Bridgewater TA7 9LF
Tel 0278 723113
(Courses P only covering classes 2, 3, 4, 5, 6, 8 and 9).

Shepton Mallet

Skilltrain Associates, Shepton Mallet
Tel 0850 592812
(Courses P only covering classes 2, 3, 4, 5, 6, 8 and 9).

Taunton

Friendberry Ltd, Stogumber, Taunton TA4 3TP
Tel 0984 56310
(Courses T + P covering classes 1, 2, 3, 4, 5, 6, 8 and 9).

South Yorkshire

Doncaster

Doncaster, Rotherham and District Motor Trades GTA, Rands Lane Industrial Estate, Rands Lane, Armthorpe, Doncaster DN3 3DY
Tel 0302 832831
(Courses T + P covering classes 2, 3, 4, 5, 6, 8 and 9).

Staffordshire

Brownhills

Road and Management Training (Ron Rogers Training), Colliers Close, Coppice Side Ind. Estate, Brownhills, Staffs
Tel 0543 452769
(Courses T + P covering classes 2, 3, 4, 5, 6, 8 and 9).

Leek

Transed Associates, Springbuck House, Leekbrook Ind. Estate, Cheadle Road, Leek ST137AP
Tel 0538 381313
(Courses T + P covering classes 2, 3, 4, 5, 6, 8 and 9).

Stafford

Leight Environmental Ltd, Dunstan Hall, Dunstan Stafford ST18 9AB
Tel 0785 712666
(Courses T + P covering classes 2, 3, 4, 5, 6, 8 and 9).

Suffolk

Felixstowe

P&O Transport Training Ltd, Dock Road, Felixstowe IP11 8JB
Tel 0394 674139
(Courses T + P covering classes 1, 2, 3, 4, 5, 6, 8 and 9).

Stowmarket

RTT Training Services Ltd, Mendlesham Training Centre, Norwich Road, Mendlesham, Stowmarket IP14 5ND
Tel 0449 766473
(Courses T + P covering classes 1, 2, 3, 4, 5, 6, 8 and 9).

Surrey

Bookham

EP Training Services Ltd, The Old Library, Lower Shott, Bookham, Surrey KT23 4LR
Tel 0372 450800
(Courses T + P covering classes 1, 2, 3, 4, 5, 6, 8 and 9).

Guildford

British Oxygen Co (BOC) Ltd, The Priestly Centre, 10 Priestly Road, Guildford GU2 5XY
Tel 0483 579857
(Courses T + P covering classes 2, 3, 4, 5, 6, 8 and 9).

Tyne & Wear

Felling

Colin Hall Assessment and Training Services, Design Works, William Street, Felling, Gateshead NE10 0JP
Tel 091-495 0066
(Courses P only covering classes 2, 3, 4, 5, 6, 8 and 9).

Van Hee Transport Ltd, William Street, Felling, Gateshead NE10 0JP
Tel 091-438 2512
(Courses P only covering classes 2, 3, 4, 5, 6, 8 and 9).

Hebburn

Tyne and Wear ADR Training, Station Road, Hebburn N31 1NY
Tel 091-430 0505
(Courses T + P covering classes 2, 3, 4, 5, 6, 8 and 9).

Newcastle upon Tyne

Tyneside Training Services Ltd, Airport Industrial Estate, Kingston Park, Kenton, Newcastle upon Tyne NE3 2EF
Tel 091-286 2919
(Courses P only covering classes 2, 3, 4, 5, 6, 8 and 9).

West Midlands

Aldridge

Westgate Training, Westgate, Aldridge, West Midlands WS9 8EZ
Tel 0922 743783
(Courses T + P covering classes 2, 3, 4, 5, 6, 8 and 9).

Birmingham

Chemfreight Training Ltd, South Birmingham College, Robin Hood Centre, Pitmaston, Hall Green, Birmingham B28 9PW
Tel 0928 580505
(Courses T + P covering all classes).

West Midlands Training Group Ltd, Granby Avenue, Garretts Green Industrial Estate, Birmingham B33 0TJ
Tel 021-784 9722
(Courses T + P covering classes 2, 3, 4, 5, 6, 8 and 9).

Kingswinford

West Midlands Training Group Ltd, Dudley Road, Kingswinford DY6 8BS
Tel 0384 401199
(Courses T + P covering classes 2, 3, 4, 5, 6, 8 and 9).

Wolverhampton

BOC Ltd (Training Centre), Lower Walsall Street, Wolverhampton, West Midlands WV1 2EP
Tel 0902 453131
(Courses T + P covering classes 2, 3, 4, 5, 6, 8 and 9).

West Sussex

Crawley

Shell UK Downstream (SUKD), Colas Ltd, Rowfont, Crawley RH10 4NF
Tel 0342 711087
(Courses T + P covering classes 2, 3, 4, 5, 6, 8 and 9).

West Yorkshire

Batley

Linkman Tankers Ltd, Nab Lane, Birstall, Batley WF17 9NH

	Tel 0924 478235 *(Courses T + P covering classes 2, 3, 4, 5, 6, 8 and 9).*
Bradford	Ellis & Everard (UK) Ltd, 46 Peckover Street, Bradford BD1 5BD Tel 0274 377000 *(Courses P only covering classes 2, 3, 4, 5, 6, 8 and 9).*
	Sandel Business Services, C/o Bradford Transport Training, Cumberland House, Tumbling Hill Street, Thornton Road, Bradford BD1 2GX Tel 0274 306457 *(Courses T + P covering classes 1, 2, 3, 4, 5, 6, 8 and 9).*
	West Yorkshire Training Group, 420 Tong Street, Bradford BD4 6LP Tel 0274 689814 *(Courses T + P covering classes 2, 3, 4, 5, 6, 8 and 9).*
Dewsbury	Haz Training Services, Woodkirk Valley Counry Club, 1172 Leeds Road, Woodkirk, Dewsbury WF12 7JR *(Courses T + P covering classes 2, 3, 4, 5, 6, 8 and 9).*
Leeds	ASC Training Services, ASC House, 54 North Street, Leeds LS2 7PN Tel 0532 453480 *(Courses T + P covering classes 2, 3, 4, 5, 6, 8 and 9).*
	Hargreaves Training Services, Unit 3, Parkside Industrial Estate, Glover Road, Leeds LS11 5JP Tel 0532 701188 *(Courses T + P covering classes 2, 3, 4, 5, 6, 8 and 9).*
Wakefield	Sandal Business Services, PO Box 85, Wakefield WF2 6XZ Tel 0924 820078 *(Courses T + P covering classes 2, 3, 4, 5, 6, 8 and 9).*

Wiltshire

Devizes	Wiltshire Transport Training Ltd, Hopton Industrial Estate, London Road, Devizes SN10 2EX Tel 0380 723712 *(Courses P only covering classes 2, 3, 4, 5, 6, 8 and 9).*

SCOTLAND

Aberdeen	Chemfreight Training Ltd, Blackness Avenue, Altens, Aberdeen AB1 4PG Tel 0928 580505 *(Courses T + P covering all classes).*
	Friendberry Ltd (in assoc with J Gilbert), Cloverhill Road, Bridge of Don Industrial Estate, Aberdeen AB2 8EE Tel 0224 825644 *(Courses T + P covering classes 1, 2, 3, 4, 5, 6, 8 and 9).*
	Sandel Business Services, Hax-Train (Scotland), Granitehill Enterprise Centre, Granitehill Road, Aberdeen AB2 7AX Tel 0224 875464 *(Courses T + P covering classes 1, 2, 3, 4, 5, 6, 8 and 9).*
	John Wood Group Training Centre, Greenhill Road, Aberdeen AB1 4AX Tel 0224 875464 *(Courses T + P covering classes 2, 3, 4, 5, 6, 8 and 9).*
Bonnybridge	LAGTA Group Training, Seabegs Road, Bonnbridge FK4 2AQ Tel 031-554 8506 *(Courses T + P covering classes 2, 3, 4, 5, 6, 8 and 9).*

Coatbridge	LAGTA Group Training, 7 Palacecraig Street, Shawland, Coatbridge ML5 4RY Tel 031-554 8506 *(Courses T + P covering classes 2, 3, 4, 5, 6, 8 and 9).*
Dundee	Tayside Road Transport GTA Ltd, Smeaton Road, Wester Gourdie, Dundee DD2 4UT Tel 0382 623261 *(Courses P only covering classes 2, 3, 4, 5, 6, 8 and 9).*
Dundonald	Scottish Express International, Olympic Complex, Drybridge Road, Dundonald, Ayrshire KA2 9BE *(Courses T + P covering classes 2, 3, 4, 5, 6, 8 and 9).*
Edinburgh	LAGTA SOS Group Training (Edinburgh) Ltd, Marine Parade, Western Harbour, Leith, Edinburgh EH6 6PB Tel 031-554 8506 *(Courses T + P covering classes 2, 3, 4, 5, 6, 8 and 9).*
Glasgow	Glasgow Training Group Ltd, 120 Crownhill Road, Bishopbriggs, Glasgow G64 1RP Tel 041-762 1461 *(Courses T + P covering classes 2, 3, 4, 5, 6, 8 and 9).* Ritchies HGV Training Centre, Hobden Street, Glasgow G21 4AU Tel 041-557 2212 *(Courses P only covering classes 2, 3, 4, 5, 6, 8 and 9).*
Gourock	Thomas Bateman Training Services, 97 Albert Dock, Gourock, Renfrewshire PA19 1NN Tel 0475 630588 *(Courses T + P covering classes 2, 3, 4, 5, 6, 8 and 9).*
Greenock	Fleet Support Transport Training, 64 Westblackhall Street, Greenock, Renfrewshire PA15 1XR *(Courses T + P covering classes 2, 3, 4, 5, 6, 8 and 9).*
Hamilton	West Midlands Training Group (North) Ltd, Units 1 & 2, Cadlow Bridge Industrial Estate, Low Waters Road, Hamilton Tel 0689 285846 *(Courses T + P covering classes 2, 3, 4, 5, 6, 8 and 9).*
Livingston	Freight Transport Association, MOTEC, Dean's Industrial Estate, Hardie Road, Livingston EH54 8AR Tel 0892 526171 *(Courses T + P covering classes 2, 3, 4, 5, 6, 8 and 9).*
Stonehaven	EJS Training, 4 Rolland Place, Drumlithe, Stonehaven AB3 2YS Tel 05694 294 *(Courses T + P covering classes 2, 3, 4, 5, 6, 8 and 9).*
Thurso	Friendberry Ltd (in assn with Thurso College), Ormlie Road, Thurso KW14 7EE Tel 0984 56310 *(Courses T + P covering classes 1, 2, 3, 4, 5, 6, 8 and 9).*

WALES

Clwyd

Sandycroft	Cameon Ltd, Allan Morris Ltd, Factory Road, Sandycroft, Deeside CH5 2QJ Tel 0244 533320 *(Courses T + P covering classes 2, 3, 4, 5, 6, 7, 8 and 9).*
Wrexham	Gatewen Training Ltd, Gatewen, New Broughton, Wrexham LL11 6YA Tel 0978 720907 *(Courses P only covering classes 2, 3, 4, 5, 6, 8 and 9).*

Dyfed

Pembroke Dock

Mainport Training, Pembroke Enterprise Centre,
Kingswood Estate, Pembroke Dock, Dyfed
Tel 0646 684315
(Courses T + P covering classes 2, 3, 4, 5, 6, 8 and 9).

Mid-Glamorgan

Treharris

Pass Go Ltd, 32/34 Cardiff Road, Quakers Yard,
Treharris CF46 5DU
Tel 0443 411019
(Courses T + P covering classes 1, 2, 3, 4, 5, 6, 8 and 9).

South Glamorgan

Barry

Barry Training Services, Holt Buildings, Powell Duffryn
Way, No 1 Dock, Barry CF6 6EW
(Courses T + P covering classes 2, 3, 4, 5, 6, 8 and 9).

NORTHERN IRELAND

Crumlin

Transport Training Services Ltd, 15 Dundrod Road,
Nutts Corner, Crumlin, Co. Antrim, Northern Ireland
Tel 0232 825653
(Courses T + P covering classes 1, 2, 3, 4, 5, 6, 8 and 9).

Index

Index of Advertisers